AF883740

The Persian Gulf

Holocene Carbonate Sedimentation and Diagenesis in a Shallow Epicontinental Sea

Edited by

B. H. Purser

With 250 Figures, 7 Plates and 3 Maps

EXTRA MATERIALS extras.springer.com

Springer-Verlag Berlin Heidelberg GmbH 1973

Bruce H. Purser
Laboratoire de Géologie Historique
Faculté des Sciences
Université de Paris Sud
91-Orsay/France

Additional material to this book can be downloaded from http://extras.springer.com.

ISBN 978-3-642-65547-0 ISBN 978-3-642-65545-6 (eBook)
DOI 10.1007/978-3-642-65545-6

This work is subject to copyright. All rights are reserved, whether the whole or part of the material is concerned, specifically those of translation, reprinting, re-use of illustrations, broadcasting, reproduction by photocopying machine or similar means, and storage in data banks.

Under § 54 of the German Copyright Law where copies are made for other than private use, a fee is payable to the publisher, the amount of the fee to be determined by agreement with the publisher.

The use of registered names, trademarks, etc. in this publication does not imply, even in the absence of a specific statement, that such names are exempt from the relevant protective laws and regulations and therefore free for general use.

© 1973 by Springer-Verlag Berlin Heidelberg
Originally published by Springer-Verlag Berlin Heidelberg New York in 1973
Softcover reprint of the hardcover 1st edition 1973

Library of Congress Catalog Card Number 72-97023. Offsetprinting: Julius Beltz, Hemsbach/Bergstr. Bookbinding: Konrad Triltsch, Graphischer Betrieb, Würzburg, Germany.

Introduction

This volume, although not an integrated synthesis, treats most aspects of Holocene sedimentation and diagenesis in the Persian Gulf, grouping 22 contributions under a single cover and in one language. Because these sediments and diagenetic minerals are comparable to those existing in many ancient sedimentary basins, their appraisal should be of value to the enlarging group of workers who interpret ancient sedimentary rocks.

The essential morphological, climatic and oceanographic factors determining Holocene sedimentation and diagenesis in the Persian Gulf are summarized in the introductory article by PURSER and SEIBOLD. These environmental controls and the overall morphology of the Persian Gulf have much in common with Shark Bay, Western Australia, described by LOGAN et al. (1970). On the other hand, the Persian Gulf is markedly different from the better known Florida and Bahamian provinces; the floor of the Persian Gulf is gently inclined from continental shoreline to bathymetric axis (80–100 m); the Bahamian province, on the other hand, is horizontal and extremely shallow (2–10 m), with very sharply defined shelf edges surrounded by deep oceanic waters. These contrasting architectural styles are related to different tectonic frames. The Persian Gulf, at least in its Iranian half, is strongly affected by a late Tertiary fold system which determines the regional outline of the basin, the inclination of its sea floor, and the orientation of its bathymetric axis; the Bahamian and Florida morphologies are only remotely influenced by underlying tectonic systems, their characteristic "shelf" morphologies being the consequence of accretion to sea level under stable tectonic conditions. Because of these contrasting morphologies, regional sediment patterns in the Persian Gulf and Bahamas are quite distinct.

Major differences in the diagenetic properties of the Persian Gulf and Caribbean Holocene sediments are related mainly to differing climatic factors. The hot, arid climate of the Middle East stimulates the formation of evaporitic minerals, including widespread dolomite. The cooler, somewhat wetter Caribbean climate does not favour evaporites, and dolomite seems to be less widely distributed. These climatic differences are probably reflected also in the rates of carbonate mineral stabilization.

In sum, the processes and patterns of Holocene carbonate sedimentation and diagenesis differ markedly from one province to another. Consequently, effective interpretation of ancient analogues requires a sound understanding not only of the modern "carbonate shelf" provinces typified by the Caribbean area, but also of the shallow epicontinental basins with gently sloping floors, such as Shark Bay and the Persian Gulf.

Of the 31 authors comprising 9 nationalities who have contributed to this volume, 23 are associated with one or other of three organizations:

The Kiel University group under the leadership of Professor E. SEIBOLD had carried out extensive surveys in the Gulf of Oman and within the Iranian parts of the Persian Gulf. Their results, published mainly in the "Meteor" Forschungsergebnisse (1969–1972) give abundant data concerning the oceanography of the basin. Detailed sedimentological and ecological studies are confined essentially to the deeper parts of the Persian Gulf. New information relating to these marly sediments is given by SEIBOLD et al., MELGUEN, and SARNTHEIN and WALGER, in this book.

The Imperial College of London group has carried out extensive research, mainly along the Trucial Coast during the period 1961–1968. This work was initiated by G. EVANS and D. J. SHEARMAN and directed subsequently by G. EVANS; its results are expressed mainly in the form of doctoral theses. Various aspects have been published by BUSH (1970), BUTLER (1969), EVANS (1970), EVANS et al. (1964, 1969, etc.), KENDALL and SKIPWITH (1968), KINSMAN (1964, etc.), MURRAY (1965, etc.), and SHEARMAN (1963, etc.). Their most significant contributions include detailed studies of the relationships between sabkha diagenesis and water chemistry, and coastal sedimentation and ecology; lagoonal Foraminifera, in particular, have been treated in considerable detail. Additional results concerning sabkha diagenesis are offered by BUTLER, BUTLER et al., and BUSH, while coastal sedimentation and ecology are treated by EVANS et al., in this volume.

Shell Research B. V. (Kon. Shell Exploratie en Produktie Laboratorium, Rijswijk, The Netherlands) has concentrated its research around Qatar Peninsula but has also carried out regional surveys throughout most of the Arabian half of the basin. Initial results were published by HOUBOLT (1957). The project was continued by A. J. WELLS in collaboration with Illing Associates, their studies of Recent dolomite, evaporites and beach rocks being published by WELLS (1962), ILLING et al. (1965), and TAYLOR and ILLING (1969). The results of subsequent research directed by B. H. PURSER are presented in this volume by EVAMY, DE GROOT, HUGHES CLARKE and KEIJ, KASSLER, PURSER, SHINN, and WAGNER and VAN DER TOGT.

The objectives of the Shell Research group differed somewhat from those of the Kiel and Imperial College groups. Being an oil company, much of Shell's research was directed towards the establishment of the basic aspects of modern sedimentation and diagenesis which would aid in the understanding and exploitation of certain hydrocarbon reservoirs; this research tends to evaluate the forest without identifying all the species composing it. The different styles of research employed by the respective university and oil company groups are apparent in their contributions to this volume; both have their virtues.

Although the greater part of research in the Persian Gulf has been carried out by the three groups mentioned above, significant contributions to this volume include those of HSÜ and SCHNEIDER of the Polytechnical Institute of Zürich, KUKAL and SAADALLAH of the Central Geological Survey, Prague and the University of Baghdad respectively, and by LOREAU (and PURSER) of the Muséum National d'Histoire Naturelle, Paris.

Six of the contributions were presented orally, and as printed abstracts, at the VIII International Sedimentological Congress (Heidelberg) in September, 1971. It should be stressed, however, that although much data relating to the Holocene sediments of the Persian Gulf has already been published, the new material presented here constitutes the first attempt to group results. This coordination enables readers for the first time to obtain an overall picture of this basin.

In conclusion, on behalf of all contributors to this volume, I would like to thank the following persons and organizations who have contributed in many nonsedimentological ways to the understanding of sedimentation and diagenesis in the Persian Gulf:
- the Rulers and Governments of virtually all states bordering the Persian/Arabian Gulf for permission to work within their territorial waters and coasts;
- H. V. DUNNINGTON and the staff of Abu Dhabi Petroleum Company for generous help to Imperial College personnel and to myself and my colleagues J.-P. LOREAU and K. J. HSÜ;
- LUDWIG HAPPEL of Munich who, as head of geological research for Shell (Rijswijk), encouraged both his own staff – including myself – and the researchers of Imperial College, London;
- Huntings Air Services for use of aerial photographs, certain of which are reproduced in this volume;
- Dr. KONRAD F. SPRINGER and his staff for their very active support.

I would also like to offer my personal thanks not only to my fellow-contributors for their cooperation in producing this symposium, but more especially to my colleagues who shared the occasional pleasures and numerous discomforts of the long months of fieldwork in the Persian Gulf.

Orsay, March 1973 — BRUCE H. PURSER

Contents

The Principal Environmental Factors Influencing Holocene Sedimentation and Diagenesis in the Persian Gulf

B. H. Purser[1] and E. Seibold[2]

ABSTRACT

The Persian Gulf is a marginal sea with an average depth of 35 m, and a maximum depth of 100 m near its narrow entrance. Its elongate bathymetric axis separates two major geological provinces - the stable Arabian Foreland and the unstable Iranian Fold Belt - which are reflected in the constrasting coastal and bathymetric morphologies of Arabia and Iran. The Persian Gulf has a gently inclined sea floor lacking "shelf edges" comparable with those of modern Caribbean carbonate provinces.

The arid, sub-tropical climate with summer temperatures attaining 50° C, and frequent winds, stimulate the formation of evaporitic minerals and the delivery of aeolian dust to the basin. Fluviatile influx is limited to the Tigris-Euphrates-Karun delta and to the mountainous Iranian coast where terrigenous sediments contrast with the relatively pure carbonates forming in the shallow seas in front of the low deserts of Arabia.

Excessive evaporation and partial isolation from the adjacent Indian Ocean provoke abnormal salinities throughout most of the basin, which attain a maximum of ca 70 $^{o}/oo$ in remote Arabian lagoons.Because the prevailing "shamal" wind blows down the axis of the gulf from the NW, most coastal environments are swept by waves and surface currents which favour the formation and dispersal of carbonate sands on the Arabian side and terrigenous material on the Iranian. Tidal currents influence sediment textures, even in the deepest parts of the gulf, while extensive rock bottoms influence the biota and skeletal composition of Holocene sediments. These are mixed with significant amounts of relict sediment, especially in the deeper parts of the basin.

INTRODUCTION

This paper introduces the reader to the Persian Gulf by outlining those regional environmental parameters which seem to influence the nature and distribution of its Holocene sediments. Its authors have refrained from treating local environmental aspects, which are discussed in individual contributions. Because this volume is essentially sedimentological, water circulation and other properties are treated only in sufficient detail to permit an understanding of sediment formation, distribution, and diagenesis; oceanographic aspects are treated in greater detail by Hartmann et al. (1971).

1. Lab. de Géologie Historique, Université de Paris Sud (Orsay).
2. Geologisch-Paläontologisches Institut, Universität Kiel.

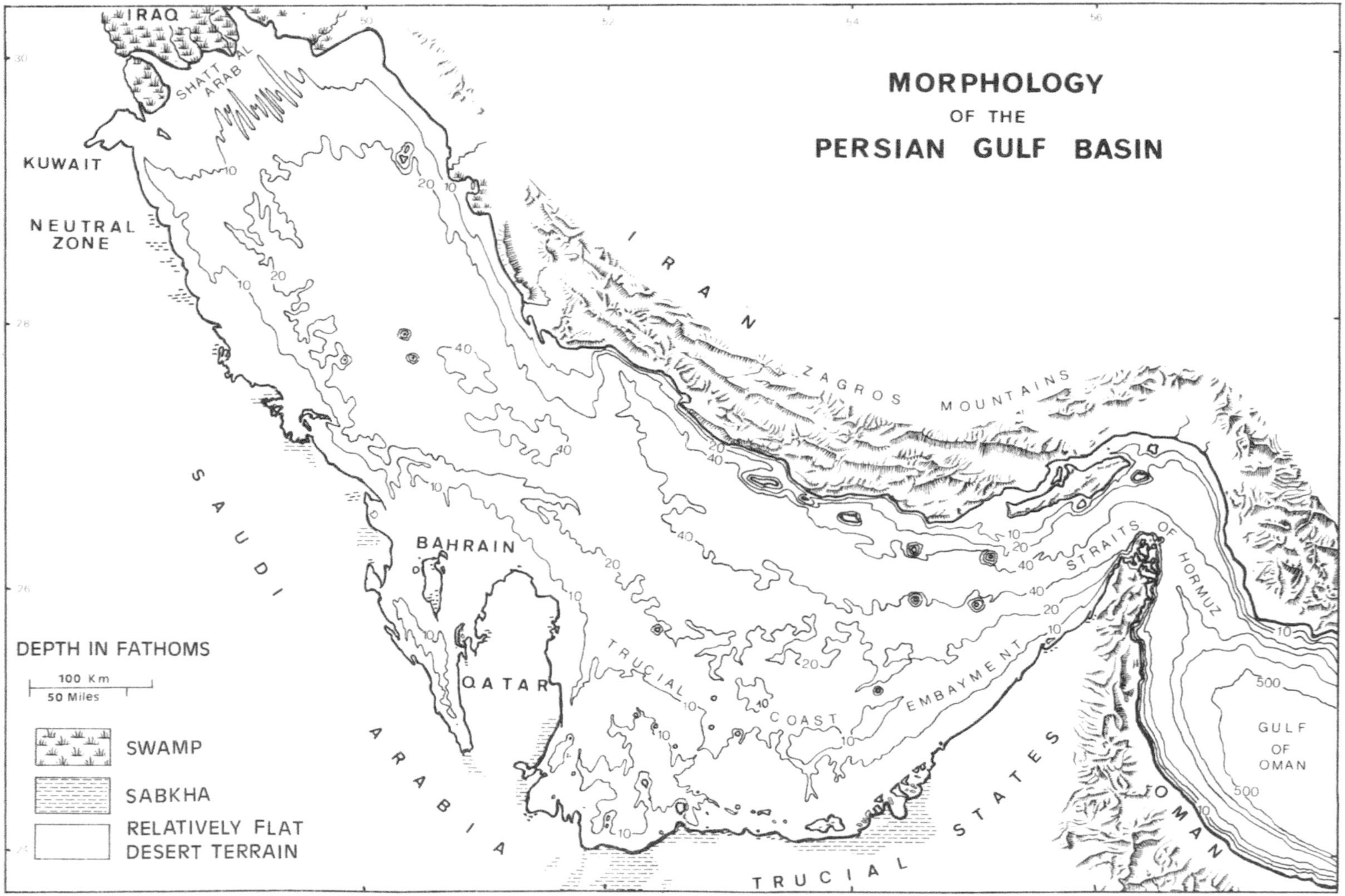

Fig. 1. Map of the Persian Gulf Region showing the principal morphological aspects of the basin and surrounding landmass (Based mainly on British and U.S. Naval maps)

This introductory paper has been prepared in collaboration with virtually all contributors to this volume, partly in an attempt to avoid needless repetition in individual texts. The authors are particularly indebted to their respective colleagues in "Shell Research" , Rijswijk, the Netherlands, and in the University of Kiel, F.R. Germany.

A. THE MORPHOLOGICAL FRAMEWORK OF THE PERSIAN GULF

The Persian Gulf (Fig. 1) is a marginal sea measuring some 1000 km in length and 200-300 km in width, covering an area of approximately 226,000 km^2. Its average depth is ca. 35 m, and it attains its maximum depth of ca. 100 m near its entrance - the Straits of Hormuz (see details in Seibold and Vollbrecht, 1969, Seibold and Ulrich 1970, and Kassler, in this volume). It is virtually surrounded by arid land and connected to the Indian Ocean only by the 60 km wide Straits of Hormuz. The entire basin lies upon the continental shelf whose margin and slope occur in the Gulf of Oman (Fig. 1).

The elongate axis of the Persian Gulf basin separates two distinct morphological provinces whose character is related closely to contrasting tectonic styles. The Arabian side of the basin constitutes part of the relatively stable Arabian Foreland flanking the Pre-Cambrian Arabian Shield (Lees and Richardson, 1940, and Lees, 1948), while the unstable Iranian area constitutes a major Tertiary fold belt (Lees & Falcon, 1952). The following outline of the major morphological units of the basin will be better understood if these fundamental tectonic differences are kept in mind.

The relatively linear character of the Arabian coast is modified by Qatar Peninsula whose presence strongly influences the marine currents and patterns of sedimentation along the SE side of the Persian Gulf. To the E. of Qatar Peninsula is a broad, shallow area (depth 10-20 m) studded with numerous shoals and salt-dome islands, many of which have a characteristic "volcanic" appearance (Kassler, Fig. 5, in this volume). An irregular bathymetric ridge, the "Great Pearl Bank Barrier" , extends eastwards from Qatar across this Trucial Coast Embayment, influencing sedimentation along the central parts of the Trucial Coast (see Purser & Evans, in this volume). This latter shoreline is characterized mainly by low, evaporitic, supratidal flats or "sabkhas" , which locally attain 10 km in width, and by extensive storm beaches in more exposed settings. Both beaches and "sabkhas" grade back into the low deserts of Arabia (Purser & Evans, Butler et al., Hsü & Schneider, in this volume).

The NW end of the Persian Gulf terminates in the Tigris-Euphrates-Karun delta, or "Shatt al Arab" which today appears to have only local effects on marine environments (see Seibold et al. in this volume).

The Iranian shores of the basin are essentially linear and rocky, with narrow coastal plains associated with the estuaries of a number of small rivers which drain into the gulf from the adjacent Zagros Mountains. This mountainous hinterland, frequently exceeding 1500 m in altitude, contrasts with the low deserts and rocky mesa topography of Arabia.

The bathymetric form of the Persian Gulf

The basin is characterized by a marked asymmetry across its axis; its floor on the tectonically unstable Iranian side of the basin is relatively steep (175 cm/km), while that on the more stable Arabian side slopes gently (ca. 35 cm/km) towards the bathymetric axis. This axis is thus situated relatively close to the Iranian coast, especially in the E half of the basin, its position and orientation being determined by underlying Iranian tectonic trends (Kassler, in this volume).

The Persian Gulf has been subdivided into a series of bathymetric provinces by Seibold & Vollbrecht (1969), illustrated in Fig. 2. A clearly defined ridge, the "Central Swell" , either related to the underlying Iranian fold system, or possibly of sedimentary origin, subdivides the Iranian half of the basin into two secondary depressions - the "Western Basin" and the "Central Basin" . The Arabian side of the Persian Gulf has been termed the "Arabian Shallow Shelf" by Seibold & Vollbrecht, mainly for oceanographic reasons. It should be noted, however, that this same area has been termed the "Arabian Homocline" by Kassler and other "Shell" contributors to this volume, the term "homocline" being chosen mainly to stress the inclined character of the Arabian sea floor. This slope, although very gentle, permits a high degree of wave agitation along the Arabian coastline and, in this respect, contrasts both morphologically and sedimentologically with the more classical "carbonate shelves and platforms" of the Bahamas and SE Florida. It should be stressed that there are no clearly defined, regional "shelf edges" within the Persian Gulf comparable to those of the Caribbean carbonate provinces - and hence no true barrier reefs or "shelf edge" carbonate sand bodies. The oceanographic, "continental shelf edge" occurs outside the Persian Gulf where it is situated well below wave base.

The secondary "basins" , "shelves" and Arabian "homocline" are further modified, on a more local scale, by a complicated system of local highs and depressions; the former attain a relief of up to 50 m and are generally rocky in character. This highly complicated relief results in complex, local sediment patterns, especially where the sea floor is above wave base (see Purser, in this volume). Precise descriptions

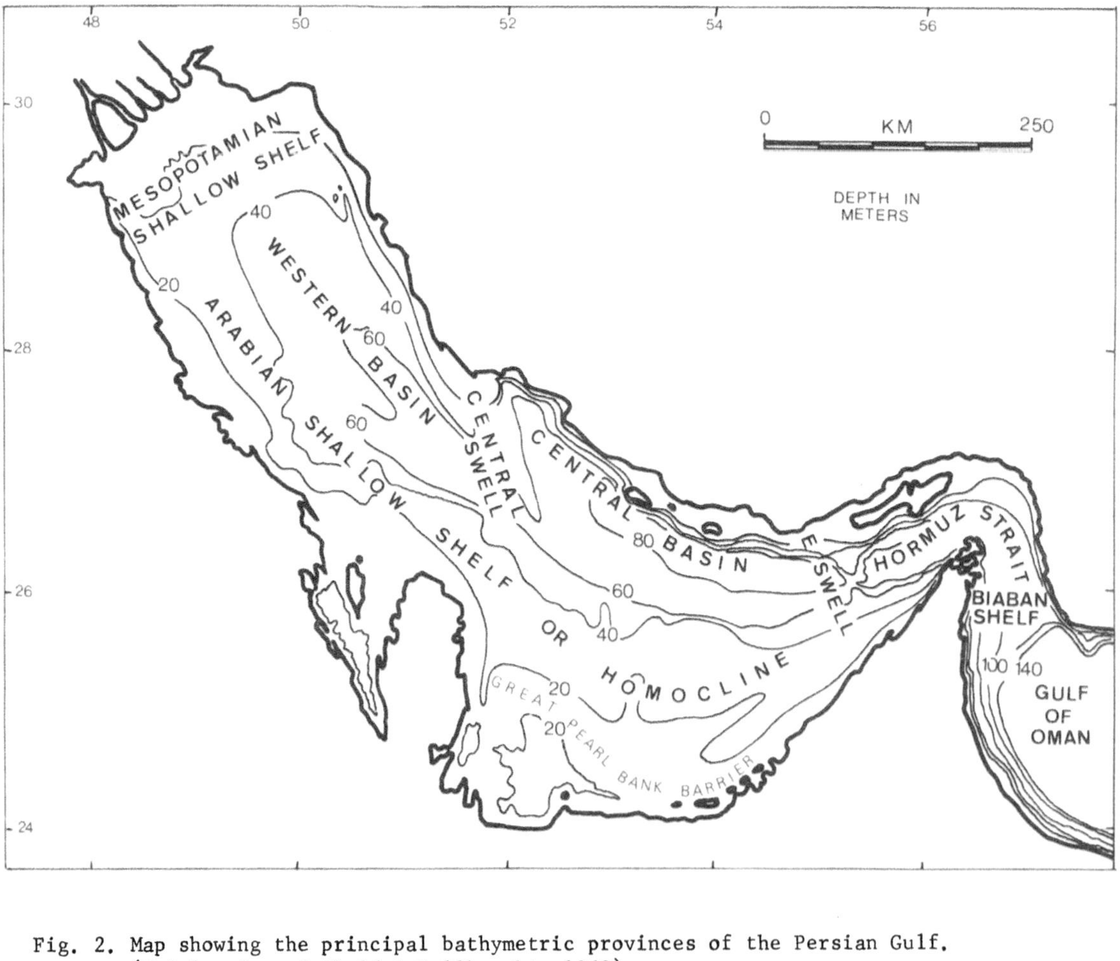

Fig. 2. Map showing the principal bathymetric provinces of the Persian Gulf. (Mainly after Seibold & Vollbrecht, 1969)

of the morphology of the Persian Gulf and its geological evolution are discussed by Kassler and by Seibold et al. in later contributions and by Seibold & Vollbrecht (1969).

B. REGIONAL CLIMATIC FACTORS

The Persian Gulf lies between the latitudes 24° - 30.30° N and has an arid, sub-tropical climate. Because it is almost surrounded by land its climate is essentially continental. As such it is characterized by marked seasonal fluctuations which, because of the absence of oceanic buffering, impart a high degree of variability and instability to the sedimentary environments within the basin. This contrasts with other carbonate provinces, including the Bahamas, where stable oceanic conditions prevail.

Strong winds characterize most desert areas including those bordering the Persian Gulf. The "shamal" blows mainly from the NW in the N parts of the gulf but tends to veer to N as one approaches the Trucial Coast in the SE. On Qatar Peninsula this wind attains Beaufort scale forces of 7-10, especially during the winter months (Houbolt, 1957). Summer months are essentially calm. In addition to their principal effects - the creation of waves and surface currents - these winds transport terrigenous sediment (including Pre-Holocene, dolomitic dust) to most marine environments. Dust storms are a well-known feature of the Mesopotamian Plain, at the N end of the gulf, and their sedimentary effects are discussed by Kukal & Saadallah in this symposium. Aeolian transport to the marine environments along the leeward coast of Qatar Peninsula is discussed by Shinn, and that along the Iranian shores by Seibold et al., in this volume.

Average annual rainfall in coastal Arabia is less than 5 cm (Evans, 1969) and fluviatile influx along the low, SW shores of the basin is virtually limited to occasional flooding of desert wadis following local storms. On the Iranian shore the mountainous relief, locally covered with snow, receives 20-50 cm of rain favouring numerous streams and small rivers.

Fluviatile influx ; while the climate of the Persian Gulf region is arid and local fluviatile run-off is low, the total catchment area has its N watershed far into the Zagros and Taurus mountains. These areas of higher rainfall supply the Tigris, Euphrates and Karun rivers which combine at the Shatt al Arab, giving a very considerable water influx. However, these rivers appear to deposit most of their sediment before entering the Persian Gulf, in tectonically subsiding marsh areas of Iraq (Berry et al., 1970). However, the smaller but more numerous streams draining the Zagros Mountain area in coastal Iran, characterized by frequent "flash-floods" , appear to deliver significant amounts of terrigenous sediment to the Iranian parts of the basin (Hartman et al., 1971, Sarnthein, 1971). Melguen, and Seibold et al. (in this volume) demonstrate that many sediment parameters can be related directly to the present Iranian river mouths. As noted already, the Arabian side of the basin, in contrast, lacks fluviatile influx; this is probably one of the fundamental reasons for the predominance of almost pure carbonate sediments throughout the Arabian half of the basin.

Summer air temperatures in Arabia frequently attain 45 - 50° C but the air temperature may decend to near 0° C in winter. The combined effects of frequent winds, high temperatures, and low precipitation result in excessive evaporation of Persian Gulf waters (124 cm/year, Butler, 1970), causing high salinities, especially in coastal areas. These factors lead to the formation of a suite of evaporite minerals, especially on the wide "sabkhas" of Arabia, discussed by Butler, Butler et al., and Hsü & Schneider in this symposium.

Oceanic influences are minimal because the Persian Gulf is almost completely surrounded by land masses. The absence of oceanic buffering permits marked variations in water temperature and salinity, as discussed in a previous section. Isolation from the adjacent Indian Ocean by the Masandam Peninsula (Fig. 1) also limits severely the delivery of planctonic elements to the Persian Gulf, the latter being confined to the vicinity of the Strait of Hormuz (see Hughes Clarke & Keij, and Seibold et al., in this volume).

C. INTERNAL ENVIRONMENTAL FACTORS

Regional currents: The high loss of water in the gulf through evaporation is not compensated by precipitation and river inflow. As a result a slow circulatory surface current flows into the gulf moving anticlockwise along the Iranian coast (Emery, 1956; Hartmann et al., 1971). Although this current has little demonstrable effect on the texture and transport of sediment (apart from planctonic organisms) it brings new oceanic water into the basin and thus plays an important rôle in determining the distribution of salinity, temperature and nutrients within the Persian Gulf waters. During the summer, surface waters attain temperatures of 36° C in the central parts of the gulf, higher temperatures being recorded in coastal areas (see previous sections). Temperatures in winter, however, may fall below 20° C. Salinity of the surface waters increases from approximately 36.6 % near the entrance to 40.6 % near the NW end of the basin. (Figs. 3, 4). Due to the combined effects of water cooling and evaporation the highly saline surface waters sink to the bottom, raising salinity and lowering the temperatures of these deeper waters (Hartman et al., 1971). Waters flowing out of the gulf follow the deeps near the Masandam Peninsula, their more saline waters with relatively high oxygen and nutrient contents being traceable beyond the edge of the continental shelf. This bottom current has significant sedimentological effects, discussed in this symposium by Seibold et al.

Tidal currents: Regional tidal currents are oriented approximately parallel to the axis of the gulf. They attain velocities of ca. 50 cm per second some 0-4 m above the bottom (Hartman et al., 1971) influencing sediment textures even in the deepest parts of the basin (see Seibold et al., in this volume). In coastal channels of Abu Dhabi, tidal velocities may exceed 60 cm per second (Evans, 1970), their bi-directional movements favouring the local development of spectacular oolitic tidal deltas discussed by Loreau & Purser in a subsequent contribution. Tidal range varies between 1.5 m on the coastal barriers of Abu Dhabi but diminishes to less than 1 m within the adjacent lagoons (Evans, 1970) and averages 0.5 - 1 m along the coast of Qatar (Houbolt, 1957).

Wind-driven currents and waves: Waves and surface currents are almost certainly the most important mechanisms of sediment transport along the relatively shallow Arabian parts of the basin, and also along parts of the Iranian coast. Because waves and surface currents are related to the NW "shamal" wind they are directed mainly towards the SE. Most of the Arabian coast and adjacent offshore environments are exposed to these winds and their sediments are frequently dominated by "high energy" bioclastic and oolitic sands. These coarse sediments can occur down to ca. 20 m in the SE parts of the Persian Gulf, suggesting that effective wave base may extend down to that depth. The waves and surface currents produced by "shamal" winds, often of several weeks' duration, together with tidal currents, tend to homogenise the surface waters to a depth of ca. 30 m. In the axial parts of the gulf, wave action may attain twice this depth (Sarnthein, 1970). Because the "shamal" winds blow down the axis of the gulf from the NW, wave base is almost certainly at its minimum along the sheltered shores of the Shatt al Arab, at the NW end of the basin.

Water depth: Water depth influences both the degree of water agitation and light intensity of the sea floor and thus, indirectly, the textures and composition of its sediments. Most of the floor of the Persian Gulf is deeper than 20 m and it is characterized by moderate to low degrees of water agitation and muddy sediments. These grade into "higher-energy" carbonate sands towards the southern, shallow, more agitated coastal environments or the tops of numerous offshore highs. Coastal sediments on the Iranian side, however, are generally fine due to terrigenous supply.

The depth distribution of algae and hermatypic corals (discussed by Hughes Clarke & Keij in this volume) suggests that the euphotic zone extends down to approximately 20 m in the southern (Arabian) parts of the gulf, but exceeds 30 m in the clearer, axial parts. Light penetration is influenced by the turbidity and seems to be more limited in depth within the more agitated waters of the Trucial Coast Embayment than elsewhere in the Gulf. Measurements by Ziegenbein (1966) have shown that light penetration in the Persian Gulf is considerably less than that in oceanic provinces such as the Bahamas and Great Barrier Reef.

Restriction: "Restriction" may be defined as a divergence in the physical and chemical properties of sea water relative to oceanic water at the same depth and latitude, and it is reflected in the diversity and type of organisms able to tolerate it (see Hughes Clarke & Keij, in this volume and "adjacent sea effect" in Sarnthein, 1971). It is most readily expressed in terms of a single measurable parameter such as salinity or temperature.

Salinity: Most of the Persian Gulf can be considered partially restricted due to the limited water interchange with the Indian Ocean through the narrow Straits of Hormuz. Surface salinities in the central parts of the gulf average 37-40‰, while shallow parts of the Trucial Coast have salinities of 40 - 50‰, rising to 60 - 70‰ in remote lagoons and coastal embayments such as the Gulf of Salwa (Fig. 3). In the axial parts of the gulf salinity increases 2-4‰ with depth (Fig. 4), as shown by Seibold (1970).

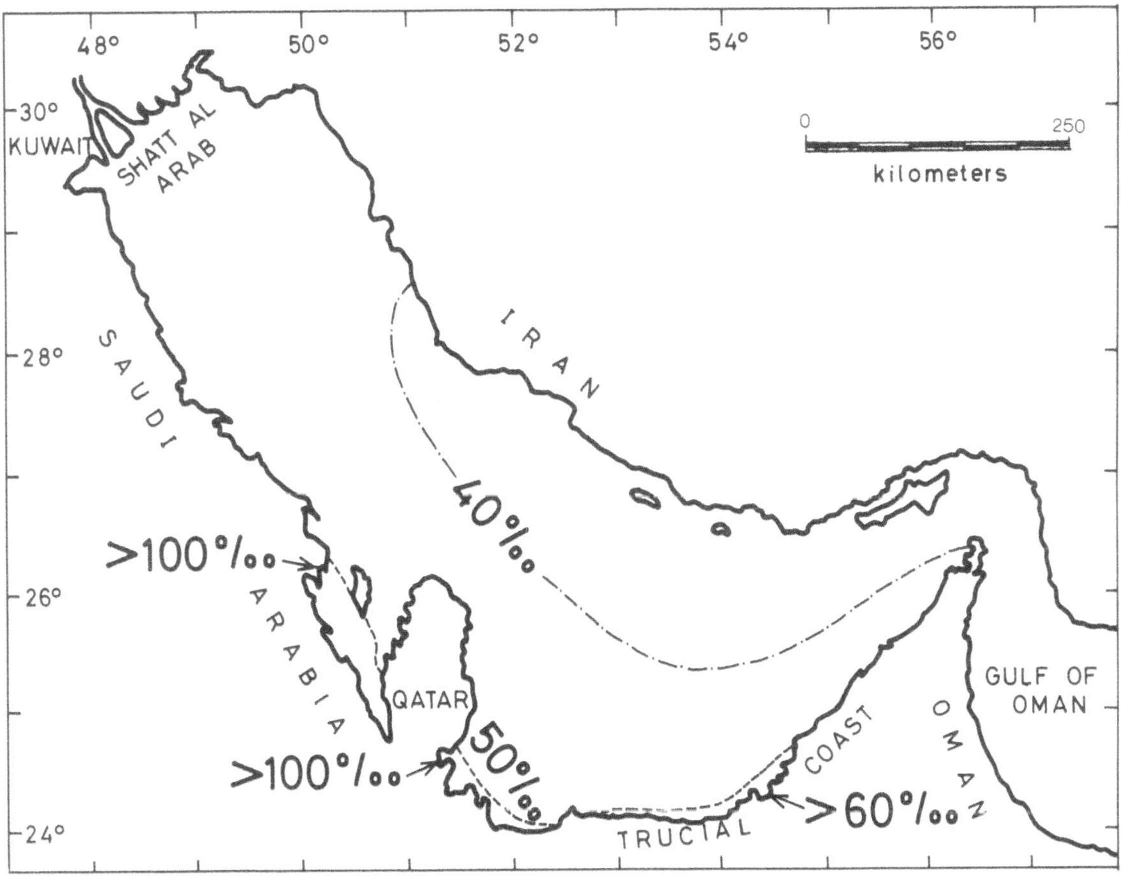

Fig. 3. Map showing major salinity trends within the Persian Gulf. (Based on Emery, 1956; Brettschneider, 1970; and Shell Research data)

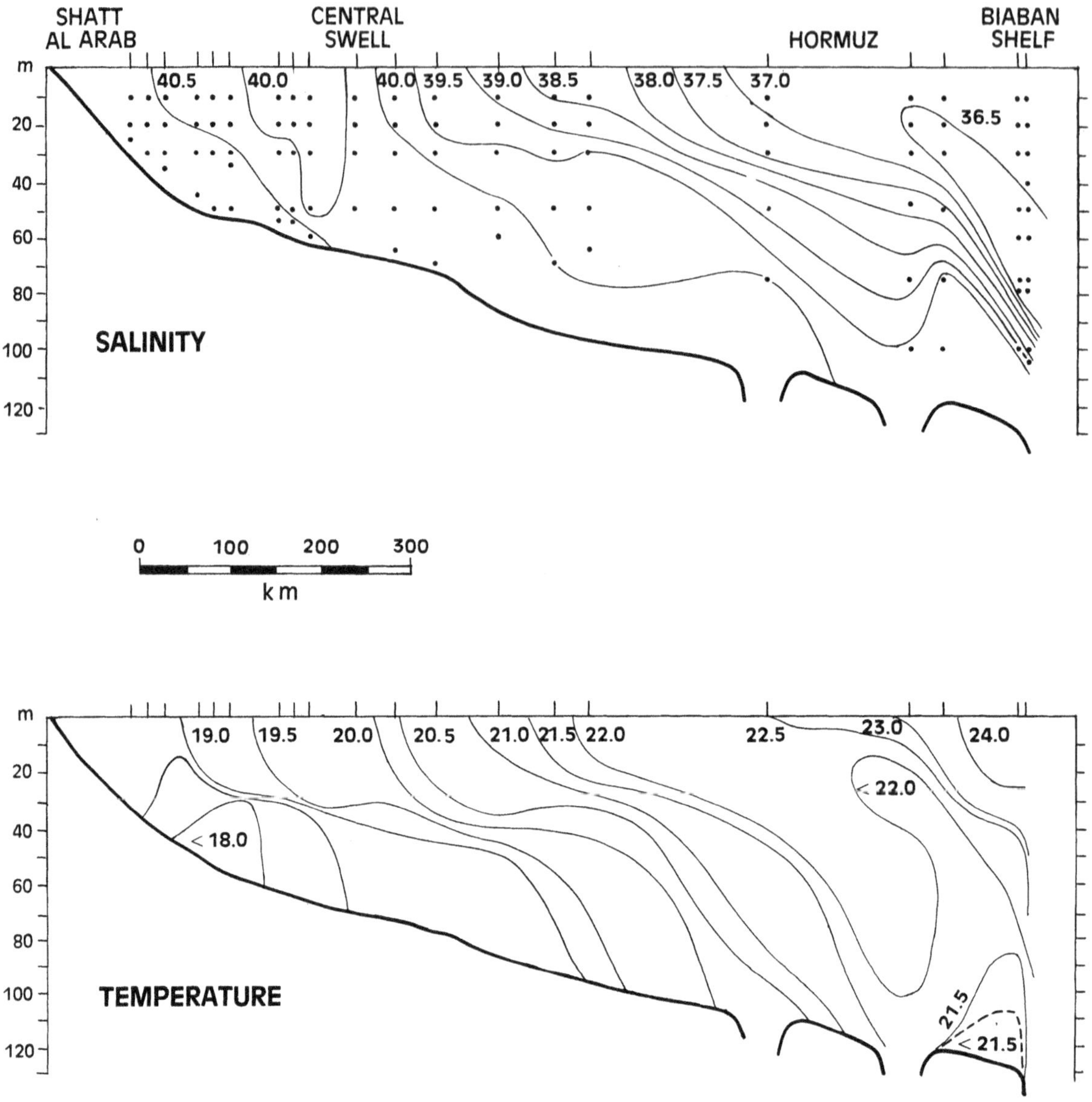

Fig. 4. Profiles along the axis of the Persian Gulf showing vertical distribution of salinity and water temperatures. (After Seibold 1970)

Temperature: The range of variation in water temperature tends to increase away from the entrance of the Persian Gulf. Temperature and salinity trends are broadly comparable, both attaining a maximum in shallow embayments and lagoons isolated from the main body of Persian Gulf water. In these geographically isolated situations water temperatures (in Abu Dhabi and Qatar lagoons) can fluctuate between 40° C in summer and 15° C in winter, seasonal variations being less marked in the deeper parts of the basin. Temperature profiles measured in the axial parts of the gulf (Fig. 4) suggest a poorly-defined thermocline (Seibold, 1970) which tends to rise from ca. 40 m near the Straits of Hormuz to ca. 20 m near the NW end of the gulf.

Nutrients: Because of the previously-mentioned pattern of water circulation, the Persian Gulf contains oxygen down to the bottom. The nutrient content, however, is low, with the exception of mixing zones that extend from the continental shelf margin to the Straits of Hormuz (Hartmann et al., 1971).

Substrate: The Arabian parts of the Persian Gulf have widespread areas of rock bottom, especially at depths of less than 35 m on the Arabian Homocline. These rock substrates are the product of Holocene submarine lithification (Shinn, 1969), but also include Tertiary limestones and dolomites. Generally, a thin layer (5-20 cm) of bioclastic sand lies on a rock pavement, the hard surface, bored and encrusted with organisms, being exposed frequently between ripples or irregular patches of sand. These widespread rock bottoms undoubtedly favour an epifaunal biota and thus influence the composition of biogenic sediment. They may be an important factor favouring the overall dominance of skeletal carbonate in the southern Persian Gulf. In the deeper, Iranian, parts of the basin, however, substrates generally are muddy and fixed epifaunas are more or less absent. Exceptionally, secondary hard substrates may be formed by skeletal debris and/or coarse relict particles in areas of low sedimentation rates. On mud bottoms off Iranian river mouths, down to depths of approximately 70 m, infaunas are more important.

Biogeography: The biogenic carbonate sediment in the Persian Gulf is characterized by the absence of certain skeletal elements known to be common in similar environments elsewhere. Significant absentees include the algae *Halimeda* and *Penicillis* and many of the larger molluscs (see Hughes Clarke & Keij, in this volume), which may be excluded from the Persian Gulf by some intolerable environmental factor. However, it is possible also that some forms have not migrated into the gulf since the Flandrian transgression, due to slow rates of spread or to the presence of migration barriers outside the Persian Gulf.

The surface sediments of the Persian Gulf, in common with most other marginal basins, are a complex mixture of modern and relict elements. While the former are essentially the product of factors outlined in the preceding text, it is evident that the relict fraction may have been produced under somewhat different conditions (Sarnthein, 1971). In other words, the composition and distribution of the surface sediments of the Persian Gulf should not be analyzed only in terms of modern environment.

The Structural and Geomorphic Evolution of the Persian Gulf

P. Kassler[1]

ABSTRACT

The origin of the present-day morphology of the Persian Gulf has been studied and is summarized in figure 3. The Gulf is a tectonic basin of late Pliocene to Pleistocene age, whose morphology is greatly influenced by the tectonic style. The topography of Iran and the Iranian coastal islands is controlled by the intense folding of the Zagros orogeny, on NW-SE to E-W trends. The much more subdued relief of the Arabian side is the result of gentler tectonic movements: Plio-Pleistocene folding, faulting and salt diapirism, superimposed on older, predominantly north-south-trending growth structures ("Arabian folds"). There is evidence over much of the Gulf of interference between Arabian and Zagros folds.

This tectonically controlled morphologic pattern was subdued by sedimentation of Pleistocene limestones, but locally rejuvenated by Quarternary tectonic adjustments.

The sea level fell by as much as 120 m during the Pleistocene, emptying the Gulf; river valleys were eroded down the slopes. The sea then cut a series of platforms, at its level of maximum retreat and at times of relative standstill during the post-glacial rise. However, in spite of these Pleistocene physiographic modifications, the underlying tectonic control of morphology is still apparent, and there is a partial correlation between bathymetric and structural highs and lows.

The topography controls the type and the thickness of the marine sediments. Sediment type is largely a function of the biological communities which give rise to skeletal material; these vary in character from shoals to depressions. Sediment thickness is shown by sparker records to be least on topographic highs, and greatest in depressions.

The Recent unconsolidated sediments are the product of the post-glacial Flandrian transgression, which, according to Fairbridge (1961), began about 18 000 years B.P. and reached its present level about 5000 years B.P. These sediments are expected eventually to smooth out the pre-Recent topography by filling up the depressions and extending over the highs. The thickest Recent sediments are found in the Gulf axis.

---

1 The essence of this paper, with P. Kassler's agreement, was presented by M.W. Hughes Clarke at the 1971 International Geological Congress at Heidelberg. P. Kassler is now with N.V. Turkse Shell, Ankara, but was with Shell Research B.V. Rijswijk in 1967-1968.

INTRODUCTION

The framework of the Persian Gulf is structural, much of the topography being closely related to this structure, with contributions from Late Tertiary and Quaternary erosion and sedimentation. The object of this contribution is to evaluate these factors and describe the resulting morphologies, which have a strong influence on the patterns of Recent carbonate sedimentation in the Gulf.

During early 1968 the author spent 6 months at the Laboratory of Shell Research B.V., Rijswijk, The Netherlands, working on a variety of material. This included a network of "sparker" profiles (total length almost 2000 km; see fig. 1), and accompanying echo-sounder and asdic records, which together with the British and American Admiralty charts, give a reasonable picture of sea-bottom morphology. The sparker used is a shallow-penetration device, the deepest reflection observed being about 30 m below sea-bottom. Other information was obtained from commercial foundation surveys of offshore oil fields. The coastal morphology of Qatar and the Trucial Coast was studied by means of aerial photographs. A number of unpublished oil company geophysical maps and reports provided background structural information.

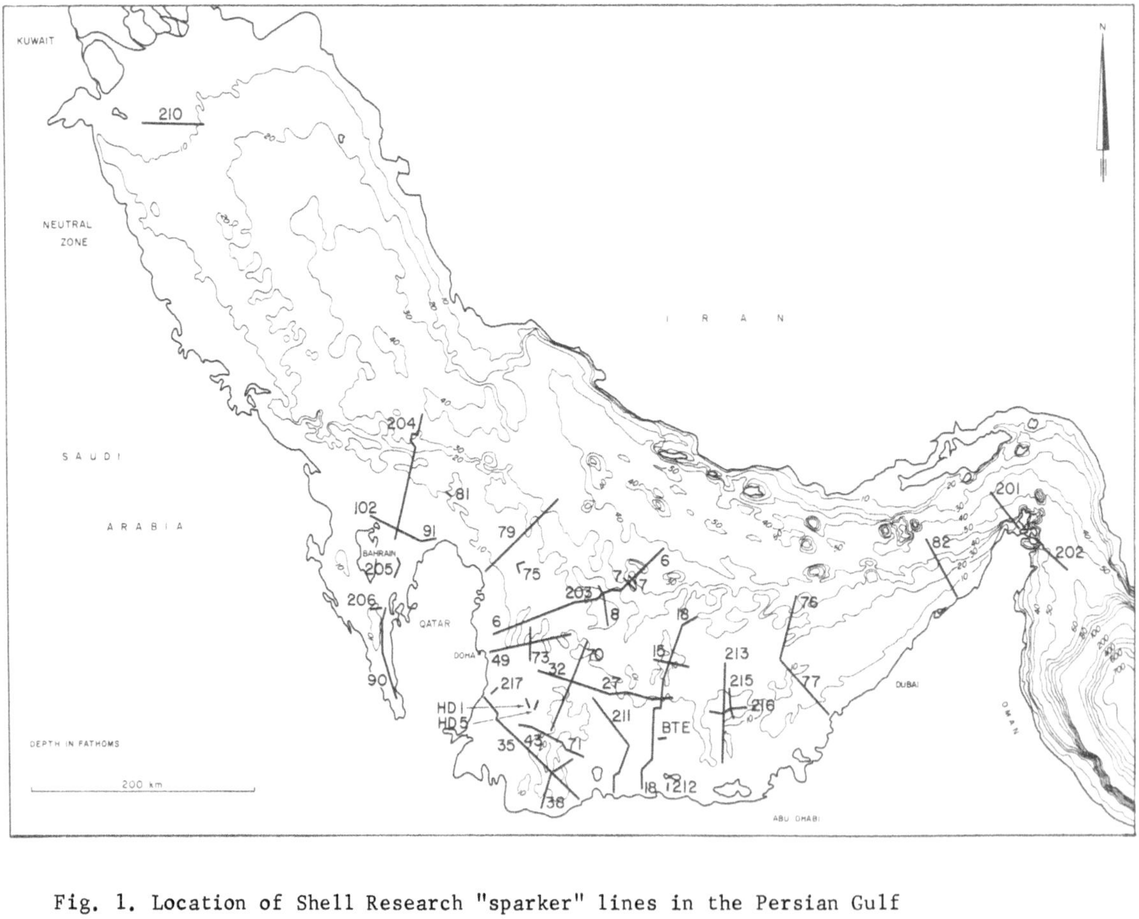

Fig. 1. Location of Shell Research "sparker" lines in the Persian Gulf

Up to 1968 a number of authors[2] had written on various aspects of the structural development and morphology of the Persian Gulf, and much of the author's information is derived from the then existing published literature, together with unpublished oil company reports. This includes work on theoretical and practical aspects of structures, sedimentology, and Pleistocene sea-level movements, and is referred to in the text. Parts of the report referring to the Gulf of Salwa owe much to discussion with A.J. Keij and G.R. Varney of KSEPL, and the work of Houbolt (1957) was an invaluable starting point for considerations of submarine morphology. Regional background information on Recent carbonate sedimentation was available thanks to the work carried out by Shell Research B.V., Rijswijk, The Netherlands on the carbonate sediments of the Persian Gulf.

GEOLOGICAL SETTING (Figs. 2 and 3)

The present-day Persian Gulf is a shallow tectonic depression nearly 1000 km long, formed late in the Tertiary in front of the rising Zagros mountain front. The basin is asymmetrical, the slope of the Arabian flank being much gentler than that of the Iranian side, and the deepest waters lie close to the Iranian shore. Depths of up to 95 fm (165 m) occur, although most of the axis is about 40 to 50 fm (74-92 m) deep. There is no true shelf margin on either flank, although on the Arabian side there is a relatively marked slope break at about 20 fm.

The difference in slope between the two flanks of the Gulf reflects the fundamentally different tectonic histories of Iran and Arabia. The low-lying, arid Arabian Coast and adjacent shallow sea floor are characterized by large, low-dipping anticlines, with NS or NE-SW ("Arabian") trends, which include most of the major oil fields of the region. Some of these anticlines, such as Dukhan and Bahrain, have probably had continuous relief, as a result of salt growth, since the Mesozoic. The Arabian coastline is itself partly structurally controlled. The anticlines of Qatar, Dukhan, Bahrain and Damman make prominent if low-lying topographic features. A number of structural observations indicate that the coasts of Saudi Arabia and the northeastern Trucial States (Hasa and Pirate Coasts) may be fault-controlled.

The Iranian coast, on the other hand, is mountainous, ridges up to 1500 m high being formed by anticlines of the Plio-Pleistocene Zagros orogeny with trends varying from NW to EW. These are huge folds with steeply-dipping flanks, the result of orogenic movements directed towards the Gulf. Thus, these structures differ radically in geometry and origin from the Arabian folds. The chain of islands in the Gulf (Kharg, Sheikh Shuaib, Qishm, etc.) form part of the Zagros foothills belt.

The interaction of Arabian and Zagros folds is apparent in the submarine topography of the Gulf. Although the sea-floor gradient is predominantly very gentle, there are some 20 islands, as well as numerous submarine highs or shoals. Most of the islands are piercement salt domes and many of the shoals are also clearly due to salt diapirism. Other shoals, however, show Arabian or Zagros trends or combinations of these.

The highs exert considerable influence on the sedimentary pattern because the biologic communities they support provide local sources of sand-size skeletal sediment, as discussed by Purser elsewhere in this volume.

The main topographic feature of the head of the Gulf is the Tigris-Euphrates delta. The present-day delta is constructed of lobes of clastic sediment up to 30 m thick, extending almost 100 km seawards from the river mouths towards the Gulf axis.

The Oman Mountains form a large, rugged range which sweeps round from NW-SE to a NS trend as it approaches the Gulf. The last major uplift of the range began in the late Tertiary and continued into the Recent. The northern plunge of the

2 As this study was carried out in early 1968, it refers only to those works either published or known to the author in the prepublication stage before September 1968.

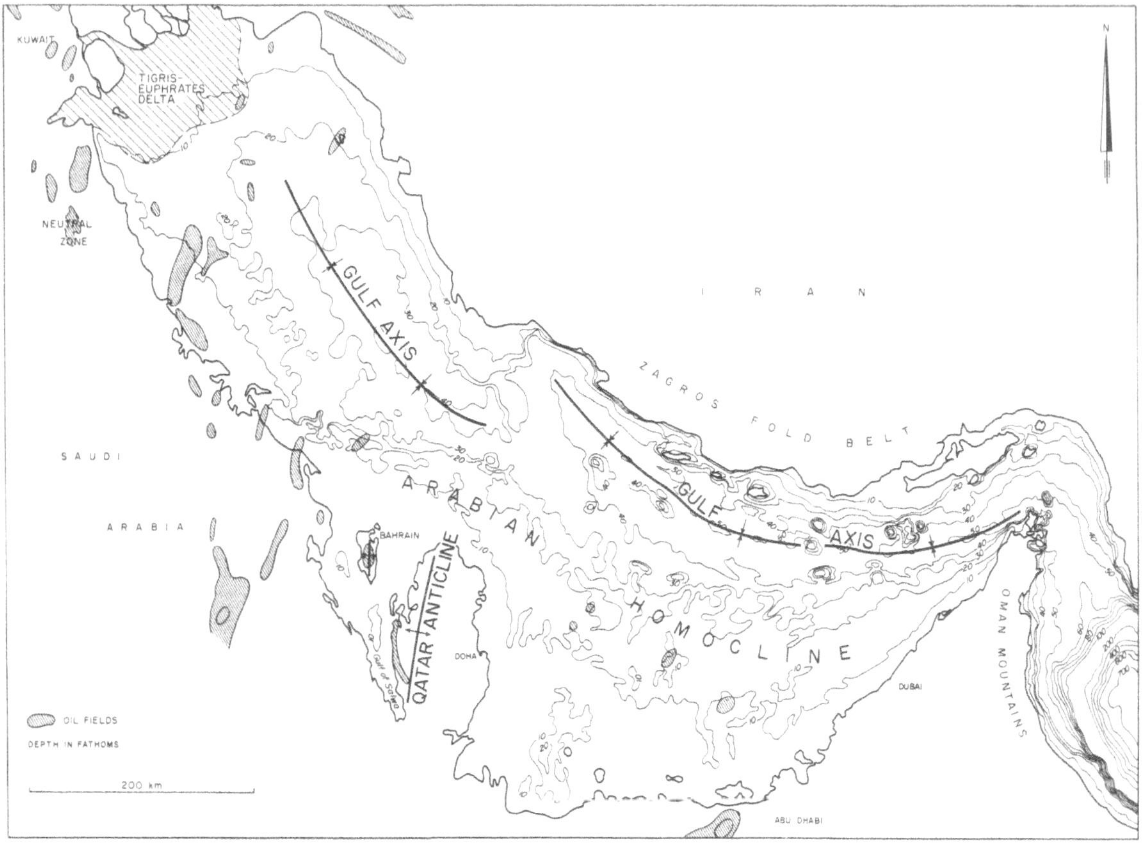

Fig. 2 Geological setting of the Persian Gulf

range into Hormuz Strait causes a severe constriction of the mouth of the Gulf, which leads in turn to a restriction of water circulation and of animal life over the entire basin. Thus, the existence of this topographic feature influences the character of the sediments over a very large area.

STRUCTURAL EVOLUTION

Pre-Pliocene movements

Growth structures

Most of the subsurface structural work done in the Persian Gulf area deals with the Mesozoic history of the major oil field growth structures (e.g. ARAMCO, 1959 and Elder, 1963). Three main categories of pre-Pliocene structural uplift are recognized:

linear growth structures having a NS or NE-SW ("Arabian") trend (e.g. Ghawar field, Saudi Arabia)

domal growth structures, often aligned along Arabian trends (e.g. Umm Shaif field, Abu Dhabi)

piercement salt domes (e.g. Yas Island, Abu Dhabi, Fig. 5)

There has been a great deal of discussion about the origin of the growth structures. The domal growth structures and piercement domes are clearly due to salt diapirism. Some oil company geologists also invoke this mechanism for the linear anticlines, whereas others (ARAMCO, 1959) believe these to be due to basement uplift.

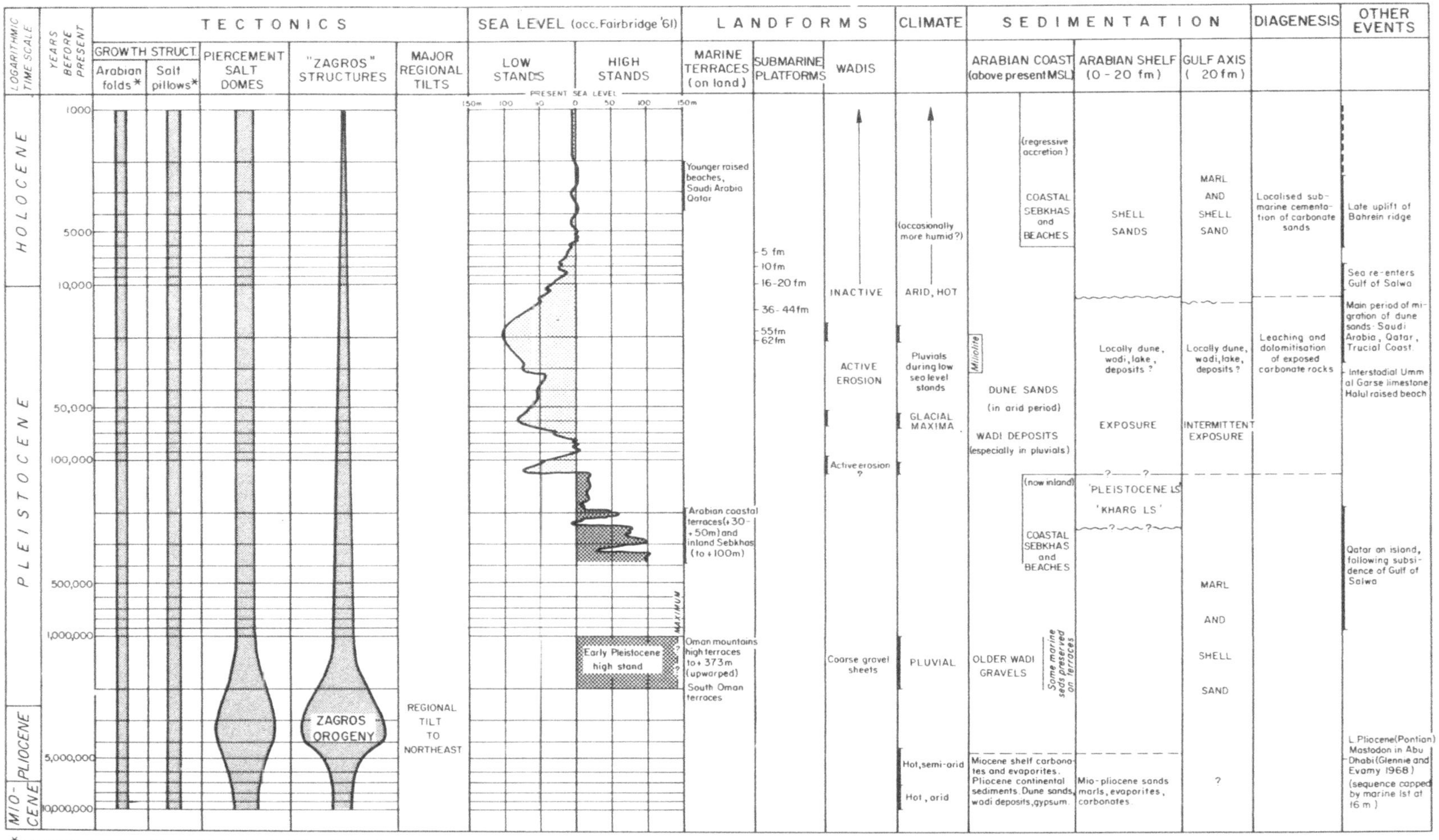

* Growth structures show local variation in time and intensity of movement

Fig. 3. Generalized late Tertiary and Quaternary history of the Persian Gulf and Arabian coastal areas (logarithmic time scale)

Fig. 4. Interpretation of the shallow structure and topography of the central Persian Gulf, based on sparker profiles and echo-sounder

Arabian growth structures have two striking characteristics: prolonged upward movement during sedimentation and low vertical relief. Very prolonged structural growth has been documented (unpublished oil company reports) as dating, perhaps, from as early as the Permian and lasting well into the Tertiary. In some cases the movement was pulsatory, in others very regular. In spite of the long periods during which uplift occured, this was very slow, as attested by the very low dips (10° maximum) found on the flanks. Growth structures are generally expressed in the stratigraphic record by subtle variations in thickness or lithology.

Pre-Pliocene movements and topography

The effects of pre-Pliocene movements have, except locally, been obscured by later tectonics and sedimentation. The Bahrain and Qatar surface structures may be exceptions to this, as they seem to owe their present form largely to pre-Pliocene movements.

Plio-Pleistocene movements - the Zagros orogeny

Folding

The most intense folding occurred on the Iranian mainland where cross-sections of the mountain range show folds with amplitudes of several thousand meters (British Petroleum, 1956). The intensity of the folding diminished markedly from the Iranian coast seawards. Even on the coastal islands, each of which is a Zagros anticline, the folds have dips of 10-20°, compared to 50° and more on the mainland (H.R. Grunau & M.F. Shepherd, personal communication).

The main axis of the Persian Gulf is a Zagros feature and was formed in Plio-Pleistocene times. In fact the axis consists of two separate synclines (see encl. 1 and fig. 4); the more easterly of these trends is situated relatively close to the Iranian coast in the Kangan area, and is separated from the north-west axis by the prominent "Jabrin Ridge" (= Central Swell, Seibold et al., in this volume), probably a large Zagros anticline. The area between the Qatar Peninsula and the Jabrin Ridge is a transverse shoal between two synclinal areas and seems to reflect uplift along the northward prolongation of the Qatar anticlinal axis.

Sparker reflections from probable Miocene levels in the Gulf of Salwa, W of Qatar (Fig. 4) suggest folding on axes oblique to the marked NS trend of nearby growth structures such as Dukhan and Awali (Bahrain). Two of these post-Miocene structures seem to be aligned with the Uwainat and Jaleha culminations of Dukhan, suggesting that the oil field was affected by Zagros cross-folds. It is easiest to visualize (as suggested by G.R. Varney in an unpublished report) the whole Salwa-Bahrain-Qatar area as a structural complex of north-south Arabian trend structures deformed slightly by NW-SE (Zagros trend) cross-folds. The same may be true of other areas of the Arabian shelf, for which little information on Plio-Pleistocene folding is available. In particular, areas of present-day topographic shoals with a "Zagros" orientation (Fig. 4) suggesting Quaternary uplift along these trends, must also be strongly suspected of having had some Pliocene folding. The area of Rig az Zakum - Abu al Abyad - Khor al Bazm in offshore Abu Dhabi is another possible example. Zagros folding has also been reported from the Kuwait area (Fox, 1959).

Salt diapirism

The Plio-Pleistocene was a period of extensive extrusion of salt domes (R. Blaser, personal communication) in the Gulf as well as on the Iranian mainland, due in part to Zagros fold movements. Some of the salt-dome islands may occur at inter-sections of Zagros folds and Arabian growth structures (a situation analogous to that described by de Sitter, 1965, pp. 314-7, from the High Atlas). Several salt-dome islands show doming of Mio-Pliocene beds around the central plug (e.g. Yas Island, fig. 5).

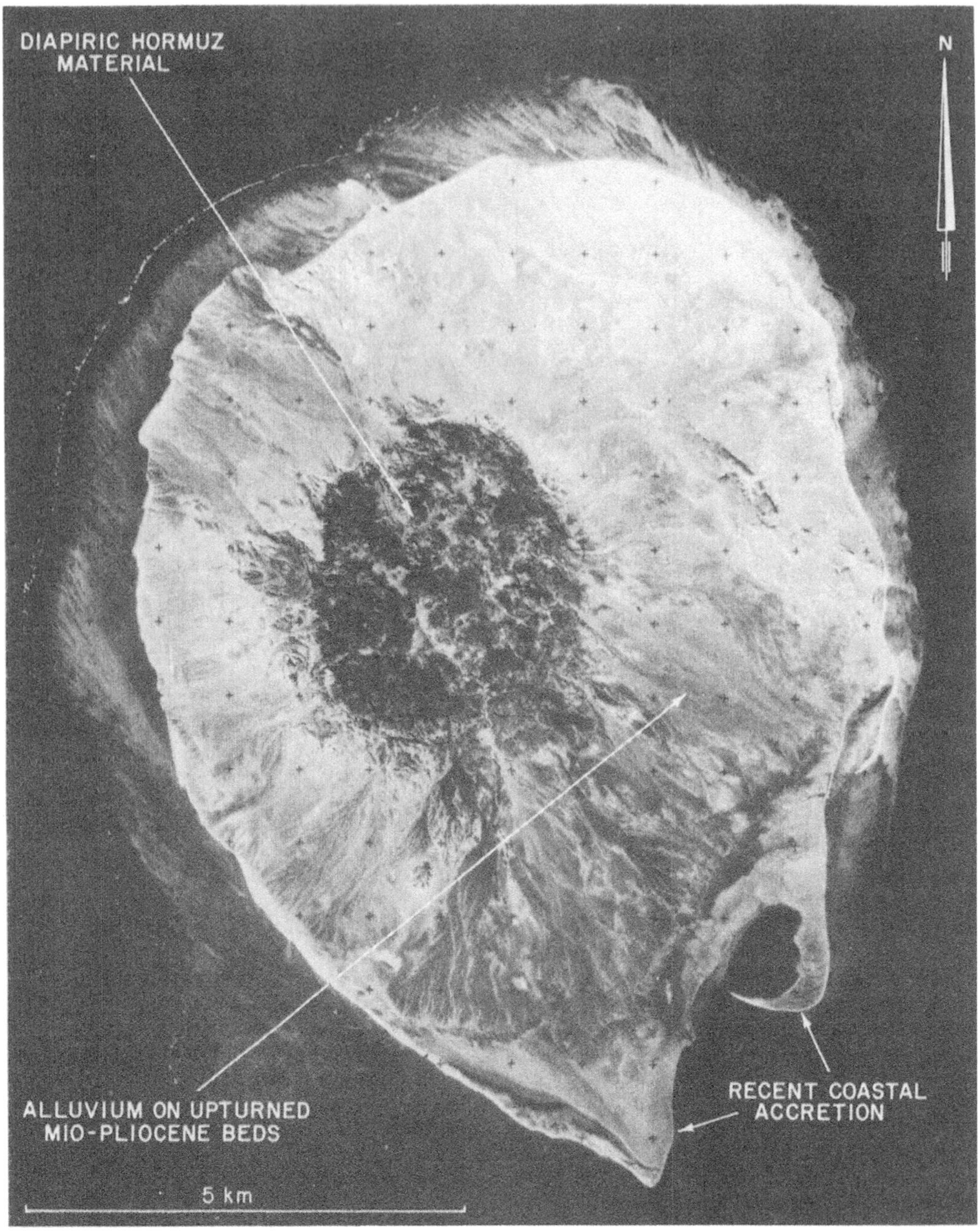

Fig. 5. Aerial photograph of Yas Island, Abu Dhabi showing piercement salt dome core of Paleozoic "Hormuz" rocks

Faulting and fracturing

"Shell" sparker profiles of the eastern Gulf of Salwa, and a photogeological study of Qatar and Abu Dhabi, provide evidence of a fault and fracture system affecting Miocene and possibly Pliocene rocks. The trend is northwest to north-northwest, similar to that of a large portion of the Saudi Arabian coastline, which may itself be partly controlled by faulting. One such fault apparently affects the southern end of the Dukhan structure (Qatar) (fig. 4), causing the axis to swing away from the main north-south trend.

In addition to the faults that have been mapped, aerial photos show certain outcrops of Miocene rocks to be affected by parallel, closely-spaced, NNW-oriented fractures which have caused an alongation of the outcrops in this direction. This elongation persists into Abu Dhabi where, for example, at Jebel Baraka it affects beds dated by Glennie & Evamy (1968) on the basis of mastodon remains as Lower Pliocene (Pontian). Here, however, the outcrop orientation can no longer be definitely identified on photos as due to fracturing, and may be confused with some erosional effect of the shamal wind.

The eastern Trucial Coast, from Abu Dhabi town to Ras al Khaima is very straight and may also lie on a fault trend complementary to that of the Hasa Coast.

Uplift of Oman Mountains

Little direct evidence is available to permit precise dating of this late Tertiary uplift, but a zone of tectonic and topographic complexity occupying the whole of the Hormuz Strait suggests a close connection with the Zagros movements; the main uplift of the Oman Mountains may, similarly, have started in the Pliocene and, as discussed in a subsequent section, continued well into the Quaternary.

Tilting of the Arabian homocline

The Arabian homocline probably had a seawards tilt prior to the Zagros orogeny. This is suggested by the regional thickening of Mio-Pliocene formations in this direction and also suggested on sparker line 204. During the Zagros movements the Gulf axis subsided, and this tilt was amplified. A number of onshore and offshore Arabian oil fields, including Ghawar (ARAMCO, 1959), have oil/water contacts tilted to the northeast.

The tilt of the Arabian land surface is also shown by the north-easterly trend of older Quaternary wadi gravels (Holm, 1960 and Powers et al., 1966), indicating a stream drainage system flowing from Central Saudi Arabia across S. Qatar and W. Abu Dhabi.

Sparker profiles show the existence of one or more tectonic hinge zones (fig. 4), along which a slight steepening of the Gulf-ward dip occurs. In some cases the present-day topographic gradient steepens slightly along these hinge-lines.

Plio-Pleistocene movements and topography

The foundations of the modern Persian Gulf were largely laid in the Plio-Pleistocene Zagros orogeny. The outlines of the Gulf, with the Zagros range to the north, the offshore islands of Iran, the subsiding axes, and the possibly fault-controlled Hasa and Pirate Coasts were formed at this time, as was the Oman Mountains uplift with its profound effect on circulatory patterns within the Gulf. The broad distribution of shoals, so fas as these are structurally controlled, is likely to have been similar to that of today, although the detailed appearance of highs and lows may have been rather different. The topographic "grain" of the Gulf, which has influenced all subsequent physiographic processes, dates back to the Plio-Pleistocene and is essentially tectonic in origin.

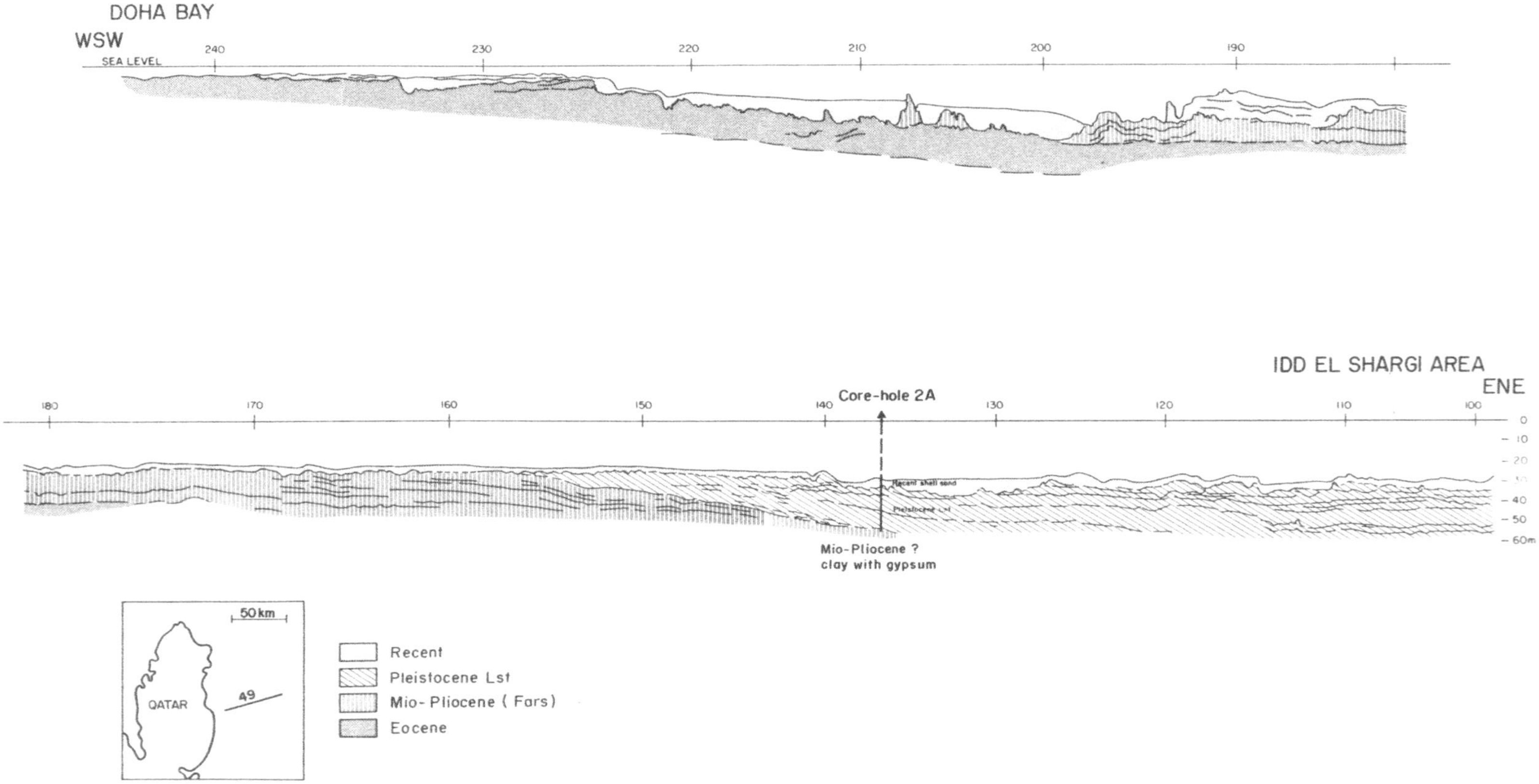

Fig. 6. Sparker line n° 49 showing the relationship between pre-Holocene basement rocks and Holocene sediment cover

Quaternary Movements

Arabian coast

Photogeological study of present-day subaerial drainage patterns on Qatar Peninsula suggests that they are partly radial around a number of low domes. The data of uplift of these features can only be suggested very tentatively as post-dating a presumed peneplanation of Qatar by a Pleistocene sea at higher eustatic level. The southernmost dome corresponds to a residual negative gravity anaomaly and is probably a salt feature. The northern group lies on the trend of the Bahrain/Pearl Bank Ridges and is probably related to Zagros trend cross-folding, although salt movements may also have had some influence.

Oman Mountains

These were uplifted considerably in the Quaternary. Lees (1928) describes a series of raised marine terraces at elevations up to + 373 m on the E flank of the range, with marine fauna (Venus, Cardium, Turritella, Cerithium, oysters); "in a state of preservation similar to that of sub-Recent raised beaches". These beaches could have been formed during early Pleistocene periods of eustatically high sea level, which may have reached + 150 m, according to observations recorded in Saudi Arabia (Holm, 1960), and elsewhere in the world (Fairbridge, 1961). However, judging from Lees's description, they are likely to be much younger. Similar terraces in the northern Oman mountains are tilted, and locally folded (K.W. Glennie, personal communication).

Arabian homocline

Evidence of Quaternary structural movements in the shallow offshore waters is derived largely from dips visible on sparker profiles. This can be ambiguous for two reasons:

- the ages of reflecting horizons are often not known;
- it is difficult to distinguish depositional dips from structural dips.

In the area E of Qatar, however, where most of the profiles were made, the reflections may emanate from a distinctive Pleistocene limestone encountered in a number of offshore wells, which is apparently affected by structural movements (Fig. 6).

The pattern of dip vectors obtained from these shallow reflectors (fig. 4) agrees fairly well with the topography and suggests a strong structural influence on morphology. The shapes of shoals and depressions east of Qatar are strongly reminiscent of those produced by folds of two interfering systems and both Zagros and Arabian trends are discernible.

The oil field structures, Rig az Zakum and Umm Shaif (offshore Abu Dhabi) are both marked by shoals. In each case, the shoal has a slightly more "Zagros" shape than the underlying growth structures, an indication of slight re-orientation of old structures by subsequent movements of a different stress-pattern.

The subsidence of the Gulf of Salwa (W. Qatar) is a Quaternary feature which postdates the "older Quaternary wadi gravels" mentioned above. G. R. Varney, (unpublished oil company report) suggests that the Bahrain Ridge (fig. 4) has been rising in the last few thousand years. Partial confirmation of this view comes from the recognition within the Gulf of Salwa of Holocene marine platforms at -5 fm and -10 fm. With the present configuration of the Bahrain Ridge, the rising Holocene sea could not have entered the Gulf of Salwa before it reached -4 fm (about 6000 years ago, according to Fairbridge, 1961). The platform at -10 fm would be about 8000 years old, however, and still earlier marine conditions are suggested by the recent dating of an oolitic sand from there at 11 000 $\pm$ 400 years old. Although the Bahrain Ridge is clearly a young structural uplift, the Holocene blockage of the Gulf of Salwa may have been accomplished partly by uplift and partly bu sedimentation (the ridge is an area of active coral reef growth). The same general arguments may be advanced for the Pearl Bank Ridge, (Fig. 4) which is probably also a Quaternary structural feature.

The most spectacular Quaternary movements in this area are those of the piercement salt domes. These attain topographic elevations of up to 161 m (Zirko Is.), well above those of nearby coastal areas. Quaternary raised beaches occur on many, if not all, of the islands. W.A.C. Russell, (unpublished oil company report) cites "Quaternary littoral deposits" at + 60 m on Das Island, at + 45 m on Zirko and at + 18 m on Ardhana and Qarnain. G.L. Nicol (personal communication) noted on Halul Island that raised beaches "with a striking similarity to present-day forming beaches" occur at up to + 35 m, and that in some cases they are tilted. Shells from one of them were dated by C^{14} and gave ages ranging from 32 700 years B.P. These dates should, according to Fairbridge (1961), correspond to the mid-Würm interstadial, when the sea level rose temporarily to about -40 m (-15 m according to Curray, 1961). Thus, Nicol argues that the present elevation of the beaches is the result of salt diapirism and not of a higher sea level.

Iranian Coastal Islands

H.R. Grunau and M.F. Shepherd (personal communication) made observations that clearly demonstrate Quaternary movements. The Kharg Limestone, which is Plio-Pleistocene according to micro- and macro-fauna, is unconformable on Tertiary formations. On the islands of Kharg, Hendurabi and Sheikh Shuaib it is arched gently over the culmination of the underlying Zagros anticline. Several coastal islands show evidence of further movement in the form of sub-Recent raised beaches.

Zagros Mountain Range

Lees and Falcon (1952), writing of the area at the head of the Gulf, describe the elevation of the Zagros as continuing into the Recent, with complementary depression of the Gulf. Episodic upwarp is shown by uplifted river terraces, incised by later down-cutting of the river. The Iranian coast is an area of active folding with drowned valleys in synclines, and tilted terraces on anticlines. Rising anticlines are shown by the authors to have deformed canals and other artifacts. A close relationship exists between shallow Quaternary uplifts and subsurface structure.

Tigris-Euphrates Delta

It has been suggested that the delta is still actively subsiding. Lees and Falcon (1952) state that although the rivers are carrying sediment into the coastal marshes, these do not seem to be building above sea level. V.S. Colter (unpublished oil company report) has observed that the Kuwait part of the delta has subsided by as much as 37 m below sea level in the last 5000 years. His view is based on the occurrence of dolomitic muds, which he takes to indicate deposition near sea level, in association with marine lime mud in a borehole at 37 m below sea level. Oysters from the marine mud were dated by C^{14} at 5080 to 5980 years old. No sedimentary structures indicating very shallow sedimentation or exposure occur and for this reason the author regards Colter's evidence as inconclusive. The presence of diagenetic dolomite indicates the flow of dolomitising waters, and has no absolute environmental significance. The presence of submarine platforms (discussed in greater detail in a subsequent section) in the Kuwait part of the delta indicates, on the contrary, relative stability at the present day. A platform at -16 to -20 fm has been mapped northwestwards from offshore Abu Dhabi to Kuwait, where the sea floor becomes more graded, suggesting that it is covered by delta sediment. Colter reports, (using sparker lines and borehole results) that the modern delta sediments in this area were laid on a more or less horizontal surface at about 115 feet below sea level. This is approximately 19 fm, or well within the depth range of the present marine platform. The age of this surface, according to correlation with Fairbridge's scale

(Fig. 9) would be about 9000 to 11 000 years B.P. This suggests Holocene progradation across a stable surface rather than filling of a subsiding depression, although slight subsidence may in fact be occurring.

Quaternary movements and topography

In general, the effect of Quaternary tectonics on the morphology is much less than that of the Pliocene movements. and has taken the form mainly of local adjustments of existing features. This is certainly true for the major uplifts bounding the basin and of most of the shoals within it. The most important exception to this generalization is the Gulf of Salwa area where relatively small vertical movements during the Quaternary caused considerable changes in the topography.

The major influence on morphology in the Quaternary, as discussed in the following section, was that of sub-aerial and coastal erosion. In many cases this may have been simultaneous with localized structural uplift.

GEOMORPHIC EVOLUTION

Post-Pliocene topography

Following the Plio-Pleistocene Zagros movements, the topography of the Persian Gulf approached its present form. At the end of the Pliocene (Fairbridge, 1961) sea level was probably about 150 m higher than present sea level. It receded to approximately its present level (which it may have reached about 100 000 years B.P.) in steps, which left their mark on the Arabian landscape as a series of marine terraces and sabkhas (now "inland" sabkhas). Such terraces are described on the Trucial Coast by Lees (1928) and in Saudi Arabia by Holm (1960). Glennie (1970) has suggested that the Oman inland sabkha of Umm as Samim represents "a relict arm of the sea left behind as a result of relative sea level changes". This view is also held by Holm (1960, p. 1368).

The Arabian coastal plain sloped to the northeast. For at least part of the time when sea level was high, the climate was wet, allowing wadis to pour a sheet of coarse alluvial gravel, originating in central Saudi Arabia, over the Saudi Arabian coastal plain and across the later site of the Gulf of Salwa to Qatar and western Abu Dhabi (Holm, 1960; Powers, 1966). Spot heights given by U.S.G.S. (1961) suggest that the sheet may be graded to a sea level of about + 90 m.

Subsequent to the deposition of these gravels the Gulf of Salwa subsided along its bounding fractures. A chain of sabkhas extending from its head around the base of the Qatar Peninsula to Khor Odaid may have been formed during a subsequent phase when Qatar was an island (with relative sea level about 30 m above its present level).

Pleistocene sedimentation

The Pliocene clastics and evaporites of the Fars Formation are overlain unconformably over a large part of the Persian Gulf by shelly and oolitic limestones of presumed Pleistocene age. The "Kharg Limestone" dated faunally as Plio-Pleistocene, is of this type. Similar limestones are described from the Trucial Coast by Skipwith (1966), from Das Island and offshore Abu Dhabi by Russell (op.cit.), and are known in foundation boreholes from offshore Qatar. This limestone is a shelly calcarenite up to 100 feet thick. No diagnostic fauna was found in it, but regional geology suggests a Pleistocene to latest Pliocene age. The lithofacies suggests a shallow-water origin, and the limestone was probably deposited at a sea level slightly lower than that of today. This may tend to favour a Pleistocene age, since the Plio-Pleistocene boundary is believed to have been a time of higher eustatic sea level.

Russell (op.cit.) describes a probably correlative limestone as forming sea-floor outcrops over much of the offshore Abu Dhabi in the form of "submarine limestone platforms .... sometimes bounded by small pseudo-scarps .... The attitude of the limestones is horizontal to sub-horizontal, jointing is rare, and bedding indistinct".

This Pleistocene limestone shows measurable thickness variations. In the Mahdan Mahzam area (E. Qatar) it varies from 25 feet thick over the oil field structure to 40 feet nearby. One core hole between the Maydan Mahzam field and Doha penetrated about 100 feet of this limestone. This relationship is interpreted as being due to a combination of infilling of an existing topography and structural growth during and after sedimentation. Late Pleistocene topography probably reflects that of the Pliocene in a subdued form.

Pleistocene retreat of the sea

During the Pleistocene glaciations, the sea level in the Persian Gulf was lower than it is at present. This is shown by the existence of submarine platforms at various levels and drowned river valleys.

Submarine platforms are discussed in a subsequent section but are mentioned at this point because they provide a clear indication that sea level has fallen at least to -62 fm (120 m), the depth of the deepest platform recognised. Furthermore, they demonstrate that there was regional tectonic stability in the Gulf (with the exception of the localized movements discussed above) throughout the latter part of the Quaternary.

These observations are of considerable interest in that they show that during the glacial maximum (about 70 000 to 17 000 years B.P. according to Fairbridge, 1961), continental conditions prevailed over almost the whole of the Persian Gulf, the sea having withdrawn to the Hormuz Strait. During maximum regression, the basin was a very large river valley carrying Tigris-Euphrates waters directly into the Gulf of Oman.

Further evidence of sub-aerial exposure of part of the present sea bottom of the Gulf comes from the work carried out by M.S. Thornton (personal communication) who shows that the Pleistocene limestone, in cores taken in a present water depth of some 37 m, has undergone leaching, probably indicative of sub-aerial exposure of the rock.

A partial return of the sea during the Würm glaciation (Gottweig (Aurignacian) Interstadial) is postulated by Fairbridge (1961) and Curray (1961). This would have been about 30 000 to 45 000 years B.P. (Fairbridge), or about 25 000 years B.P. (Curray), and sea level reached -40 m (Fairbridge) or -15 m (Curray). One dated sample from the Persian Gulf tends to support Curray's view of this event. It is a marine limestone from Umm al Garse NE of Qatar, dated by C^{14} at 26 700 years, which was collected 22 m below present sea level. A dated raised beach on Halul Island (Nicol, op.cit, dates ranging from 32 700 years B.P. to more than 44 300 years) also falls in this general range.

The term "miliolite" is used by field geologists in the Trucial Coast area to describe a group of carbonate-cemented sand dunes, frequently including marine shell debris and ooliths. That the miliolite was deposited during a period of lowered sea level is demonstrated by the fact that aeolian bedding within it extends below present sea level (Skipwith, 1966). Recent C^{14} dating of samples by Imperial College (London) gives ages of 20 000 to 30 000 years B.P. (G. Evans, personal communication). Thus the miliolite may consist, in part, of re-worked marine deposits of the interstadial mentioned above.

Subaerial erosion: structurally controlled Pleistocene river valleys and residual land forms

Bathymetric contouring based on the Admiralty charts together with information from sparker and echo-sounder profiles, suggests the existence of a network of channels which are interpreted as drowned river valleys. These are most apparent in areas of steeper topographic slope, such as the steps between submarine terraces, and often cannot be distinguished on the platforms themselves except where the base-level of paleo-channel erosion is below that of the platform. The density of the channels depends partly on the variable degree of bathymetric control. Nevertheless, the scale of development of the whole system suggests that it formed at a time when the climate was wetter than it is today.

The general relationship between the drowned river valleys and the submarine platforms is inferred to be that the period of the maximum development of the rivers or wadis coincided with maximum retreat of the sea during the Pleistocene glaciation. The post-glacial rising sea cut platforms in a topography already formed by fluviatile erosion (Fig. 7). The situation could have been more complex since pluvial periods may have occurred during the rise in sea level causing rejuvenated erosion. It is significant, however, that channels can be mapped down to the deepest platform (in the Gulf of Oman), indicating that channels were formed at a single eustatic sea level.

The drainage pattern is dendritic, and from the Arabian coast radiates predomininantly towards the NE down the structural slope of the basin. There was, however, some internal drainage into local depressions situated south of the Bahrain/ Pearl Bank Ridges.

Figure 7 shows the interpreted effect of rising structures on the drainage pattern. Locally around the structures the drainage would be radial, while regionally the drainage systems would flow towards synclines, swinging aside to avoid the highs. Thus structural highs would become residual erosional highs. Some parts of the Gulf, particularly offshore Abu Dhabi (Umm Shaif and Zakum structures) come close to the idealized picture shown in figure 7. An analogous situation occurs in the prominent drowned valley system of the Musandam Peninsula (Lees, 1928), where the old river valleys radiate from the plunging nose of the Oman Mountains uplift.

Figure 4 summarizes evidence for the coincidence of topographic and structural highs in the central Persian Gulf; it suggests that a considerable number of topographic highs show evidence of structural origin. Some have dip reversals on sparker records, others seem, from seismic or gravity interpretations, to overlie deeper structures. The shape and trend directions of many highs are strongly suggestive of structural influence. Areas to the northwest of Qatar (offshore Saudi Arabia and Neutral Zone) seem to show little effect of localized Quaternary uplift in their drainage patterns which are dominated by the regional tilt of the sea floor.

A number of bathymetric highs may be of purely erosional origin having only indirect relation to structure. One may expect, for example, an arc of cuestas around Qatar Peninsula, marking the up-dip edge of the harder Miocene rocks. However, no indication was found in the Gulf of the "inverted-structural-relief" type morphology, in which topographic relief is provided by rings of cuestas round synclines. This is characteristic of a more mature stage of landscape formation, such as that in western Abu Dhabi (fig. 4). The present submarine landscape of the Persian Gulf was probably formed over a few thousand years of subaerial exposure and is morhologically juvenile.

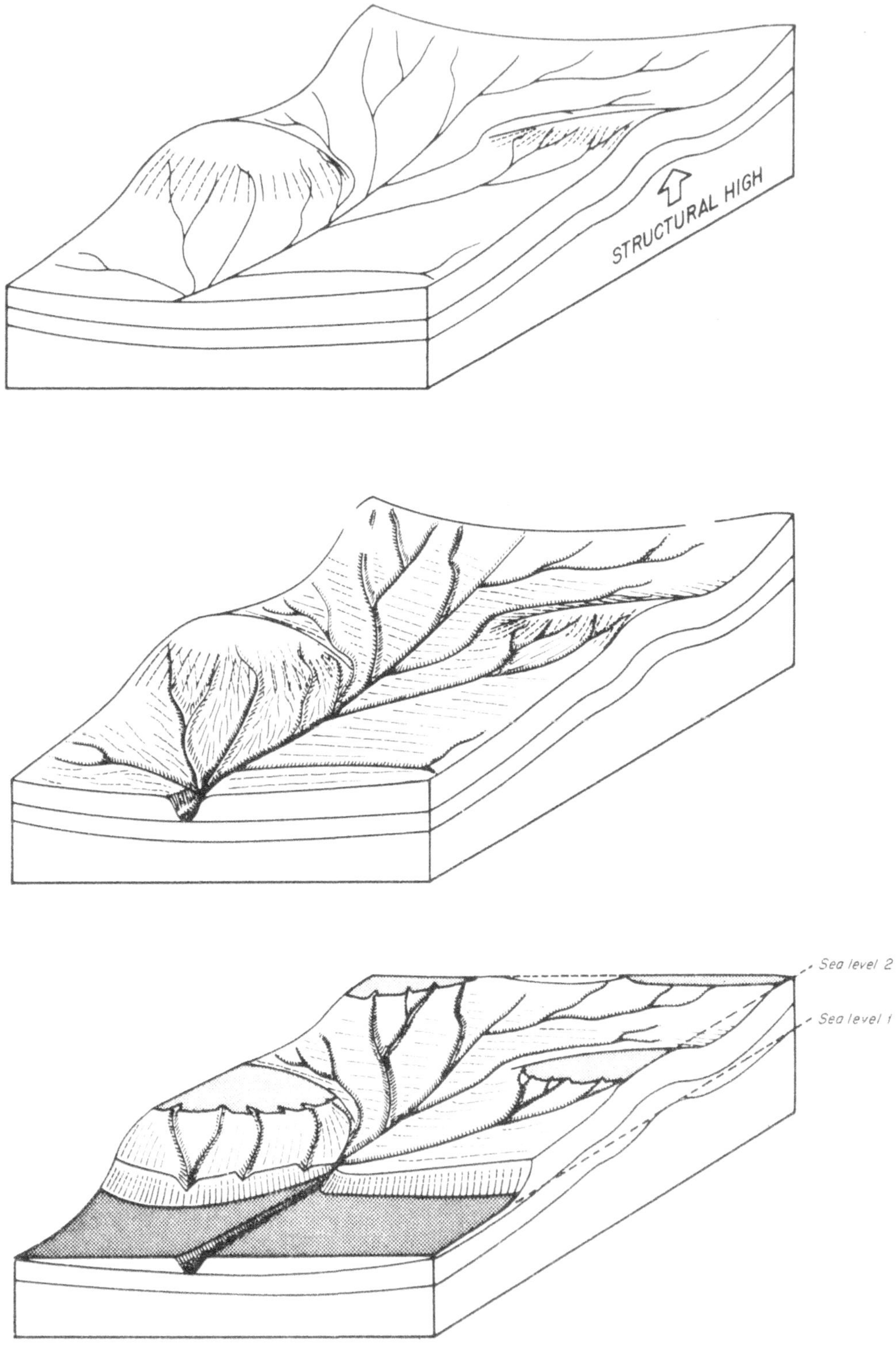

Fig. 7. Block diagrams showing interpretation of physiographic evolution of submarine morphology in the Persian Gulf

Late Pleistocene and Holocene submarine platforms

The submarine platforms are interpreted on the basis of sparker profiles, echograms and bathymetric charts as erosion surfaces, which are subhorizontal or dipping gently seawards, their regularity depending on the hardness of the rock and the amount of sediment cover (see the topographic profiles on fig. 8).

Fairbridge (1961) discusses the mode of formation of such platforms and concludes that "the base-level of sub-aerial or intertidal erosion is mean low-tide level, and it is to this level that a contemporary marine platform approximates ; it is to this level that coral reefs grow" (p. 140).

Several submarine platforms in the Gulf are recognized and described by Houbolt (1957), who regards them as indicating stages in the post-glacial rise of sea level. He considers they were formed in pairs, each temporary sea level being marked by a surface of accumulation and by a surface of abrasion. This is no doubt the case today along the Arabian coast, the result of some 5000 years of static sea level and accompanying regressive coastal accretion. However, it is questionable whether these conditions are comparable with those in late Pleistocene and early Holocene times when the sea level was rising rapidly (Curray, 1961; Fairbridge, 1961; Cullen, 1967). In the circumstances, erosion seems likely to have been the dominant process of platform formation. The platforms are therefore regarded as surfaces cut by the sea, during pauses in the post-glacial rise in sea level, at levels corresponding to mean low-tide level at the time. Thus,the step from one platform up to the next level is a fossil coastline.

The deepest platform recognized is at -62 fm (120 m) (Fig. 9). This level occurs only in the Gulf of Oman and the Hormuz Strait, both areas of questionable tectonic stability. However, the next level, at -55 fm (100 m), occurs in both the Gulf of Oman and the Persian Gulf, at constant level. In addition, -62 fm agrees with Curray's (1961) estimate of the sea level at the peak of the Würm glaciation. Fairbridge (1961) believes that this level should be -55 fm. The broad agreement of the Persian Gulf levels with those of these authors, as well as the constancy of individual terrace elevations within the Persian Gulf, suggests that both the NE Gulf of Oman and the Persian Gulf are tectonically stable today.

The exact timing of the late Quaternary, post-glacial sea level rise is still uncertain (see Curray, 1961; Fairbridge, 1961; Cullen, 1967). However, there seems to be agreement that the sea level began to rise about 17 000 to 20 000 years B.P., and that it reached its present level some 5000 years ago, in a series of rapid advances separated by stillstands, this rise being known as the "Flandrian transgression". The steps in the transgression are marked throughout the world by drowned shore lines and marine platforms.

Six main platform levels are defined, mainly from studies of the Arabian parts of the Persian Gulf (Fig. 9)

- 5 fm (9 m)
- 10 fm (18 m)
- 16 - 20 fm (29-37 m)
- 36 - 44 fm (66-80 m)
- 55 fm (100 m)
- 62 fm (120 m)*

* Apart from the 30 m terrace, these levels do not correlate closely with the stillstands recently proposed by Sarnthein (1971) mainly on the basis of distribution of relict sediments. This may be due to the fact that Kassler's levels are essentially erosional while Sarnthein's are mainly depositional. Ed.

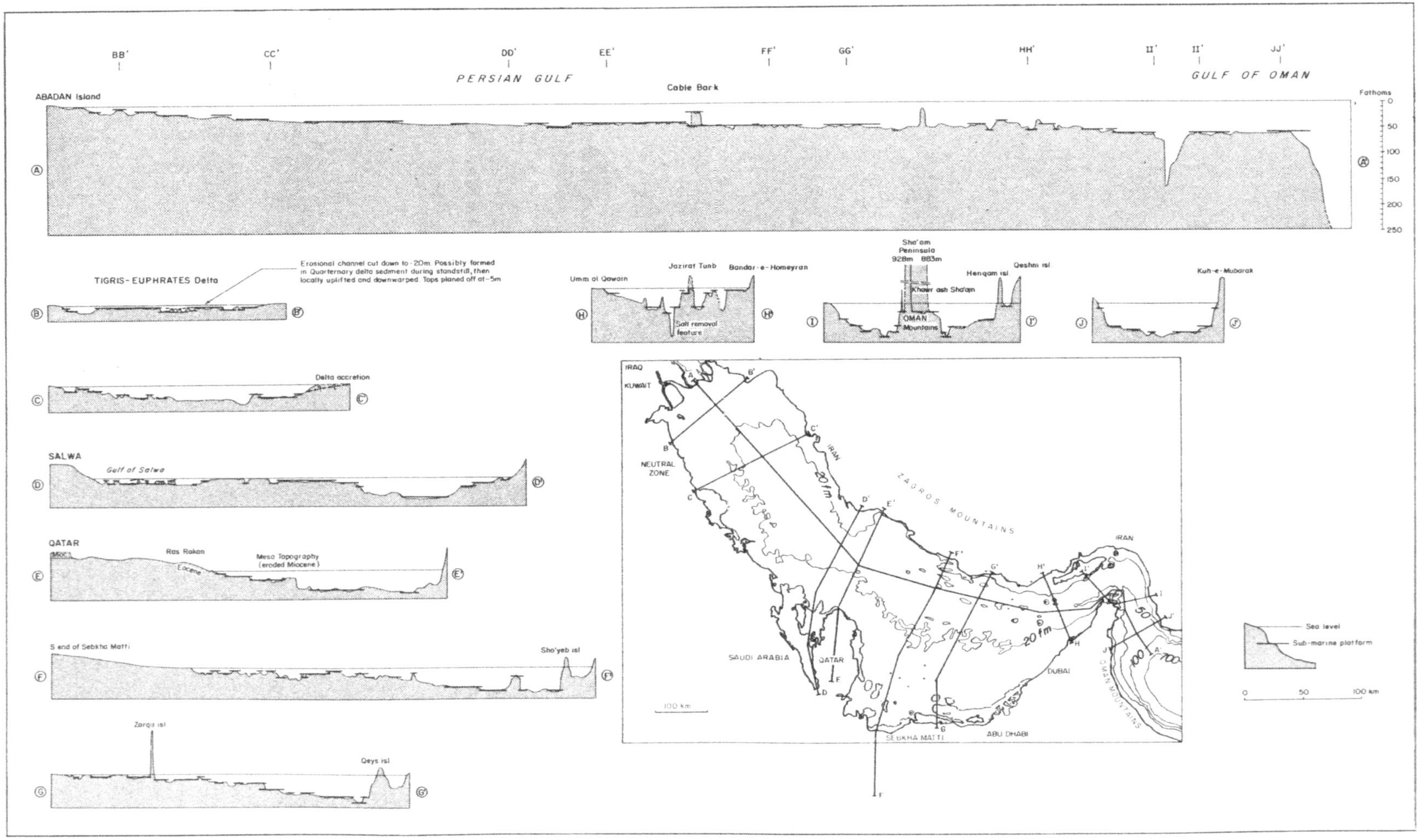

Fig. 8. Regional bathymetric profiles of the Persian Gulf

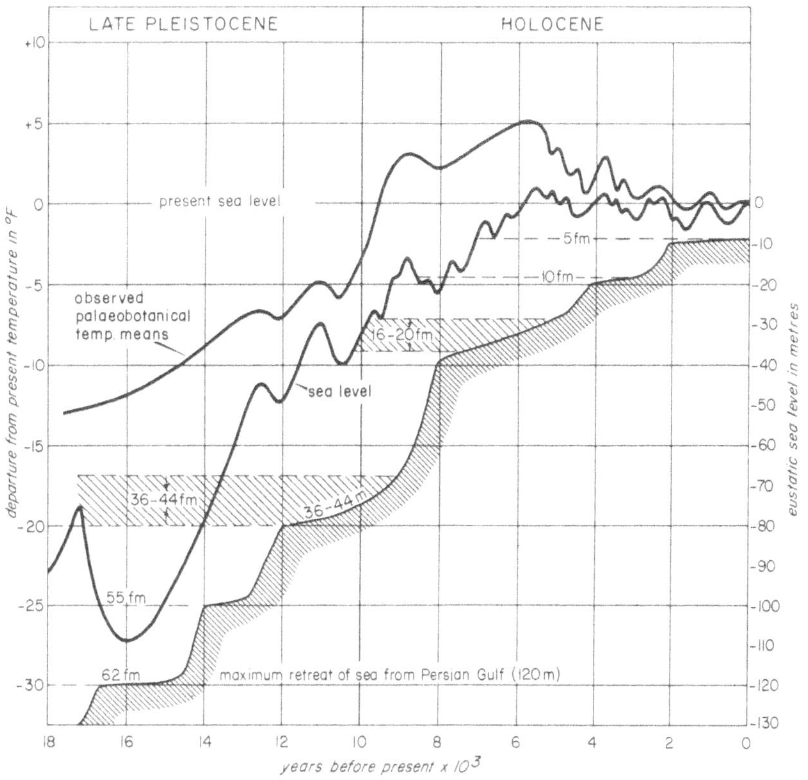

Fig. 9. Persian Gulf submarine platforms and late Pleistocene and Holocene sea level and temperature changes. (based partly on Fairbirdge, 1961)

There is, at the moment, no direct evidence of the age of these platforms. Figure 9 shows one method in which they can be related to Fairbridge's generalized late Quaternary sea-level curve and this correlation has been used as a working hypothesis.

Figure 7c shows theoretically how an older landscape, whose morphology is controlled by structure and drainage, might be affected by the cutting of two submarine platforms. It can be seen that the general pattern of highs and lows still exists in a subdued form but that all but the deepest stream valleys are obscured on platform surfaces (as is also shown by bathymetric contouring).

The submarine topography in the area east of the Qatar Peninsula is dominated by the shallower platforms. The study of these platforms confirms the possibility that the desert sands of Abu Dhabi could have been blown across from Saudi Arabia by way of Qatar and the Great Pearl Bank area (E.A. Shinn, personal communication), up to about 8000 years ago (assumed age of the 10 fm platform). Investigations in the

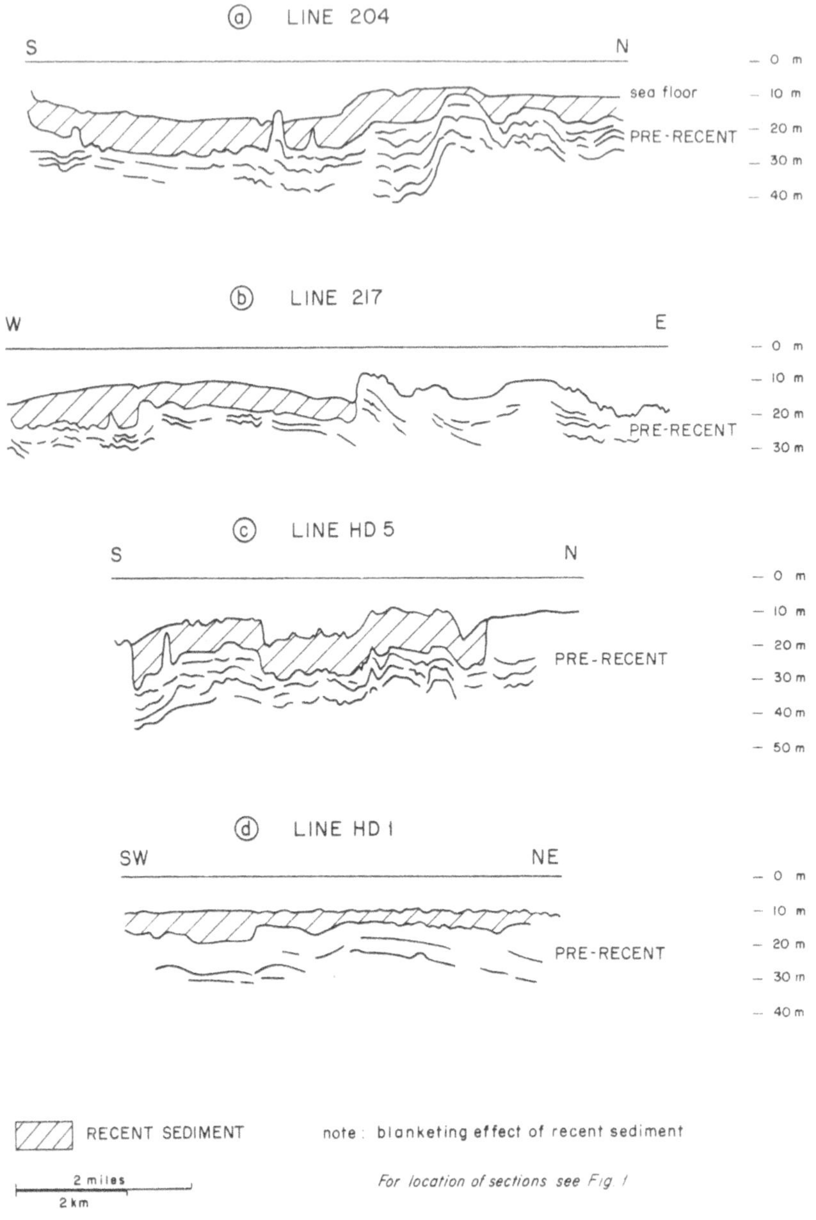

Fig. 10. Profiles illustrating the mode of Holocene sediment accumulation, based on "sparker" results

Gulf of Salwa also point to Holocene migration of aeolian dunes across an area that is now marine; sparker profiles show topographic mounds suggestive of sand dunes, with a steeper flank to the south. These rest on a surface identified as the 10 fm platform and are themselves truncated by the 5 fm platform. If, indeed, these highs are drowned dunes, the date of the last dune migration across the Gulf of Salwa would be approximately 7000 to 8000 years ago. A.J. Keij and G.R. Varney (unpublished oil company report) state that the Recent carbonate sediments of the Gulf of Salwa are underlaid by uncemented quartz sands.

Effects of late Pleistocene and Holocene marine sedimentation

A very general impression of the distribution of unconsolidated marine sediments in the Persian Gulf is based on foundation boreholes and sparker profiles. In the shallow water areas (those most intensively studied by Shell Research), sediment thickness varies up to a maximum of about 10 m, although large areas have less than 2 m. Greater thicknesses seem to occur in the deeper water towards the Gulf axis, perhaps reflecting a supply of clastic material from the Zagros range, as well as a shorter interruption of sedimentation during the Pleistocene fall in sea level.

The sediments described as "unconsolidated" in this context, post-date the last glacial fall in sea level. Their age thus ranges from about 18,000 years to present day and they can therefore be classified as latest Pleistocene to Holocene. These products of the Flandrian transgression overlie older rocks, either Pleistocene limestones or Pliocene Fars clastics, with a marked unconformity. (The "unconsolidated" sediments may contain submarine lithified layers of the type described by Shinn, 1969, which can be distinguished on petrographic criteria from the older rocks.)

Sparker profiles indicate that, volumetrically, most sediment in the Gulf accumulates in topographic depressions, thus tending to smooth out the relief (Fig. 10). In deeper parts of the Gulf, this process may have effectively obscured the older relief; on the Arabian homocline, however, many rocky highs protrude through the sediment. The origins of these depressions are various. The largest constitutes the axis of the Persian Gulf, which is a structural depression. Several large topographic lows in offshore Abu Dhabi, seaward of Sabkha Matti, appear also to be structural synclines. On a more local scale, drowned Pleistocene wadis may constitute sediment traps, such as that around the southern edge of the Rig az Zakum structure in offshore Abu Dhabi. Both local and regional depressions are characterized by mud sedimentation. Depressions on the Arabian sea floor are filled with relatively pure carbonate mud, while that constituting the Gulf axis contains a considerable amount of impure carbonate muds, or marls.

From these observations it can clearly be seen that the pre-Recent rocky topography exercises a fundamental control on the character and thickness of the sediments overlying it. The sedimentary sequence thins over topographic highs and thickens towards inter-high areas, as well as towards the deeper water in the tectonically controlled basin centre. Shoal sediments are coarser grained than those in the depressions. This is analogous to the situation found in a number of fossil carbonate sequences.

It has been assumed in the foregoing discussion that all rocky highs in the Persian Gulf are of pre-Recent age, thus ignoring the possibility of carbonate build-up, either by reef growth or by "mounding" and submarine lithification. In fact little evidence is available to show that build-up has created appreciable topographic relief in the present-day Persian Gulf, and it is possible that the Flandrian transgression was so rapid that upward growth was unable to keep pace with the sea level. Sparker records indicate a virtual absence of sediment on the crests of most Arabian highs and, when present, it seems to be unconformable on older residual land forms.

Results of geomorphic processes

It has been shown that the fundamental morphologic framework of the Persian Gulf was formed in the Pleistocene tectonic phase, and that Quaternary tectonics modified this in detail but not in substance. This may also be true of the geomorphic processes described above. The net effect of these seems to have been to reduce the relief associated with the Pliocene tectonic structures by a combination of erosion of highs and sedimentation in the depressions, without fundamentally altering its distribution; synclines having positive topographic relief have not been found.

The relative importance of the different geomorphic processes discussed is very difficult to assess. One might guess that on a regional scale the cutting of submarine platforms affected the morphology more than the river valley erosion which preceded it. If sedimentation continues at the present rate under the existing conditions of relative tectonic quiet, it should succeed in obliterating most of the pre-Recent morphology in the near geologic future.

CONCLUSIONS

1. The morphologic framework of the Persian Gulf is tectonically controlled in Iran by the intense Plio-Pleistocene Zagros folding,and towards Arabia by late Pliocene to Pleistocene folding, faulting and diapirism superimposed on a much older pattern of growth structures. The subsiding axis of the Gulf is a Zagros syncline whose formation was accompanied by a regional north-easterly tilt of the Arabian homocline.

2. This structurally controlled morphology was partly subdued by sedimentation of the Pleistocene Maydan Mahzam and equivalent limestones in offshore areas.

3. Renewed,or continued tectonic activity uplifted some structures during the Quaternary, but caused only local modifications of the morphologic pattern. These Quaternary movements affect a wide variety of areas and types of structure.

4. Sea level fell by a maximum of 120 m during the Pleistocene, at which time the sea retreated to the Strait of Hormuz leaving the entire Persian Gulf exposed. Rivers eroded structurally-controlled valleys down the flanks of the basin. Tectonic uplifts escaped erosion in some cases, and remained as topographic highs. Other highs are purely erosional in origin.

5. At its level of maximum retreat, and during its subsequent rise to its present level, the sea cut a series of platforms in the pre-existing topography.

6. The pre-Recent morphology which resulted from these events and processes is regarded as predominantly tectonically controlled, the Quaternary erosion and sedimentation having had no more than a modifying influence. There is some correlation between structure and bathymetry, and shapes of topographic features show interfering Zagros and Arabian trends.

7. This morphology exerts regional and local control on patterns of sediment accumulation. Shallow water and deeper water topographic highs and topographic depressions each have characteristic associations of sediment type and distribution.

8. The present-day sedimentary cycle is likely to show a pattern of thinning over topographic highs and thickening towards inter-high depressions and the basin centre. Sedimentation has so far not masked the underlying topography; presumably it will gradually blanket and obscure the pre-Recent morphology.

Organisms as Producers of Carbonate Sediment and Indicators of Environment in the Southern Persian Gulf

M. W. Hughes Clarke[1] and A. J. Keij[1]

ABSTRACT

The principal carbonate producing and environmentally significant organisms are discussed in a general manner aimed at facilitating the interpretation of certain fossil assemblages. The mode of life, effects of restriction (salinity and temperature), light, and biological breakdown of the various organisms is described in order to understand their distribution within the basin and the manner in which they contribute to the sediment mass.

Particular attention is devoted to the Foraminifera whose distribution and abundance (foram number) are analysed in terms of bio- and thanatocoenose. Maximum numbers of individuals, locally exceeding 20,000/ gram of sediment, are attained in certain restricted coastal lagoons and at the foot of the 36 m terrace. In the former areas (lagoons) species of Peneroplis frequently exhibit aberrant growth forms.

Increasing restriction seems to have clearest effect where salinities attain 50 gm/litre, this barrier coinciding with the disappearance of most perforate Foraminifera, red algae, corals, echinoderms, and many other groups. The most common organisms in the Persian Gulf are illustrated photographically.

I. INTRODUCTION

As part of the project studying Recent carbonate sediments in the Persian Gulf, carried out by Shell Research BV in the years 1965 - 67, a general investigation was made of the organisms concerned in producing and characterising the sediment. The aim of this investigation, and of the whole project, was to understand only the broad relationships and principles in order to gain a better understanding of ancient carbonate sediments. For this reason, the degree of detail with which the studies of the organisms were carried out was strictly prescribed and no attempt was made to identify and name all the organisms that were observed and collected. Formal identification was therefore limited to those forms that were numerically or environmentally significant, and most attention was paid to micro-organisms; in particular the Foraminifera, as they are so widely used in oil company stratigraphic work. This paper, with these limitations, is a general survey of our observations and supports the sedimentological studies presented elsewhere in this volume.

[1] Shell Research B.V., Rijswijk, The Netherlands.

Methods

Most of the material used in this study was obtained from bulk sediment samples, i.e. Van Veen grab samples for the most part,with lesser numbers of core (including box-core) samples. In total, including the material utilised by Houbolt (1957), samples from over 4000 stations were studied. A relatively limited amount of sampling specifically for living organisms was carried out using three types of dredges and a plancton drogue. In shallow areas near coasts and on shoals, some collecting and observation by diving was carried out.

Evaluation of the floral and faunal composition of the bioclastic fraction of the sediments was based mainly on the distinction of morphological groups in a manner permitting comparison with thin sections of ancient limestones, i.e. "high-spired" gastropods,"small"lamellibranchs, etc. These basic observations are recorded on regional profiles given in the accompanying article by Wagner and v.d. Togt*. These very generalised methods have the disadvantage that the taxonomic information necessary for detailed local studies is not recorded; to have undertaken a complete determination of the faunas of the Persian Gulf would have required considerably more time than was available for this project, especially as many undescribed species, particularly ostracods and micro-mollusks, are present. The resulting generalisations are essentially regional and should be applied in that manner. Local studies could have been made mainly by subdividing the associations defined.

For the micro-organisms, a standard sample treatment procedure was carried out on the dry sediment samples. Fifty grams of dried 'grab' sample were washed over sieves and the fraction larger than 75µ retained. The Foraminifera within this fraction were counted and the 'foram number' (number of specimens per gram of sediment) was calculated. In the fraction larger than 150µ, the major foraminiferal groups were determined; on maps only large and somewhat artificial assemblages were plotted. These included: arenaceous, imperforate, smaller and larger perforates, together with their various admixtures. This broad grouping was employed partly because it gave a reasonable comparison with the associated sedimentological studies based mainly on thin sections. A number of samples containing living micro-organisms were preserved for subsequent study using alcohol and mercuric chloride. This allowed better determination of the actual life habitat of the more important indicator organisms; again,most attention was concentrated on the Foraminifera.

Previous work

At the time of completion of our studies, relatively few publications on the fauna and flora of the Persian Gulf existed, the principal contributions being those of Melvill & Standen (1901) on the mollusks,and of the Danish expedition (echinoderms, algae, fishes, sea snakes) (1939 - '49). However, as sedimentological research in carbonate provinces almost invariably involves a study of the organic remains that constitute the bulk of the sediment, the pioneer studies of Houboult (1957) and Sugden (1963), followed by those of Evans (1966), Kendall (1966), and Skipwith (1966), contain much pertinent information. The publications of Murray (1965 - '70) specialise in the taxonomy of the Foraminifera of the Trucial Coast. Nonbiological studies include the publications of Emery (1956), who contributed to the understanding of the hydrography of the Gulf, an aspect basic to the understanding of ecology in the area, and Pilkey & Noble (1966) who evaluated, in a general manner, the distribution of carbonate and noncarbonate minerals in the Holocene sediments of the Persian Gulf.

---

* Appended at the back of the volume.

II. PERSIAN GULF ORGANISMS IN RELATION TO ENVIRONMENTAL PARAMETERS

The major modes of life

It is necessary firstly to consider the different modes of life. Planktonic faunas are scare in the sediments of the Persian Gulf. Planktonic Foraminifera occur in significant quantities only near the entrance of the Gulf in the Straits of Hormuz, their distribution being discussed by Seibold et al. elsewhere in this volume. Pteropods, on the contrary, occur throughout the Gulf, but various forms are limited to the entrance, or to the deeper marine parts of the basin. Coccoliths can be particularly frequent in muds on the proximal homocline* (plate 1A).

Nectonic animals play an insignificant role in sediment production and seem to have very limited value as environmental indicators. Otoliths of fish are rare in spite of an abundant fish fauna.

Benthonic life is abundant in the Persian Gulf, its character being strongly influenced by the nature of the substrate upon which it lives. On hard rock, live a multitude of organisms requiring a firm substrate on which to attach their tests, shells or colonies. Others use the rock as a medium into which to bore for protection (sponges, algae, mollusks). Rock areas generally carry a rich and diversified organic community, the character of which is determined by the other prevailing environmental factors, such as current, light, etc.

In sediment bottoms, an entirely different type of organic community is found. Many animals burrow in the sediment, either for protection or for food, while others move over the soft surface. Certain higher plants occur in the shallower areas and may form thick carpets of vegetation, binding the sediment by their extensive systems of rhizomes.

In this text, organisms that live in or on sediment are termed 'infauna', and those connected with hard substrates, or living on the fronds of weeds, are termed 'epifauna'. Extensive rock bottoms are a characteristic of the Arabian parts of the Persian Gulf, being partly outcropping pre-Holocene rocks and partly recently cemented sediment, (Shinn, 1969) and epifaunal habitats are common.

The infaunas of the muddy, central parts of the Gulf are strikingly poor in living forms, in spite of seemingly constant temperatures and salinities. On the irregular surface of the Arabian homocline, with its rather complex pattern of muds and sands, infaunal life is relatively abundant, although not particularly varied, down to depths of about 40 m. In the muddy areas, infaunal gastropods and lamellibranchs, together with ostracods and forams, are relatively common (foram numbers ranging between 1000 and 10000) as are certain burrowing echinoids. Solitary, infaunal corals occur in the deeper parts of the basin, generally at depths greater than 10 m.

Among the epifauna, corals are abundant on most hard substrates; whilst they occur at all depths, they are most prolific on rock highs and near the coast at depths less than 15 m, where they help to build patch or fringing reefs. This reef community is dominated by melobesioid algae similar to the genus Neogoniolithon, together with various octocorallid and scleractinian corals, sponges and serpulid worms. Oysters (sensu lato) are abundant locally and the byssate forms have been exploited by the pearl fishing industry for centuries. Various gastropods and regular echinoids feed on these epifaunas and activate the breakdown of the skeletons to sand- and mud- sized sediment.

In offshore areas, transport of epifaunal debris into surrounding deeper environments with infaunal associations is common, but generally limited to short distances, unless an encrusting organism is carried on detached, floating seaweed, as is the case with certain bryozoans.

---

* For definition and discussion of this term see introductory article by Purser and Seibold. Ed.

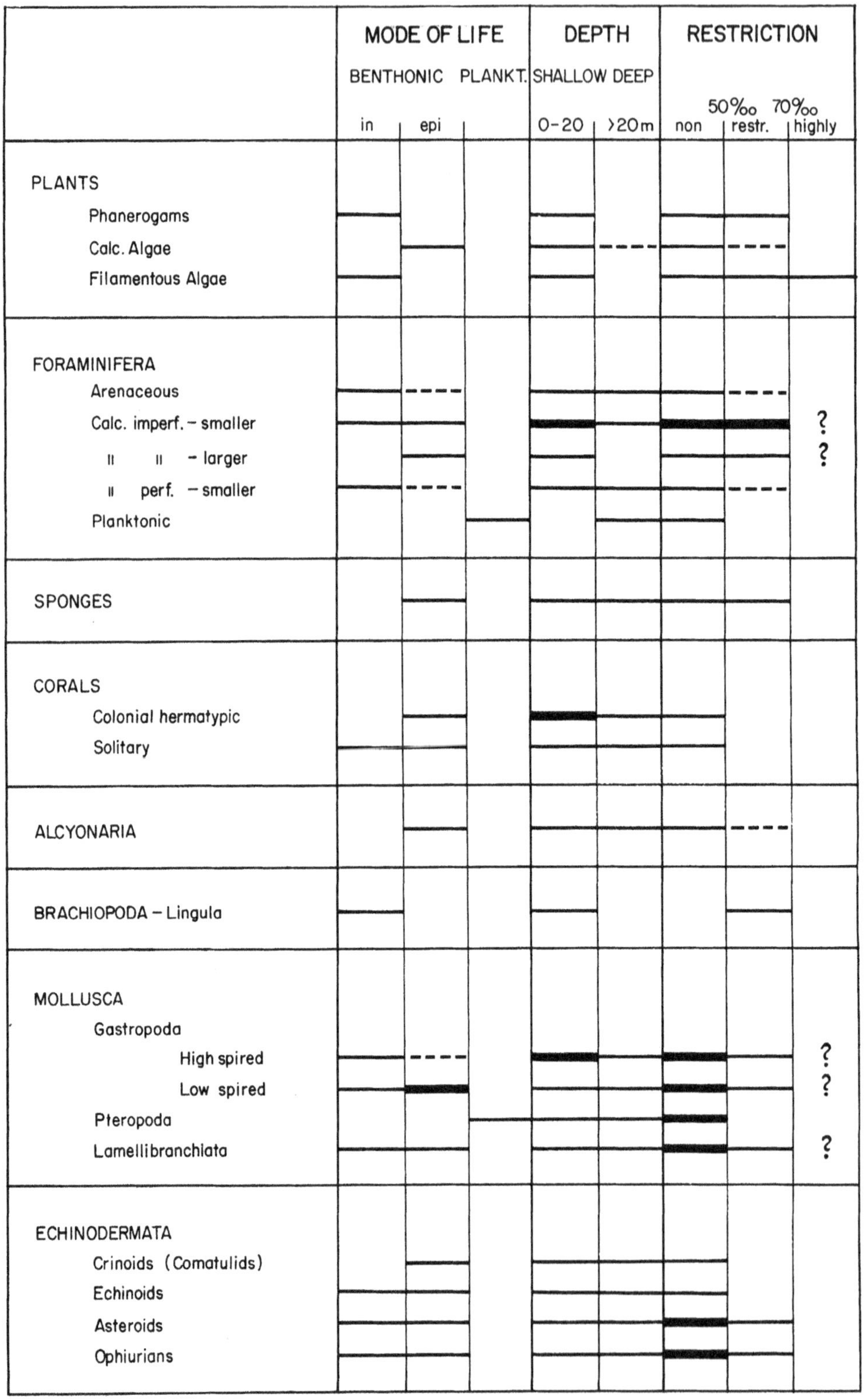

Fig. 1. Schematic representation of relationships between major plant and animal groups and substrate, depth and salinity

In the shallower areas of the Arabian coastal zone and around emergent offshore highs, significant amounts of reef detritus and other epifaunal debris are transported laterally to fill lows and to accumulate as coastal spits, beaches, and sediment tails, described by Shinn, Purser, and Purser and Evans, in this volume.

The relationships between the major animal and plant groups and their substrates are summarized in figure 1.

The effects of restriction

The concept of 'restriction' is widely used among geologists and palaeontologists without any precisely agreed definition, but always with the meaning that a peculiarity in the fossil assemblage suggests that the environment was in some way 'abnormal' . The peculiarity in the assemblage is usually that the species diversity is less than would be expected in that particular sediment grade and age interval, but can also include features as 'dwarfing' or aberrant growth forms.

Lowered species diversity and other peculiarities in the Persian Gulf organisms is associated with a decreasing interchange between the water body concerned and the main mass of the Persian Gulf water. In the Persian Gulf studies, we have chosen to quantify this abnormality in the water solely with reference to the salinity, but in no way do we imply that salinity is the only factor that prevents the development of the full species compliment of organisms.

In the Persian Gulf, no animal species has been observed to make its appearance as the result of increasing restriction. Moreover, the number of individuals does not necessarily decrease as a result of restriction; the most restricted environments are often exceedingly rich in plant and animal life made up by relatively few species. Foraminiferal numbers in certain restricted lagoons around Qatar Peninsula may exceed 30,000,the sediment then being composed almost entirely of imperforate forams. In contrast, sediments containing few recognisable skeletal remains often occur in nonrestricted areas of the marine homocline. This is particularly true for the lamellibranch muds (Wagner and v.d. Togt, this volume) where the lack of Foraminifera and other organisms is probably due to such environmental factors as unfavourable substrate, sedimentation rate,or poor light penetration.

The progressive disappearance of certain faunal and floral groups within the Persian Gulf permits an arbitrary three fold subdivision of the degree of restriction of the basin. Whilst these changes are measured against salinity, it is certain that the faunal character is also very much dependant upon the range of diurnal or seasonal temperature fluctuations, which, however, have been measured in very few localities. In general, in the Persian Gulf there is seemingly a fairly close parallelism between increasing salinity and increasing temperature fluctuation as both characterize rather shallow water.

The three subdivisions are:

The 'normal marine environment' , salinities up to ca. 50 ppm

Salinities in this range occur throughout the greater part of the Persian Gulf. The associated fauna and flora is an impoverished Indo-Pacific one lacking many characteristic and volumetrically important elements, e.g. the lamellibranch *Tridacna* (the giant clam), the gastropod *Lambis* , and the algae *Halimeda* and *Penicillis*. The reasons for such absences are not understood, especially as many of these organisms occur in the adjacent Gulf of Oman and Red Sea. However, it should be clear from the (salinity) data given by Purser and Seibold (this volume) that the bulk of the 'normal' Persian Gulf water is already slightly altered, salinity being everywhere over 39 ppm within the Straits of Hormuz. Within the range of the "normal" environment, it was observed that at approximately 45 ppm salinity, especially at the entrance to the land-locked Gulf of Salwa, several important groups disappear. They include perforate Foraminifera such as *Operculina* , *Heterostegina* , and *Amphistegina* , gastropods of the genera *Strombus* , *Conus* , *Xenophora* , lamellibranchs such as *Ervillea* and all pectinids, and all echinoids with the exception of the infaunal *Clypeaster*.

At approximately 48 ppm there is a very marked change in the faunas and floras of the Persian Gulf. Above this point all coral, alcyonarians, echinoids, and melobesioid algae disappear, together with most of the calcareous perforate and arenaceous Foraminifera.

The 'restricted' environment, salinities 50 to ca. 70 ppm

In many coastal areas where salinities exceed ca. 50 ppm, faunas are dominated by imperforate Foraminifera and gastropods. This dominance may, however, be the indirect result of the widespread synsedimentary lithification of the Holocene sediments rather than the water chemistry; the rocky substrates support a prolific growth of brown algae which are the favoured habitat of many imperforate Foraminifera and certain Bryozoa.

These hard substrates apparently are unfavourable for the majority of lamellibranchs, but are the sites of an abundant gastropod community. *Cerithium*-type gastropods are abundant particularly on shallow subtidal and intertidal sediments, but also on the flora on the rocky substrates. They are particularly in evidence on wide tidal flats where both salinity and temperatures are extreme. These faunas have been studied in areas with salinities up to ca. 65 ppm where they seemed to show only minor variation.

The 'highly restricted' environment, salinities exceeding ca. 70 ppm

Little is known regarding the faunas inhabiting restricted lagoonal areas with salinities in excess of 65 ppm. Certain isolated lagoons, including the N side of Khor Odaid in SE Qatar, seem to be faunal deserts, containing little more than cyprideid ostracods. These same lagoons are the sites of gypsum precipitation, with measured chlorinities exceeding 50 ppm.

The effects of temperature

Temperature is generally admitted to be one of the main factors determining the presence or absence of organisms. The offshore waters of the Persian Gulf fluctuate in temperature from approximately 20° C in winter to over 32° C in summer. The sea floor in the central parts of the Gulf has a more constant temperature, which is above 20° C throughout the year in most areas while in shallow coastal areas water temperatures may drop below 15° C in winter and rise above 35° C in summer. These extreme fluctuations are especially reflected in the nature of the reefs in the Persian Gulf. It is generally agreed that prolific coral growth does not occur where water temperatures drop much below 20° C. The fringing and patch reefs of the Gulf lack many of the species and the spectacular luxuriance of the adjacent Indo-Pacific area. The large proportion of dead corals in the Persian Gulf reefs may be the result of the winter temperatures intermittently falling to the lethal level and emphasises the relatively unfavourable nature of these environments.

The effects of light

The nature of organic communities is influenced by the intensity of the available light, which controls the floral elements. The base of the photic zone in a given latitude is determined essentially by the turbidity of the sea. In shallow seas, such as the Persian Gulf, suspended sediment in the water column is considerable and light penetration correspondingly reduced. Whilst the light penetration in the

Pacific and Carribbean allows luxuriant growth of flora, and also that fauna directly associated with plants (e.g. herbivore gastropods, corals with algal symbionts, etc.), down to more than 40 m, in the Persian Gulf this comparable zone reaches barely to 20 m. The authors use the relative abundance of light-dependant, benthonic organisms as the main factor in dividing the marine environments into two simple categories:

'Shallow marine' : recognised by the presence of one or other of the following groups; micro, blue-green algae, responsible for most of the finely bored peripheries on skeletal grains.
Calcareous, and other algae.
Hermatypic corals (which include most reef-building corals).
Large perforate Foraminifera and certain imperforate groups, e.g. Peneroplidae.

'Deep marine' : areas with the sea bed below the well-lit zone and therefore lacking the elements listed above (unless they are transported or relict, which is often the case in the Persian Gulf).

These two categories are also coincidentally recognisable with other parameters, e.g. sand sediment traction is largely restricted to the 'shallow marine'
The object in making this subdivision is to facilitate the interpretation of ancient carbonate environments, and thus such categories must be relative and not involve absolute depth figures.

III. BIOLOGICAL BREAKDOWN OF SKELETONS

The breakdown of organic carbonate skeletons begins during the life of the organism, various boring algae and fungi, mollusks, echinoids and fish being the main agents of destruction. Their activities reduce the rigidity of the organic skeletons and facilitate their breakup by wave and current action.

Perhaps the most spectacular example of organic disintegration occurs on the numerous coral-algal reefs. These are generally the habitat of great numbers of regular echinoids, of which *Echinometra mathei* is the most common; individual coral heads frequently support ten or more echinoids (plate 1B). Coral heads in all stages of destruction were observed. This destruction, while probably mainly the result of echinoid attack, is also caused by many other organisms, including boring algae and sponges (which perhaps are the food of the echinoids), boring mollusks, and parrot fish. The process clearly results both in the continued modification of the reef and in the rapid production of important amounts of carbonate sand and mud. Field experiments with *Echinometra* carried out by E.A. Shinn (pers. comm.) show that each echinoid can produce ca. 0.5 g of medium grained lime sand per day when browsing on algal-infested, dead *Acropora* fronds.

A second scale of organic attack is evident on the surface of individual carbonate sand grains which, under the microscope, are seen to be perforated by a system of superimposed, minute borings. These seem to be made mainly by micro-algae, but could also be fungal. Although less striking than echinoid attack on reefs, this algal boring is present on virtually all skeletal grains in the shallow marine environment, especially where the sediments are relatively free of mud. Continued boring results in the modification both of the shape and size of the individual grains, the process probably being more effective than mechanical abrasion.

IV. PRINCIPAL GROUPS OF ORGANISMS IN THE PERSIAN GULF

Algae

Calcareous and noncalcareous algae are locally abundant at depths of less than 10 - 15 m. Red, Melobesioid, calcareous algae are the binding element in the coral rubble slopes and dead reefs. On the offshore shoals they form characteristic skeletal oncoids, consisting either of slightly irregular nodules or branching structures. Melobesioid algae apparently do not tolerate conditions in the areas with salinities higher than 50% and are absent in the Gulf of Salwa. The brown and green algal weeds are important as hosts for rich epifaunal associations. In the Gulf of Salwa and Trucial Coast lagoons, the common occurrence of the heavily calcified stems and coronas of the dasyclad alga Acetabularia characterises the sediment in those areas with a salinity higher than 50ppm.

Blue-green, filamentous and unicellular algae are widespread throughout the shallow marine environments of the Persian Gulf and play an important role in sedimentation.

They occur as:

Stromatolitic algal mats in the protected intertidal areas of the Trucial Coast and Qatar. The diversity in morphology of these mats is illustrated in several publications, including that of Kendall & Skipwith (1968). No subtidal stromatolitic mats have been observed, although the filamentous algae are common.

Mucilagenous algae are widespread in inter- and shallow subtidal environments. These algae occur as a thin (several mm), semi-transparent, fragile sheet on the sediment surface, especially in moderate to low energy environments. They are seemingly more abundant on sandy sediments which they stabilise and thus possibly facilitate the synsedimentary lithification of these sediments.

Unicellular algae inhabit the surfaces of most carbonate sands in the Persian Gulf, including the oolite sands, their presence often imparting a vague greenish tint to the sediment. Some types of minute algae perforate the surface of grains and rock substrate by chemical processes producing a dense system of micro-borings. These are a characteristic feature of most shallow marine and intertidal sediments of the Persian Gulf.

Foraminifera

The distribution of three large groups of Foraminifera has been plotted on a regional scale (fig. 2), i.e. arenaceous, calcareous imperforate, and calcareous perforate Foraminifera. The first group comprises species of Textularia together with rare Clavulina and encrusting Placopsilina . The smaller calcareous imperforates are dominated by miliolids, the larger imperforates by Peneroplis and Sorites ; the smaller perforates are a heterogeneous group including rotaliids, Elphidium , Cibicides , Discorbis , Cancris , etc., while the larger perforates include Operculina , Amhistegina and Heterostegina .

Many Foraminifera are infaunal, such as Bolivina , Bulimina and Operculina , of which some prefer a mud substrate, others muddy sand. An important observation is that a surprisingly large number of the living Foraminifera are seen to be epifaunal. The most obvious are those found cemented to rock or shells (Homotrema , Acervulina , Placopsilina). However, many others are loosely attached to the large weeds that grow on rock areas, e.g. Amphistegina , Cibicides , many miliolids, Peneroplis , Sorites etc. These genera, although their tests are always found in the sediment, did not live in the sediments and in reality constitute part of the epifauna on nondepositional (rock) areas. Their tests have been transported by currents or dropped to the bottom when the plant substrate on which they rafted perished.

A diagrammatic cross-section (Fig. 3) illustrates where the more obvious or important Foraminifera live. The diverse, and particularly the epifaunal, habits favoured by many have obvious implications in any attempts to quantitatively survey and sample the living Foraminifera.

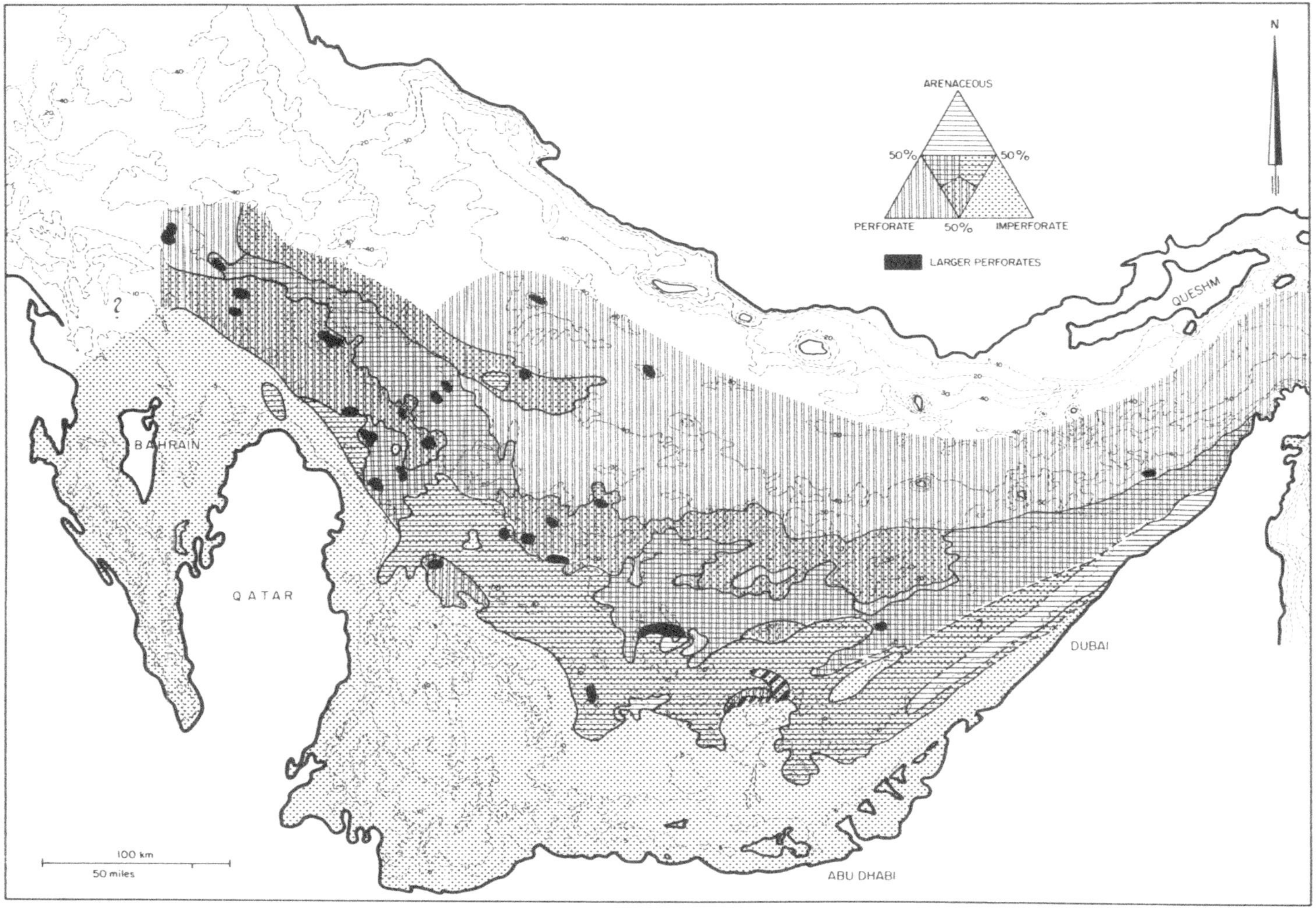

Fig. 2. Distribution of the principal foraminiferal assemblages, SE Persian Gulf

Scattered planktonic Foraminifera occur in the deeper water east of Qatar, but only in the Strait of Hormuz area, and immediately to the west of it, do they form a significant portion of the dead Foraminifera assemblages. Only species of Globigerinoides penetrate into the Gulf as far as Qatar, Orbulina and Globorotalia remaining near the entrance. The cause of the rapid decline in number within the Gulf is unknown.

Regional foram. distribution:

The imperforate assemblage is predominant in the coastal zone and in the restricted Gulf of Salwa. Offshore, especially at ca. 15 m depth, arenaceous Foraminifera become more numerous and a mixed imperforate-arenaceous assemblage occupies an irregular belt extending eastwards from the N end of Qatar Peninsula (see fig. 2).

An arenaceous assemblage locally becomes dominant on Umm Shaif high, the proximal parts of the Arabian marine homocline in the E parts of the Trucial Coast Embayment, and around Zirko Island. The imperforate Foraminifera are replaced progressively by perforate and arenaceous forms with increasing depth and decreasing restriction, a mixed arenaceous-perforate assemblage occupying a broad, irregular belt extending across the outer parts of the Trucial Coast Embayment down to depths of ca. 55 m.

A perforate assemblage dominates in the argillaceous lamellibranch muds occurring in the centre of the basin, and locally within the lamellibranch muds somewhat nearer the coast. The large perforate assemblage has a localised distribution, being restricted mainly to the tops of highs situated on the outer parts of the homocline. A generalised distribution map of the larger foraminiferal genera is given in figure 4.

Foraminiferal numbers:

The foram number was plotted on a regional scale (fig. 5) using units of 0 - 100, 100 - 1,000, 1,000 - 10,000, and > 10,000 specimens larger than 75μ per gram of dry sediment. In general, high energy sediments have very low foram numbers, whilst moderate to low energy sediments have numbers in the order of 1,000 - 10,000.

Highest foram numbers occur in two contrasting environments:

(a) A narrow zone bordering the outer edge of the 20 fm. (36 m) terrace. The abundance of foraminiferal tests, especially small perforates, in this relatively deep, open marine environment may be due to one or more factors, including high rates of reproduction or low rates of sedimentation. These forams, virtually none of which were found living, often occur in fine sediments rich in molluskan debris. This debris accumulates near the foot of the 20 fm. (36 m) terrace and grades basinwards into argillaceous lamellibranch muds, and landwards into lamellibranch muddy sands accumulating on the 20 fm. terrace. The abundant small foraminiferal tests is in equilibrium with the mud grade (smaller than 63μ) of the total sediment. Thus it is probable that these Foraminifera have been transported from the shallower areas and concentrated by sedimentary processes at the foot of the 20 fm. terrace (see fig. 6).

(b) Exceedingly high foraminiferal numbers also occur in somewhat restricted coastal embayments where the dominantly imperforate assemblages may have foraminiferal numbers that exceed 20.000. These high numbers are due, at least in part, to the abundance of miliolid and peneroplid Foraminifera which live on brown algae covering the rocky substrates in these coastal environments. However, the main factor is clearly the lack of any alternative sediment.

The concentration of Foraminifera in the sediments of the Persian Gulf is clearly influenced both by ecological and sedimentary factors; the foraminiferal number of any given sample has little environmental value unless both factors are considered.

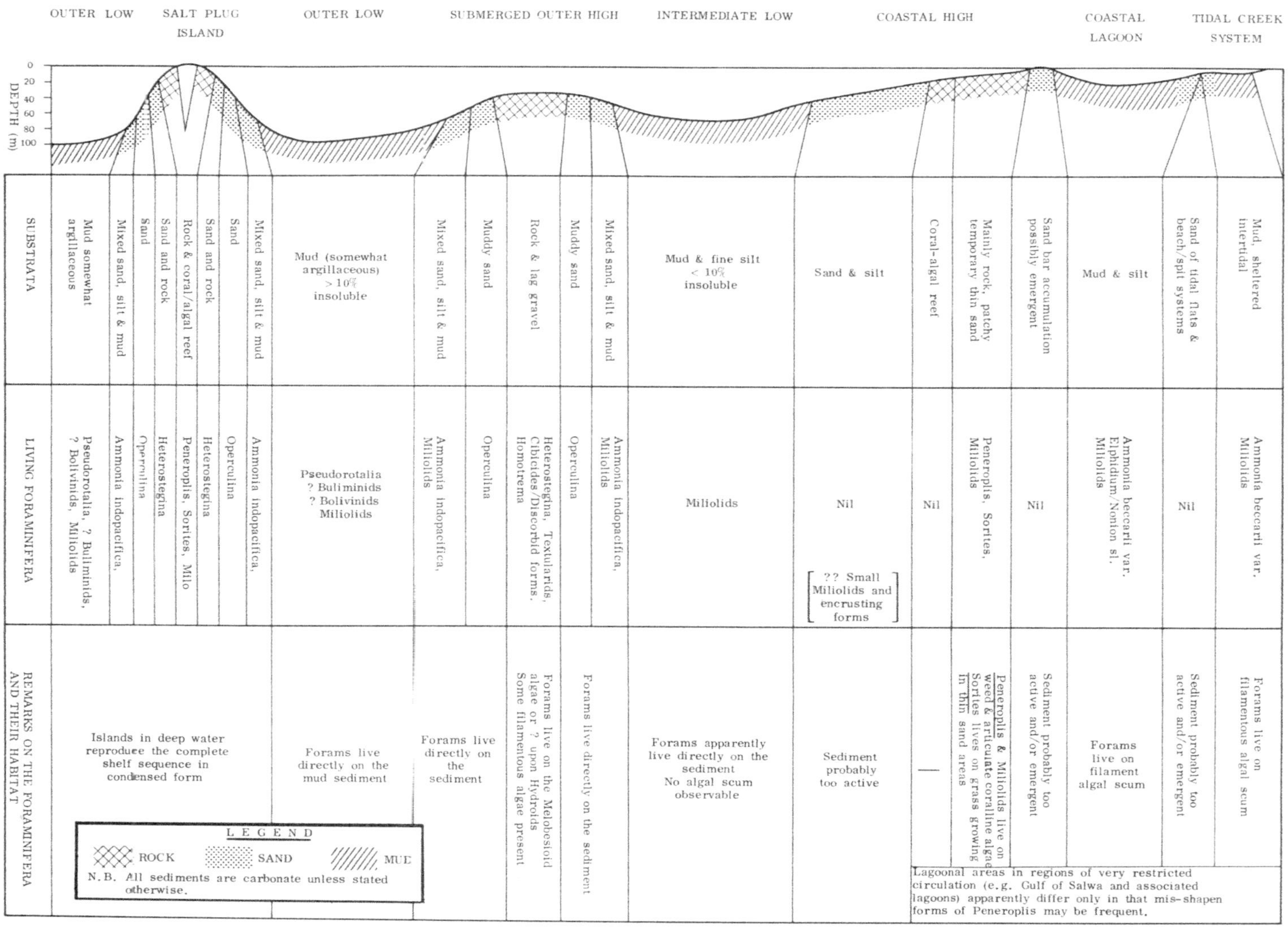

Fig. 3. Schematic cross-section illustrating distribution of living foraminifera (from field observations) Southern Persian Gulf

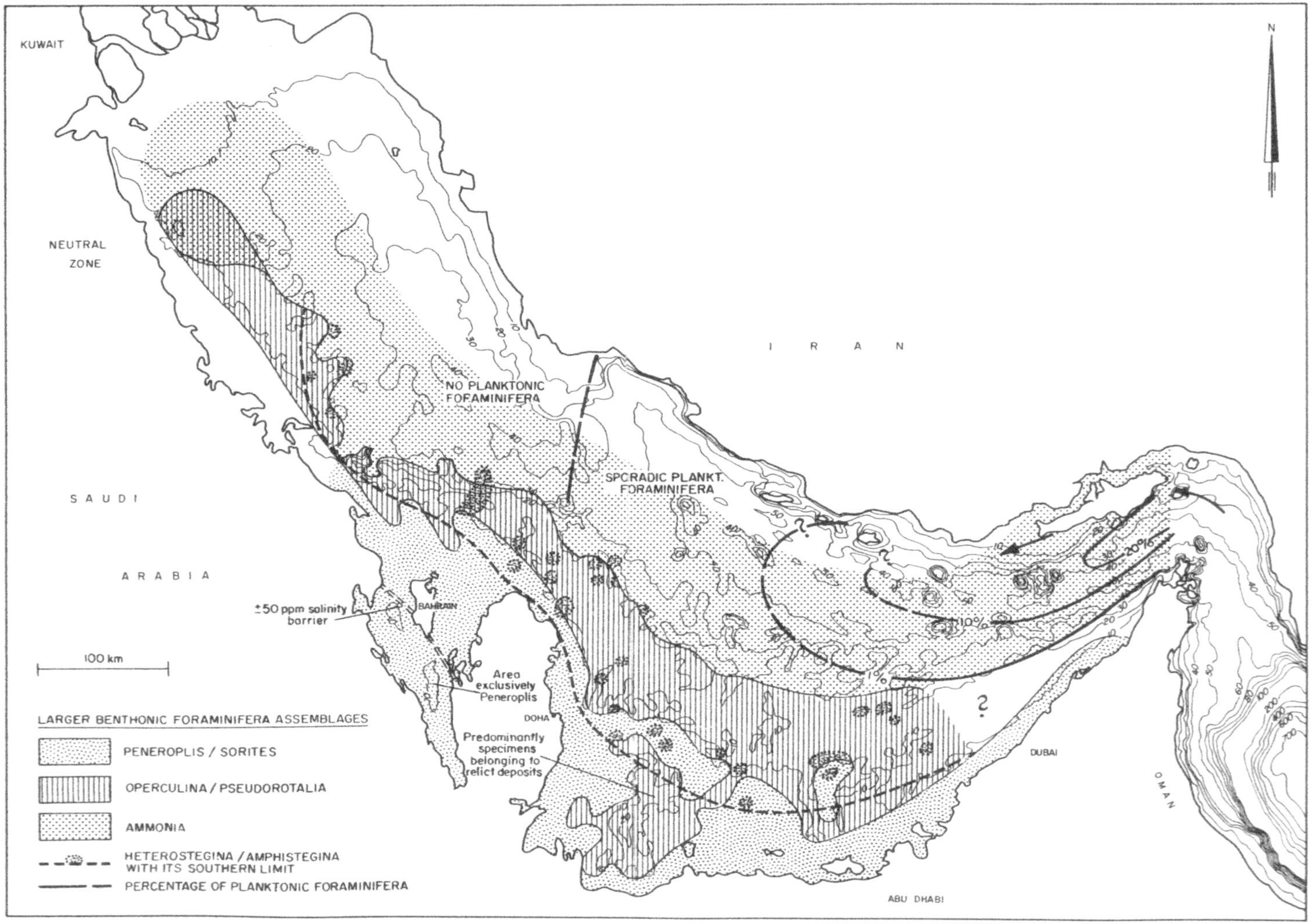

Fig. 4. Distribution of larger foraminifera in the S Persian Gulf

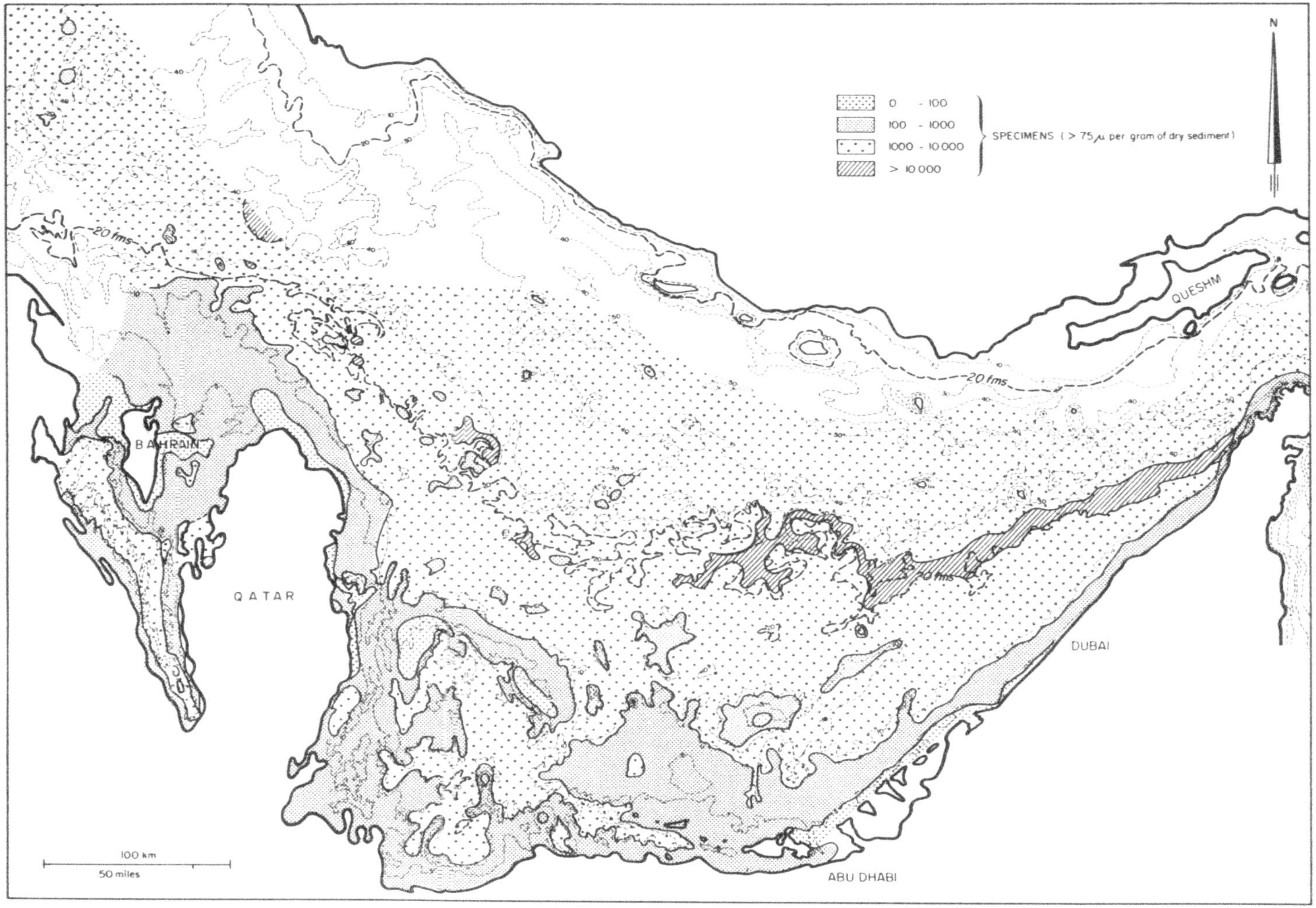

Fig. 5. Distribution of foraminiferal numbers, SE Persian Gulf

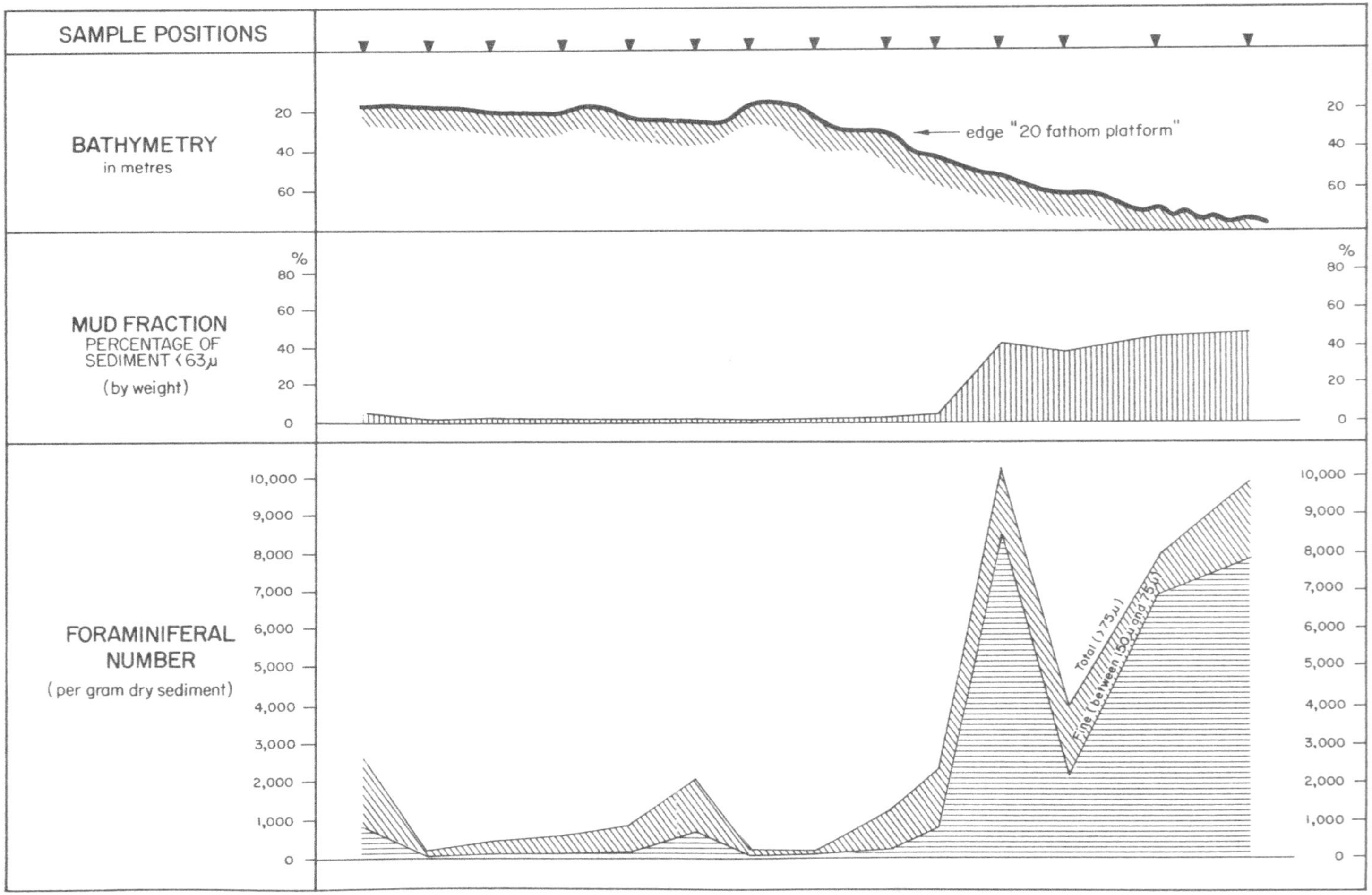

Fig. 6. Relationships between foraminiferal number, mud fraction and bathymetry on section II of Houbolt (1957), N Qatar

Aberrant growth in Peneroplis:

Of the two species of Peneroplis identified in the coastal environments, up to 5% of the larger speciments (> 400μ) commonly exhibit growth aberrations. However, in the restricted environments (salinity 50 - 70‰) in the lagoons around Qatar and in western Abu Dhabi, the proportion of such irregular growth variants increases and they may constitute 30% of the larger forms (plate 2). This growth aberration is probably a response to the great variability of the climate in these environments, but, whatever its origin, it appears to constitute a valuable marker for such a degree of restriction.

Corals (plate 3)

Although corals are widespread in the Persian Gulf, both as solitary and colonial forms, the number of genera and species is somewhat limited.

The branching Acropora , the massive Porites and the brain coral Platygyra are the commonest types. A striking form occurring in the deeper water is the mushroom-shaped Turbinaria . Small solitary corals were found attached to various bioclastic particles offering sufficient substrate. The solitary, infaunal corals Heterocyathus and Heteropsammia , which live in symbiontic association with the worm Aspidosiphon , are widespread, but are particularly abundant on the Oman side of the Strait of Hormuz. In the deeper water muds, a flabellid species (? Placotrochus) was often found in small numbers.

Kinsman (1964) reported that the coral Porites in the Trucial Coast lagoons had its maximum salinity tolerance at 48 ppm and its maximum temperature limit at approximately 42° C. Studies at the eastern entrance of the Gulf of Salwa confirm that corals disappear where the salinity exceeds 50 ppm.

Octocorallia are locally abundant where hard substrate is available, preferably at depths exceeding 10 m. They are absent in the Gulf of Salwa. In dredge samples, gorgonians were frequently recovered infested with brittle stars.

Worms

The calcareous tubes of Spirorbis are often abundant on plants in shallow marine environments. Serpula is common in coarse sediment where it encrusts rock or skeletal fragments. Non-skeleton-secreting worms are common burrowers and disturb bedding, tending to homogenise the sediment. Certain polychaete worms are active borers, their activity helping to weaken and break carbonate skeletons to finer sediment.

Sponges

Sponges are locally abundant, especially in coarser sediment and on rock bottoms. In addition, juvenile clionid sponges are important borers of calcareous skeletons and produce a system of micro-cavities which weaken the shell structure. These animals are particularly abundant in parts of the restricted Gulf of Salwa.

Brachiopoda

A species of infaunal Lingula was collected at several stations in restricted environments in the Khor al Bazm and Gulf of Salwa. It lives in muddy molluscan sand at depths of 4 - 16 m where salinities vary between 56 - 60 ppm. No previous record of the occurrence of Lingula in supersaline environments is known in the available literature. However, it is known that Lingula tolerates other extreme conditions such as foetid muddy sediments or brackish water.

Bryozoa

Zoaria of encrusing Bryozoa, such as Thalamoporella and Parasmittina , are locally abundant in the sediment. They live on the stems and fronds of brown weeds which detach and float to beaches, or even inland during periodic flooding of the sabkhas. These algae can be transported to particularly remote restricted environments, such as the south of the Gulf of Salwa, where they disintegrate and deposit their encrusting Bryozoa.

The total bryozoan fauna is qualitatively poor. Articulated species seem unable to withstand salinities higher than 50 ppm; only one such species was found sporadically alive in the Gulf of Salwa. However, several living encrusting types were noted in these restricted areas.

Ostracoda

Ostracoda are present in almost every sample, the amount varying according the sediment type. Very rich assemblages occur in nearshore muds, but very few occur in the mobile sands. Species of Loxoconcha and Xestoleberis are abundant at localities rich in growing weed, i.e. Qatar coastal lagoons and the Gulf of Salwa. A relatively varied assemblage of some 20 species persists at the S end of the Gulf of Salwa where salinities exceed 60 ppm. In the inner Khor Odaid lagoon where a chlorinity of over 50 ppm was measured, living Cyprideis was found, the only other life being masses of filamentous algae growing on a thick gypsum crust.

Molluska (plates 4 and 5)

Mollusks are particularly common in the Persian Gulf. In general, the gastropods are predominant in the lagoonal environment and the lamellibranchs on the offshore marine homocline.

In the lagoons, species of Cerithium , Bullaria and Mitrella , together with the thin shelled lamellibranch Lucina , are common. In the intertidal zone of sand beaches, Cerithium occurs in droves. In the 'high energy' areas the epifaunal lamellibranchs Chama , Spondylus , several Arca species, the oyster Pinctada and the mussel Brachyodontes live together with corals and coralline algae. The mobile coral-algal sands contain the small reddish-purple Ervillea purpurea in abundance, and its shell fragments often give this sediment a typical reddish appearance. The more protected muddy sands carry a Timoclea layardi-Circe-Glycimeris-Dentalium assemblage, whilst the muds are characterized by small and heavy valves of Corbula , Phacoides and Nucula.

It is striking that all large species of Indo-Pacific mollusks are absent; they include species of the lamellibranch Tridacna and the gastropods Cassis , Cymbium , Lambis , Voluta and the larger species of Conus and Cypraea . The only large lamellibranch that occurs is the infaunal Pinna bicolor , which lives three quarters buried in the muddy sands of lagoons or shallow shoals, often in dense populations.

Fragile Pteropoda occur in the deeper water muds near the entrance to the Gulf. Diacria barely penetrates the Gulf, but the long beaked shells of Cavolinia longirostris are typical of the basin centre muds. The extremely fragile tubes of Creseis occur nearer the shore.

Echinodermata (plates 6 and 7)

Echinoids, seastars and brittlestars are common in the Gulf. The echinoids belong to a limited number of species; the epifaunal,black spined Echinometra is abundant on coral rubble and rock outcrops from the low tide line down to 20 m. Diadema clusters in depressions and crevasses of rock and coral. Rare specimens of the cidarid Prionocidaris baculosa were dredged from rocky bottoms. The small, regular

Temnotrema is widespread at all depths and was often collected alive, attached to the concave underside of empty mollusk valves. The regular Temnopleurus lives gregareously in shallow water on sandy bottoms covered with marine plants. Species of the burrowing genus Clypeaster are the commonest echinoids in the gulf, two species living at depths exceeding 20 m. The commonest, Clypeaster humilis lives between the low tide line and 35 m depth, often in association with Echinodiscus auritus . These forms are infaunal, burrowing to some 10 cm in the surface layers of sands and muddy sands. In the deeper muddy sands and muds species of Metalia and Lovenia are very active burrowers, often occurring in abundance. In the deeper, argillaceous, carbonate muds near the entrance to the Gulf, empty tests of an Echinocyamus species are common.

Spines and fragments of tests of echinoids, and the ossicles of arm skeletons of sea and brittle stars occur in small quantities in all samples, provided the salinity of the water is not higher than 48 ppm. In the lagoonal areas of Khor al Bazm and the Gulf of Salwa, where the salinity exceeds 50 ppm, two species of sea stars (an Asterina and an Asteropecten) and one species of brittle star were found living, but echinoids were absent from these same areas.

Free-swimming comatulid crinoids were found in dredge samples at various localities, especially where the bottom was rocky or the sediment very coarse. No stalked crinoids were found.

The relationships between the principal groups of organisms and their environments are summarized in figure 1.

V. CARBONATE PRODUCING ORGANISMS

A relatively small number of organisms dominated by the mollusks, and followed by the calcareous algae, corals, Foraminifera, Bryozoa, and echinoids, produce the bulk of the skeletal material that constitutes most of the Holocene sediments of the Persian Gulf. Certain of these groups have a patchy distribution and may be dominant locally where their ecological requirements are fulfilled. They include the coral/algal reefs and pearl oyster banks, which are dependant on hard substrates in relatively high energy environments. Certain lamellibranchs are locally dominant; epifaunal forms such as the Chama , Spondylus , Pinctada group occur in reefs, while a group of infaunal species, which includes species of Corbula , Nucula, and Phacoides , are common in muddy environments.

Foraminifera are particularly abundant in two distinct environments; the imperforate miliolids and Peneroplidae predominate in shallow, protected embayments and lagoons where they may constitute nearly all of the sediment; the perforate Amphistegina - Heterostegina group are exceedingly abundant on certain offshore shoals at moderate depths.

Various other groups contribute lesser amounts of sediment, but they may be characteristic, and are therefore important, as environmental indicators, as seen in the foregoing outline of the major groups.

The carbonate sediments in the southern Persian Gulf therefore reflect the broad ecological limitations of the environments upon the organism assemblages. The different sediment types defined in the article by Wagner and v.d. Togt (in this volume) show this relationship.

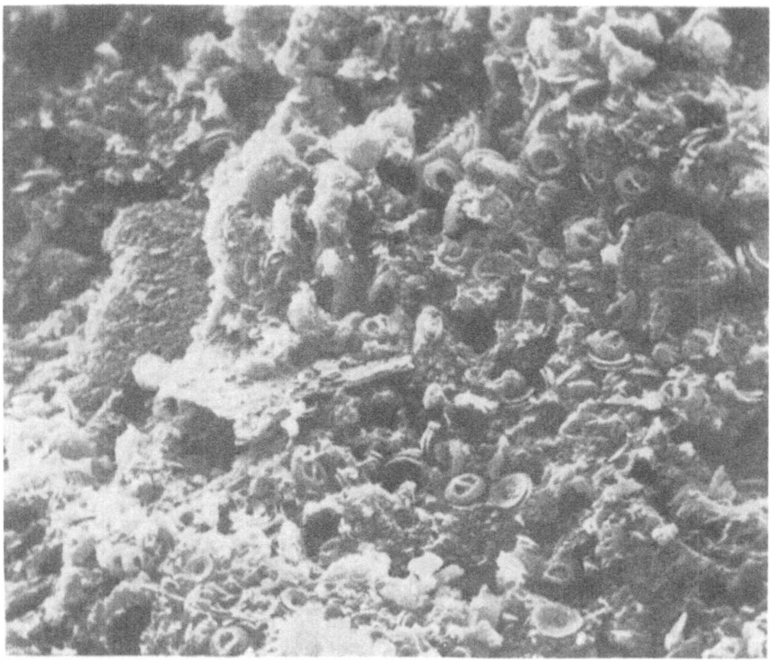

Plate 1A. Carbonate mud in depression on proximal homocline, approx. 85 km offshore Sabkha Matti, western Abu Dhabi.
Note that some 70% of the sediment consists of a single coccolith species. The remaining sedimentary material is unidentifiable, but is largely broken aragonitic skeletal debris.
Sample station T.319, depth 27 m.(TNO REM Foto No. 19786)

Plate 1B. Underwater photograph showing numerous individuals of Echinometra mathei browsing over dead coral, mostly of the genus Acropora

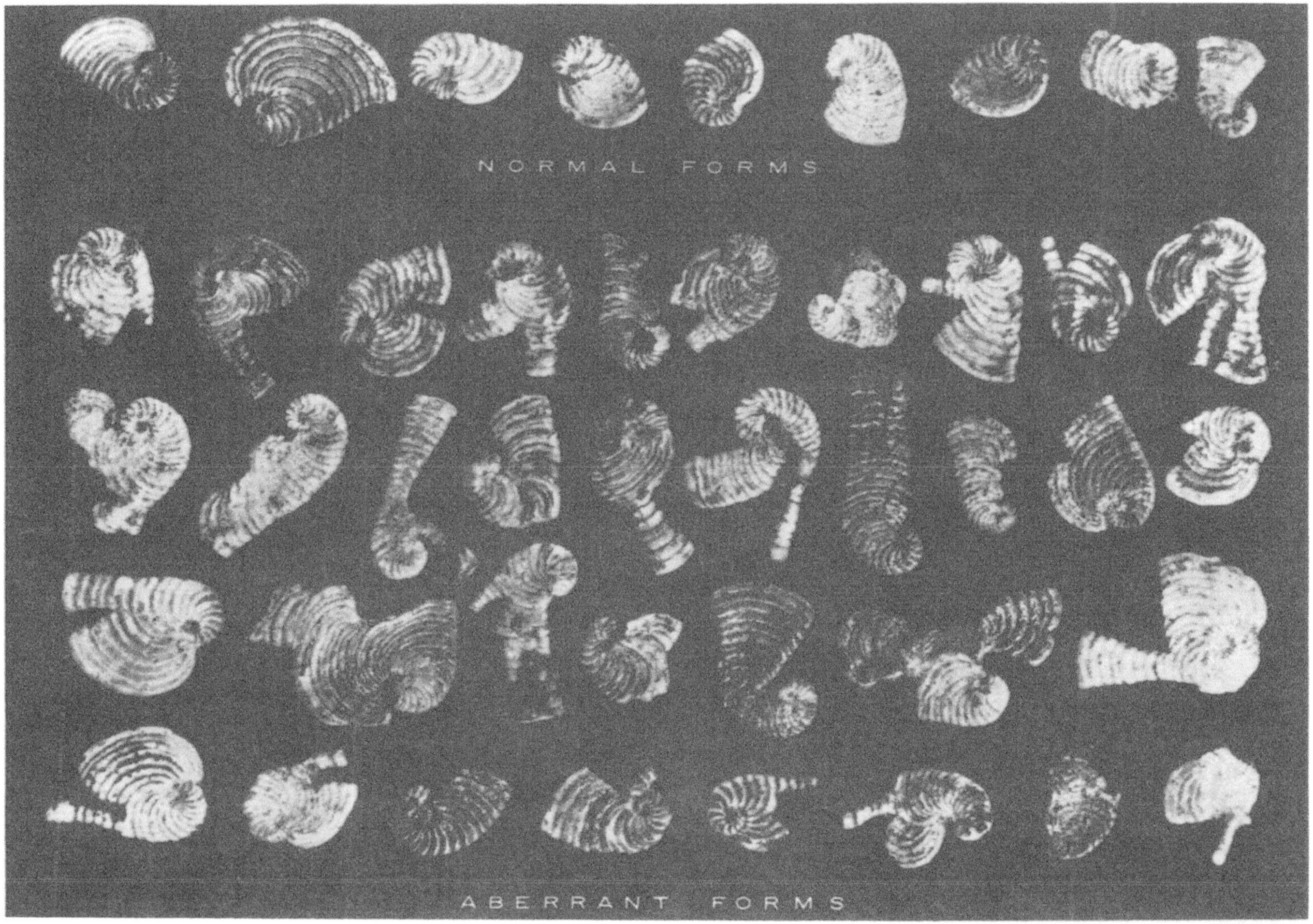

Plate 2. Normal and aberrant growth forms in a species of <u>Peneroplis</u>. Dead specimens from sediment sample W 1298 collected in Dohat al Hussein (W. Qatar) where salinities range over 60 ^{o}/oo. (x 9)

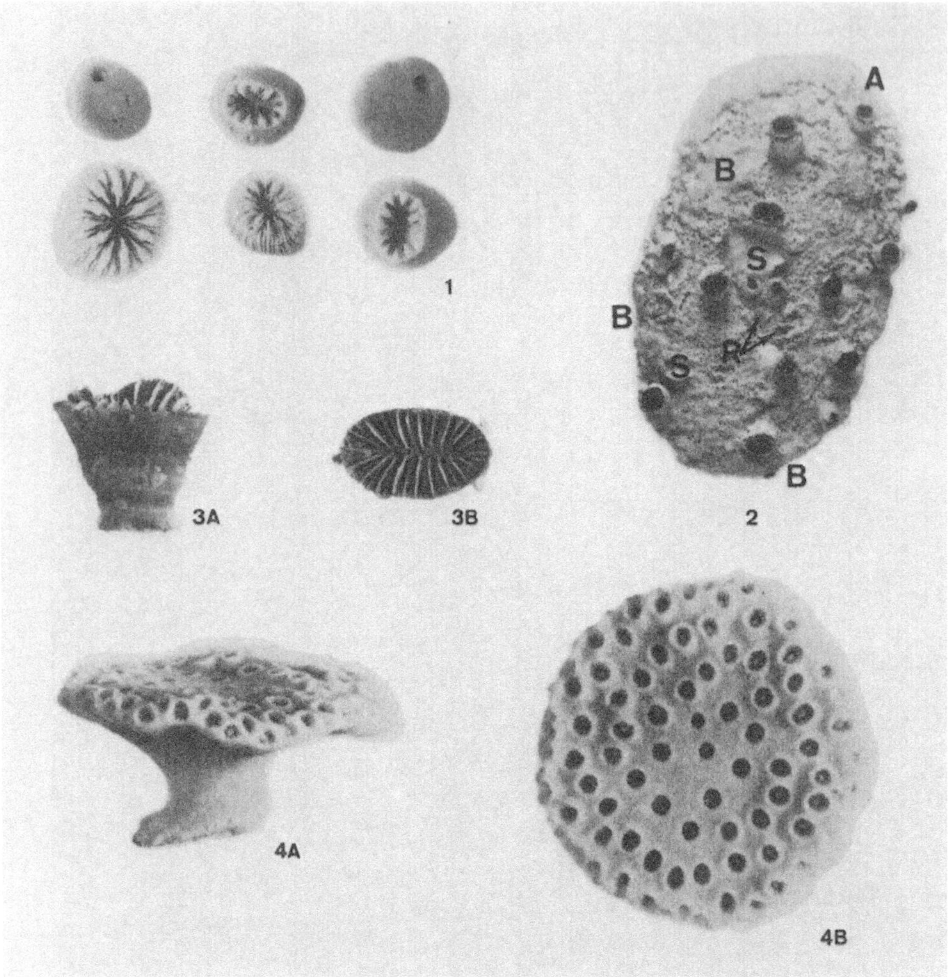

Plate 3

1. Infaunal solitary corals of the genera Heteropsammia and Heterocyathus, whose base encloses the tube of a sipunculid worm. (x 1.2)
2. Solitary epifaunal corals attached to lamellibranch valve. The latter completely bored by clionid sponges and overgrown by Algae (A), Bryozoa (B), Spondylus (S), serpulid worms and the foraminifer Acervulina (R). (x 0.8)
3. Solitary flabellid coral belonging to infauna of deeper water muds. A: side view, B: top view (x 1.2)
4. Turbinaria (x 0.8), side view (A) and from above (B), showing mushroom-shaped growth, probably reflecting adaptation to reduced light conditions at the depth where this species lives, i.e. greater than 15m

Plate 4. Shallow-Marine Mollusks (x 0.75, except no. 19 and 20)
1: Pecten aff. senatorius L; 2: Clycymeris cf heroicus (Melvill and Standen); 3: Pinctada radiata; 4: Circe scripta (L); 5: Clycymeris maskatensis (Melvill); 6: Spondylus sp.; 7: Conus keatii Sowerby; 8: Cypraea turdus; 9: Cardium sp.; 10. Venus cf calophylla Hanley; 11: Arca sp.; 12: Timoclea layardi (Reeve); 13: Ervillea purpurea Desh.; 14: Murex cf turbinalis Lmk.; 15: Xenophora corrugata (Reeve); 16: Strombus plicatus (Röding); 17: Strombus decorus persicus Swainson; 18: Bullaria sp.; 19: Pinna bicolor (x 0.35); 20: Ceritium sp. (x 0.35); 21: Fusinus sp.; 22: Turritella maculata Reeve; 23: Dentalium sp.

A

B

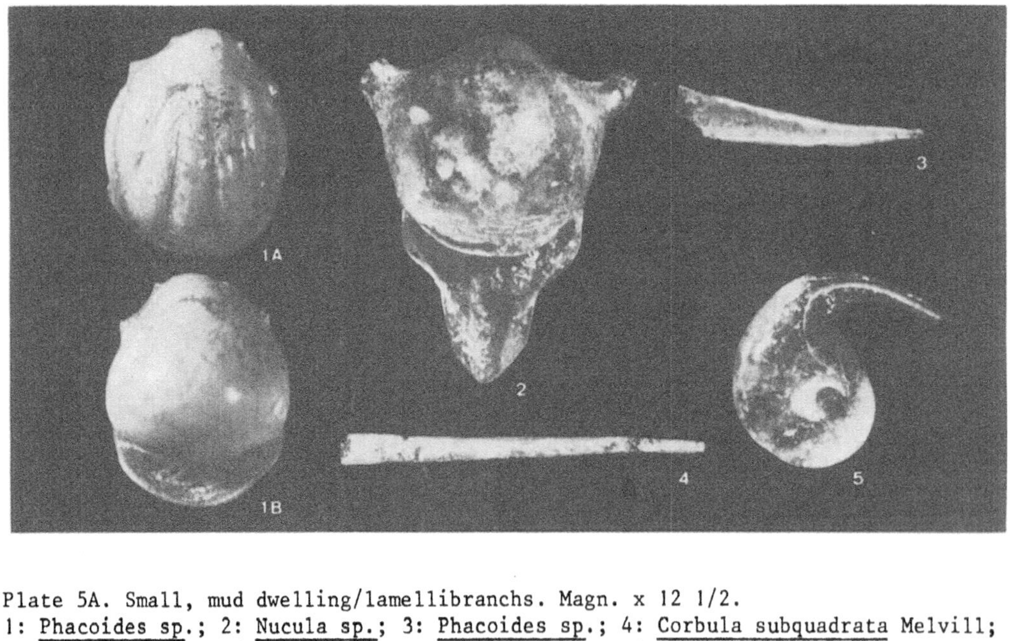

Plate 5A. Small, mud dwelling/lamellibranchs. Magn. x 12 1/2.
1: Phacoides sp.; 2: Nucula sp.; 3: Phacoides sp.; 4: Corbula subquadrata Melvill; 5: Lima juglandula (Melvill); 6: Arca requiescens Melvill

Plate 5B. Pteropoda
1: Diacria quadridentata (Lesueur); a: dorsal, b: ventral side (x 12 1/2); 2: Cavolinia longirostris (Lesueur) (x 6); 3: Creseis virgula (Rang) (x 25); 4: Creseis acicula (Rang) (x 12 1/2); 5: Limacina inflata (d'Orb.) (x 25)

Tests of Diacria and Cavolinia occur mainly in deeper marine areas, the remaining types accumulating in many shallow and deep marine parts of the Gulf

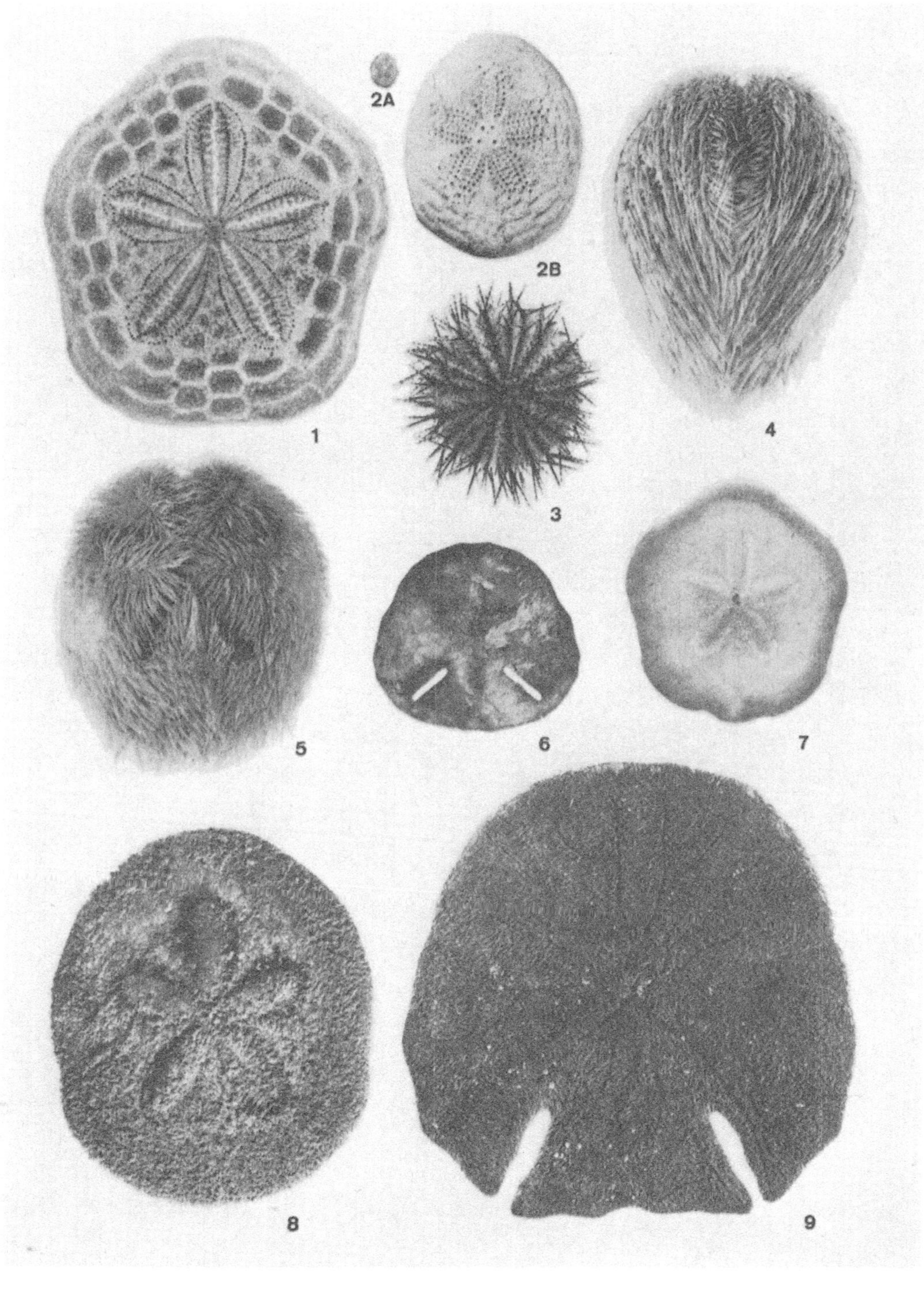

Plate 6. Infaunal echinoids (x 0.75, except no. 2)
1: Clypeaster reticulatus (L); 2: Echinocyamus crispus (Mazzetti) magn. a: x 0.75, b: x 4.5; 3: Temnopleurus toreumaticus (Leske); 4: Lovenia elongata Gray; 5: Metalia persica (Mortensen); 6: Echinodiscus bisperforatus (Leske), juvenile; 7: Clypeaster rarispinus (de Meyere); 8: Clypeaster humilis (Leske); 9: Echinodiscus auritus (Leske), ventral side

Plate 7. Epifaunal echinoids and sea stars (x 0.75)
1: Test of Diadema setosum Gray, a form common on reefs; 2: Juvenile Diadema; 3: Temnotrema scillae (Mortensen), a: dorsal, b: ventral side; 4: Echinometra mathaei (Blainville) a: test with, b: and without spines; particularly abundant on reefs; 5: Prionocidaris baculosa (Lamarck), note abundant epifaunal serpulid tubes on the spines, together with Homotrema (arrows); 6: Asterina cephea iranica Mortensen; 7: Astropecten sp.

Holocene Sediments and Sedimentary Processes in the Iranian Part of the Persian Gulf

E. Seibold, L. Diester, D. Fütterer, H. Lange, P. Müller, and F. Werner[1]

ABSTRACT

Sections 2-7 summarize investigations of the regional grain size distribution, carbonate content, sand, silt and clay fractions, colors in the fraction < 0,1 micron, content of organic carbon and nitrogen, and sedimentary structures with biogenic cavities. Results are combined in Fig. 27, showing facies distribution and sediment transport direction, and in Fig. 28 showing the regional distribution of relative sedimentation rates. Sedimentary processes and the history of the post-glacial transgression are discussed. Finally, the Iranian Side of the Persian Gulf is regarded as 1) a model area for the formation of a fully marine, marly, fine grained molasse, 2) a sedimentation model for a marginal sea within the arid climatic zone, and 3) an area for the study of interbedded limestones and marls.

INTRODUCTION

A team of geologists and micropaleontologists from the Geological-Paleontological Institute of Kiel University, and physical, chemical and biological oceanographers from Kiel and Hamburg took part in the International Indian Ocean Expedition which visited the NE part of the Persian Gulf with R.V. "Meteor" from March 3 to April 24, 1965 (Dietrich et al 1966). The expedition itself, as well as the evaluation of the material collected, was supported by the German Research Society.

The aim of the expedition was to investigate a marginal sea in an arid climate with regard to its hydrography, morphology, sedimentology and micropaleontology, and to compare the results with observations recorded in the Baltic, a marginal sea in a humid climate (Seibold 1970, Seibold et al. 1971). Of particular interest were the characteristics, distribution, source and division of the post Pleistocene marl sediments, and the study of the Holocene rise in sea level and its effect on morphology and sedimentation.

1) Environmental framework

The morphological, climatological and hydrological features influencing sedimentation and sediments are briefly mentioned in the Introduction of this volume. More details about the Iranian side of the Persian Gulf are to be found in Seibold & Vollbrecht 1969, Seibold & Ulrich 1970, Sarnthein 1970, Hartmann et al. 1971.

1 Geological-Paleontological Institute of Kiel University.

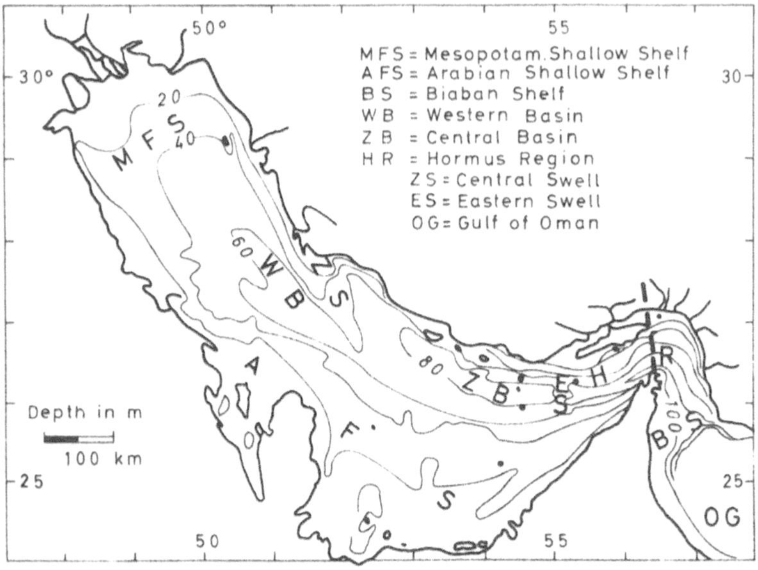

Fig. 1. Simplified bathymetric map of the Persian Gulf (After Seibold & Vollbrecht 1969)

2) Grain size distribution

(Details see Hartmann et al., 1971)

The main trend in the grain size distribution pattern is an increase in coarse grained fractions with water depth; in part, a parallel increase takes place with increasing distance from land, except in nearshore areas. In the Western Basin tongues with increased fine grain fractions are parallel to the coast, whereas in the Central Basin they are more or less perpendicular to the coast. The shallows are characterized by coarse grain sizes (Fig. 2). Generally, however, the sand content (> 63 micron) of the entire area is less than 50 %.

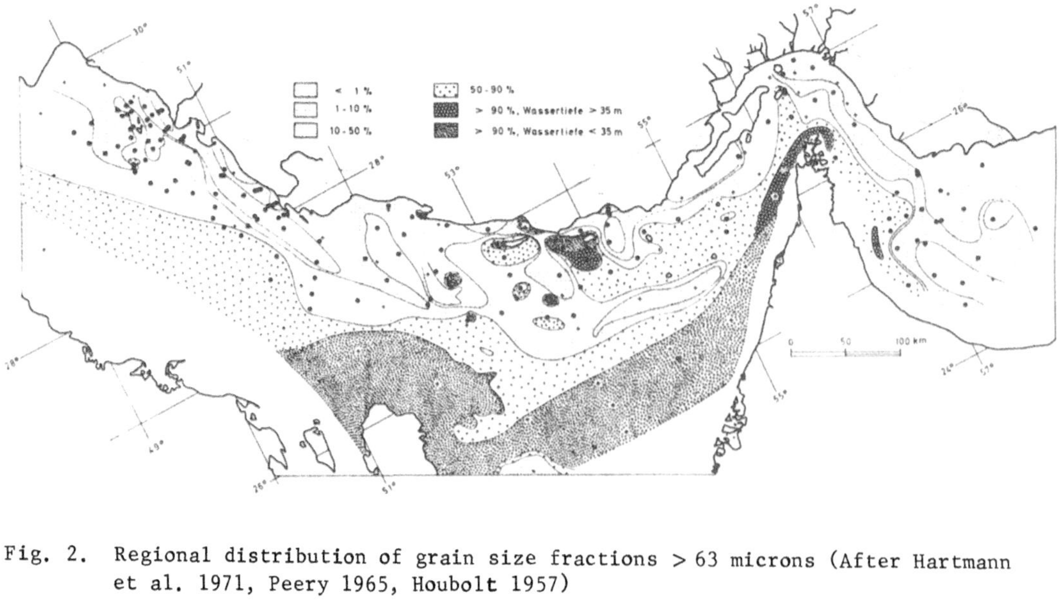

Fig. 2. Regional distribution of grain size fractions > 63 microns (After Hartmann et al. 1971, Peery 1965, Houbolt 1957)

3) Carbonate Content

(Details see Hartmann et al., 1971, Sarnthein 1971a)

Except for Iranian coastal areas the sediments nearly always contain more than 50 % carbonate minerals (Fig. 3). "Marl" is therefore the dominant sediment type. Carbonate content reaches 80 % towards Arabia and in shallows. Comparison with the grain size distribution map (Fig. 2) shows that an increase in carbonate content generally parallels an increase in mean grain diameter and vice versa. This fact alone suggests that, besides local terrigenous carbonate supply and carbonate relict sediments, organisms are the primary contributors of carbonate to the sediments; many fossil marl sediments surely had similar origins.

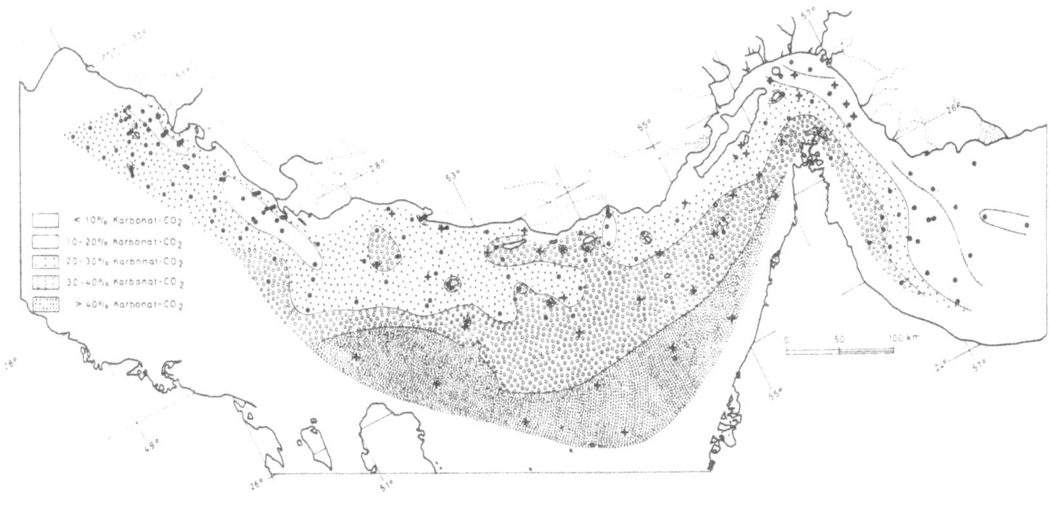

Fig. 3. Regional distribution of the carbonate-CO_2 contents of total samples (After Hartmann et al., 1971, Peery 1965, Houbolt 1957)

4) Coarse fraction (> 63 microns)

4.1) Description of components
(Details see Sarnthein 1971a)

The main constituent groups of the coarse fraction are a) faecal pellets and lumps, b) non-carbonate mineral components, c) Pleistocene relict sediments, and biogenic d) benthonic and e) planktonic components. Samples from the main sedimentary environments show the following compositions:

Table 1. The composition of samples from the principal sedimentary environments in the Iranian part of the Persian Gulf

Sample type	1) Clayey Marl (with few coarse grains)		2) Calcareous Marl (rich in coarse material)		3) Calcareous Marl		4) Calcarenite (coarse)		5) Clay (for comparison)	
Distribution	Near shore (River mouths)		Near shore with limited terrigenous inflow, and offshore in basins		Off river mouths (in shallow water)		Shallows and Island Slopes		Continental Slope Gulf of Oman	
Carbonate-CO_2 %	18.15 %		32.3 %		23.4 %		41.7 %		10.6 %	
Fraction 0.06-2.0 mm % / > 2 mm	0.16	0	57.4	1.92	37.12	0.05	89.65	7.33	14.74	0
Subdivision:	%		%	%	%	%	%	%	%	
Foraminifera:										
Calcareous	41.3		11.25	-	1.45	-	6.9	0.35	1.06	
Arenaceous	0.5		4.65	-	1.75	-	0.9	-	-	
Planktonic	-		0.06	-	-	-	-	-	28.8	
Sponges	Tr.		-	-	-	-	-	-	-	
Corals	-		Tr.	-	-	-	-	0.14	-	
Bryozoa	-		0.16	-	Tr.	-	1.2	0.25	-	
Mollusks:										
Benthonic	22.3		40.9	49.0	1.1	92.2	7.5	7.9	Tr.	
Planktonic	5.4		3.05	0.12	Tr.	-	0.25	-	Tr.	
Crustacea:										
Balanus	0.1		Tr.	1.5	Tr.	-	0.1	0.12	-	
Decapods	0.9		3.95	0.6	0.2	7.8	0.9	0.2	Tr.	
Ostracods	5.45		1.1	-	0.5	-	Tr.	-	0.1	
Echinoids	14.9		1.05	0.3	0.45	-	0.3	Tr.	Tr.	
Ophiurids	0.05		0.85	-	0.3	-	0.3	-	-	
Fish remains	1.65		0.1	-	Tr.	-	0.12	Tr.	0.55	
Plant remains	1.15		Tr.	-	0.3	-	-	0.1	0.1	
Calc.Algae	-		-	-	-	-	Tr.	2.2	-	
Biogenic Relict Components	-		10.7	16.3	-	-	68.6	38.8	-	
Miscellaneous: Carbonate and Non-carbonate Grains	6.3		22.0	32.2	93.8	-	12.7	50.0	69.2	

(Examples from Sarnthein 1971a)

Fig. 4. Regional distribution of plant remains. Plotted values =

$$\frac{\text{percentage of plant remains in the sand fraction}}{\text{percentage of plant remains, benthonic Foraminifera and mollusks in the sand fraction}} \times 100$$

(After Sarnthein, 1971)

Fig. 5. Regional distribution of the plankton:benthos ratio in the Foraminifera. Plotted values =

$$\frac{\text{percentage of planktonic Foraminifera in the sand fraction}}{\text{percentage of total Foraminifera in the sand fraction}} \times 100$$

(After Sarnthein, 1971)

4.2) Regional distribution of components
(Details see Sarnthein 1971a)

The "rule" of independent availability of coarse fraction component groups is essential for an understanding of the formation of the marl sediment: creation of the main component groups takes place more or less independently and transport sorting has no drastic effects, at least on a regional scale. This is suggested by the facies distribution and also by the simultaneous occurrence of both living and dead benthonic Foraminifera faunas (Haake, Lutze, unpublished).

a) Higher percentages of faecal pellets and lumps can be correlated with large amounts of fine sediment and presence of organic C. No discernable change takes place in carbonate minerals as a result of digestion and faecal pellet formation.

b) The non-carbonate sand components orginate mainly from the Iranian rivers (well-sorted fine sands), from relict minerals (poorly sorted medium sand) on basin bottoms and from reworked salt dome material (medium sand to gravel) in localized areas. Wind transport produces only a minimal "background value" of very fine sand.

c) The main part of the coarse, poorly sorted, relict sand particles was deposited during the Late Pleistocene and forms a quasi-autochthonous cover over wide areas where little recent sedimentation takes place and where bioturbation has mixed the relict and modern fractions.

d) Benthonic organisms are the most important contributors of carbonate to the sediments of the Persian Gulf, whereas in the Gulf of Oman planktonic forms play the most important rôle.

The areas influenced by the inflow from Iranian rivers can be recognized by an abundance of ostracods, benthonic Foraminifera, and land-plant fragments which are transported along the bottom like sand particles. They have been found up to 30 km offshore (Fig. 4). Some species of benthonic Foraminifera are indicators of water depth. For example Quinqueloculina crassicarinata Collins occurs predominantly at a depth of 30 m (Haake 1970). Some of the depth zones characterized by a particular fauna become somewhat deeper as the mouth of the Gulf is approached. This may indicate dependency on water properties (Haake, Lutze, unpublished). On the continental slope, benthonic Foraminifera decrease in number at water depths greater than 200 m; this decrease is even more pronounced in the benthonic mollusks.

Wave base seems to be reflected in a maximum concentration of echinoderm fragments within the sediment (15-50 m, depending on fetch, Sarnthein 1970). Unrelated to sediment character, echinoid remains dominate the Western Basin and the Central Swell whereas ophiuroids are dominant in the Central Basin.

The relationships between fauna and substrate are numerous:
The shallow rocky bottoms of the shallow submerged highs and island slopes (sediment type 4, Table 1) are free of coral reef formation except for a weak indication of reef building west of the island of Kais (- 16 m). The larger Foraminifera, such as Heterostegina sp. populate such bottoms. Field studies and laboratory experiments have shown that this group of Foraminifera alone may contribute 150 g $CaCO_3/m^2$/year. This corresponds to 10 cm/1000 y (Lutze et al. 1971).

Hard-bottom sediments (type 2), such as are predominant in the basins, are characterized by an abundance of epibenthonic organisms such as solitary corals, Bryozoa, mollusks, especially gastropods and barnacles. Hard parts of worm tubes are also present.

Soft-bottom sediments may be subdivided into two groups according to their respective population. In areas with high sedimentation rates, epibenthonic organisms and mollusks are generally less frequent (sediment type 1). In areas where low sedimentation rates prevail, mollusks increase in frequency. They are able to generate a type of hard bottom with their own shell remains.

Fine sand bottoms (type 3) where sediment is being moved, as in nearshore, shallow water areas, are characterized by the absence of epibenthonic organisms.

e) Planktonic Foraminifera (as sediment particles) decrease gradually in frequency as compared to benthonic forms with increasing distance from the Indian Ocean. The ratio P/B is 50:50 near the continental shelf margin, a typical figure throughout the world. The ratio decreases rapidly towards the Persian Gulf and falls to 5 % in the Central Basin (Fig. 5, Sarnthein 1971a, Lutze "Adjacent Sea Effect", unpublished).

The maximum values are clearly displaced towards the Iranian coast, a direct indication of surface water inflow. Most, if not all, of these planktonic Foraminifera are passively transported from the Indian Ocean and do not live in the Gulf itself.

A similar regional distribution pattern (Fig. 6) is shown by the ratio planktonic/benthonic mollusks. In contrast to the P/B ratio of forams, this ratio is dependent mainly on water depths within the Persian Gulf. According to direct observations of plankton hauls and indirect methods such as $^{18}O/^{16}O$ ratios of shell carbonates (Hoefs & Sarnthein 1971), these planktonic Pteropods and Heteropods are indigenous to the Persian Gulf. In a thick water column the total amount of sediment constituted by their shells should be higher than in a thin one. By comparing different regional situations Sarnthein (1971a, 1972a) developed a method of estimating sedimentation rates from these ratios (Fig. 28).

Fish remains (bones, scales, teeth, otoliths) accumulate on the continental slope near the Gulf of Oman and at river mouths along the Iranian coast, the result of increased nutrient content and a related increase in plankton.

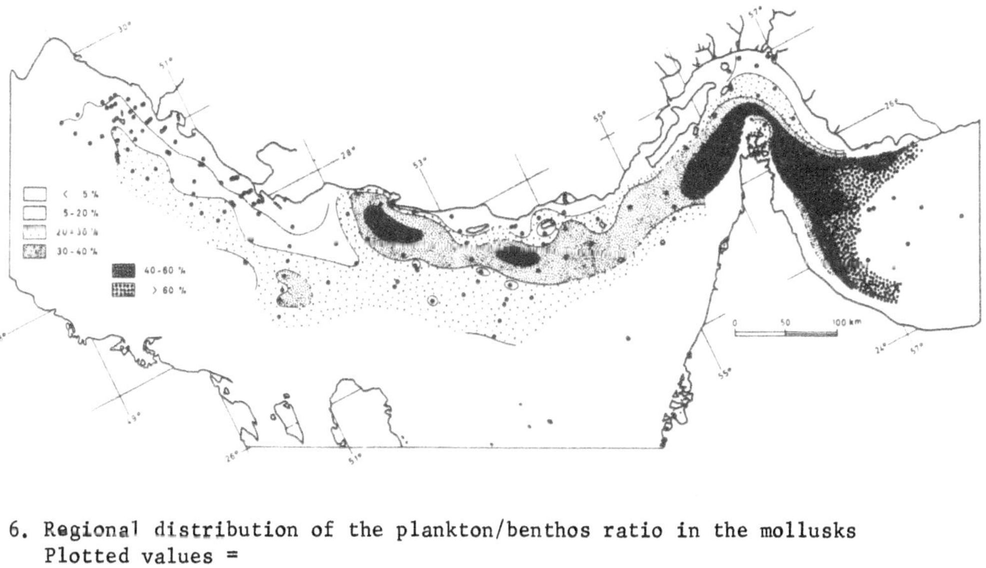

Fig. 6. Regional distribution of the plankton/benthos ratio in the mollusks
Plotted values =

$$\frac{\text{percentage of planktonic mollusks in the sand fraction}}{\text{percentage of planktonic + benthonic mollusks in the sand fraction}} \times 100$$

(After Sarnthein 1971)

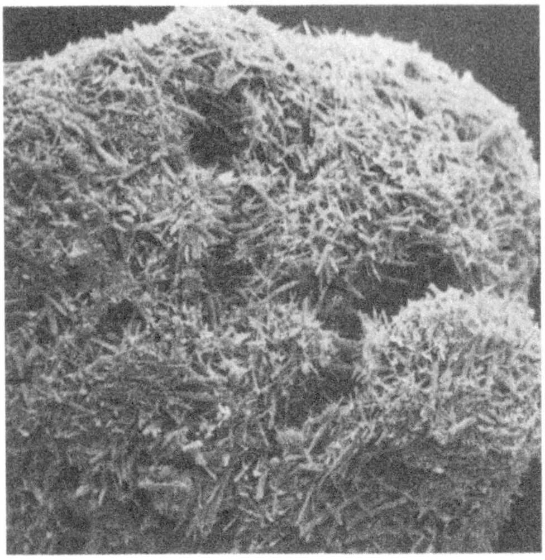
7

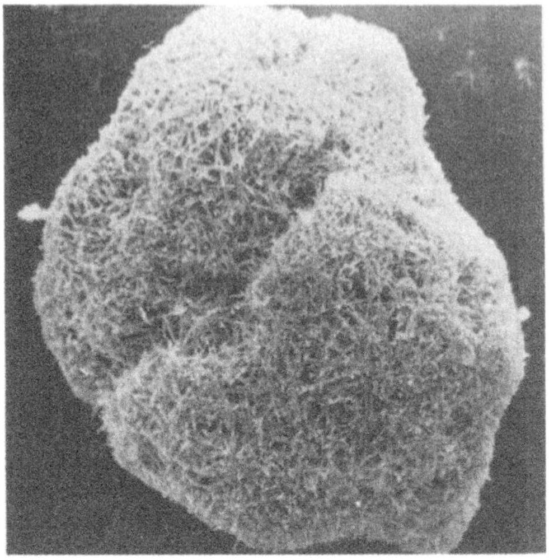
8

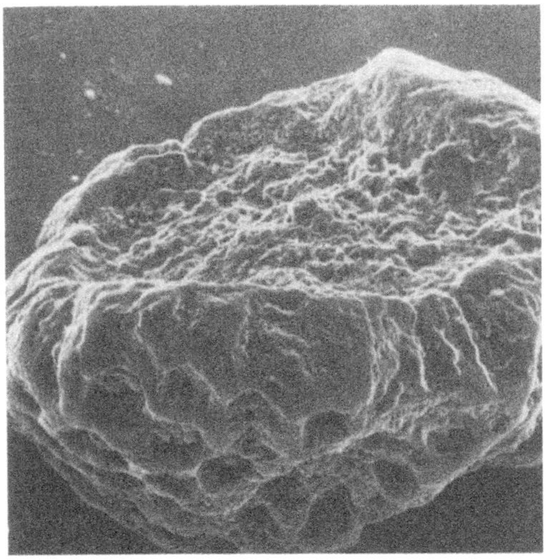
9

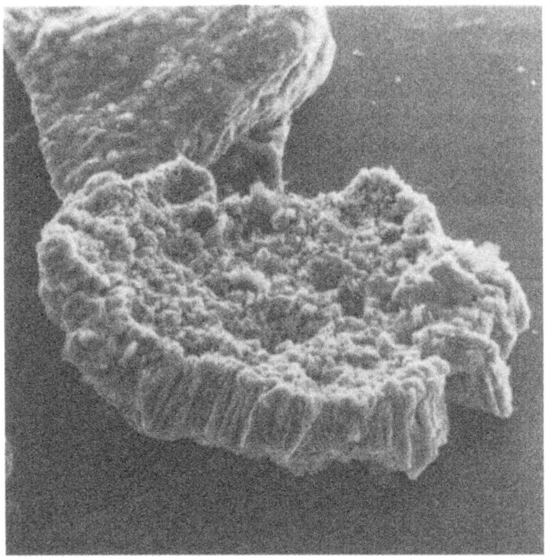
10

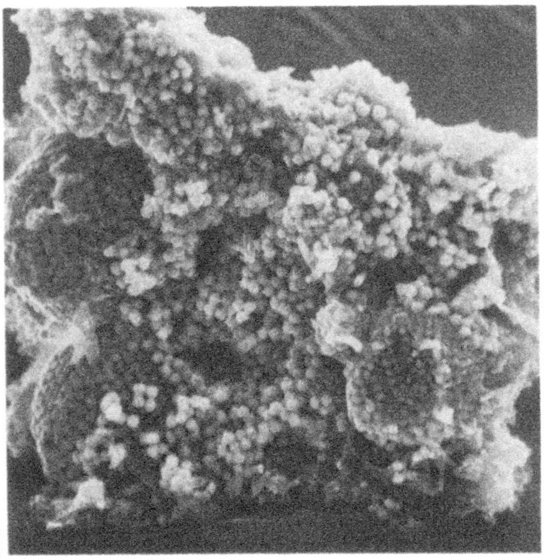
11

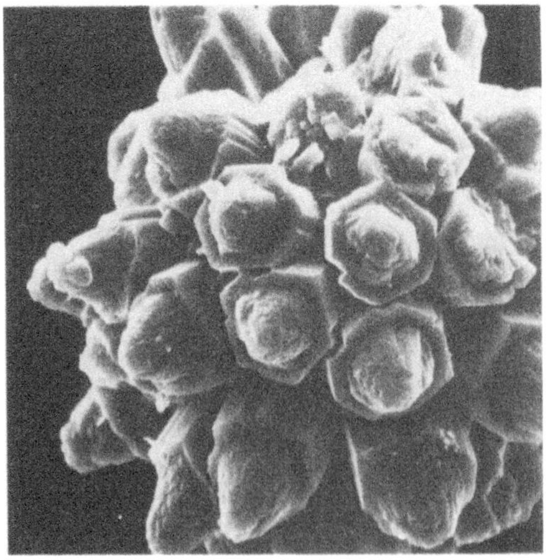
12

Legends to Figs. 7 - 12 on the following page

5) Fine fraction (< 63 microns)

5.1) Quantitative component analysis of the silt fractions (63 - 2 microns) (Fütterer)

a) The marly sediments on the Iranian side of the Persian Gulf show very high percentages of fine grained fractions; for example, in 100 out of 144 samples, the silt comprised more than 40 % of the total sediment. Therefore a detailed examination of the silt fraction, mostly by scanning microscope, is of major importance for the quantitative evaluation of the total sediment. It supplements investigations of the clay size fraction smaller than 2 microns by x-ray analysis, and that of the sand size fraction larger than 63 microns by light microscope. An inherent difficulty in such a study is that only the surface morphology can be illustrated. Characteristics such as colour, surface polish, transparency or opaqueness, which are used in coarse grain analysis, are not available. Therefore a catalogue has been prepared for exact determination of the single particles and this is being continously appended. For this purpose artificial silt has been made from different groups of organisms, feldspars, quartz, micas, and various limestones. For scanning electron microscope investigation the silt is separated by sedimentation into fractions from 2 to 6, 6 to 20, and 20 to 63 microns. This separation should be performed with great care, and, in particular, the dispersing agent must be chosen carefully: if, for example, Sodium Pyrosphosphate is used, some artifacts will be obtained.

Silt is scattered on a slide and overlapping photographs taken by the scanning electron microscope are grouped into a mosaic. The magnification depends on the size of the fraction; 500 to 1000 times for the 20 to 63 microns fraction, 2000 times (6 to 20 microns) and 5000 times (2 to 6 microns). Depending on the density of their distribution some 600 to 1000 particles per fraction were analyzed.

b) The determination of biogenic and non-biogenic particles is not usually difficult. Within the Persian Gulf, however, the surface sediments are often mixed with Pleistocene aragonitic material (Fig. 7). A single grain of this material consists of clearly-defined needles up to 20 microns long. This grain is very similar to a recent fragment of *Halimeda tuna* from the Adriatic Sea, (Fig. 8).

c) The determination of terrestrial biogenic detritus is possible only with larger particles and complete nannofossils. Fig. 9 shows a fragment of a fossil Foraminifera characterised by the massive, compact recrystallized nature. In contrast, a fragment from a Recent Foraminifera (Fig. 10) has well preserved shell structure.

d) The third point is the determination of authigenic minerals, such as pyrite. In the 6 to 20 micron fraction, pyrite may attain about 2 %. Very often, Foraminifera are completly filled with framboidal pyrite (Fig. 11).

Fig. 7. Grain of aragonitic mud, Persian Gulf. Width of photo 50 microns

Fig. 8. Particle of *Halimeda tuna* from the Adriatic Sea, Recent, with no significant difference from Fig. 7. Width of photo 50 microns

Fig. 9. Fragment of fossil Foraminifera with its typical massive texture; river sample, Rasnaband, Iran. Width of photo 50 microns

Fig. 10. Fragment of Recent Foraminifera with well preserved crystallites; Persian Gulf. Width of photo 50 microns

Fig. 11. Authigenic pyrite (framboidal pyrite) from an infilling of a Recent Foraminifera; Persian Gulf. Width of photo 50 microns

Fig. 12. Spicule of Recent ascidian; Persian Gulf. Width of photo 20 microns.
Fig. 7-12 Stereoscan microphotographs (Fütterer)

e) Of special importance is the determination of typical "silt organisms". These are organisms, especially coccoliths, or fragments, whose size is limited essentially to the silt fraction. They contribute a maximum of only 5 % of the 6 to 20 microns fraction. Another specific group are spicules of Ascidiae (Fig. 12) which may constitute up to several percent of a given grain size fraction.

Normally, the biogenic components of the silt fraction consist of fragments which can be determined only by their ultrastructure and which are very similar within several different groups of organisms. In contrast to coarse grain analysis (Sarnthein 1971a) where several dozen different grain types can be distinguished, morphological silt analysis can, at the moment, only distinguish some fifteen.

5.1.2) Preliminary results

Whereas biogenic components in the sand size fraction represent 50 - 100 % of the total number of particles, in the silt size fraction they represent only 20 - 60 %. The proportion of organisms that produce silt size particles is minor, as is that of authigenic mineral particles. Some details are illustrated in Fig. 13 which shows samples with high silt content from different environments.

In contrast to the sand fraction in the Persian Gulf, in which the main components are biogenic, the silt fraction in most samples consists of non-carbonate detrital grains (Samples 1117, 1116, 1177). Sample 1071 from the Gulf of Oman is similar to 1116 in that it contains a relatively high percentage of biogenic material in the fraction 20-63 microns; however, in the Persian Gulf this consists mainly of mollusks and benthonic Foraminifera whereas in the Gulf of Oman planktonic Foraminifera are more prominent. In both areas the silt fraction generally contains more non-carbonate detritus than carbonate detritus.

5.1.3) Mineralogical analysis of silt fractions
(Details see Hartmann et al., 1971)

Generally, the carbonate content in the three silt size fractions from the Persian Gulf which were investigated increased with increasing grain size, indicating the importance of biogenic carbonate constitutents. With increasing distance from shore the calcite/aragonite ratio decreases in all silt fractions, because of decreasing terrigenous calcite and increasing biogenic aragonite and relict aragonite mixed by bioturbation. Accordingly, no significant supply of either terrigenous or shallow marine precipitated aragonite is added from the Iranian side of the basin. The calcite/aragonite ratio in the silt fraction again increases sharply from the continental shelf margin towards the Gulf of Oman as a result of a decrease in aragonite-secreting benthic organisms and in increase in planktonic Foraminifera with calcitic tests.

Whereas the Central Basin of the Persian Gulf is characterized, along the entire coastline, by an increased dolomite content in the 2-6 micron size fraction, areas of increased dolomite in the 20-63 micron size fraction are concentrated near river mouths; dolomite and proto-dolomite are generally brought in by those rivers in whose drainage system dolomitic rocks are found.

5.2) Mineralogical analysis of the clay fraction
(Details see Hartmann et al., 1971, Esteoule et al., 1970)

5.2.1) Terrigenous material in the finest fractions consists of (in approximate order of decreasing abundance) calcite, quartz, dolomite, feldspar, chlorite, illite, palygorskite, swelling minerals and a small amount of kaolinite. Some of these minerals can be directly related to the geology of the hinterland. Our observations indicate that palygorskite is not currently forming in the Persian Gulf.

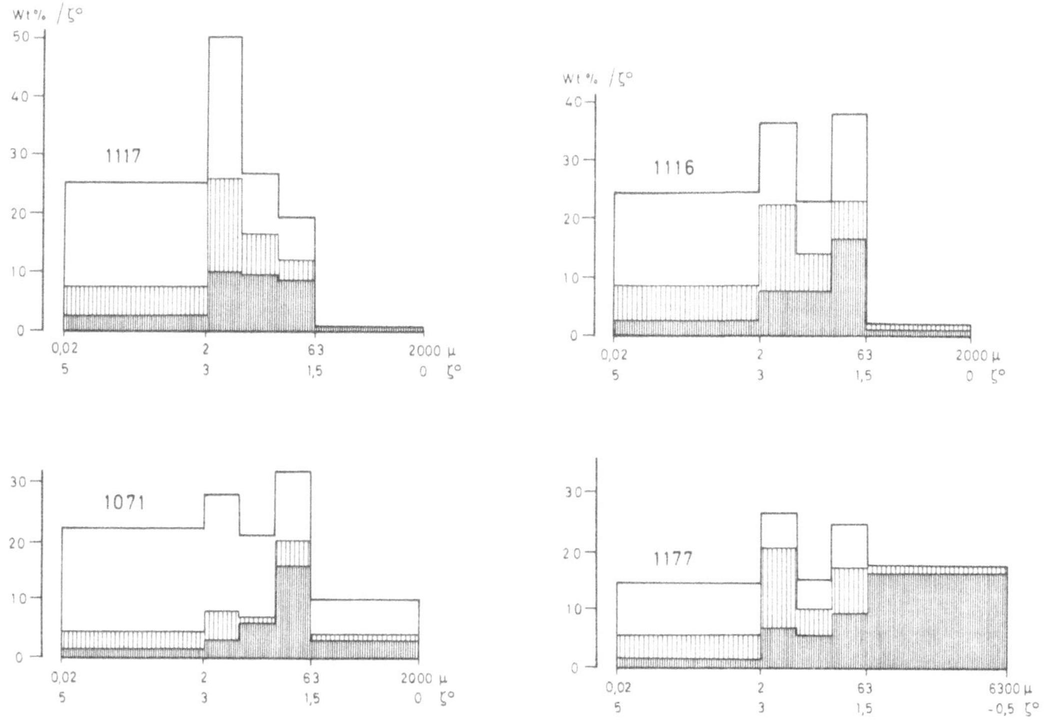

Fig. 13. Histograms of components of silt fractions (Fütterer) Sample 1117 = Western Basin, high sedimentation rate; 1116 = Western Basin, lower sedimentation rate; 1177 = Eastern Slope of Central Swell, very low sedimentation rate; 1071 = Gulf of Oman, 1100 m water depth low sedimentation rate, for comparison.

Legend:
Grain size distribution of total sample = outer line
white = detrital noncarbonates,
hatched = detrital carbonates,
closely hatched = biogenic carbonates
0.02 - 2 -micron fraction extrapolated, > 63 micron after Sarnthein 1971 (Fütterer)

5.2.2) <u>Colors (Lange)</u>

Differences in color can be distinguished more clearly in the finest clay fraction (ca. < 0.1 micron) than in the total sediment sample. The regional color distribution (blackish olive-green to light red) outlines the areas of influence of individual rivers. The majority of the clay samples have grayish, olive-green hues. The darkest olive-green is found in the Gulf of Oman. Reddish hues are concentrated nearshore and especially in front of river mouths, the high percentage of red components therefore decreasing away from land. The red coloration is dependent upon the rock type eroded in the borderland.

Throughout large areas of the Gulf the mineralogical composition of the clay fraction remains constant irrespective of sediment color. Changes in color from reddish to greenish indicates a destruction of the red components which takes place either during marine transport (20-40 km) or, more likely, through slow reduction of the uppermost sediment after deposition. High sedimentation rates do not allow sufficient time for complete reduction and in such areas red sediment is preserved at greater sediment depths. In the Persian Gulf reddish sediments are therefore correlated with high sedimentation rates.

6) Content of organic carbon and nitrogen

(Details see Hartmann et al., 1971)

The sediments of the Persian Gulf contain generally between 0.5 and 1, or at the most 2 %, organically bound carbon (Fig. 14). The total nitrogen content (less than 0.3 %) shows a similar distribution. Nevertheless, there are regional differences in the C/N ratios (Fig. 15) whose interpretation, although posing many problems, suggest that:

a) they are independent of the rate of organic production or rate of sedimentation of organic material - with the exception of sediments with very low content of Recent or sub-Recent organic material;

b) they are not noticeably influenced by redox conditions in the water and the sediment;

c) they are lowered by nitrogen-bound,or adsorbed in lattices of,terrigenous minerals especially in the case of sediments with low organic contents found near the coasts and,

d) (Müller) increase noticeably with depth in the upper layers of the sediments (up to 50 cm), probably due to a preferred decay of nitrogen-rich components of the organic material.

Fig. 14. Regional distribution of organic carbon content (After Hartmann et al., 1971)

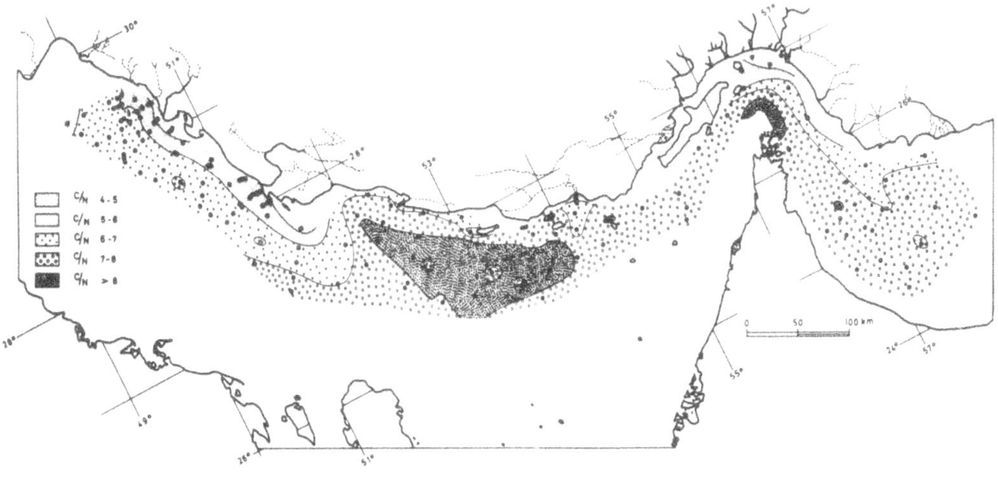

Fig. 15. Regional distribution of C_{org}/N-ratios (After Hartmann et al., 1971)

7) Structures (Diester)

The Holocene sediments show the following textural types:
1. homogenous material, 2. finely laminated material, 3. sediment which shows traces of bioturbation: a) pockets or lenses of shells in a fine groundmass, b) sharply defined burrows, c) mottled structure, 4. shell layers with orientation of the shell long axes - an indication of current influence.

7.1) Sorting

Holocene cores from the Strait of Hormuz, the Central Basin and the Central Swell can be subdivided into 2-3 coarser-grained and 2-3 finer-grained intervals. The following aspects suggest sorting: changes in the median of the sand fractions are paralleled by changes a) in the fine fraction content, b) in the median of each biogenic component and c) in the biogenic composition.

a) Fig. 16 shows the correlation between the median of the sand fraction and the fine fraction content in the cores. The sand fraction contains almost none, but the fine fraction consists almost entirely of terrigenous material. Decreasing fine fraction content in each core coincides with increasing median values of the biogenic sand fraction.

b) The medians of each biogenic component, as well as the medians of the intermixed

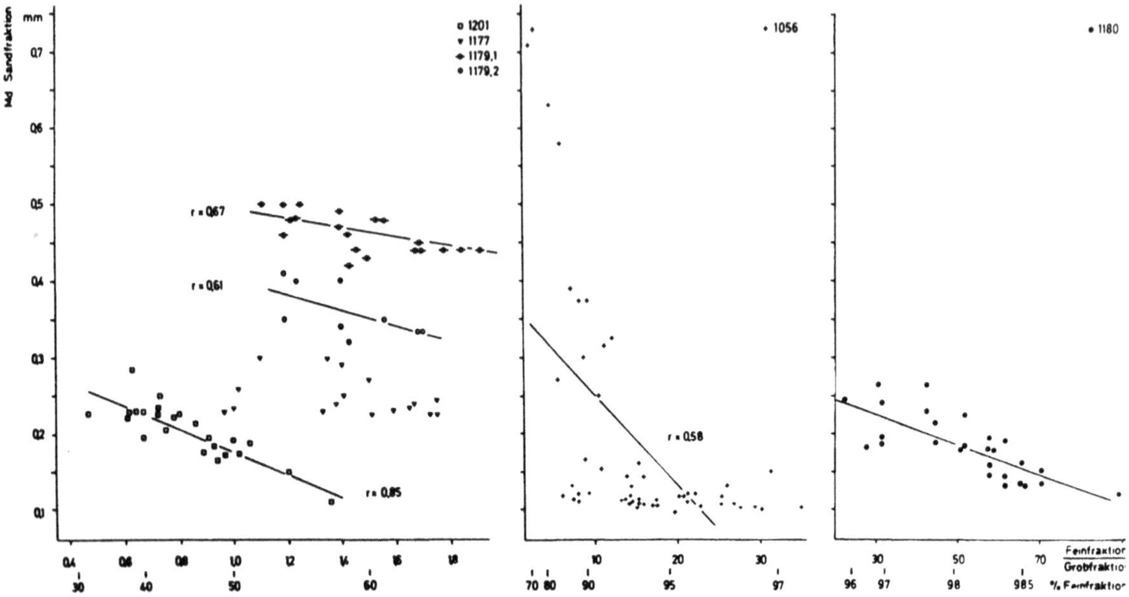

Fig. 16. Median of the sand fraction versus fine fraction content in layers of Holocene cores (Diester)

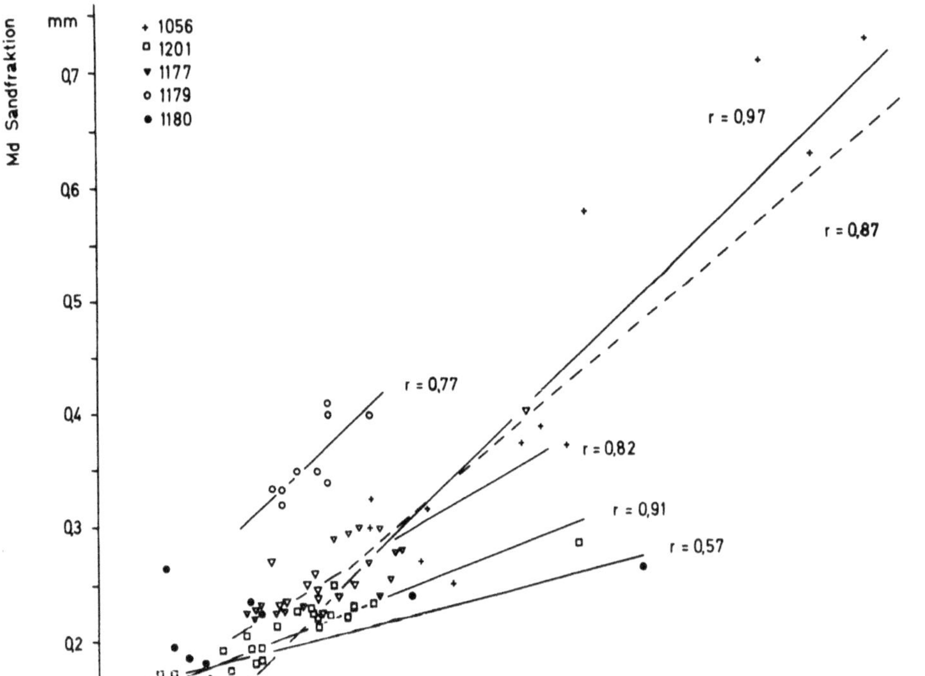

Fig. 17. Median of the sand fraction versus median of pteropods in layers of Holocene cores (Diester)

portion of the underlying relict sediment (coarse shell debris, ooids, lithified glauconite and aragonite lumps) increase with an increase in the sand fraction median. Fig. 17 illustrates the correlation between the sand fraction medians and the medians of the pteropods. This correlation is valid not only within each core, but also when all the cores are considered as a whole. The pteropods are an example of planktonic, and hence substrate-independentorganisms. All other biogenic groups show the same dependence on grain size.

c) The percentage composition of the biogenic components is also dependent on grain size.

With increasing median values of the sand fraction the proportions of benthonic mollusks to benthonic Foraminifera, of decapods to ostracods, of benthonic to planktonic Foraminifera, of benthonic to planktonic mollusks, of pteropods to planktonic Foraminifera, of Globigerina (except Orbulina) to Orbulina also increase. Experimental measurements of the sedimentation rates of various biogenic particles (Fig. 18) show that, in the described ratios of various biogenic components to each other, the particles with the smaller equivalent diameter decrease in relationship to those with larger equivalent diameters in proportion as the grain size of the sand fraction increases.

The median values for the coarser layers increase from top to bottom in the cores. The sedimentation rates (calculated in 2 cores) for the total sediment, for the terrigenous material and for each individual biogenic component are higher in the upper portion of the core than in the lower portion where larger median values suggest increased reworking.

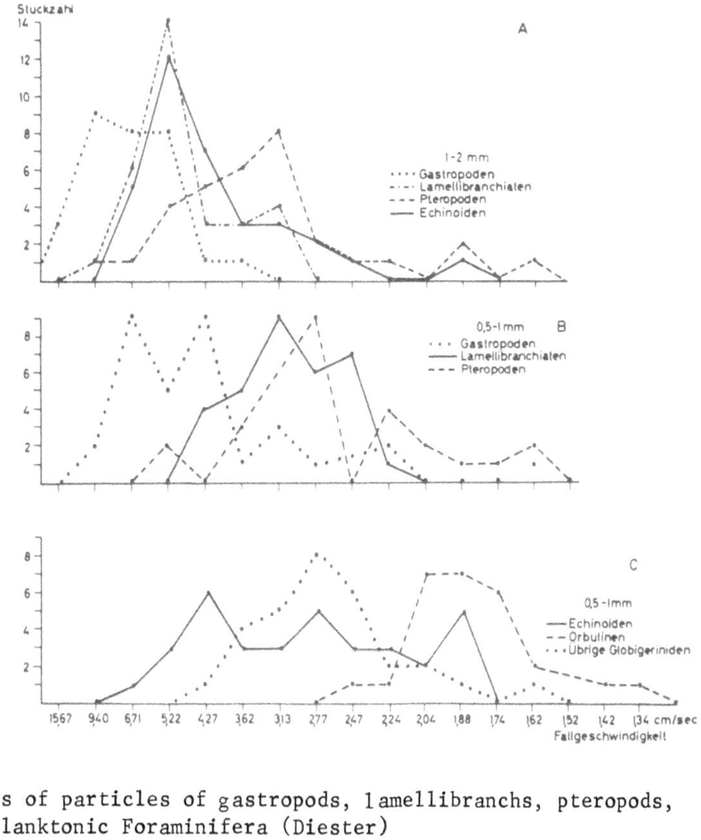

Fig. 18. Sedimentation rates of particles of gastropods, lamellibranchs, pteropods, echinoids and of planktonic Foraminifera (Diester)

Conclusions

1) Many results of the coarse fraction analysis of the Holocene cores admit to no other explanation than sorting, although it is very surprising that there has been sorting in some very fine-grained marls, with only 2 % sand content. In the water depths examined (between 40 and 100 m) this must be attributed to the influence of tidal currents on bottom sediments.

2) The quantity of terrigenous material transported into the Persian Gulf by rivers increases from the beginning of the Holocene to the present. In this general trend 2 periods are intercalated which have more terrigenous supply and 2 periods with lesser supply. Assuming that the quantity of terrigenous material is controlled by climate (rainfall), 2 humid and 2 dry periods can be distinguished in the Holocene.

7.2) Pleistocene aragonite muds

In some cores, the Holocene marls are underlain by Late Pleistocene aragonitic muds. This mud is very fine grained (> 50 % < 2 microns) and poor in fossils (0.5-1.8 % biogenic particles in total sediment). The sand fraction consists almost entirely of white lumps, ca. 0.1 mm in diameter, composed of aragonite needles, or of detrital minerals of the same size. The aragonite mud was probably not formed in situ. The water depth at time of formation was at least 12 m but not more than 35 m.

The sorting of the sediment (predominantly of the fine sand), the absence of larger biogenic particles and of pellets, ca. 0.2-0.5 mm in diameter, which are typical of Recent and Pleistocene areas of aragonite formation, and also the sedimentological conditions near the sampling points, indicate a transport of aragonitic mud from the area of formation in very shallow waters.

7.3) Biogenic cavities (Werner)

X-ray radiography (Fig. 21, 23, 24), structural casts (Fig. 19), thin sections (Fig. 22) and direct observation (Fig. 20) reveal wide-spread biogenic cavities in the Persian Gulf sediments. These include the interior of mollusk and Foraminifera shells as well as a rich inventory of burrows. The largest ones are branching Decapod burrows with diameters of up to several cm and extending more than 1 m down into the seabed (Fig. 19, 20). These burrows may be open and sometimes their walls are reinforced with mollusk shells, mostly with the concave side against the wall (Fig. 21). Other cavities of this Decapod type are filled with laminated sediment during use (Fig. 22) either due to the animal's activities or passively by falling sand grains. Such stratification is an indication of rapid sedimentation of sand in these traps. Very often assemblages of well preserved mollusk shells are found in the form of nests (Fig. 23), probably collected by the inhabitants as objects which hindered their digging. A large pore space should be preserved here.

These large burrows contrast with the branching systems of very small tubes made (width 0.1 to 0.3 mm) by other animals (Fig. 24). These networks of tubes are sometimes so dense that they can have lengths of 1 m condensed into 1 cm^3; this can provide a good drainage system for pore water. The inner sides of the walls are generally mottled with minute pyrite crystals.

Because these cavities, and other types of burrows, occur within a wide range of water depths and facies, most of them are of only minor importance as facies indicators. They should, however, be considered with respect to vertical sediment transport, the possibility of water circulation within the sediment, the preservation of mollusk shells,and other diagenetic processes.

Open cavities are found not only at the sediment surface, but also down to a depth of several meters in the sediment, without deformation. Normally, however, the number of open cavities in a given core decreases with depth. In order to obtain a more quantitative picture in some cores, open and filled burrows were counted (Fig. 25). Figure 26 shows that a slight decrease in incidence of open burrows parallels sediment depth. However, overburden pressure can be excluded as a cause because the shear strength of the core material is high enough everywhere to prevent compaction, and horizontal tubes are not deformed in any sediment examined down to 4 m. Possibly this trend is the result some of the tubes being filled in from above during sedimentation.

8) Facies distribution (see Sarnthein 1971a, additional aspects in Melguen and Sarnthein & Walger, this volume)

The sediments of the Iranian part of the Persian Gulf can be grouped provisionally into different facies divisions based on grain size, sand components, carbonate contents etc. (Table 1):

fine grained, clayey marl facies
(near river mouths with some subfacies types (Melguen) or
coarse grained, high carbonate marl facies rich in relict sediment
(basin bottoms poor in Holocene sediments) etc. Fig. 28)

9) Sedimentary processes

As shown in the previous chapters, analysis of surface sediments results in a better understanding of various sedimentary processes, keeping in mind known morphological, climatological and hydrological features.

9.1) Tidal currents, which have been calculated and measured, act more or less parallel to the Gulf axis, and rework sediments. This is proven by sorting effects in the sand fraction (Chapter 7.1). The regional net transport (at least for the sand fraction) resulting from these bipolar currents, however, may be small. This is illustrated by facies boundaries, the limited dispersion away from the coast of river mouth sediments with higher contents of minerals typical for the given hinterland, of plant and vertebrate remains, ostracods, etc.

9.2) Wave base, important for sedimentological investigations, this may range from 40 m (near islands) to about 70 m on the flanks of shallow submerged highs. The submerged highs within the Central basin, where terrigenous supply is minimal, consequently have sediments which are distinctly different from those of the deeper basin area: extremely high percentages of coarse grains, high carbonate content, low total organic carbon and nitrogen content, and low C/N ratios. A widespread, vertically restricted zone of maximum sedimentation of echinoid and ophiuran skeletons can also be related to the base of wave action. Depths of this zone are 12-14 m on the lee side of islands, but may attain 20-30 m in less protected settings, depending upon the sample location in relation to fetch. (Sarnthein 1970).

9.3) Wind-induced longshore currents together with the action of waves and tidal currents results in a net transport in the Western basin (Fig. 27) which extends out along the Central Swell. This is shown in the coarse fraction > 63 microns, by total carbonate content, carbonate in fine fractions < 2 microns, 2-6 microns and 20-63 microns, by calcite:aragonite ratios in the fractions 2-6 and 20-63 microns, by quartz:dolomite ratios in the fraction 2-6 microns, and by the quantitative component analysis of the coarse fraction.

9.4) The slope from the Iranian coast into the Central Basin is generally steeper than that into the Western Basin. The Central Basin also has interspersed islands and shallow highs which, together with the steeper slope, tend to complicate sediment distribution parallel to the coast, except in very shallow waters.

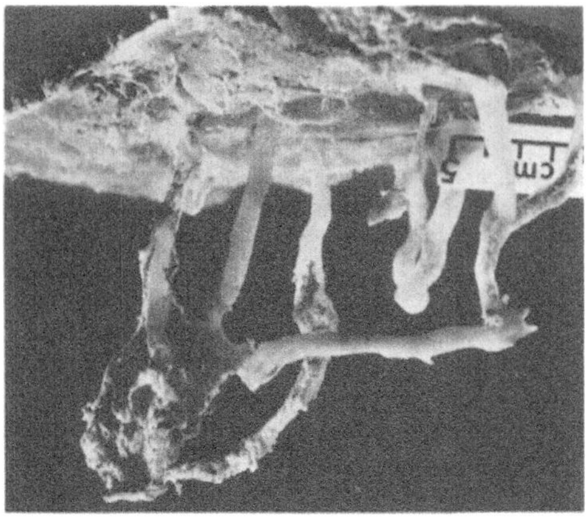

19

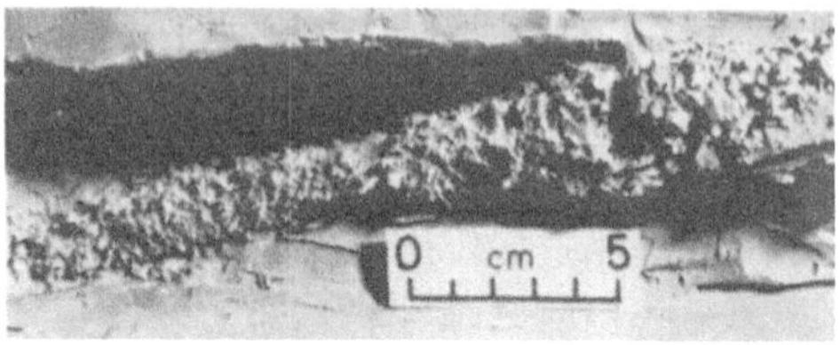

20

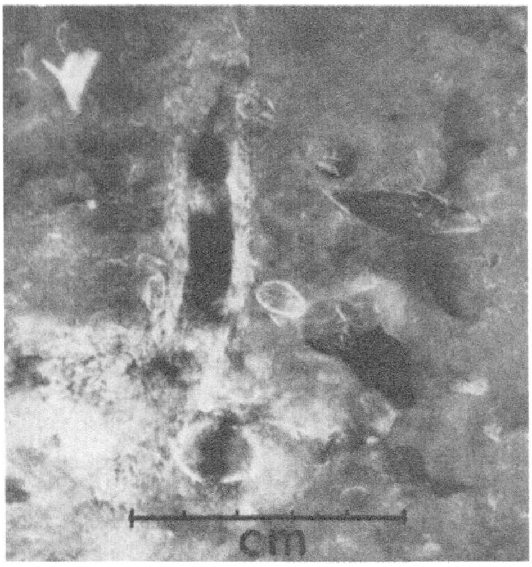

21

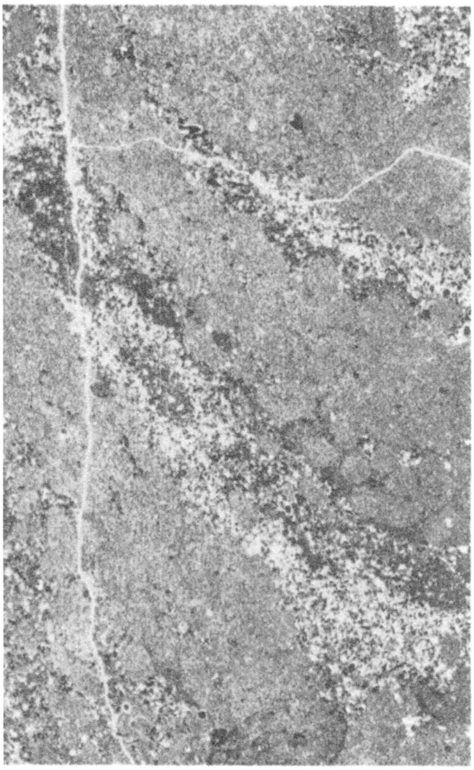

22

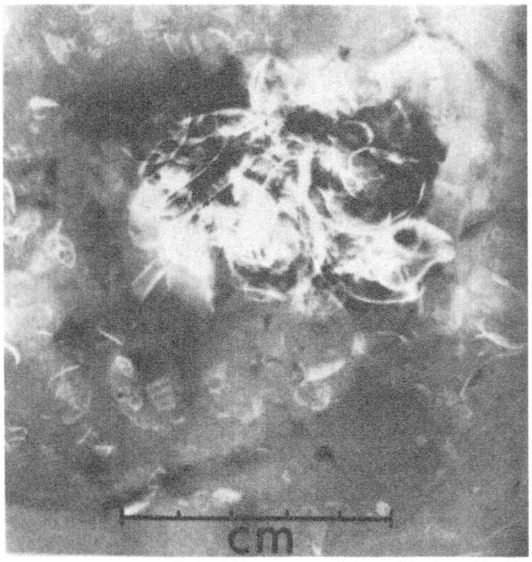

23

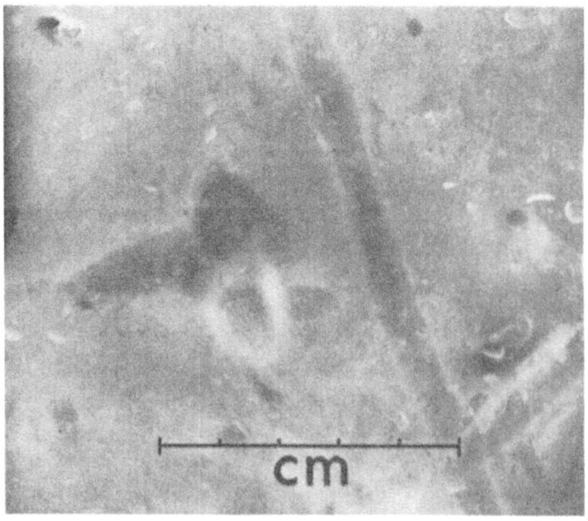

24

Legends to Figs. 19 - 24 on the following page

The spatial distribution of the coarse fractions and the quantitative variations of their components seem to indicate sediment transport perpendicular to the basin's long axis and along the steepest gradients well into the Central Basin, down to depths of 80-100 m (Fig. 27). Perhaps gravity currents ("flash floods") caused by the infrequent but heavy rainfalls in the arid Zagros mountain area are responsible for this type of sediment transport.

9.5) The inflowing surface current, diverted to the Iranian side, transports planktonic organisms into the Persian Gulf, as shown by the decreasing ratio of planktonic to benthonic Foraminifera and of planktonic Foraminifera to planktonic mollusks in the sediments between the continental shelf edge and the interior of the Gulf. The current axis can be mapped from the sediment based on the maximum values of the ratio of the median grain size of Globigerina/median grain size of total coarse fraction (Sarnthein 1971 b). This current also brings organic matter from the Indian Ocean into the Persian Gulf. Nutrients are added by "fresh" upwelling waters of the Gulf of Oman. Both nutrients and organic matter diminish very rapidly as the water moves into the Persian Gulf, this depletion resulting in generally low organic carbon content in the sediments within the gulf. Furthermore, west of the Central Swell the content of organic carbon is lower than east of it, if samples of similar grain size distribution are compared.

9.6) The outflow carries well oxygenated water over the bottom of the Persian Gulf and the resulting oxidation may further decrease the already low content of organic matter in the sediments.

In the Masandam Channel and in the Biaban Shelf Channel, the outflowing bottom current prevents deposition of fine material and transports sediment particles well beyond the continental shelf margin. The path of the outflowing Persian Gulf water is expressed by higher carbonate contents both in total samples and in individual size fractions, higher contents of aragonite and dolomite in individual size fractions and a non-carbonate mineral maximum in the sand fraction in the sediments of the Gulf of Oman.

9.7) According to the regional distribution of the clayey marl and calcareous marl facies (Table 1, types 1 and 2), the terrigenous supply is dominated by Iranian rivers. The supply from Euphrates/Karun, at present, is confined to the area north of 29° N.

A generalization of sediment transport directions is illustrated in Fig. 27.

Fig. 19. Structure cast of branching Decapod burrows. Central Basin, low sedimentation rate. Sediment surface at top, width of photo 18 cm (Werner)

Fig. 20. Decapod burrow filled with shells. Length more than 2.80 m. Western Basin, high sedimentation rate. (After Melguen, Thèse Université Rennes 1971)

Fig. 21. Radiograph of a Decapod burrow lined with mollusk shells. Western Basin, low sedimentation rate (Werner)

Fig. 22. Thin section of part of a Decapod burrow filled with sand and pellet layers. Western Basin; low sedimentation rate of total sediment. Width of photo 1.5 cm (Werner)

Fig. 23. Radiograph of a mollusk shell nest. Sediment depth 180 cm. Western Basin, high sedimentation rate (Werner)

Fig. 24. Radiograph showing branching systems of very small tubes. Central Basin, low sedimentation rate. (Werner)

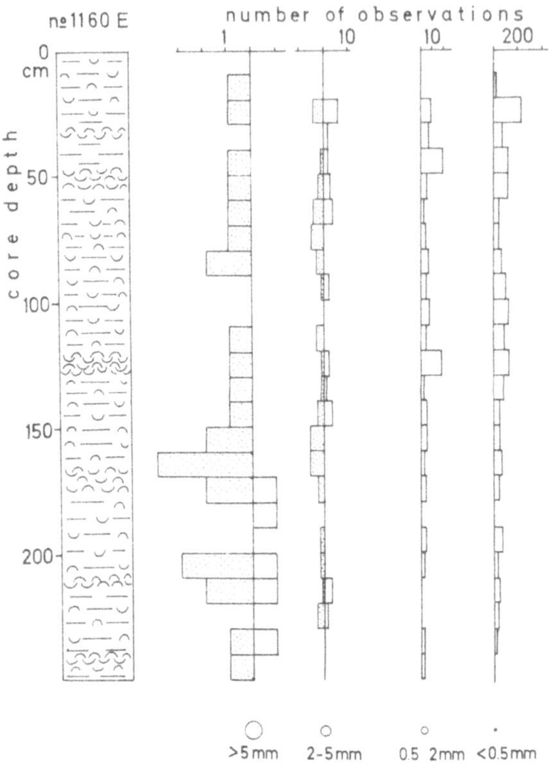

Fig. 25. Diagram showing the distribution of filled (dotted) and open burrows of different diameter classes (number of burrows counted in sections of 10 cm vertical length, 11 cm width and 2 cm thickness). Western Basin, high sedimentation rate, marl with various amounts of shells. (Werner)

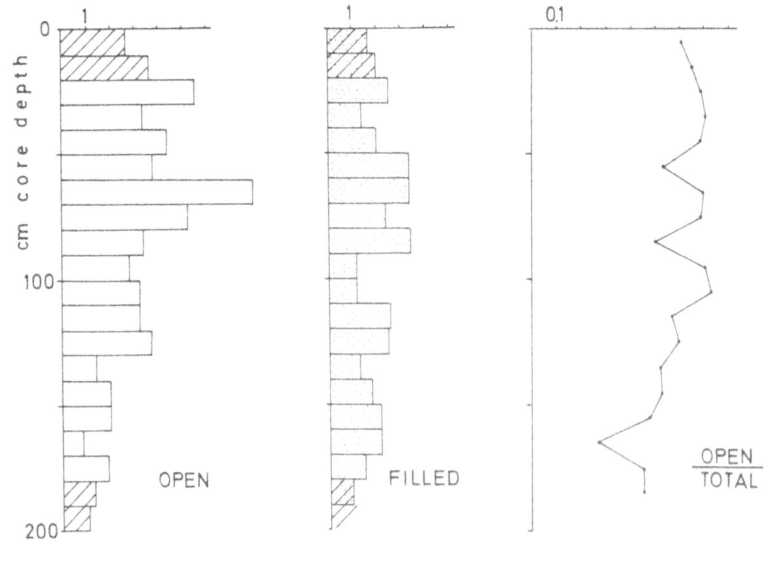

Fig. 26. Diagram as in Fig. 25. Mean values of filled and open burrows of the size classes 5 and 2-5 mm, 6 cores from Western and Central Basin. Hatched = core sections technically disturbed. (Werner)

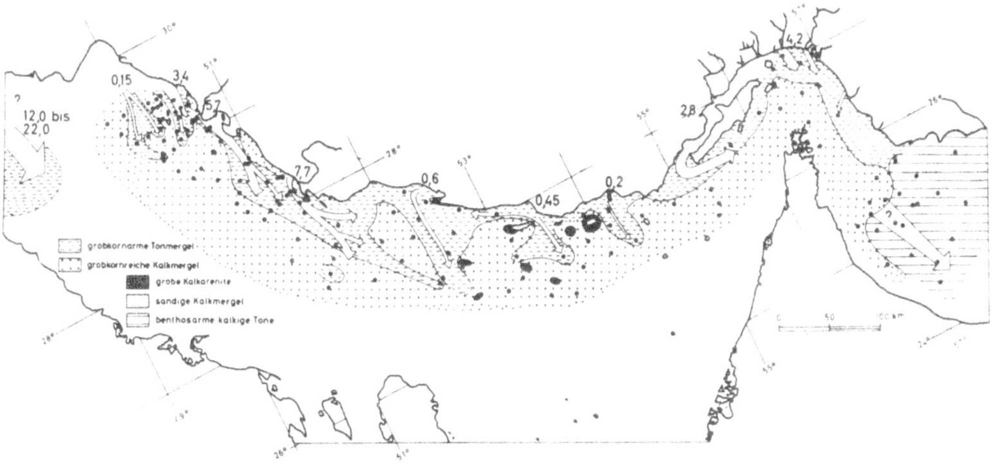

Fig. 27. Facies distribution, sediment transport directions and material balance. Plotted values = estimated yearly sediment supply from rivers (million m^3) Main facies types: 1) Clayey marls (with few coarse grains) 2) calcareous marls (rich in coarse material) 3) coarse calcarenite 4) calcareous marls (with fine sands) 5) Clay (with few benthonic remains) (After Sarnthein 1971)

Fig. 28. Regional distribution of relative sedimentation rates determined on the basis of planktonic mollusk frequency. Values in the Gulf of Oman uncertain. Approximately indices 1 and 2 = 0.5-2 cm/1000 years, 2-10 = 1-10 cm/1000 y etc. Index 500 = more than 500 cm/years sedimentation rate (After Sarnthein 1971)

10) Sedimentation rates

Holocene terrigenous sediments extend 20-30 km seaward into the Western Basin and about 25 km onto the Biaban Shelf. Relative sedimentation rates, determined on the basis of planktonic mollusk frequency (Sarnthein 1971a), show exceptionally marked regional differences. These can usually be related to river mouths or offshore morphology. (Fig. 28).

Aeolian influx is difficult to asses from our data. However, it probably is of minor influence on the Iranian Gulf side and may add, at the most, a few centimeters of fine sediment per 1000 years. Aeolian supply in the N parts of the Persian Gulf is discussed in greater detail by Kukal and Saadallah, in this volume. An example of different sedimentation rates of terrigenous material and planktonic and benthonic remains of some groups of organisms based on quantitative component analysis of the coarse fraction and on ^{14}C-Dates is given in Fig. 29 (Diester). The cored sequences represent about 9000 years, and give an average rate of total sedimentation of 7 cm/1000 y (Core 1177 from the Central Basin), 7 cm/1000 y (1201 from the deeper Hormus Region), 41 cm/1000 y (1056 from the Eastern Biaban Shelf with high terrigenous supply). Sedimentation rates are generally lower in the bottom parts of the cores and higher in the upper parts.

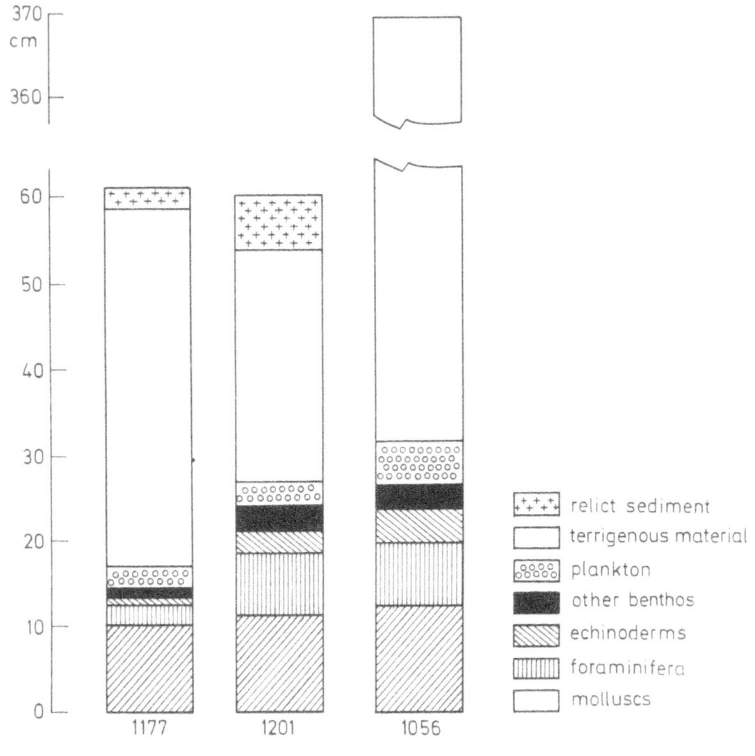

Fig. 29. Thickness (cm) of terrigenous material (white), relict sediment (crosses), plankton (circles), benthos (Echinoids, Foraminifera, mollusks- and others (black) in Holocene cores representing about 9000 years (Diester)

Direct observations and culture experiments with larger Foraminifera (Heterostegina) collected from shallow submerged highs and island slopes in the Central Basin show that these Foraminifera, with the help of symbiontic algae, may produce a calcite skeletal sequence of several cm/1000 years. This extremely high production may be a modern analogue of the formation of some Nummulitic or Fusulinid limestones, requiring less sorting influence than previously supposed (Lutze et al., 1971).

11) History of the post-glacial transgression

(Details see Sarnthein 1971a, 1972b)

As mentioned in preceeding sections, relict sediment particles occur in many surface sediments of the Persian Gulf, especially in areas with low sedimentation rates. They are mixed vertically by bioturbation and moved laterally (Diester) by current reworking. Therefore, the post-glacial history of the Persian Gulf is important for an understanding of Recent sedimentation. The following results are preliminary, being based mainly on indirect conclusions and not on cores, due to the lack of vibro-corers able to penetrate coarse sands or cemented material during the expedition in 1965.

These relict, mainly aragonite components, show an intricate facies pattern reflecting the paleogeography of carbonate tidal environments during the transgression. Studies of the sea floor morphology and the investigation of these relict components also indicate periods of sea-level stillstand (Fig. 30).

The deepest sub-fossil particles,formed originally in very shallow waters, (ooliths, reef material) occur near the continental shelf margin in water depths of 105-125 m. At the time of their formation the Persian Gulf was essentially a dry, flat area crossed by an ancient Shatt-River.

Between 100 and 65 m water depth "polymict coquinas", (mollusks, serpulids, bryozoans, Foraminifera, some individual corals), in the shallower parts covered by unlithified aragonite mud, reflect a rapid transgressive migration of intertidal environments. West of the Central Swell a temporary lagoon was formed and filled with aragonite mud and some terrigenous deltaic sediments.

Transgression stillstands are indicated at 64-61 and 53-40 m by coarse, frosted quartz and ooid concentrations embedded in lithified aragonite mud. These formations are interpreted as drowned strand dunes associated with a ridge and trough system of dunes. (Seibold & Vollbrecht 1969 believed that similar morphological features found in shallower depths represent presently active,or fossil mixed forms resulting from tidal current erosion and accumulation). A possible third standstill lies at about 30 m.*

The Late Pleistocene climate was probably more arid than today, as indicated by substantially reduced amounts of Iranian river sediments, and widespread aragonitic muds now absent on the Iranian side of the basin.

The analysis of coarse grain components in cores (Diester 1970, Melguen 1971) revealed variations in several parameters which reflect more than local changes in sediment supply or distribution. They suggest a general increase in fluviatile sediment during the Holocene and also superimposed regional variations in river supply. These changes may possibly be correlated with European climatic variations (Diester 1970).

12) General applications

1) The Iranian side of the Persian Gulf may be regarded as a model area for the formation of a fully marine, marly, fine grained <u>molasse</u> :

a) It represents a final stage geosyncline which is being filled as a foreland to post-orogenic Zagros folding.

---

* These depths do not coincide with the regional terrace levels discussed by Kassler in this volume. The discrepancy may be explained by the erosional origin of the terraces and the sedimentary origin of the relict sands. (Editor)

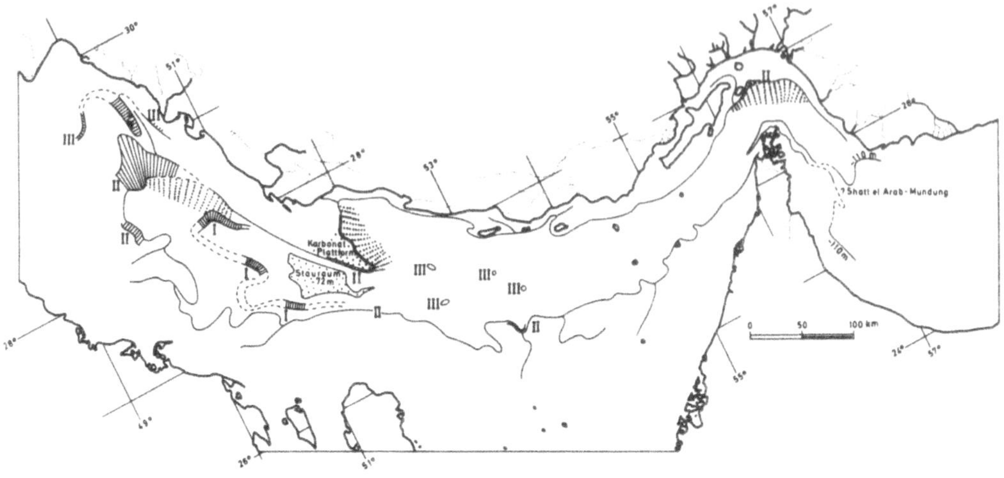

Fig. 30. Paleogeographical evolution during late and post Pleistocene times. The 110 m line shows the approximate maximum of sea level fall. The horizons I, II and III represent suggested phases of stillstand during the sea level rise - at 61-64 m, 40-53 m and approx. 30 m depth. Dotted surfaces represent areas with (relict) aragonite mud sedimentation. (After Sarnthein, 1971)

b) The frequently reddish and grey-green sediments are erosion products of the uplifted Zagros folds. The sediment fans thin towards the edges and, in many cases, can be traced to particular river mouths.

c) The sediment transport in this shallow sea, although essentially parallel to its axis, can also be at an angle to the Gulf axis, or normal to the coast.

2) The Persian Gulf as a whole is a sedimentation model for a marginal sea within the arid climatic zone. It differs from seas within humid zones in many ways : circulation pattern, water layering, bottom water characteristics, sediment character (content of carbonate, evaporites, organic carbon, metals, sharpness of biofacies boundaries) and lagoonal conditions.

3) Non-carbonate material is brought to the Persian Gulf almost exclusively by lateral rivers from the Zagros Mountains. Variations in the drainage pattern, caused by climatic or tectonic changes, therefore are expressed directly in sediment cores, although somewhat masked by partial reworking. Analysis of the area near the Iranian coast increases our understanding of the origins of ancient interbedded limestones and marls.

Classification of Modern Marl Sediments in the Persian Gulf by Factor Analysis

M. Sarnthein and E. Walger[1]

ABSTRACT

The marly bottom sediments in the northern half of the Persian Gulf were studied in 170 grab samples. Eighty two variables were catalogued for the gravel and sand fractions. Thirty eight of them were analyzed by comprehensive Q- and R-mode factor analysis in addition to previously evaluated conventional ratio plots (Sarnthein 1971). A cross-check evaluation of R- and Q-mode factor patterns allow the definition of a series of mappable facies units: benthos-poor calcitic clay (deep water), foram-mollusk-echinoderm marl (outer delta), foram-ostracod marl (delta axis), fine sandy calcareous marl (inner delta), molluskan coarse-grained marl (off-delta shelf plain), "relictiferous" coquina marl (shelf plain), large foraminiferal coquina arenite (banks), marly "relictiferous" coquina arenite (starved shelf). The main source of carbonate in these marls is terrigenous. Consequently in regions of high sedimentation (delta), the total carbonate content of the sediment is low because autochthonous biogenic carbonate and relict carbonate are diluted in these areas by high portions of lower carbonate material.

INTRODUCTION

170 surface samples from the northern portion of the Persian Gulf (Meteor-cruise 1965, Dietrich et al. 1966) presented the opportunity for an extensive study of the sediment type "modern marl" by a quantitative analysis of the fraction coarser than 63 µ. This study was organized in a manner similar to those previously carried out on pure carbonate sediments (e.g. Imbrie & Purdy 1962). According to Hartmann et al. (1971) these sediments are mixtures of clay and lime with carbonate contents of generally 20 - 90 %, justifying the term "marl" (classification according to Füchtbauer 1959).

In a first approach the large amount of data was evaluated by conventional ratio plots of single variables, or groups of them, to find correlations between these and single definite qualities of the depositional environment (Sarnthein 1971, see also Seibold et al., this volume). The present study is an attempt to sum up the total information contained in the various parameters of all samples by condensing them into a few types of sedimentary facies, here using factor analysis as one of the instruments offered in the tool kit of multivariate methods.

On the one hand the term "facies" is used as a petrographical concept reflecting certain assemblages of coarse fraction components (R-mode analysis), on the other hand as a collective term for sediments from easily recognizable types of regional environments with natural discontinuities between them - these may also be

1. Geological-Paleontological Institute of Kiel University.

recognizable, through analogous criteria and methods, as depositional environments in fossil marl sediments (Q-mode analysis). Because of its specialized approach this paper supplements that of Seibold et al. (this volume). A detailed introduction to the environment, hydrology and geological setting of the Persian Gulf is given by Purser and Seibold (this volume).

LABORATORY METHODS

For the component analysis, the sand fraction (0.063-2.0 mm) from each of 170 bottom samples was split into five subfractions (-1-0; 0-1; 1-2; 2-3; 3-4 Φ°). Five hundred to eight hundred grains were counted in each subfraction. The grains were catalogued in up to 40 Recent and relict grain-type categories (e.g. Foraminifera, corals, Bryozoa, mollusks, relict oöids, pellets, etc.). The individual percentages from each subfraction were then added together after being weighed to correspond to the actual weight-percent of the sand sub-sample which they represented. The final quantitative result from a particular sand fraction was then considered as being a mixture of both weight and volume. The gravel fraction was counted separately and the values calculated as weight percentages (details of method and petrography in Sarnthein, 1971).

OUTLINE OF THE COMPUTATIONAL METHOD

As is shown in the pertinent literature (e.g. Harman, 1960, Ueberla, 1968, Lebart & Fenelon, 1971), a data matrix consisting of N_v variables which are observed in M_s samples may be imagined to define either M_s points in a Euclidean space of N_v dimensions ("variable space") or to define N_v points in such a space of M_s dimensions ("sample space"). In the example discussed below, $N_v = 37$, $M_s = 170$. Unfortunately our capacity for visualization of spatial relationships is confined to only three dimensions. However, given the case of a sufficient number of significantly linear (or at least monotone, approximately linear) trend relations between the observed variables, it would be possible to find a "subspace" of markedly fewer dimensions - hopefully four, or still better three or less - in which the cloud of the points given may be contained without flattening it unduly. The objective of factor analysis, as understood here, is to determine such a subspace of as few dimensions as possible in the given data matrix.

More exactly, this point of view characterizes the model of a "main-components analysis". Yet here we prefer to retain the term "factor analysis" because it seems to be the more widely known. This may be justified by the expectation that, in analysing as many variables as are given in the present example, the differences between the results of these two methods become very small. "Factor measurements", a specific type of variable within the model underlying factor analysis s. str., are not considered here.

The process of determination involves an orthogonal rotation of the system of coordinate axes combined with a certain proportional stretching of the corresponding scales so that the coordinate values of the given points in relation to the new axes fall within the range - 1 to + 1.

For reasons which cannot be discussed here, the new coordinate values are called "factor loadings" ; the whole matrix containing the factor loadings as elements is designated "factor pattern" . Consequently, the new coordinate axes are called "factor axes" , and the space they determine, "factor space" . Variances estimated within the new system are also renamed: they are called "communalities" .

The projection process leading from the sample space to the corresponding factor space, which contains the points representing variables, is called "R-technique" or "R-mode" ; the resulting matrix of new coordinate values accordingly

"R-mode factor pattern" . The analogous process starting from the variable space is called "Q-technique" . It yields the "Q-mode factor pattern" containing the new coordinate values of points representing samples.

If a factor analysis is successful, the resulting factor space has a small number of dimensions. In practice, however, the cost of obtaining this desired effect is nearly always the loss of some portion of the information contained in the original multidimensional data matrix. A measure of the amount of the total original information recovered in the factor space is the ratio of total communality (i.e. total variance extracted by the factor space) to total variance in the original space.

Once a factor space is laid bare, one may perform further rotations within it without destroying the special properties which are essential for the primary solution. This freedom can then be used to find the position of a secondary system of factor axes, which is best adapted to the shape of the point swarm. This facilitates its final evaluation.

In the present study the so called "VARIMAX-procedure" was always used for making such a secondary (orthogonal) rotation.

In the case of the R-mode analysis reported below, the raw data matrix contains a great number of percentage values and, in addition, the water depth at the sampling site. Water depth is a variable of quite a different type and scale. In such a situation it is inevitable that the scales of the variables must be adjusted for both position and range to make them sufficiently comparable. This is achieved by normalizing the variables to mean values 0 and standardizing them to variances 1.

In the case of the Q-mode analysis, water depth as a variable was abandoned. The data matrix then contained only percentage values, that is, variables of the same type throughout.

However, as implied by the special form of Q-technique used in this study, the data matrix was subjected to a certain normalization process in this case also; the sum of squares of the corresponding variable values for each sample adjusts to 1. This follows from the use of the so called "cosinus theta coefficient" (for details see Imbrie & Purdy 1962, Imbrie 1963, Harbaugh & Demirmen 1964) as a measure of similarity between the samples.

Since in the Q-technique the samples play the same role as the variables play in the R-technique (and vice versa), a computer program of appropriate capacity was needed to process a data matrix with > 170 "variables" for the present study. As none of the machines hitherto available to the authors provides a core storage capacity of more than 32 K words, a suitable program had to be developed. It was written in pure FORTRAN IV, avoiding the use of any machine-dependent language elements. It has now been implemented with a capacity of max. 172 variables as a chain job on an IBM 360/50 under DOS with 32 K words core storage installed at the Institut für medizinische Dokumentation und Statistik of Kiel University.

Because a factor space has only a few dimensions, it can be plotted in a comprehensible number of two dimensional projections. Such projections are the plots discussed below.

As a Key for the interpretation of these plots, the following relations may be considered.

In the R-mode factor space a small distance between two points signifies that the variables they represent have a high linear correlation; for two points having the same separation the greater their distance from the origin of the factor axes, the higher the degree of correlation.

In the Q-mode factor space the analogous situation can be interpreted as signifying that the samples represented by the points are characterized by sets of variable values, which, in respective pairs, are sufficiently similar.

It must be realized that points may be close together in one of the 2-dimensional projections, but an appreciable distance apart when seen in a different projection plane.

Points which are situated on, or near, a factor axis on which they have a high loading represent those variables which, in the rotation process of the coordinate system, exerted a high influence on the position of the new system in relation to the original one.

Consequently, points which are situated near the origin of the factor axes in all the 2-dimensional projections had correspondingly little influence on the rotation process.

When the points were combined into groups, the aim was to make the groups as isometric as possible. In some cases, however, rather elongate ones are inevitable. If such a chain of points is situated near a single factor axis, this signifies that the main common property of the variables represented by these points is a negative one: they are maximally independent of all those variables which have appreciable loadings on one, or more, of the other factor axes.

GROUPING OF MARL COMPONENTS

In Persian Gulf marl sediments the regional abundance of a certain constituent particlesis determined primarily by the "rule" of independent availability (Sarnthein 1971). It states that different component groups are deposited independently of each other. Their sediment quantities are mainly controlled by the conditions of their primary production (e.g. ecology), whereas joint redeposition and sorting, as shown by Diester (1972), have more local importance. Consequently the abundance of one constituent particle has the effect of increasing, or decreasing, the percentages of all other components within a sample. This complicated interplay of mutual dilution can be most successfully bypassed by grouping together components with similar abundance profiles, as was expected from an R-mode factor pattern (Fig. 1).

33 of 77 sand- and gravel-sized grain types originally differentiated were used as variables in the present factor analysis. Thirty three relict grain types were combined into 5 main groups as they were shown to influence present-day facies patterns only collectively (Fig. 1). Similarly, many small skeletal groups in the gravel fraction were lumped together as "total skeletals" . These variables jointly form a closed percentage system. However the number of 33 variables is believed to be high enough to dampen spurious correlations.

In order to characterize the position of the groups of components more exactly, the weight percentages for each of three grain size fractions and for the carbonate content were added as variables, as was the water depth of each sample (data from Hartmann et al. 1971).

If the various factor patterns of the 4 first factor axes in Fig. 1 are compared (cumulative total variance approx. 41 %), 7 main and 2 subgroups of marl components can be separated. They agree, in general, with relationships familiar from previously plotted conventional ratios of single constituent particles (Sarnthein 1971).

Factor axis I seems to join primarily the three variables of grain size distribution. Relic sediments (group 7 a and b), especially the skeletal ones (highest factor loadings), control the component grouping in the range of coarse grain sizes and high carbonate values, whereas sand sized ostracods and calcareous Foraminifera (group 3) govern the grouping in fine grained sediments; these are, for the most part, delta deposits (Fig. 2). Only the exact separation of the relict fraction allows an adequate evaluation of the genuinely Recent sedimentary features.

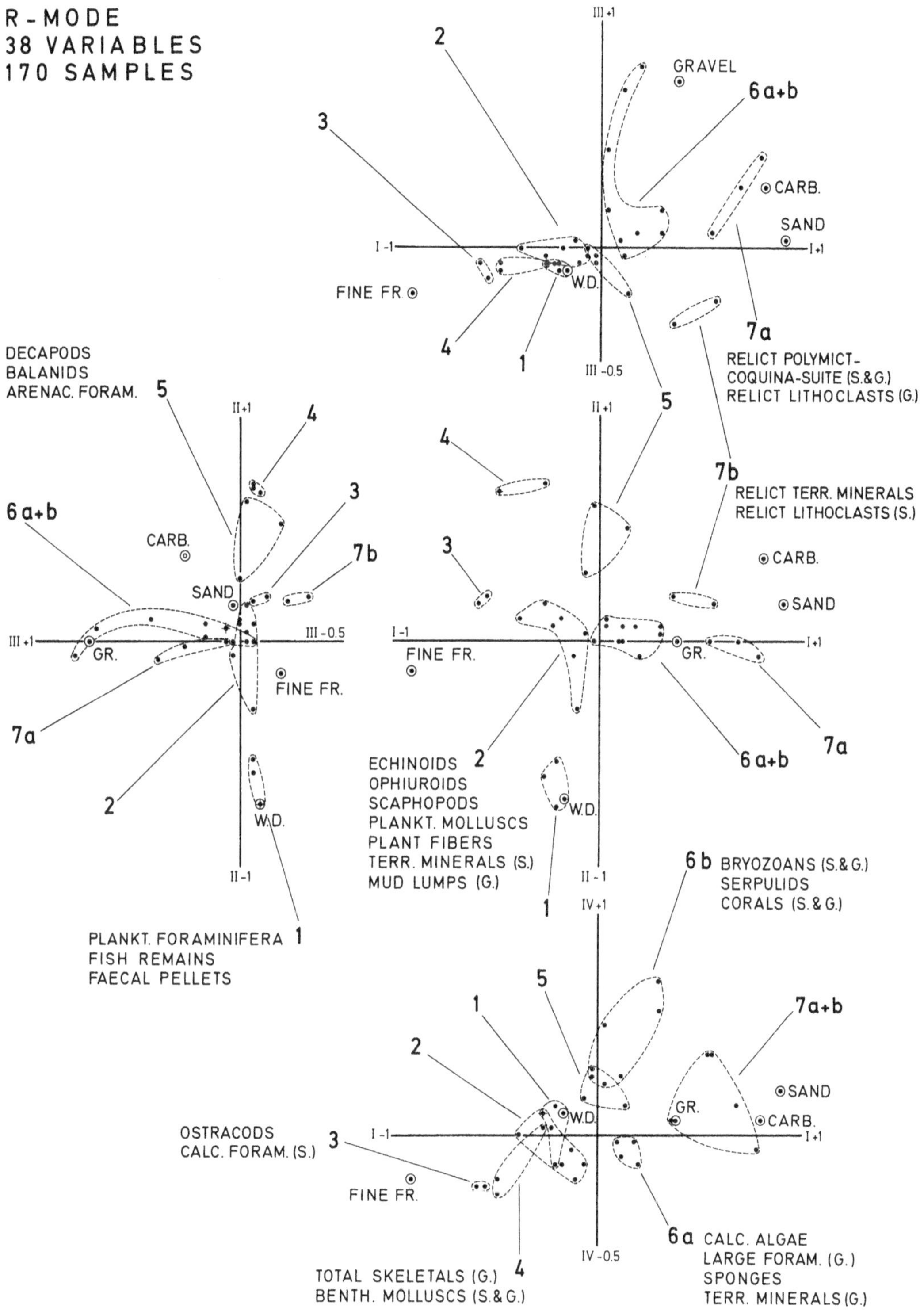

Fig. 1. R-mode factor pattern. Numbers of component groups refer to text and Figs. 4-9. Abbreviations: Carb. = carbonate; Fr. = fraction; G. = gravel sized; GR. = gravel fraction; S. = sand sized; W.D. = water depth; ⊙ encircled dots = additional characterizing variables

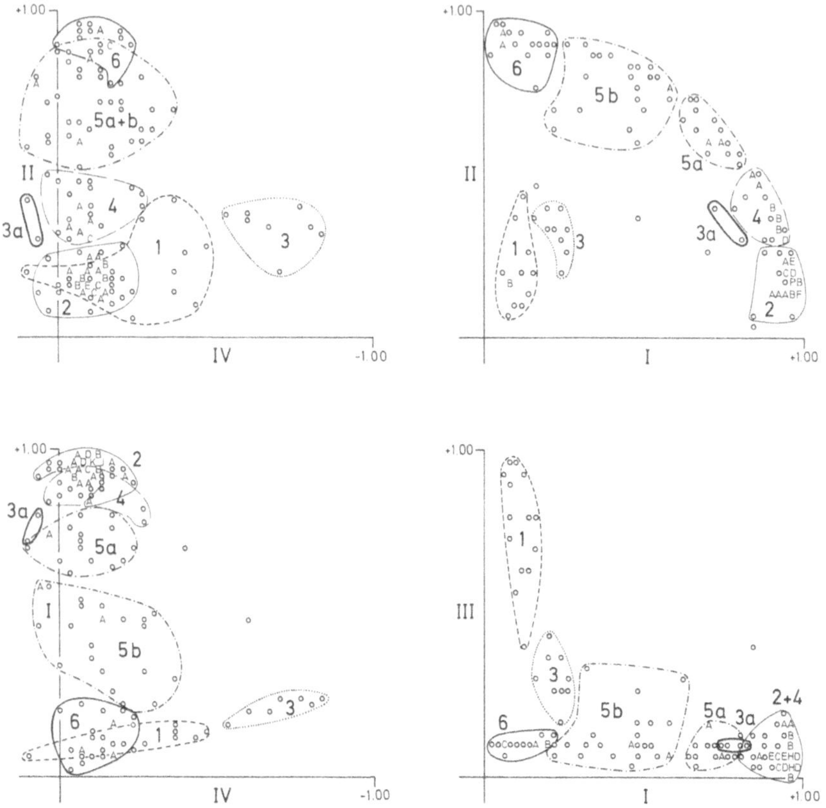

Fig. 2. Q-mode factor pattern based on 37 variables, 170 samples. Numbers of sample groups (facies units) refer to text and Fig. 3. Capital letters signify overprint values: A = 1 overprint, etc., Roman figures = factor axes

Group 1 - high abundance of planktonic Foraminifera, fish remains and faecal pellets - correlates directly with water depth and together they determine the negative end of factor axis II. On its positive end one finds group 4 (benthonic mollusks of sand and gravel fraction, gravel-sized "total skeletals") and, with some minor influence, group 5 (balanids, decapods, arenaceous Foraminifera). Accordingly they both have their optimum development in shallow water regions, group 4 in somewhat finer, group 5 in coarser sediments.

Group 6 comprises a larger closed assemblage of grains of fixed epibenthos skeletals. It correlates to some degree with the coarse grain variables along factor axis I, but is independent of water depth (factor axis II). Along factor axis III the subfraction of calcareous algae (Lithothamnium) and corals extends the group strikingly in proportion as the gravel fraction increases. This indicates a strong negative correlation with all terrigenous supply, here mainly represented by the position of the variable "fine fraction". If one considers factor axis IV, group 6 can be split into two subgroups: 6 a, on the negative side of factor axis IV, embraces calcareous algae, large Foraminifera (Heterostegina, Amphistegina), sponges and gravel sized terrigenous minerals, which are typical of sediments from shallow banks (Lutze et al. 1971, Sarnthein 1971). Coarse terrigenous minerals are here derived from local salt diapir erosion. Subgroup 6 b is controlled by Bryozoa and serpulids which have the highest loadings on factor axis IV, and which generally indicate starved sedimentary environments.

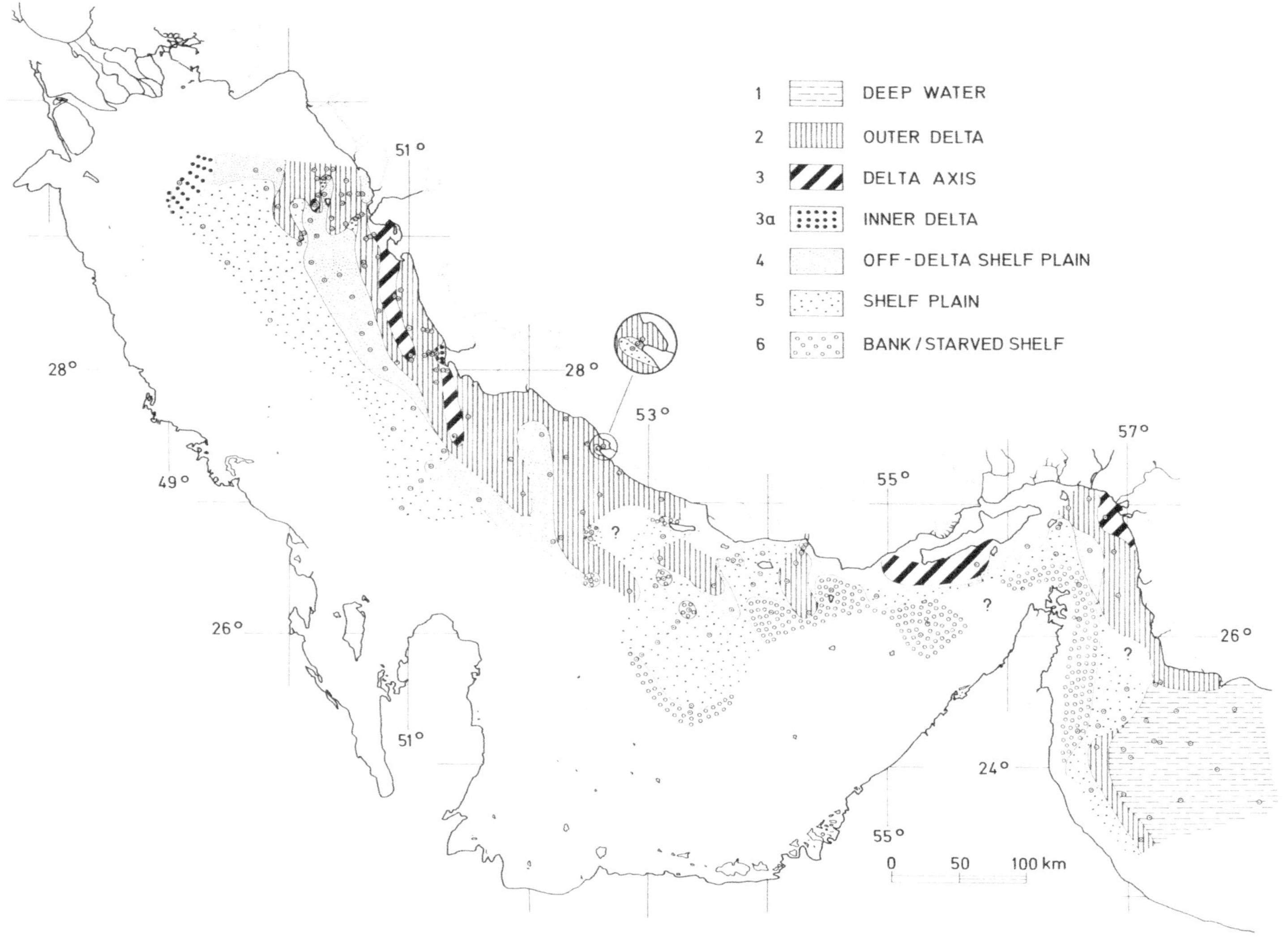

Fig. 3. Facies distribution in the Persian Gulf. Numbers refer to Fig. 2

The remaining components of group 2 (Echinoderms, planktonic mollusks ...) show only a slight influence on the given 4 factors apart from a lower correlation with the fine fraction variable and group 3; a similar deltaic position is assumed for group 2. This is supported by the delta-related abundance of plant fibers and (fine) sand sized minerals quartz, mica, carbonate rock detritus (Sarnthein 1971). The latter also correlate well with deep water-group 1 - an indication of a further relationship between groups 2 and 1 (compare the presence of "Outer Delta Facies" at the continental slope, Fig. 3).

Further quantitative data on the frequency of the component groups will be discussed later in connection with the structure of the Q-MODE sample groups, i.e. the units of marl facies.

Finally, it should be stressed that in the sample area examined, grain sizes and water depth show no significant correlation, but a correlation is found between coarse grain sizes and carbonate content (Hartmann et al., 1971, see also Seibold et al., this volume).

The factor pattern shows that higher carbonate contents in modern marl samples from the Persian Gulf are derived primarily from the adjacent relict component group 7 and, to a minor extent, from Recent benthonic skeletals in groups 4, 5 and 6 (see Fig. 10).

SEDIMENTARY FACIES UNITS

When 170 samples from the Persian Gulf are grouped by Q-mode analysis, the factor space of the first 4 factor axes (Fig. 2) extracts approx. 91 % of total variance (68/14, 4/6/2, 6 %). This corresponds to the major part of observed information contained in the matrix of 37 variables. Water depth, as a variable, was removed from this matrix to avoid an undesired *a priori* weighting of samples. The fine fraction (< 63 μ) variable was also removed as it is also reflected in the sum of sand and gravel fraction variables and could possibly overstress the importance of grain size data. Relict oöids were added as a variable in its place.

Comparing the factor patterns of Fig. 2, the sample points are classified into eight distinct groups, some of which again overlap in the projection of one or other factor axis. Only 4 samples remain isolated. They can then be considered equally related to most of the other groups.
Proceeding from group 2, which forms the end-member of factor axis I, one observes a gradational suite through group 4, 3 a, 5 a and 5 b ending at sample cluster 6, which constitutes the end-member of factor axis II. Strong intra-group correlation in all six groups is shown by their great distances from the origin. Factor axis III sets off sample group 1, factor axis IV group 3.

Mapping the eight discrete sample groups (Fig. 3) made it possible to define them as sedimentary facies units. Indeed, they give a sensitive and reliable reflection of regional environmental limits (for difficulties with facies 6, see following section) To place the facies boundaries in areas with only sparce sample coverage, additional information was used: grain size data from Houbolt (1957) and Peery (1965), bathymetry and sediment echographs (colored map in Seibold & Vollbrecht 1969 and Hughes Clarke and Keij, this volume) and even a Gemini exposure (near 55° E of Gr., e.g. Bodechtel & Gierloff-Emden 1969). In the following section an attempt will be made to characterize the facies units and their relationships by the previously delineated R-mode component groups as well as by grain size and carbonate data. However, only frequency ranges can be stated, as a Q-mode group is always determined by the interplay of all the variables. Excessive facies splitting has been avoided purposely.

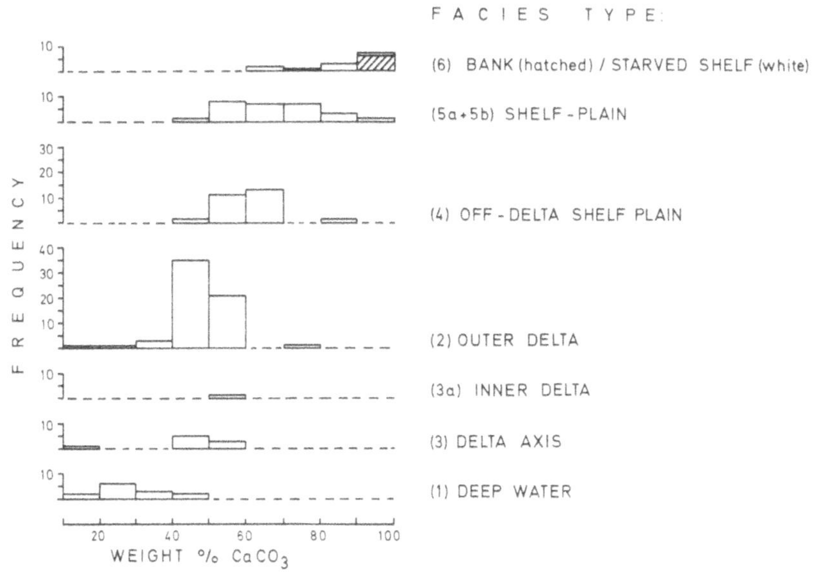

Fig. 4. Frequencies of carbonate contents

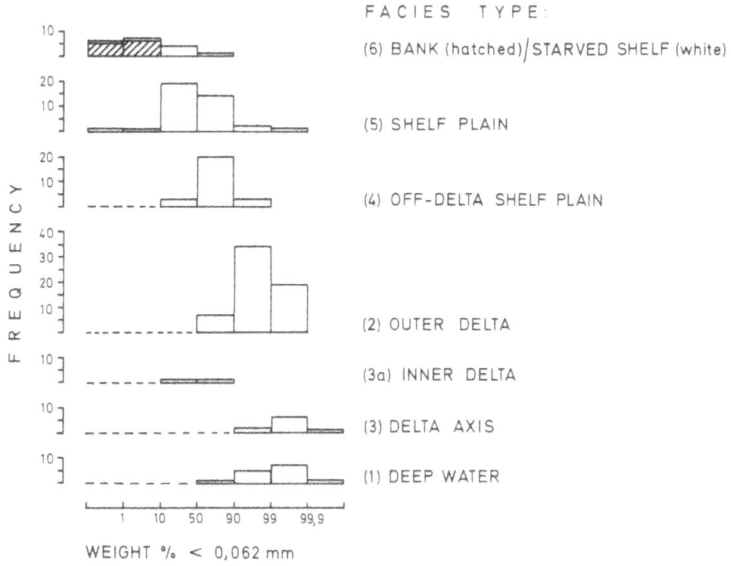

Fig. 5. Frequencies of coarse fraction percentages

(1) Deep water facies:

Samples of the deep water facies are found in the Gulf of Oman at water depths below 200 m (maximum 1500 m). Their main properties are a low carbonate content and a small coarse fraction averaging 0.1 - 10.0 % (Fig. 4 and 5). It is marked by the almost complete deficiency of mollusks and by the abundance of grain types of R-mode group 1 (Fig. 6 and 7): faecal pellets, planktonic Foraminifera and fish remains. Hence, from the petrographical point of view, the facies may be called "Benthos-poor, calcitic clay". Affinities to the various types of delta facies (2, 3) are obvious, including their spatial closeness at the continental slope. They are emphasized locally by high percentages of terrigenous sand minerals which are delivered by the outflowing current from the Persian Gulf and can be traced for roughly 100 km into the southern Oman Gulf (Sarnthein 1971). This continental shelf-derived sediment load generally results in high sedimentation rates and thereby prevents the formation of Globigerina ooze. The abundance of faecal pellets might correspond to the high organic carbon values reported there by Hartmann et al. (1971). Similar facies types were reported from the western continental slope off India (Schott et al. 1970), the Gulf of Aden (Einsele & Werner 1972) and the continental slope off Portugal (Kudrass 1972).

(2) Outer delta facies:

Off the mouths of several smaller Iranian rivers sample group 2 forms wide lobate delta areas, down to 100 m water depth. These extend 25 to 50 km away from, and perpendicular to the coast, and up to 150 km along their axes of sediment transport. The sediments are poor in coarse fraction and differ from deep water calcitic clays in that they contain twice as much carbonate (average 40 to 60 %, Fig. 4 and 10), and, even more important, in a drastic change in the coarse fraction composition. As shown in Fig. 6, R-mode groups 2 and 3 are dominant, with calcareous Foraminifera and ostracodes in the one, and primarily Echinoderms and variable amounts of plant fibres, terrigenous sand and planktonic mollusks, in the other R-mode assemblage (details in the distribution of single components see Sarnthein 1971). In border zones farther away from facies 3 and near facies 4, benthonic mollusks become abundant, representing the beginning of secondary hard bottom formation (Fig. 7); all other components are only poorly represented. Generally the term "Foram-mollusk-echinoderm marl" may be appropriate (for a more sophisticated subdivision of this facies see Melguen, this volume). The occurrence of sample group 2 in some sections of the uppermost continental slope (200 - 400 m) does resemble delta-like sedimentation.

(3) Delta-axis facies:

Along the axis of the outer delta lobes, off the major rivers, certain restricted zones are defined by the samples of group 3. They were well separated, without transition, from group 2 by the Q-mode factor patterns. However, some difficulties arise if one tries to distinguish them quantitatively from group 2 by using conventional data. The carbonate contents - 40 - 60 % - and the considerable portions of R-mode group 2 in the sand fraction in both facies, are within the same range (Fig. 4 and 6). A specific difference is found in a somewhat smaller coarse fraction, averaging 0.1-1.0 %. Increases are also found in plant fibres and terrigenous sand minerals (locally up to 60 % in sand fraction, especially mica) within R-mode group 2, an increase of the mean abundance of Foraminifera and ostracods, i.e., R-mode group 3 (40-50 %, maximum: 60-70 % of sand fraction, Fig. 6). Consequently, the petrographical term "Foram-ostracod-marl" was chosen for the delta-axis sediments. They are of particular importance for the exploration of the sedimentation processes of the Persian Gulf, as discussed in a subsequent section.

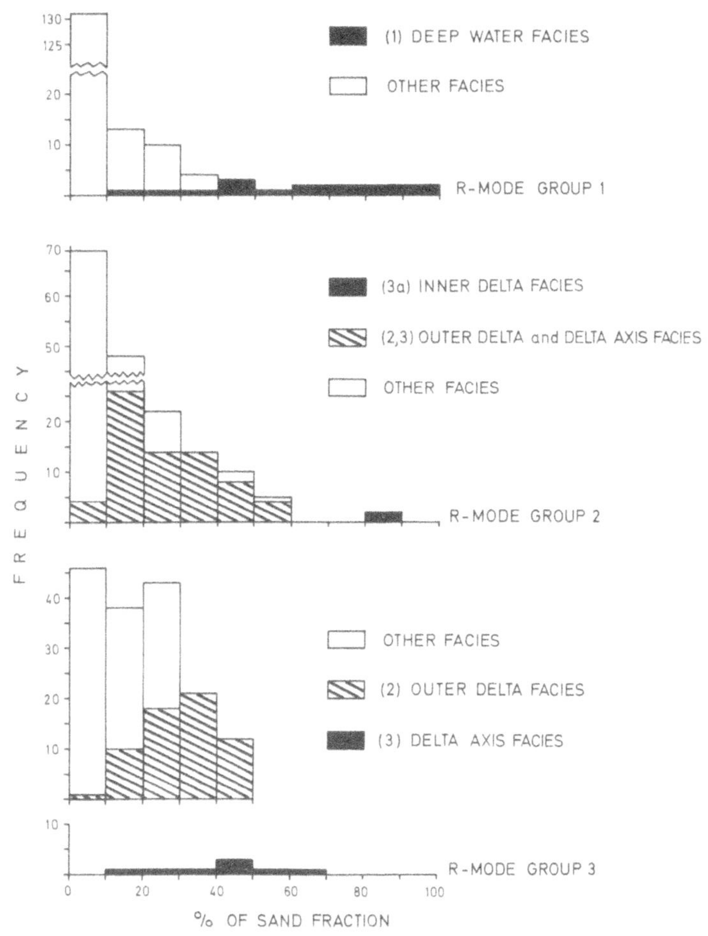

Fig. 6. Frequencies of R-mode component values in sand fraction :
R-mode group 1: faecal pellets, planktonic Foraminifera, fish remains.

R-mode group 2: echinoderms, scaphopods, planktonic mollusk, plant fibres, terrigenous sand sized minerals, mud lumps.

R-mode group 3: Calcareous Foraminifera, ostracods

(3a) Inner delta facies:

This group is represented by only two samples, both typically resembling deltaic facies. One is situated off the mouth of the river Mund, the other in the Euphrates delta. Characterized by high percentages of fine sand in an abundant coarse fraction (35-70 %, Fig. 5), they cannot be confused with any other facies. Terrigenous sand minerals, primarily quartz and carbonate rock detritus (80-90 %) displace all other grain types in the sand fraction, extending the trend of the Delta axis facies (Fig. 6). A somewhat elevated carbonate content (Fig. 4) leads to the definition of these sediments as fine sandy calcareous marl. The sedimentological significance of this facies will be discussed in context with the other delta facies. The inner delta facies has equivalents within the relict sediment facies, which locally indicate fossil deltas of the Euphrates at lower sea levels (Sarnthein 1972).

(4) Off - delta shelf plain facies:

The NE delta-lobe sediments of the Persian Gulf end abruptly with the samples of the off-delta shelf plain which spreads along the weakly inclined gulf axis. The exact boundary,as drawn,corresponds to the sudden appearence of sediments with a larger polymodal coarse fraction and higher carbonate content (now 50-70 %, Fig. 4 and 5). This change can be observed also in sediment echographs (Seibold & Vollbrecht 1969) and is caused by significantly lowered sedimentation rates (Sarnthein 1971, Diester 1972). With lowered sedimentation rates new grain types gradually appear, including fixed epibenthic skeletals and some relict grains (Fig. 8 and 9). The latter were mixed in by bioturbational movement which sometimes exceeded 2 m vertical displacement (Sarnthein 1972 b). The portion of deltaic component group 2 at the same time shrinks to less than 10 %, as shown in Fig. 6.

However, the position of sample group 4 in the Q-mode facies pattern of Fig. 2, as well as two of its main particle groups, still reflect strong affinities with the outer delta facies. This is shown by only a slight increase in benthonic mollusks to averages of 20 to 50 %, and a slight decrease of R-mode group 3 (< 30 %, see Fig. 6c and 7). Although difficult to explain, this results in an additional zone, roughly 25 km wide,of deltaic influence on the shelf plain. "Molluskan coarse-grained marl" is suggested as a descriptive facies name.

(5 a and b) Shelf-plain facies:

Samples of Q-mode group 5 a and b cover most of the up to 100 km wide plains along the bathymetric axis of the Gulf extending to the continental shelf break. Abundant relict grains, fixed epibenthic skeletals, tests from arenaceous Foraminifera, decapods and balanids are characteristic of this facies (Fig. 8 and 9). Consequently, there are essentially higher contents of carbonate and coarse fraction, compared to facies 1 to 4 in Fig. 4 and 5. Benthonic mollusk shells generally do not comprise more than 1/3 of coarse fraction (Fig. 7). Components typical for deep water or delta sediments are rare (Fig. 6). Very small, but persistent "background" values of 0.2 to 0.5 % quartz in the sand fraction may be correlated with aeolian sedimentation.

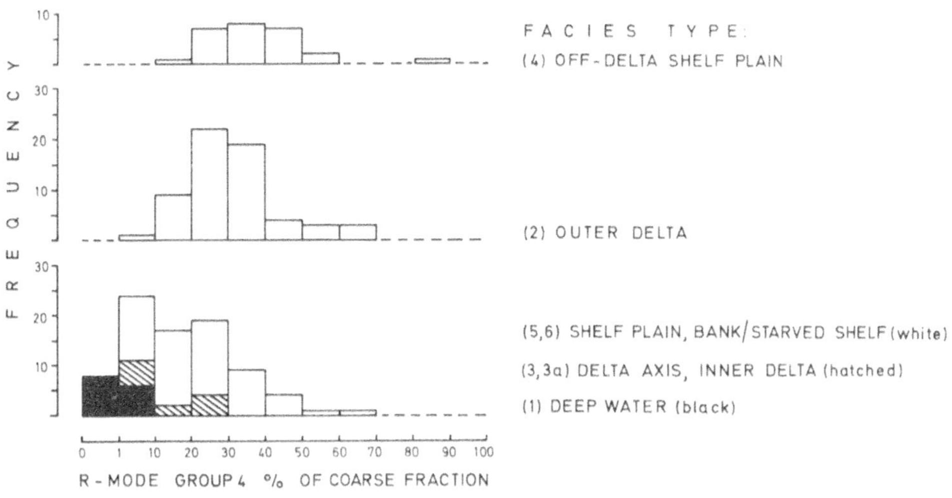

Fig. 7. Frequencies of sand sized benthonic mollusk and gravel sized total skeletals (R-mode group 4)

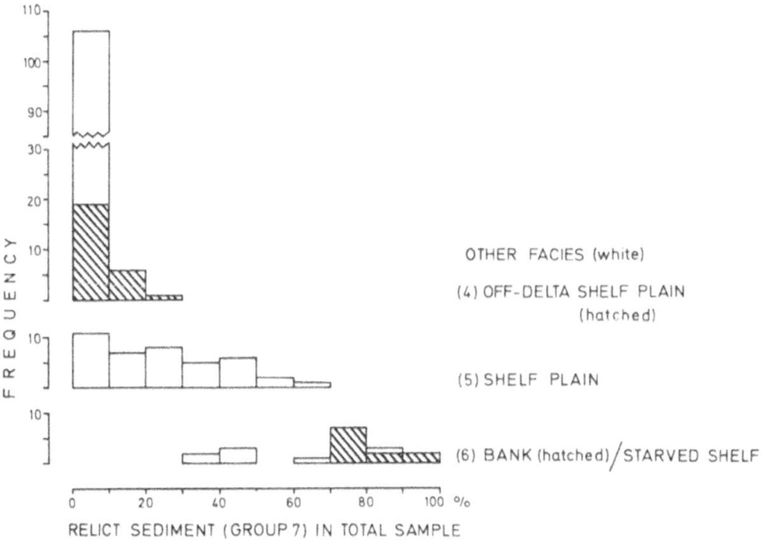

Fig. 8. Frequencies of relict sediment (R-mode group 7) in total sample

This "Relictiferous coquina marl" forms secondary hard bottoms, but no really Recent cementation processes could be observed. To some extent it corresponds to the "Coralligène" of the Mediterranean; both cover zones with low sedimentation rates of a few centimeters per 1000 years (Sarnthein 1971). These facies are generally inherited from the anomalous Pleistocene sea level changes; the facies of the relict fraction alone is a vivid reflection of the history and subfossil conditions of the Post-glacial transgression (Sarnthein 1972 a). Q-mode facies 5 a and 5 b are separated only by gradational differences in composition and were therefore condensed into one group.

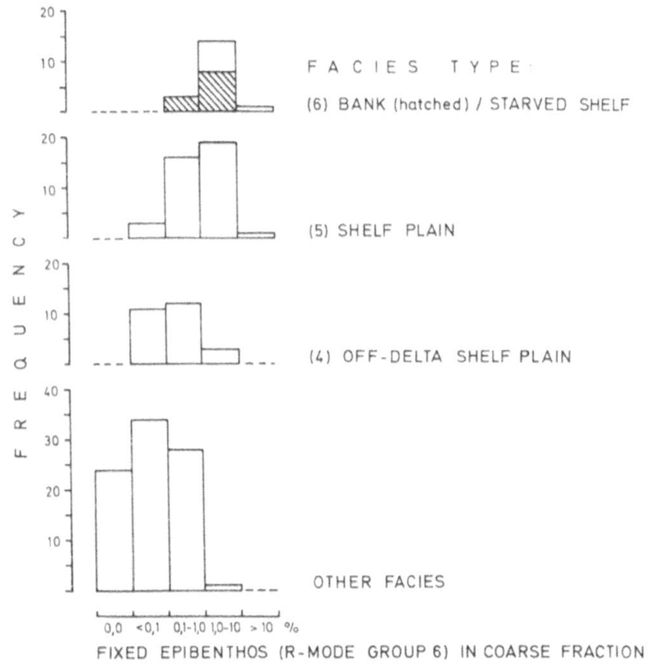

Fig. 9. Frequencies of fixed epibenthic skeletals (R-mode group 6) in coarse fraction

(6) Bank/Starved shelf facies:

Unfortunately, Q-mode group 6 in present analysis lumps together similar sediments from two different environments: the shallow zones on top of banks as well as the roughly 100 to 200 m deep areas also marked by minimal terrigenous influence, the "starved shelf". As shown by Figs. 4, 5, 8 and 9, this facies may be typified by up to 100 % coarse grain-, and/or carbonate fraction, and the abundance of relict grains and skeletals of fixed epibenthos. However, these percentages, with the exception of the grains of fixed epibenthos, are generally somewhat lower in the starved shelf sediments (60-90 %). Furthermore, R-mode component group 6 b seems to be more typical of the deeper starved shelf sediments, whereas R-mode component group 6 a definitely characterizes the shallow water subgroup (larger Foraminifera, sponges, lithothamnium algae, gravel sized terrigenous minerals). Therefore, in the Persian Gulf, one should distinguish between a "Large foraminiferal-coquina arenite" from shallow banks and a "Marly relictiferous coquina calcarenite" from the deeper starved shelf regions where there have been minimum sedimentation rates for about 12000 years. Such conditions are also reflected by larger amounts of fixed epibenthic organisms which are sensitive to any terrigenous influx. At some places there are traces of serpulite formation, while modest reef components have occasionally developed near the banks. However, real reef-building activity in shallow water zones is prevented by excessively low winter temperatures in the NE Persian Gulf (see Seibold et al., this volume).

DISCUSSION

The attempt to present a facies classification under two complementary aspects - sedimentary environment and petrography - creates several problems which merit discussion. These include the inter-relations in the delta facies complex, the distribution and origin of carbonates in a marl sediment, and the general experience of applying the classification methods outlined above.

Surprisingly, a fine sandy, calcareous marl and a foram-ostracod marl, two rather different sediments, form the connecting inner delta and delta-axis zone. Both obviously mark the main path of sediment transport from the river mouth into greater depths of water. At their common boundary the sand content drops seawards rather abruptly from about one-half to one-hundredth of a sample. This phenomenon is more dependent on the availability of the respective silt grain sizes than on hydrodynamic conditions as it is also observed within sample profiles from constant water depth (8 m) and wave exposition. Accordingly, it may be another expression of the general problem of silt deficiency preventing the formation of the transitional zone which would otherwise be expected. The peculiar concentration of deltaic sediment along narrow "delta axes" must bear some relation to the dispersal mechanisms. Apparently bodies of water very rich in suspended matter flow near the bottom as they leave the river mouth (compare density measurements of Ziegenbein 1966). Within the zone of wave activity, i.e. 20 to 30 m water depth (Sarnthein 1970), their well defined transport directions may be controlled by the steadily blowing (Shamal-, Kaus)winds (Hartmann et al. 1971). The directions are also in general agreement with the tidal current pattern. With increasing water depths, especially below wave base, gravity transport probably becomes more important. The abundance of Foraminifera and ostracods in this facies apparently results from the favourable food-rich environment and rapid burial preventing postmortal reworking (Diester 1972, Seibold et al., this volume).

The R-mode factor pattern shows that the relict sediments are primarily responsible for the highest carbonate contents in marl samples from the Persian Gulf. Further contributions are made by biogenic and terrigenous material. This three-fold origin of carbonate in marls is analyzed again in Fig. 10. The fine fraction carbonate, mostly calcitic, was grouped as terrigenous supply. Only 8 % of 152 plotted samples contain more than 50 % biogenic carbonate, and only 10 % contain more than 50 % (mainly aragonitic) relict carbonate. In the remaining 82 %, i.e. the great majority of samples, the main part of the carbonate fraction is terrigenous, although, in consequence of its comparatively lower original carbonate content the delta input accounts for a relatively low carbonate fraction in these particular 82 % of the samples.

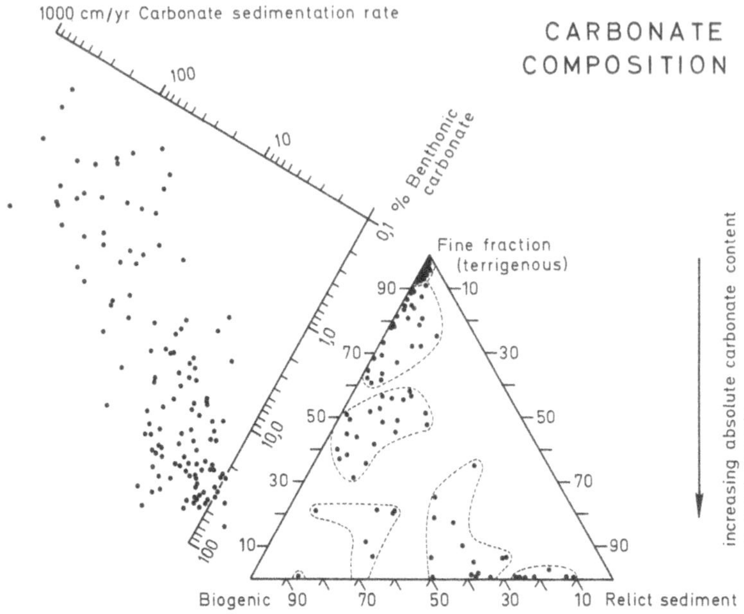

Fig. 10. Frequencies of main carbonate fractions, compared with carbonate sedimentation rates. (Erratum: 1000 cm/yr should be 1000 cm/1000 yrs)

Another aspect of carbonate sedimentation is shown on the upper left of Fig. 10 (inspired by Haake, Lutze, Röttger 1971). As the benthonic carbonate fraction increases, the relict and total carbonate contents in the total samples also increase (compare Fig. 5). However, the corresponding sedimentation rates of carbonate decrease proportionately from 100 - 500 cm to 1 - 5 cm over 1000 years (sedimentation rates from Sarnthein 1971).

To summarize, these observations indicate, firstly, an approximately constant biogenic carbonate production everywhere in the Iranian parts of the Persian Gulf falling within the range of 0.5 to 5 cm in 1000 years (see Seibold et al., this volume).

Secondly, with respect to sedimentation rates, terrigenous supply is demonstrated to be by far the most important source of marl carbonate. Thus, the abundance of carbonate in Persian Gulf marls is governed by the hereditary carbonate mineralogy of the sediments of the hinterland. Therefore, although appearing contradictory at first glance, marl samples with low absolute carbonate content characterize sediments with a relatively high rate of overall carbonate formation, and vice versa.

Carbonate content and grain size distribution are certainly important variables in assigning a sediment to a facies unit. The Q-mode factor pattern has shown, however, that no coordinate can be identified, even approximately, with such simple variables, effective identification requiring a complex assemblage of sedimentary components. This involves all the pros and cons of the method as a multivariate procedure, for example:

- A single variance of only local significance will be eliminated.
- The same analysis will:
  - fuse closely related but basically different facies units (e.g. bank and starved-shelf facies);
  - uncover meaningful relationships between various sample areas (e.g. outer delta and uppermost continental slope);
  - clearly separate sediment complexes giving the impression of coherence (e.g. outer delta and delta-axis facies).
  - The underlying mechanisms for the classification could not be completely elucidated by conventional ordination methods.
- The frequency distribution of the variable values within different facies units may be obtained.
- The resulting facies pattern is independent of the author's prejudice, e.g., some of our facies units agree well with those worked out separately for one distinct delta complex in the Persian Gulf by an "analyse factorielle des correspondences" ("contingency factor analysis") by Melguen (this volume; compare also Lebart & Fenelon 1971).

Comparative experiments with the same data matrix using other multivariate methods are in progress.

ACKNOWLEDGEMENTS

This paper owes much to the informative discussions held with the other members of the Persian Gulf working group of the Geological-Paleontological Institute of Kiel University, and particularly Professor E. Seibold for his continuous support of the data processing work. Miss Melguen kindly made available the publication by Lebart & Fenelon. Professor G. Griesser generously made computing time available on the machine installed at his institute where M. Jainz spent much time implementing the program. The computer time needed for the development of the program was provided by the computing center of the university. We should like to express our sincere thanks to all of them.

Correspondence Analysis for Recognition of Facies in Homogeneous Sediments off an Iranian River Mouth

Marthe Melguen[1]

ABSTRACT

Six sediment cores from the submarine delta of the Rud Hilla River in the northern part of the Persian Gulf consist of fine grained, homogeneous Holocene marls. The coarse (> 63 μ) fraction varies from 0.3 - 3.5 %. The cores are 2 - 4 m long and were taken in water depths of 8 - 56 m. In spite of the great similarity and homogeneity of the cored sediments, correspondence analysis (an extension of factor analysis) of the coarse fraction reveals the presence of four distinctive sedimentary facies: (1) a minerogenic facies, 10 km from the estuary; (2) an ophiuroid-ostracod facies near a lateral margin of the delta, 12 - 15 km from the estuary, (3) a benthic foraminiferal-molluskan facies, in the central part of the delta 20 km from the estuary, and near its seaward margin 120 km from the estuary, (4) a gastropod-epibiotic facies, in an area of relatively slow sedimentation on the border of the delta, 90 km from the estuary. A seventh core, taken near the seaward margin of the delta of the Rud Hilla River, penetrated homogeneous, aragonite-rich mud of late Pleistocene age. Correspondence analysis of the sand fraction of the Pleistocene sediments leads to the definition of two facies that can be readily compared with the facies identified in the Holocene cores.

I. INTRODUCTION

The Rud Hilla River flows out of Iran into the northern Persian Gulf (fig. 1). Seaward of the estuary, a submarine delta extends to the SE, parallel to the coast (Sarnthein, 1971; fig. 2). The delta is approximately 120 km long and 40 km wide. Seven gravity cores, 2-4 m in length, were taken in the area of the delta from the German oceanographic vessel "METEOR" during the 1964-1965 International Indian Ocean Expedition (Dietrich et al., 1966). Six of the cores are very similar in appearance and consist of fine grained, apparently homogeneous, Holocene sediments (fig. 3). The seventh core (1115 ; fig. 1), 2 m long and taken on the outer edge of the delta, contains whitish, late Pleistocene (approximately 12 500 - 10 000 years b.p.), aragonite-rich mud underlying Holocene green marls (fig. 3).

The purpose of the study reported in this paper was to investigate sedimentary history in the Rud Hilla delta area by looking for, and defining, different facies[2]

---

1 Geologisch-paleontologisches Institut und Museum der Universität Kiel
40-60 Olshausenstrasse, 23 KIEL
Present adress: Centre Océanologique de Bretagne, B.P. 337, 29 N BREST, France.

2 "assemblage of physical and organic characteristics ... that indicate conditions of deposition" (Moore, 1949).

within the cores (Melguen, 1971).[1] The apparent homogeneity of the sediments indicated that this would require a method of considerable sensitivity. Shepard and Moore (1954) suggested that: "coarse-fraction studies appear to provide one of the best means of differentiating modern environments, and give important evidence of distances from shore, and water depths. It is thought that the method could be applied to help differentiate environments of ancient sediments". Sarnthein (1971) utilised modifications of Shepard and Moore's procedures for his study of sand fractions of surface sediment samples from the Persian Gulf. The author, proceeding from a similar approach, has used correspondence analysis (an extension of standard factor analysis) for the statistical processing of the data derived from detailed microscopic analyses of coarse (> 63 μ) fractions.

The results show that it is possible to identify distinctive facies within the six homogeneous Holocene cores from the Rud Hilla delta and thereby to gain some insight into recent changes in the history of sedimentation in this part of the Persian Gulf. Identification of the late Holocene facies also permits a better recognition of differences between the Holocene sediments and those of the sequence containing late Pleistocene sediments that were cored on the south-western margin of the delta (core 1115; fig. 1).

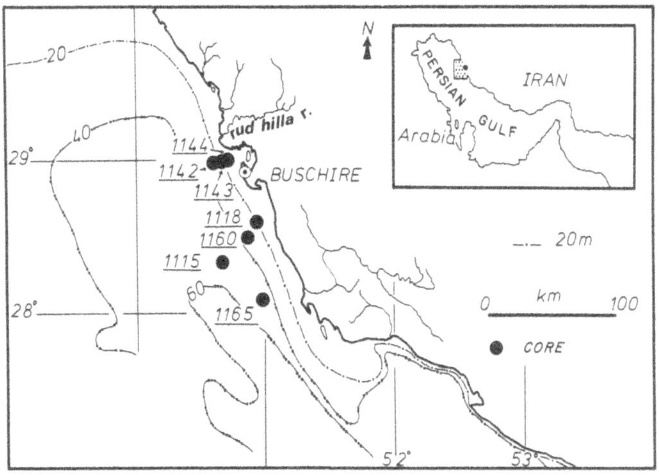

Fig. 1. Location of gravity cores taken in the Rud Hilla delta area in the northern part of the Persian Gulf

1 Full results are to be published in "Meteor" Forsch. Ergebnisse.

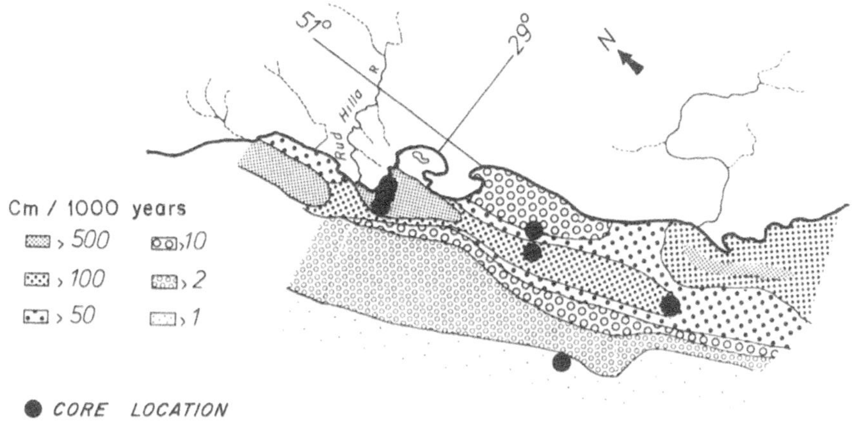

Fig. 2. Map showing approximate sedimentation rates in the Rud Hilla delta area, northern Persian Gulf (after Sarnthein, 1971)

II. ANALYTICAL METHOD

Sampling the seven cores (table 1) at about 2-7 cm intervals throughout their length provided 139 specimens for study. The sand fractions (2 000 - 63 μ), separated by wet sieving, constitute 0.3 - 3.5 % by weight of total sediment.

Each sand fraction was divided by dry sieving into five grain-size classes: 2 000 - 1 000 μ, 1000 - 500 μ, 500 - 250 μ, 250 - 125 μ, 125 - 63 μ. Six hundred and ninety five subsamples were thus obtained. Between 800 - 1 000 particles within each subsample were identified under the binocular microscope (Plas et al., 1965) and each particle was assigned to one of 33 different components (e.g. pteropods, echinoids, biotites) within various categories of components (e.g. total benthic components, total terrestrial components; fig. 4 - 5).

The information generated by the above procedures provides the basis for the recognition of facies in the cored sediments. The array of data is large and well suited for statistical treatment.

Several authors have used factor analysis in paleontological and sedimentological studies (Imbrie et al., 1962, 1964; Briggs, 1965; Klovan, 1966; Osborne, 1969; Kelley et al., 1969; Allen et al., 1970; Beall, 1970). In factor analysis, clouds of observation points in variable space (or of variable points in observation space) are examined by use of privileged axes (factorial axes or factors) which best define the approximate shape of the cloud.

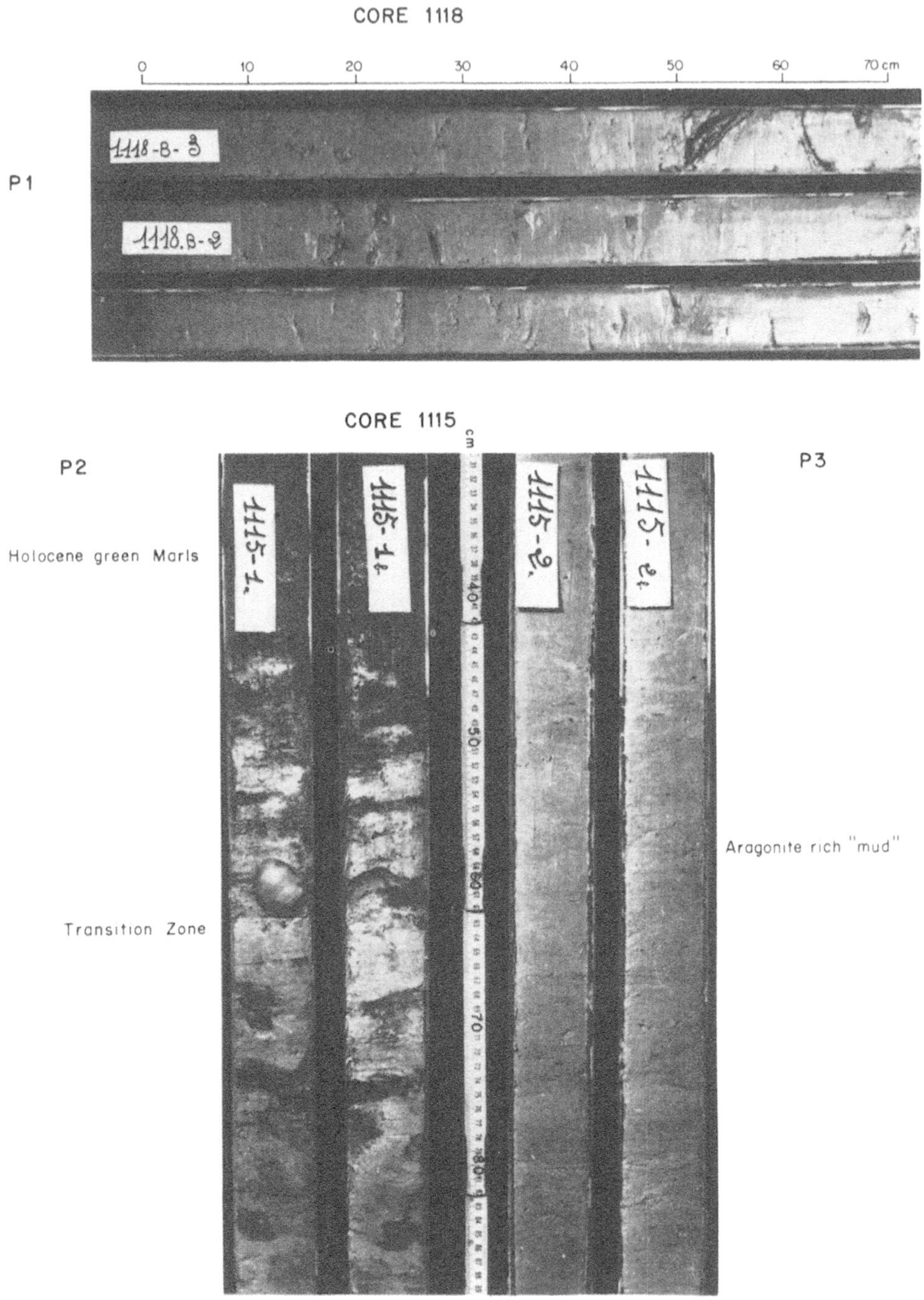

Fig. 3. Photographs of sections of cores 1118 and 1115; delta region of the Rud Hilla River, northern Persian Gulf

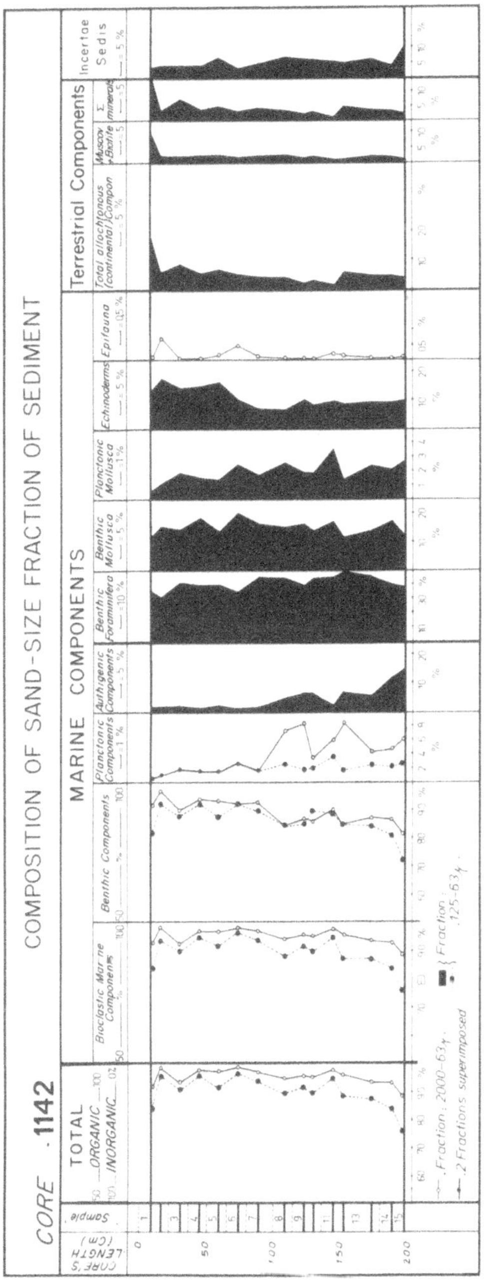

Fig. 4. Composition of sand-size fraction in sediment core 1142; Rud Hilla delta, northern Persian Gulf. The percentages are percentages of total sand fraction

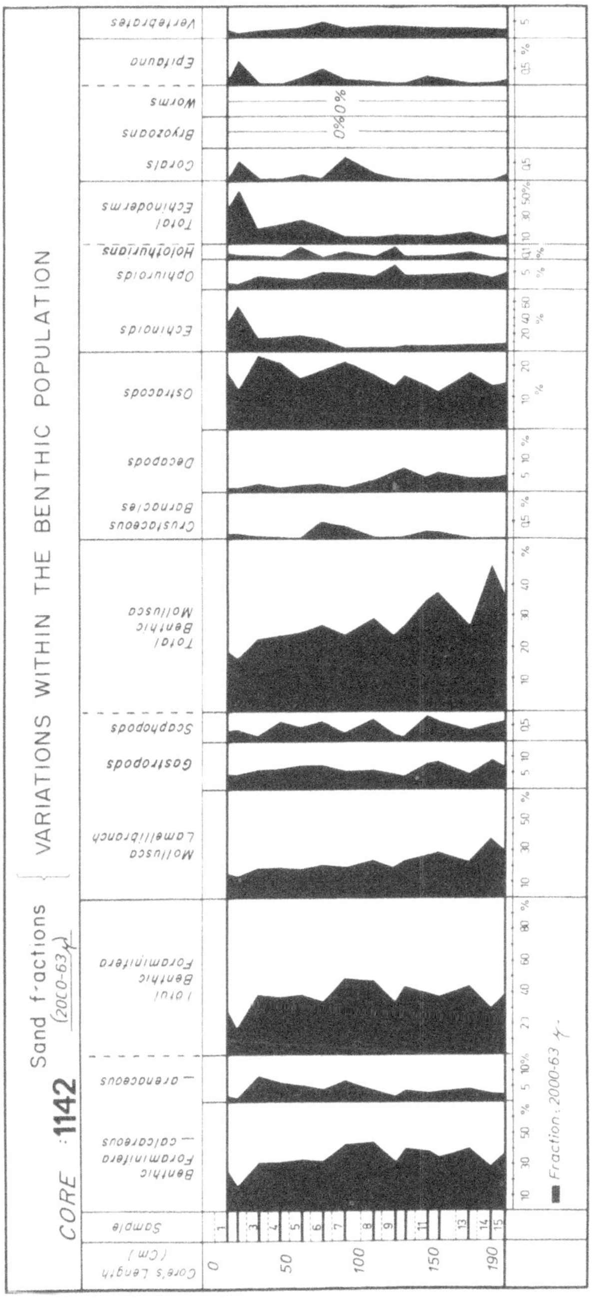

Fig. 5. Distribution of sand-size benthic components of the sediments of core 1142; Rud Hilla delta, norther Persian Gulf. The frequencies are calculated as percentages of the total benthic population

The author has used correspondence analysis, which is Benzecri's generalization of factor analysis (Cordier, 1965; Benzecri, 1970, 1971; Le Bart et al., 1971). In correspondence analysis, observations and variables (which need not be continuous as in factor analysis) play a symmetrical role. Possible proximities between lines and columns of a table can be measured simultaneously. Correspondence analysis is adapted to frequencies. It leads to a global view of proximities among observations (i.e. sand fractions), among variables (i.e. components) and among observations and variables. Consequently, correspondence analysis is particularly useful for the type of problem with which this paper is concerned because it permits a characterization of facies that takes numerous relationships into account simultaneously (Melguen, 1972).

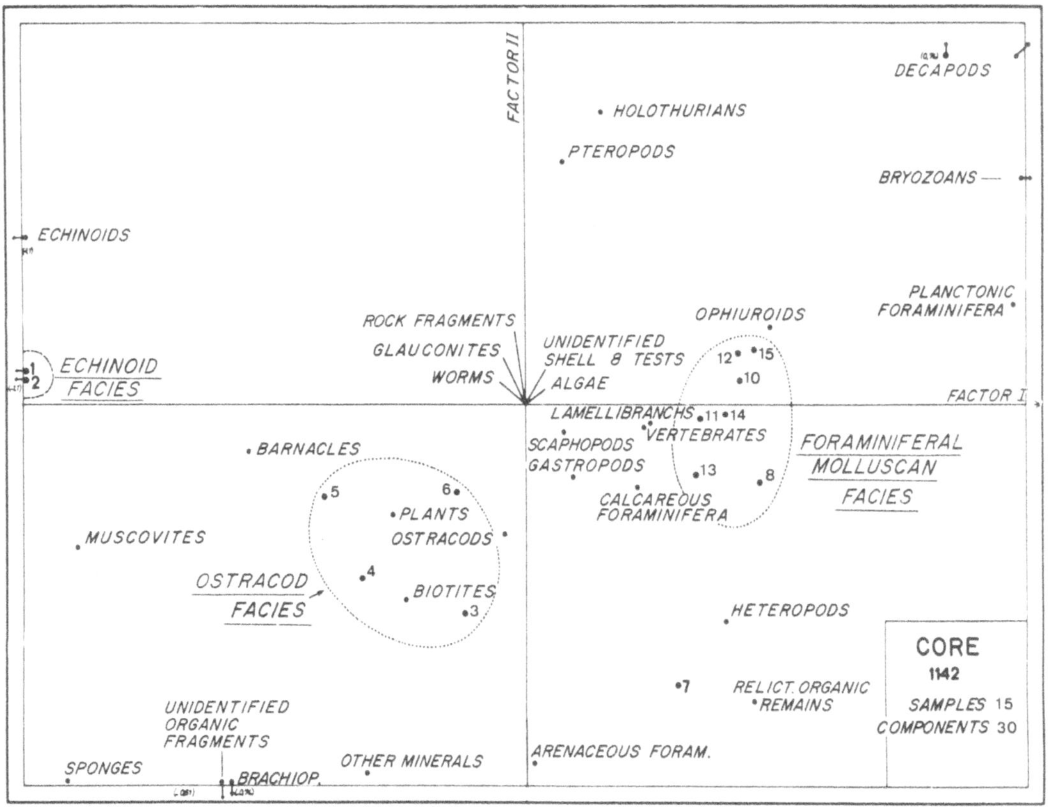

Fig. 6. Sedimentary facies determined by correspondence analysis of sand fraction data for core 1142; Rud Hilla delta, northern Persian Gulf. The proximities of sand fractions and components are shown with respect to factors I and II. Three subfacies are identified: Echinoid facies (samples 1, 2); ostracod facies (samples 3, 6); and foraminiferal-molluscan facies (samples 8, ... 15)

III. RECOGNITION AND DISTRIBUTION OF FACIES

Six facies (I - VI) of Rud Hilla delta sediments are recognized by correspondence analysis (fig. 6, 7 and 8). They can be grouped into 3 late Holocene facies (I, II, III), which include a total of 9 subfacies; 2 Pleistocene facies (V - VI); and one transitional, early Holocene facies (IV).

a) Late Holocene facies

The late Holocene facies (I, II, III) are represented by the sediments occuring today approximately along the axis of the Rud Hilla submarine delta, from the estuary towards the SE (cores 1144, 1142, 1160, 1165). Their distribution along factor I (fig. 7) is as follows:

- facies I: minerogenic[1] facies, comprising a subfacies of quartz[2],feldspar, followed by a subfacies of muscovite and biotite.

- facies II: comprising three subfacies:
  ophiuroid-plant subfacies
  ostracod subfacies
  echinoid subfacies

- facies III or benthic foraminiferal-molluskan facies, comprising four subfacies:
  dominantly foraminiferal subfacies
  foraminiferal-molluskan facies
  dominantly molluskan lamellibranch subfacies
  gastropod-epibiotic subfacies

Factor I (fig. 7) may be considered as an approximate measure of the distances of core stations to the estuary along the axis of the delta. The composition of the sand changes with increasing distance from the estuary: micas replace quartz and feldspar; frequency of plant debris increases; muscovite, initially more prominent than biotite, gives way progressively to biotite (fig. 10). The terrigenous content of the sand fractions of the cores ranges from 80 % near the estuary to less than 1 %, at the SE extremity of the delta (fig. 10).

Benthos becomes increasingly abundant as the proportion of terrigenous detritus diminishes and hence, presumably, as the environmental conditions for benthic life improve (fig. 9 & 11). The relative proportions of different benthic components reflect their adaptability to different microenvironments. The Foraminifera and ostracods, for example, are well adapted to the estuarine conditions, whereas the organisms such as the gastropods seem to prefer areas where sedimentation rates are slower (fig. 2 & 11).

b) Pleistocene facies

The sequence of facies along factor I (Fig. 7) corresponds not only to the sequence of Holocene facies that is found with increasing distance from the estuary of the Rud Hilla River but also to the sequence of Pleistocene facies in core 1115. From the bottom towards the top of core (factor I; Figs. 7 & 8) are found successively: (1) a minerogenic facies of quartz and various minerals, (2) an ophiuroid facies (3) an ostracod facies, (4) a foraminiferal-molluskan facies, (5) a gastropod and epibiotic facies. Thus,things have developed as if, since the end of the Pleistocene, station 1115 had successively occupied situations similar to those now respectively characteristic of stations 1144, 1143, 1142, 1160 and 1118 (Fig. 9). The Holocene and Pleistocene facies are not, however, identical. Factor II (Fig. 7), interpretation of which is complex (section IV.b), indicates the differences: fragments of resedimented rocks, micas and plants not as abundant at station 1115; epifauna (worms, bryozoans) and pellets, very abundant at station 1115.

---

1 with minerals as major components.
2 The Rud Hill River today drains mainly Tertiary and Quaternary sandstone marls, limestones and evaporites. (Sampo, 1969).

IV. SEDIMENTARY HISTORY

The evidence of the identity and distribution of facies in the cores from the Rud Hilla delta makes it possible to attempt a reconstruction of sedimentary history in the area. The author has, during the course of this study, maintained the view that observations of the more recent sediments are a necessary starting point for the interpretation of the Pleistocene facies, and the late Holocene sediments are

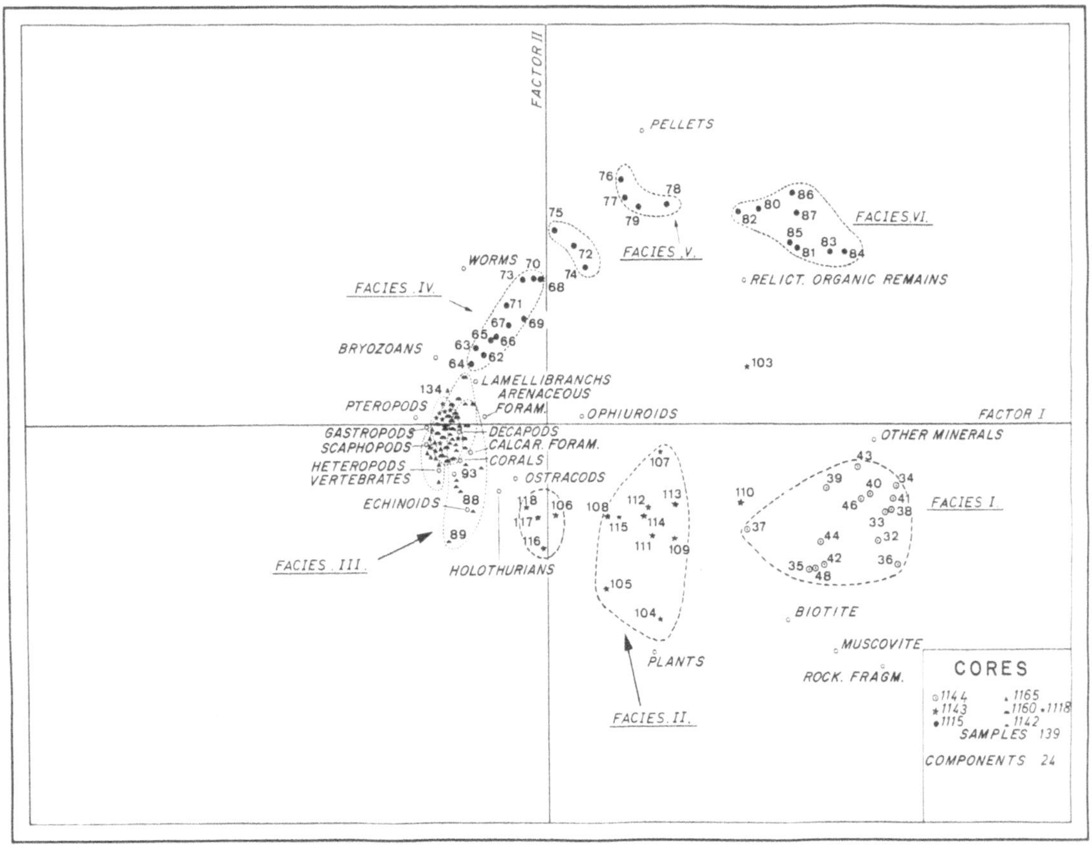

Fig. 7. Sedimentary facies determined by correspondence analysis of sand fraction data for 7 cores from the Rud Hilla delta in the northern part of the Persian Gulf. The matrix was constituted by the frequencies of components in the total sand fraction of each sample. The proximities of sand-fractions and components are shown with respect to factors I and II. Analysis of the proximities leads to the identification of 6 facies: 3 late Holocene facies (I, II, III), 2 late Pleistocene-early Holocene facies (V, VI), 1 transitional facies (IV)

discussed first.

a) Late Holocene

Sedimentation rates seem to be the main control of facies distribution in the Rud Hilla delta environment. Tentative estimates of sedimentation rates for the depositional periods represented by each subfacies (table 3) are based on a series of C 14 datations (table II) and on other evidence of present day sedimentation patterns in the area (Sarnthein, 1971).

The depositional history of the delta has tended to be characterized by an increase in the rate of sedimentation since approximately 6 000 B.P.. The increase is accompanied at stations 1142, 1143, 1118 (Fig. 9) by an evolution of sedimentary sub-facies that includes: (1) the appearance of the minerogenic sub-facies at the top of core 1143, (2) an upward change from foraminiferal sub-facies to ostracod sub-facies, to echinoid sub-facies in core 1142, and (3) an upward change from gastropod and epibiotic sub-facies, to molluskan (lamellibranch) facies, to foraminiferal-molluskan sub-facies in core 1118.

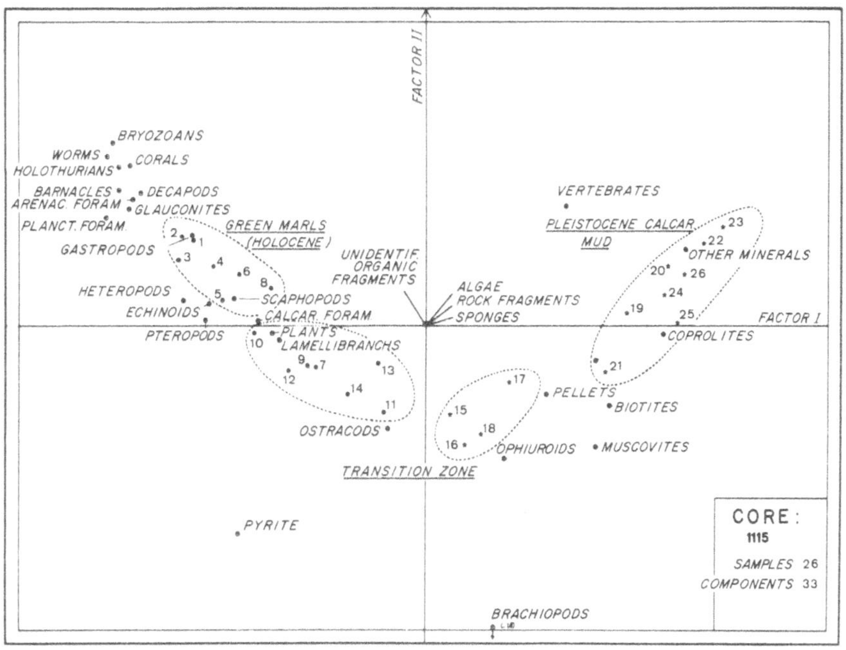

Fig. 8. Sedimentary facies determined by correspondence analysis of core 1115; Rud Hilla delta, northern Persian Gulf. Along factor I (from right to left in the diagram, that is, from bottom to top of the core, the following facies are recognized: (1) late Pleistocene facies, (2) transition facies, namely ophiuroid sub-facies and ostracod sub-facies, (3) early Holocene facies including gastropod-epibiotic sub-facies

The increased rate of sedimentation seen at stations 1143, 1142, 1118 (Fig. 9) cannot be traced across the entire delta. Core 1160 (Fig. 9) is almost homogeneous from top to bottom, sedimentation over the past 6 000 years being represented principally by one sub-facies (facies III c). Sedimentation was interrupted by three minor depositional events associated with slightly higher sedimentation rates, and with accumulation (perhaps by decapods) of shell fragments (subfacies III b).

In addition to the question of sedimentation rates, there have evidently been changes of sediment distribution over parts of the delta. Core 1165 (Fig. 9) exhibits a sequence that is evidently the inverse of core 1160: the estuarine subfacies at the bottom (subfacies III a) is followed by a foraminiferal-molluskan subfacies (subfacies III c), suggesting a decrease in the rate of sedimentation. It seems that sediment distribution on the Rud Hilla delta during the late Holocene was displaced towards the center of the gulf: this would explain why stations 1160 and 1165 have not been affected by the increase in the rate of sedimentation.

Core 1118 (Fig. 9), the facies of which reveals a marked vertical evolution, suggests an increase of the sedimentation rate. It should, however, be noted that core 1118 was taken near the landward margin of the delta. Sedimentary evolution in this area is dependent on a terrigenous supply derived not only from the Rud Hilla River but also transported, probably by currents, from the adjacent coast.

Increases in the rate of sedimentation on the Rud Hilla delta during the last 6 000 years could be the result of any one, or more, of the following processes: (1) the general increase in rainfall which affected Iran during the Holocene (Bobek, 1963, 1969; Butzer, 1966; Diester, 1972; Van Zeist and al., 1963: Van Zeist, 1967); (2) varying distribution of rainfall; less frequent but heavier precipitation would result in more violent run-off and consequently in an increased supply of terrigenous material; (3) destruction of vegetation leading to increased erosion; (4) displacement of the course of the Rud Hilla River, perhaps related to irrigation activities. Whatever the various causes may have been, the increased rate of erosion and sedimentation accelerated the progradation of the delta towards the central part of the gulf, together with migration of facies in the same direction.

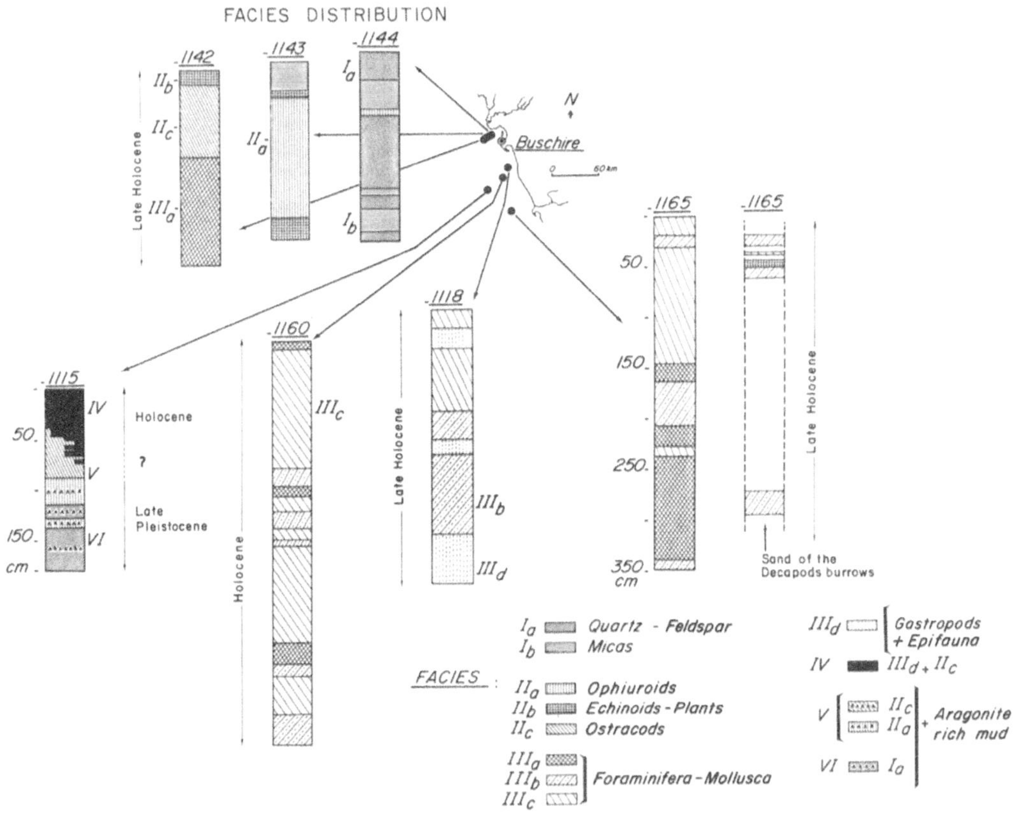

Fig. 9. Distribution of facies in 7 cores taken in the delta area of the Rud Hilla River in the northern Persian Gulf. All the cores, except 1115, have the appearance of extreme homogeneity

b) Late Pleistocene-early Holocene

Some interpretation of sedimentary history in the present Rud Hilla delta area near the end of the Pleistocene (approx. 12,500-9,500 y. B.P.: table II) can be based on the study of the sediments penetrated at station 1115 (Fig. 1), where the Holocene is relatively thin.

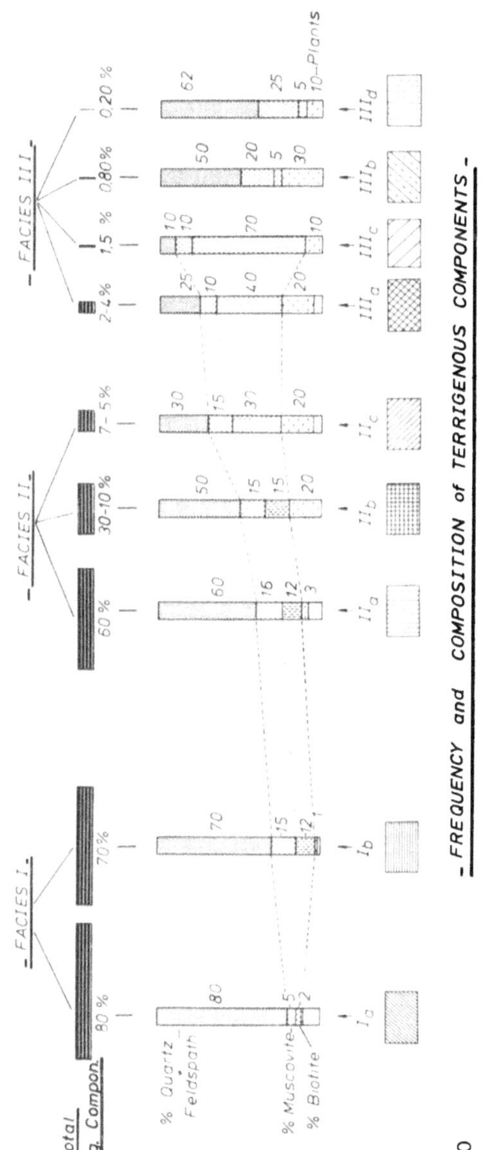

Fig. 10

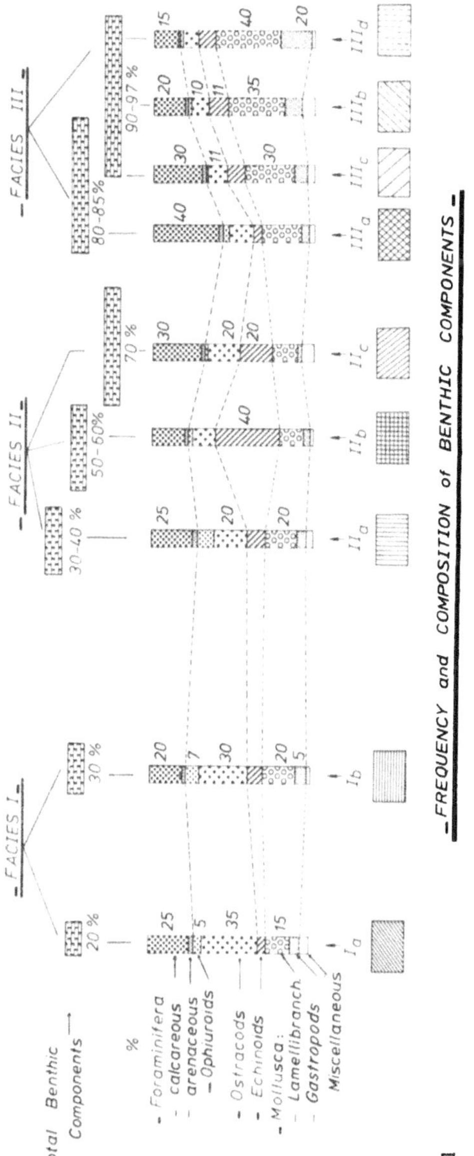

Fig. 11

Fig. 10. Nature and frequency of the terrigenous components of the sand fractions identified in the cores taken in the delta area of the Rud Hilla River, northern Persian Gulf

Fig. 11. Nature and frequency of benthic components of the sand fractions (corresponding to each type of facies or subfacies), identified in the cores taken in the delta area of the Rud Hilla River, northern Persian Gulf

The main contrasts between the Pleistocene and Holocene sediments (section III) could have resulted from several environmental differences. The deposition of whitish, aragonite-rich mud may have reflected more favourable climatic conditions (Melguen, 1971). The smaller accumulation of micas and plants may have been a consequence of increased turbulence in the overlying waters (Paremancblum, 1966). A lower sea level would have effectively placed station 1115 close to land and to the estuary of the Shatt-al-Arab river, now situated at the northern extremity of the Persian Gulf (Fig. 1), thus influencing the depositional environment at station 1115 and the provenance and transportational history of the accumulating sediments.

The relative importance of various environmental changes notwithstanding, the upward evolution of facies within the Pleistocene sediments corresponds quite closely to the successive changes of late Holocene facies that occur with increasing distance from the sedimentary source (section III). Core 1115 was taken in 56 m of water, about 90 km from the Rud Hilla River estuary, and beyond the present margin of the delta, whereas the Pleistocene facies recognized in the core (Fig. 9) suggest deposition in about 8 - 25 m of water and about 10 - 20 km from an estuary. Assuming that the estuary was that of the Shatt-al-Arab river, the sediments cored at station 1115 thus appear to represent increasing distance of the station from the estuary near the end of the Pleistocene.

The abrupt change from late Pleistocene, aragonite-rich deltaic muds to a green marl facies, apparently deposited in a water depth of 40 - 60 m, suggest that the early Holocene transgression was very rapid. The deposition of the green marls is estimated to have begun about 9 500 y. B.P. (Diester, 1972). The change may represent a sedimentary discontinuity although the pattern of transgressive sedimentation off Buschire, deduced from the facies analysis, and confirmed by the presence of benthic foraminiferal species that indicate water depths (Lutze, personnal communication), is in satisfactory agreement with established eustatic curves that show (1) a slow rise of sea level from 50 to 38 meters 12 000 to 9 000 years (B.P.) (Mörner, 1966) and (2) a rapid rise of sea level from 38 to 20 m about 9 000 years B.P. (Mörner, 1969; Curray, 1961; Jelgersma, 1961; Schofield et al., 1964; Cullen, 1967).

The termination of green marl deposition is marked by the appearance of a gastropod-epibiotic facies indicating only a very slight deltaic influence.

CONCLUSION

Detailed analysis of sand fractions and statistical treatment of the data by correspondence analysis reveals the presence of distinctive assemblages within apparently homogeneous sediments in the Rud Hilla delta area of the Persian Gulf. The different facies represented by the assemblages comprise three late Holocene (< approx. 6 000 y. B.P.) facies (including a total of 9 subfacies) and three facies of late Pleistocene-early Holocene age. The Holocene sedimentary history is marked by an increase of sedimentation rate, a progradation of the Rud Hilla delta and, therefore, by a migration of facies.

The clear recognition of the identity of the late Holocene facies permits definite comparisons to be made with the earlier facies. The evolution of facies within the Pleistocene sediments corresponds quite closely to the successive changes of late Holocene facies that occur with increasing distance from the sedimentary source. The main contrasts between Pleistocene and Holocene facies could have resulted from differences in climatic and hydrodynamic conditions, and in the nature of the sedimentary source. Correspondence analysis seems to be a tool of considerable promise for quantitative interpretation of the histories of deltaic sedimentary sequences.

ACKNOWLEDGEMENTS

I thank all those who assisted me during the course of this study. My thanks are most particularly addressed to Professors Seibold and Boillot who suggested that I undertake the research, and for whose guidance and advice I have been very grateful. Dr. Sarnthein introduced me to sand fraction analysis and Mr. Kerbaol to correspondence analysis. Professor Metivier, Miss Le Menn, Mr. Le Faou assisted me with the computer programming. I am very grateful to Dr. H. D. Needham for his generous help and advice on the improvements of the manuscript. The Centre National pour l'Exploitation des Océans provided financial support.

TABLE 1. Location and description of sediment cores from the Rud Hilla delta

CORE N°	STATION N°	LOCATION		WATER DEPTH	CORE LENGTH	DESCRIPTION
1144	M 330/3	29° 01'N	50° 42'E	8 m	2 m	brownish, homogeneous mud[+]
1143	M 330/2	29° 00'N	50° 41'E	15 m	2 m	brown-greenish, homogeneous mud
1142	M 330/1	29° 00'N	50° 39'E	23 m	2 m	greenish, homogeneous mud
1118	M 309	28° 36'N	50° 56'E	21 m	3 m	greenish, homogeneous mud
1160	M 345	28° 32'N	50° 54'E	29 m	4 m	greenish, homogeneous mud
1165	M 348	28° 07'N	51° 00'E	46 m	4 m	greenish, homogeneous mud
1115	M 306	28° 22'N	50° 40'E	56 m	2 m	3 distinctive units : - 0 - 40 cm : greenish sediment with large shell fragments. - 40 - 90 cm : intermediate units to interfingering green and white muds. - 90 - 180 cm : very fine and homogeneous whitish mud.

+ The term "mud" is used for sediment containing 99 % of particles finer than sand-size (< 63 μ).

TABLE 2. C 14 Ages[1] of sediment from the Rud Hilla delta area, Persian Gulf

CORE N°	SAMPLE LEVEL	AGE (years B.P.)
1160	0 - 7 cm	975 ± 260
	30 - 35 cm	1560 ± 175
	87 - 96 cm	2335 ± 120
	127 - 135 cm	3340 ± 195
	231 - 241 cm	2450 ± 170
	333 - 338 cm	5080 ± 185
	380 - 390 cm	5840 ± 200
1165	0 - 10 cm	1970 ± 140
	187 - 197,5 cm	2250 ± 135
	208 - 218 cm	2195 ± 215
	281 - 291,5 cm	2625 ± 155
	302 - 307,5 cm	3060 ± 115
	315 - 322 cm	3740 ± 90
	322 - 329 cm	4255 ± 245
1115[2]	Base	approx. 12,500

(1) C 14 ages of total sediment or of shells (determined by Dr. Geyh of the Laboratoire für Bodenforschung, Hannover).

(2) Age refers to a second correlatable core taken at the same station.

TABLE 3. Relationship of facies to environmental parameters in the Rud Hilla delta

FACIES	DISTANCE FROM THE ESTUARY OR THE COAST †	WATER DEPTH	CURRENT VELOCITY	SEDIMENTATION RATE
I a - I b	15 km	8 m	< 50 cm/sec †	≃ 4 - 5 m / 1 000 years
II a	15 - 17 km	8 - 15 m	"	≃ 2 - 4 m / 1 000 years
II c	19 km	25 m	"	≃ 1,5 - 2 m/ 1 000 years
III a	19 km	25 m	"	≃ 1 m / 1 000 years
III b	18 km+	21 m	"	≃ 0,80 m / 1 000 years
III c	80 - 120 km+	21 - 46 m	"	≃ 0,80 - 1 m / 1000 years
III d	18 km+	21 m	"	< 0,80 m / 1 000 years

† Near-bottom currents velocity off Buschire (after Peery, 1965).

Aeolian Admixtures in the Sediments of the Northern Persian Gulf

Zdenek Kukal[1] and Adnan Saadallah[2]

ABSTRACT

Composition of the sediments of the northernmost parts of the Persian Gulf was compared with that of the fallout from dust storms over Iraqi territory. The results of this investigation indicate the presence of considerable quantities of aeolian material in certain Persian Gulf deposits. The following points support this conclusion: a. High rate of sediment accumulation from recent dust storms b. Similar grain size: carbonate relationship in marine and dust storm sediments, c. Presence of "pseudosand" grains in both marine and dust storm deposits.

SEDIMENTS FROM THE NORTHERNMOST AREAS OF THE PERSIAN GULF

Surface sediments from different sub-environments along the Iraq coast were sampled and analyzed for mineralogical composition, grain size and carbonate content. Fig. 1 shows the areas sampled: the delta plain opposite the port of Fao and the bay sediments near the port Um-Qasir on the Iraq-Kuwait frontier.

The samples from the delta plain were collected from various intertidal sub-environments, including mud flats, sand flats, shallow channels and rudimentary sand and shell bars, and from the adjacent shallow areas down to depths of 3 m.

The Um-Qasir samples were collected from the gently sloping beach and the muddy, bay sediments down to depths of 3 m.

GRAIN SIZE OF SEDIMENTS

Analyses of grain size showed that the Md of both sets of samples lay within the silt range (between 4 phi and 6 phi). Thus, the silt fraction is much the more prominent in both environments. Apart from the prevailing silt fraction all samples contained variable amounts of fine sand and small amounts of clay. The cumulative curves all have the same shape, only the variable sand content causing them to tend toward finer or coarser grades (see Figs. 2, 3). The phi standard deviation (according to Inman's formula) of delta plain sediments ranges from 1.2 to 1.6, that of Um-Qasir bay sediments from 1.7 to 2.1.

---

1 Central Geological Survey, Prague, Czechoslovakia.

2 College of Science, University of Baghdad, Iraq.

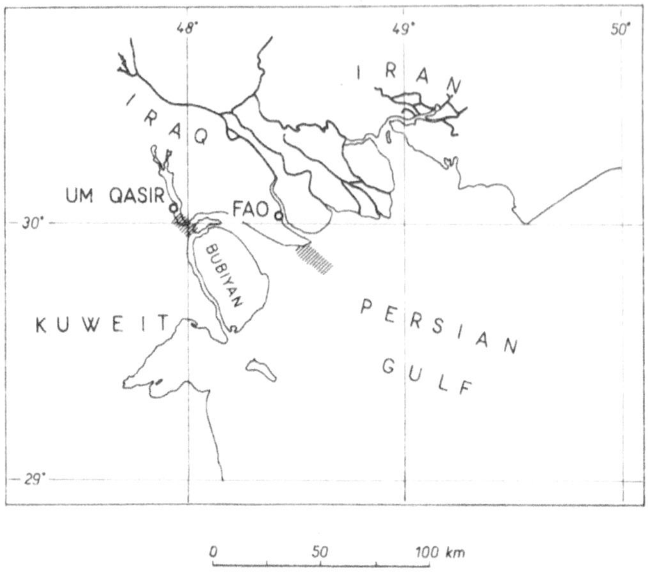

Fig. 1. Map of the NW parts of the Persian Gulf showing areas studied

COARSE FRACTIONS

As shown in Table 1, the delta plain samples were richer both in carbonate material (shell fragments and carbonate aggregates) and in mica and glauconite. Bay samples, on the other hand, were richer in quartz and other terrigenous material; the amounts of pyrite and glauconite were extremely low.

Table 1

Coarse fraction composition (< 4 phi) of delta plain and bay		
Component	delta plain	bay
Quartz	16.0 %	48.5 %
Clastic minerals	20.4 %	26.5 %
Gypsum	14.0 %	12.0 %
Rock fragments and carbonate aggregates (including pseudosand)	32.0 %	8.0 %
Mica	3.0 %	8.7 %
Shell fragments	10.0 %	1.5 %
Glauconite	2.0 %	0.5 %

The significant amounts of gypsum in both environments are due to river and wind transport from the nearby soil. The presence of aeolian material is also indicated by the high percentage of pseudosand grains. The term "pseudosand" includes heterogeneous grains the size of sand particles, which consist of firmly cemented aggregates of clay and carbonate (term introduced by P. Buringh and C.H. Edelman, 1955). Gypsum may also occur as a cementing material. Such grains are transported and rounded, together with other sand grains. After deposition and induration, however, they can be reconverted to clay or carbonate. Large amounts of pseudosand grains occur in Iraq soils (P. Buringh, 1960), in Miocene and Pliocene sediments of Iraq (Z. Kukal-J. Al-Jassim, 1971) and in aeolian suspension and fallout from dust storms (Z. Kukal & A. Saadallah, 1970).

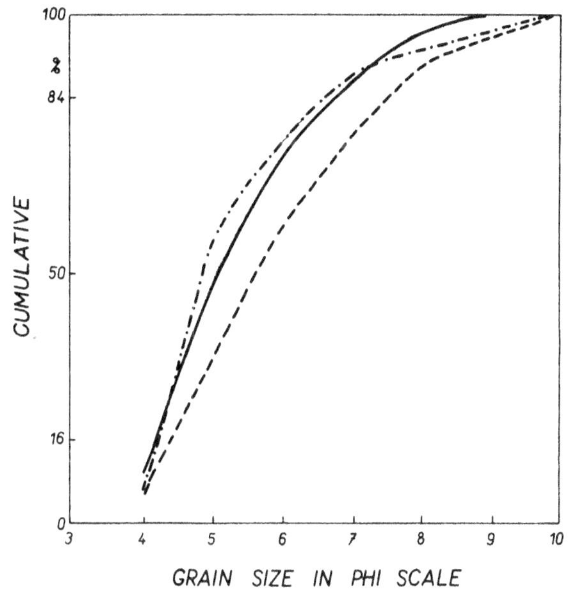

Fig. 2. Cumulative curves of delta plain (Fao) sediments
a) Outer margin of delta plain (full line)
b) mud flats within delta plain (dashed line)
c) sediments from shallow channels (dotted line)

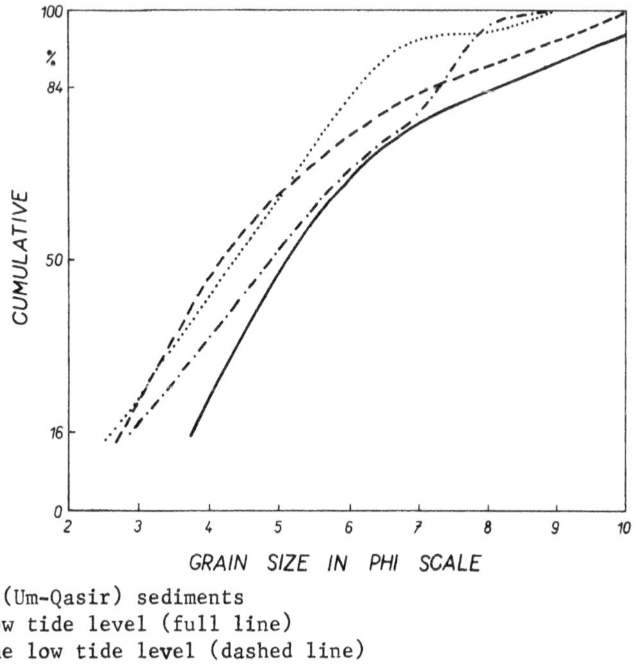

Fig. 3. Cumulative curves of bay (Um-Qasir) sediments
a) One meter below the low tide level (full line)
b) 30 m landwards from the low tide level (dashed line)
c) 60 m landwards from the low tide level (dot-and-dashed line)
d) Beach berm (dotted line)

CARBONATE CONTENT

The inter-relationship between grain size and carbonate content of deposits is shown in Figs. 4 and 5, which illustrate a marked bimodal distribution of carbonate related to grain size fractions. The first peak can be observed in the silt fraction (4-6 phi) in which the carbonate fraction attains 60 per cent. In some sediments the carbonate fraction is considerable also in the sand fraction. In some samples a second, sharply defined peak, indicates the presence of carbonate even in finest grades (8-9 phi). Grain size fractions containing the highest amount of carbonate correspond to the grain size fractions most abundant in aeolian suspension and fallout as described below.

FALLOUT FROM RECENT DUST STORMS

Dust storms are a very common event in arid and semi-arid countries. In Iraq NW winds strongly prevail over the year (M.J. Khalaf, 1957). Dust storms, however,

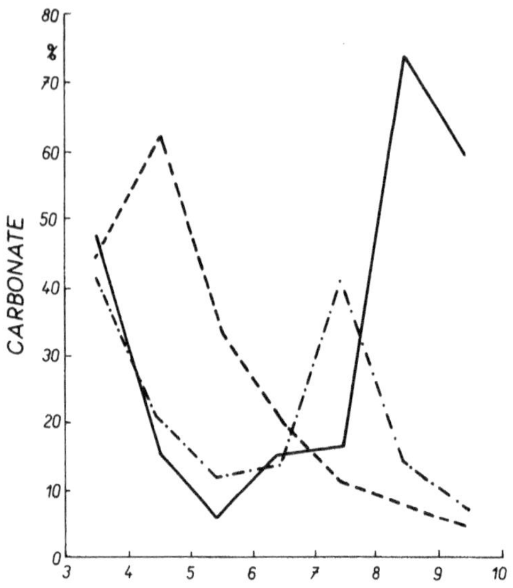

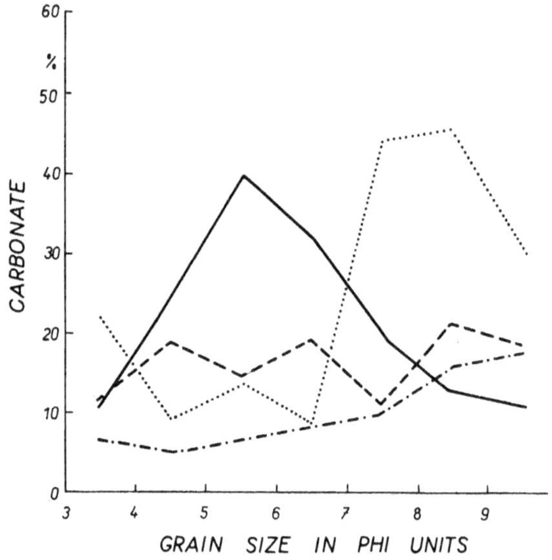

Fig. 4. Diagram illustrating the relationship between grain size and carbonate content in Fao sediments (for the location of samples see Fig. 2)

Fig. 5. Diagram illustrating the relationship between grain size and carbonate content in Um-Qasir sediments (for the location of samples see Fig. 3)

are due to the local whirl winds which are able to carry silt and sand up to 5,000 m above the ground. Due to the general movement of air masses toward the southeast the airborne suspension in usually shifted from Western deserts of Jezira or from the Euphrates valley toward the Baghdad area,and from the Southern desert towards the Shatt-al-Arab valley and northern areas of the Persian Gulf.

The occurrence of dust storm days in Baghdad and Basrah throughout 1967 (compared with the average distribution of dust storm days over the last 24 years) was as follows (according to Annual Statistical Abstracts for 1967):

Table 2
Dust storm days in Baghdad and Basrah

	Baghdad in 1967	Average for the last 24 years	Basrah in 1967	Average for the last 24 years
January	1	1.1	0	0.3
February	1	2.1	0	0.6
March	1	2.5	2	1.0
April	4	2.3	4	1.3
May	3	2.3	3	1.4
June	1	2.2	0	2.9
July	1	3.6	4	3.1
August	0	1.6	3	1.7
September	0	0.6	1	1.2
October	3	1.3	1	0.8
November	1	1.1	0	0.3
December	2	0.8	1	0.1
Total	17	21.5	19	14.7

As shown in Table 2, the distribution of dust storm days throughout the year is irregular; they are most frequent in spring and early summer. There are only slight differences in the quantity and seasonal distribution of dust in central (Baghdad) and southern (Basrah) Iraq.

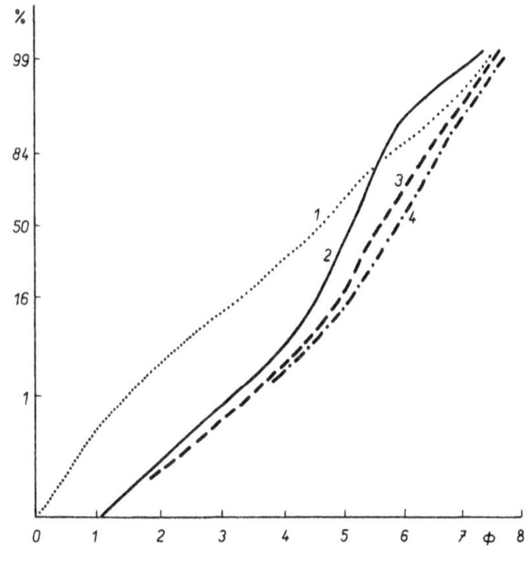

Fig. 6. Cumulative curves of the fallouts from the dust storm on April 8, 1968 (Baghdad). 1. Material deposited between 10 and 11 a.m., 2. Between 11 and 12 a.m., 3. Between 12 a.m. and 1 p.m., 4. Material deposited between 1 p.m. and 2 p.m

In order to measure the composition and rate of aeolian deposition, the products of two dust storms (April 8 and 20, 1968) were sampled and investigated (the details of the method are described by Z. Kukal - A. Saadallah, 1970). The results of grain size analyses are illustrated in Fig. 6. Although there are some differences in sand content of the samples, more than 90 per cent of the fallout consisted of silt. Greater amounts of sand appear to settle only during the first stages of the dust storm.

The mineralogic composition of the coarser fraction is as follows (arranged in order of estimated importance): calcite grains, lump aggregates, polished and rounded quartz, partly angular quartz, gypsum aggregates, chert grains, muscovite, and plant tissues. The amount of pseudosand was estimated to be 15 per cent of the total sample of the fallout from the first dust storm while gypsum grains make up 10 per cent of the sand fraction of this sample. The amount of carbonate is very high in the Iraqi dust storm sediments, 69.5 per cent and 66.1 per cent being determined in the fallouts from the two Baghdad dust storms. As illustrated in Fig. 7, the amount of carbonate generally increases within the finer fractions. Only in the coarsest fractions (< 1 phi) is the carbonate content unexpectedly high. Very similar results were established for Euphrates flood plain sediments (K.H. Al-Habeeb, 1969). The high amount of carbonate in the coarsest fractions indicates the presence of local clastic carbonate according to this author.

In order to estimate the rate of deposition of the aeolian material the amount of fallout from the individual dust storms was determined. During the two Baghdad dust storms 1.22 mm and 0.61 mm respectively of predominantly silt-sized sediments were deposited (for details see Z. Kukal - A. Saadallah, 1970). If there are 21 dust-storm days per year in Baghdad, the thickness of dust storm sediments deposited within one year should be in the order of 2.1 cm. (neglecting contemporaneous and subsequent erosion). Somewhat lower values can be expected for southern Iraq and also for the northernmost parts of the Persian Gulf.

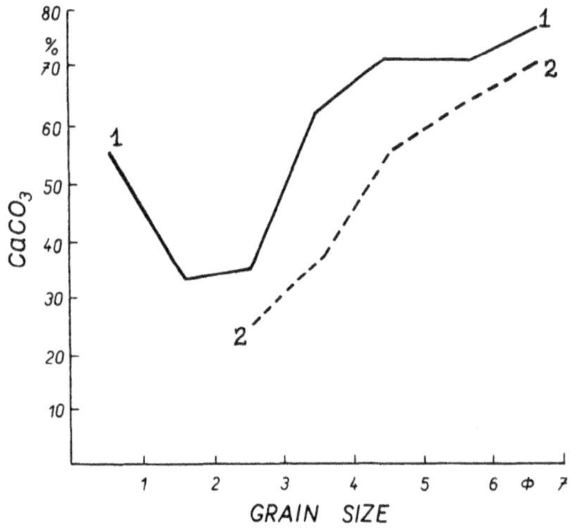

Fig. 7. Relationship between carbonate content and grain size in the total sediment deposited:
1) during the dust storm on April 8, 1968.
2) during the dust storm on April 20, 1969

DISCUSSION

The role of aeolian deposition has not been fully evaluated in spite of thorough investigation of the Holocene sediments in the Persian Gulf (see this Symposium). This paper does not give a full account of aeolian sedimentation but simply draws attention to this problem, stressing the probable importance of aeolian processes in the Persian Gulf. Three main points tend to confirm this assumption:

- the striking similarity between the carbonate content and grain size inter-relationship in Recent marine sediments in the N parts of the Persian Gulf and those of fallout from dust storms.
- the significant amount of pseudosand and gypsum grains encountered both in Persian Gulf sediments and in dust storm sediments. Their source includes the entire area of flood plain soils of the Mesopotamian Plain.
- an incredibly high rate of deposition from dust storms (2.1 cm/year). At least part of this sediment must be deposited in the northern areas of the Persian Gulf, thus contributing a considerable aeolian admixture to the total sediment mass.

A great deal of material relating to intensive aeolian sedimentation over the countries surrounding Persian Gulf has been published recently (Buringh, 1960, D.H. Yaalon & Ginzburg, 1966, Smalley & Vita-Finzi, 1968, Kukal & Saadallah, 1970).

Future research:

Further research on the aeolian sedimentation in the Persian Gulf is urgently needed. This could be carried out by direct measurement of fallout on vessels, by quantitative determination of suspension in the water column or by comparison with Recent marine sediment composition. In particular, the inter-relationship between grain size and carbonate content might offer a clue to the recognition of the aeolian fraction.

Holocene Sediment Types and Their Distribution in the Southern Persian Gulf

C. W. Wagner[1] and C. van der Togt[1]

ABSTRACT

A simplified classification of the Holocene sediments based on textures and grain type results in fourteen major units, twelve of which are essentially carbonate in composition. A brief description and photographic illustration of these units, together with the sedimentary and diagenetic processes which have contributed to their formation, is designed to give the reader a broad but valid impression of Persian Gulf sediments.

The distribution of the fourteen sediment units throughout the Arabian parts of the basin, although complicated by numerous local bathymetric highs and depressions, is relatively simple. Because the Arabian sea floor slopes progressively from a windward shoreline to the basin center there is increasing protection from wave action towards the center of the basin. As a result sediments grade from skeletal, oolitic and pelletoidal sands (and muds in coastal lagoons) and fringing reefs, through an irregular zone of compound grain sands, into widespread skeletal muddy sands, and finally into basin center muds. These simple relationships vary laterally around the Arabian side of the gulf. Lateral variation is dependant upon orientation of the regional slope with respect to the prevailing NW wind-driven waves, angle of slope, and presence or absence of regional, structurally based barriers.

I. INTRODUCTION

A. General

On first examination, samples of the surface sediments of the Southern Persian Gulf impress one with the monotony of their overall composition in that they are very largely made up of skeletal debris. However, a second more detailed look reveals an almost infinite variety within these skeletal sediments.

Several factors are involved in any attempt to subdivide such sediments: the number and nature of the samples taken (cores, grabs), their spacing, and the size of the area investigated. Except for several areas of rather close sampling (e.g. on, and around, local highs including Bu Tini or Halat Dalma), the majority of the lines extend from the Arabian coast towards the central part of the Gulf.

Although a considerable number of cores, and dredge samples were collected (some 550), the majority of the samples were taken with a 'Van Veen' grab (some 3200, including some 550 utilized by Houbolt, 1957). All of these were examined with

[1] Shell Research B.V., Rijswijk, The Netherlands.

binocular microscope to ensure that a reasonable measure of the true texture was obtained (no resorting in the sample bags, etc.).

Full sieve and pipette analysis for grain size was carried out on relatively few samples (some 300), but (in keeping with our prime interest in the interpretation of ancient carbonates) a somewhat larger number (some 500) were impregnated with plastic, and polished slabs and/or large thin sections prepared. Analyses for total insoluble residues (some 600) and carbonate mineralogy (some 70) were carried out on sufficient samples mainly to characterise the defined sediment types.

B. The method of subdivision

In describing the sedimentary fabric of ancient carbonates, it is standard practice within the Shell Group to use the Dunham classification system (see Dunham, 1962). This descriptive system identifies depositional texture on the basis of grain or mud support (with 'grains' being larger than, and mud particles smaller than 20μ) and then identifies the fabric by reference to the dominant particle(s), producing compound terms such as 'crinoid grainstone' , 'ooidal packstone' , 'radiolarian wackestone' , and 'coccolith mudstone' .

In our first inspection of the Persian Gulf sediments, an attempt was made to utilise the Dunham classes and thus preserve direct compatibility, but two factors prevented this. Firstly it became clear that the 20μ grain-size separation used by Dunham was neither practical nor meaningful in the Persian Gulf sediments, as the silt fraction (63 - 20μ) was essentially inseparable from the mud fraction (sensu stricto < 20μ) in sediment distribution. Secondly, in most of the muddy sediment samples obtained with a grab it is not always easy to distinguish grain from mud support, i.e. to distinguish a packstone from a wackestone. For these reasons, the Dunham terms were not used for the Persian Gulf sediments, being replaced by a simple threefold textural subdivision:

'sands'	: sediments with < 10% of particles < 63μ
'muddy sands'	: sediments with 10-50% of particles < 63μ
'muds'	: sediments with > 50% of particles < 63μ

These textural terms were then qualified by the dominant grain type(s) to produce compound terms. Twelve distinctive types of carbonate sediment were recognised and labelled in this way. Two non-carbonate sediment types that have a distinctive distribution were also defined. The bound coral-algal masses forming the small areas of reef cannot be termed sediments and are not treated here but constitute distinctive carbonate rock types that would fall under Dunham's term 'boundstone'.

II. THE MAIN TYPES OF SEDIMENT IN THE SOUTHERN PERSIAN GULF

A. Grouping of the sediment types

Fourteen types of sediment have been defined and their distribution illustrated by means of profiles and a map appended at the back of this volume. The sediment types are discussed in the following order:

Carbonates

Sands

Type no.	1	Lamellibranch sand
"	2	Compound grain/lamellibranch sand
"	3	Large perforate-foraminiferal sand
"	4	Coral/algal sand
"	5	Imperforate foraminferal/pelletoidal sand
"	6	Gastropod sand
"	7	Ooidal sand (oolite)

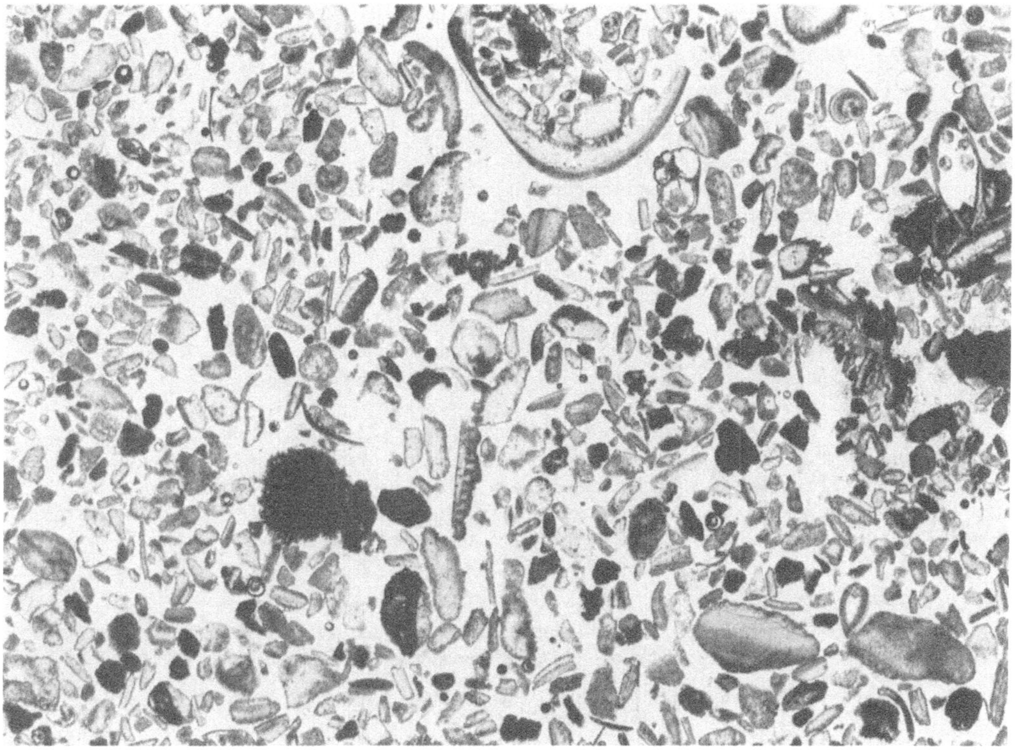

Fig. 1. Lamellibranch sand - Thin section of sample T 666, 10x

Note: Micritization of skeletal grains; some grains exhibit only a micritized periphery, others have been nearly completely micritized, such that the original skeltal structure becomes obscured.

The composition is very monotonous, the greater part of the grains being lamellibranch fragments.

Bulk carbonate composition:

Aragonite	85%
Mg calcite	10%
Calcite	5%

Sorting: moderate

Foraminiferal number: 200

Grain size:

> 2 mm	7%
2 - 0.063 mm	90%
< 0.063 mm	3%

Bulk insoluble residue: 0.5%

Foraminiferal composition:

Calcareous imperforate	15%
Calcareous perforate	15%
Arenaceous	70%

Muddy sands

Type no. 8 Lamellibranch muddy sand
" 9 Gastropod muddy sand

Muds

Type no. 10 Lamellibranch mud (< 10% insol.)
" 11 Argillaceous lamellibranch mud (> 10% insol.)
" 12 Imperforate foraminferal/gastropod mud

Non-Carbonates

Type no. 13 Quartz sand (including muddy quartz sand)
" 14 Sedimentary gypsum (precipitated in coastal lagoons)

Twelve of the fourteen sediment types are illustrated on the following pages by one or two thin section photomicrographs of the plastic impregnated sediment. For each illustrated sample the following parameters are also given.
N.B. These parameters are those of the individual samples and are not to be taken as mean or average values for all samples of that sediment type.

(a) Bulk carbonate composition: determined by means of an X-ray spectrometer (the results of all such analyses are collected in table I).

(b) Sorting: estimated visually as either 'good' , 'moderate' or 'poor' .

(c) Foraminiferal number: the number of foraminiferal tests (larger than 75μ) per gram of dry sediment.
(see Hughes Clarke & Keij, this volume)

(d) Grain size: determined by sieve analysis.

(e) Bulk insoluble residue: residue after digestion in 2 Normal hot HCL.

(f) Foraminiferal composition: the relative proportions (in the fraction > 75μ) of the three major foraminiferal groups 'calcareous imperforate', 'calcareous perforate' and 'arenaceous' (see Hughes Clarke & Keij, this volume).

B. Description of the sediment types

SANDS

Type no. 1 Lamellibranch sand (Fig. 1)

(a) Sedimentological description

The lamellibranch sand consists of rounded to angular, moderately sorted, lamellibranch fragments of medium sand size. Compound grains form a minor constituent. Mud matrix average 4% of the sediment. Bulk insoluble residue averages 1%, and does not exceed 5%.

(b) Faunal/floral elements

Most skeletal grains are lamellibranch shell fragments, small whole shells generally being present. Recognisable lamellibranchs belong mainly to the species *Ervillea purpurea*. Coralline algal debris is often associated with the lamellibranch fragments, but is subordinate. Foraminifera, echinoid fragments and small gastropods form minor constituents.

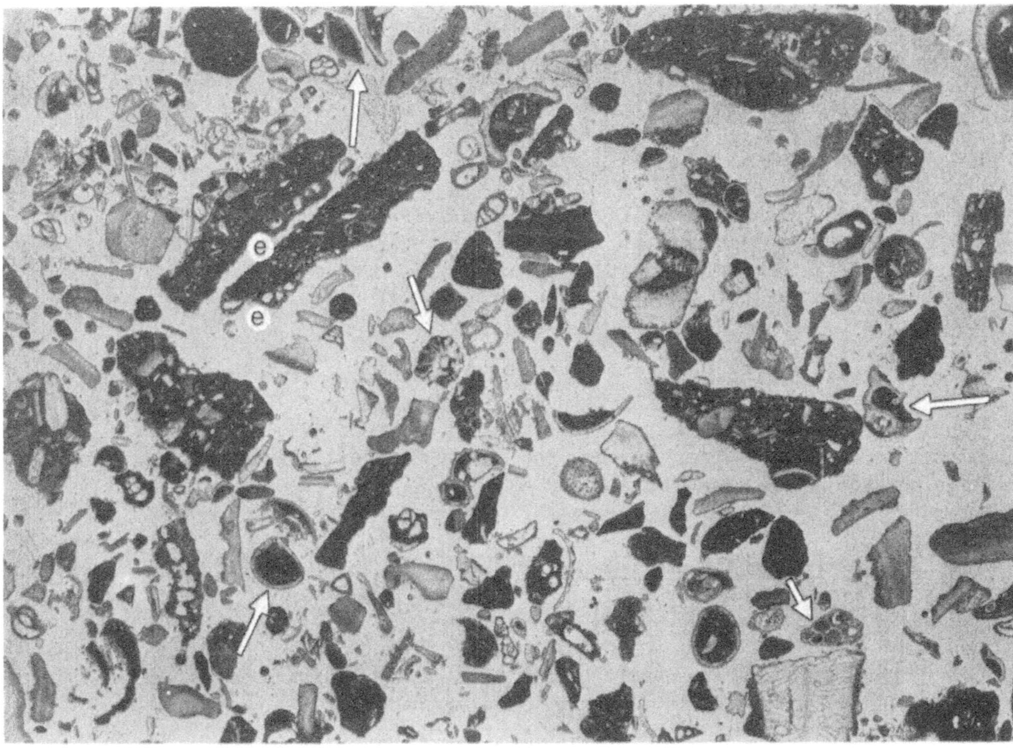

Fig. 2. Compound grain/lamellibranch sand - Thin section of sample T 101, 10x

Note: Faint micritized periphery on nearly all skeletal grains.

Infill of cryptocrystalline carbonate in interior of mollusks (arrows)

Greater variety in skeletal material than in the Lamellibranch sand (Type 1)

Presence of encrusting Foraminifera on certain compound grains (e)

Bulk carbonate composition:

Aragonite	60%
Mg calcite	30%
Calcite	10%

Grain size:

> 2 mm	11%
2-0.063 mm	81%
< 0.063 mm	8%

Sorting: moderate

Foraminiferal number: 360

Bulk insoluble residue: 0.5%

Foraminiferal composition:

Calcareous imperforate	47%
Calcareous perforate	1%
Arenaceous	52%

(c) Synsedimentary diagenetic features

Most skeletal grains have been algally bored, but still retain traces of their primary structure. Some articulated valves are filled with fibrous aragonite crystals. However, this infilling is more characteristic of the compound grain/lamellibranch sands (type 2). Compound grains are relatively rare.

(d) Relation to environment

The recognisable lamellibranchs are infaunal (Hughes Clarke & Keij, this volume) and may be regarded as indigenous or semi-indigenous elements. The minor coralline algal debris is apparently exotic, as these algae require a hard substratum and are derived from adjacent hard-rock areas, including local offshore highs.

C^{14} bulk dating indicates that the lamellibranch sands are not older than 2120 $\pm$ 150 years, and thus represent Holocene accumulation. The lamellibranch sands occur in shallow marine, nonrestricted, high-energy environments. The distribution also seems to depend upon the submarine topography, i.e. the relative absence of hard-rock substratum.

Type no. 2 Compound grain/lamellibranch sand (Fig. 2)

(a) Sedimentological description

The compound grain/lamellibranch sand consists of angular skeletal fragments and angular to rounded compound grains, the sediment being coarse and moderately sorted.

Mud matrix averages 4% of the sediment, while bulk insoluble residue averages 1% and does not exceed 3%.

(b) Faunal/floral elements

Skeletal grains are mainly lamellibranchs, both fragments and whole shells. Small gastropods, larger Foraminifera (mainly *Operculina*), smaller Foraminifera and echinoid debris also occur.

(c) Synsedimentary diagenetic features

Nearly all skeletal grains exhibit a thin, algally bored periphery. Many shells are infilled with very fine, cryptocrystalline carbonate or fibrous aragonite crystals. Compound grains consist of lithified skeletal material similar to the loose sediment and therefore possibly lithified in situ. They may, however, also be pieces broken from larger scale lithified units and transported into the area.

Most particles have a reddish-brown colouration, which is very typical for this sediment type.

(d) Relation to environment

C^{14} bulk dating* of the reddish-brown components of the sediment show that their age ranges from 2460 $\pm$ 140 to 3020 $\pm$ 160 years, indicating that the sediment is the product of present sea level. These relatively old dates suggest, however, that sedimentation is slow relative to that of other sediment types. The compound grain/lamellibranch sands occur in shallow-marine, nonrestricted, moderate-energy

* See Fig. 18 and Table I for available C^{14} datings.

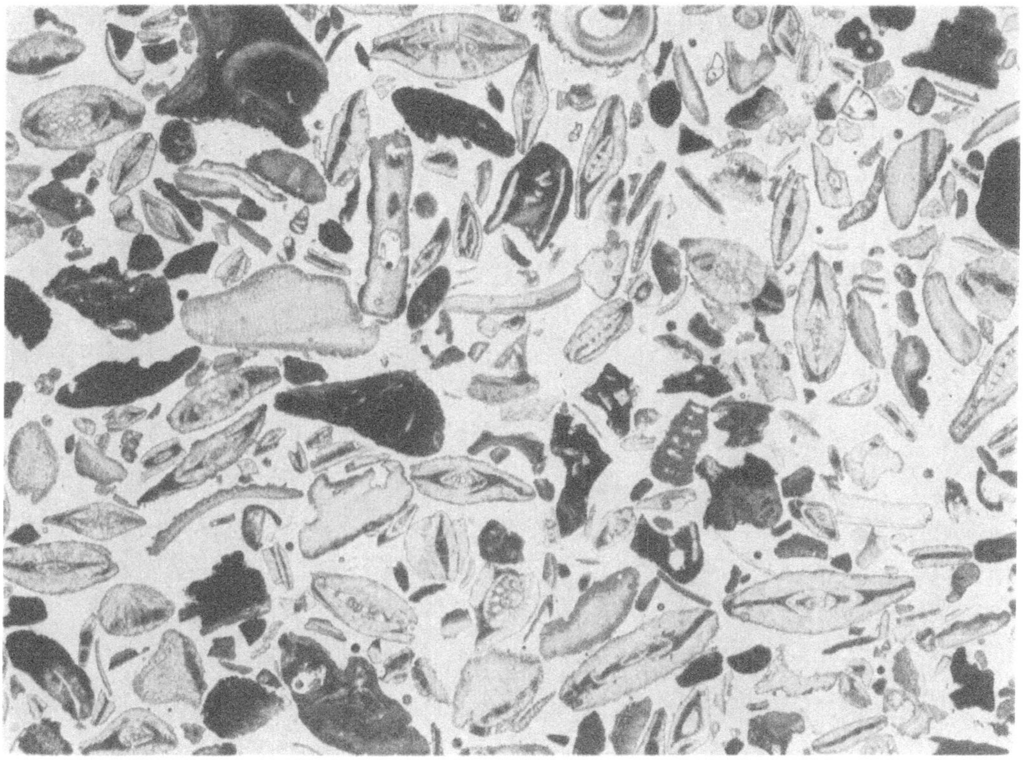

Fig. 3. Large perforate foraminiferal sand - Thin section of sample T 1139, 10x

Note: Strong degree of micritization of some particles in contrast to faint micrite periphery of most skeletal grains.

Monotony in skeletal composition (majority of particles represent whole tests of perforate large Foraminifera).

Cryptocrystalline material in foraminiferal chambers.

Bulk carbonate composition:

Aragonite	10%
Mg calcite	90%

Grain size:

> 2 mm	3%
2-0.063 mm	95%
< 0.063 mm	2%

Sorting: moderate

Bulk insoluble residue: 1.2%

Foraminiferal composition:

Calcareous imperforate	13%
Calcareous perforate	72%
Arenaceous	15%

Foraminiferal number: 400

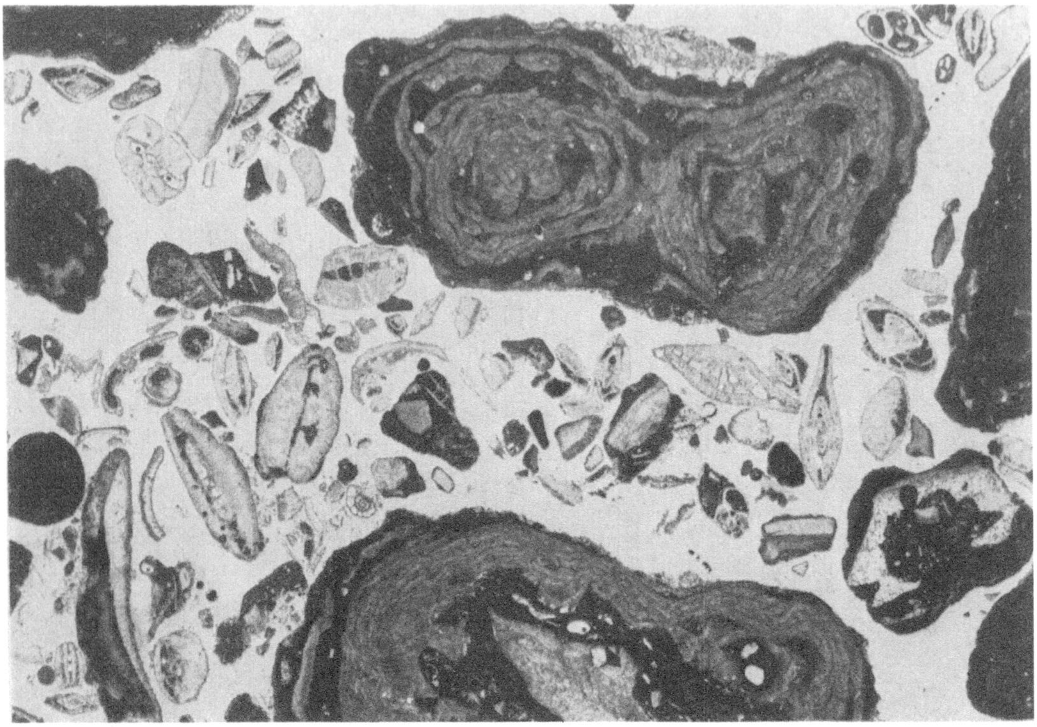

Fig. 4. Large perforate foraminiferal sand - Thin section of sample T 66, 10x

Note: Large oncoids consisting of various types of melobesioid algae which have been bored by sponges and are sometimes encrusted with various Foraminifera, worms, etc.

Cryptocrystalline carbonate material in foraminiferal chambers.

Relatively large number of compound grains.

Bulk carbonate composition:

Aragonite	30%
Mg calcite	65%
Calcite	5%

Foraminiferal number: 600

Grain size:

> 2 mm	35%
2-0.063 mm	61%
< 0.063 mm	4%

Bulk insoluble residue: 1.1%

Foraminiferal composition:

Calcareous imperforate	18%
Calcareous perforate	61%
Arenaceous	21%

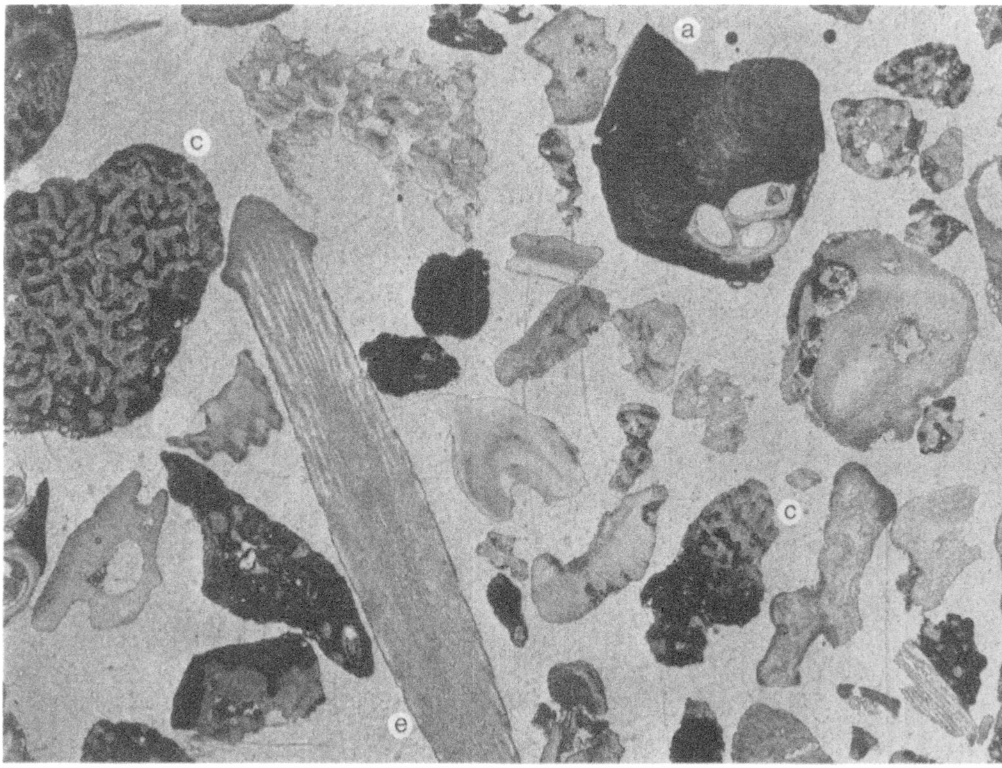

Fig. 5. Coral/algal sand - Thin section of sample T 284, 10x

Note: Skeletal elements for the greater part of epifaunal/floral origin (c = coral, a = coralline algae, e = epifaunal echinoid spine)

Presence of coral debris indicating relatively limited sediment transport (as it is easily broken down into mud-sized material).

Bulk carbonate composition:

Aragonite	65%
Mg calcite	30%
Calcite	5%

Sorting: poor

Foraminiferal number: 40

Grain size:

> 2 mm	19%
2-0.063 mm	79%
< 0.063 mm	2%

Bulk insoluble residue: 0.4%

Foraminiferal composition:

Calcareous imperforate	17%
Calcareous perforate	78%
Arenaceous	5%

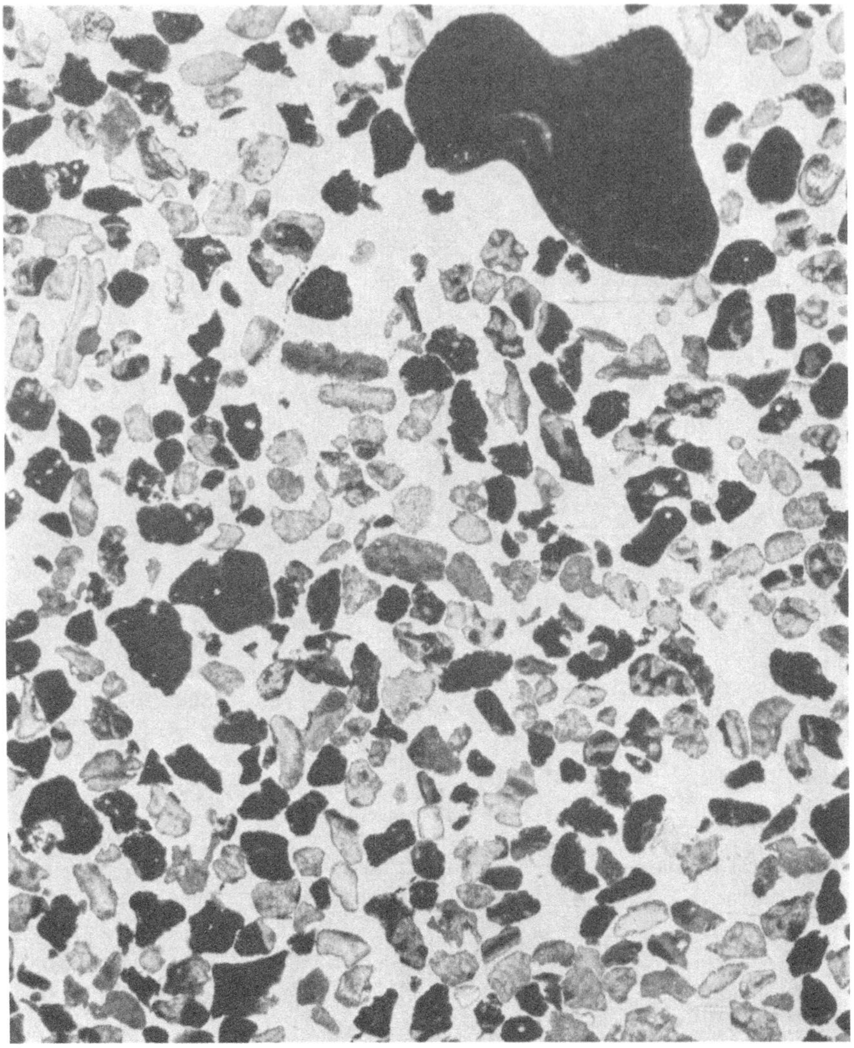

Fig. 6. Coral/algal sand - Thin section of sample T 741, 10x

Note: Coralline algal fragments bored by sponges. Only faint development of micrite peripheries on grains.

Bulk carbonate composition:

Aragonite	50%
Mg aragonite	50%

Sorting: moderate

Foraminiferal number: 30

Grain size:

> 2 mm	11%
2-0.063 mm	86%
< 0.063 mm	3%

Bulk insoluble residue: 0.2%

Foraminiferal composition:

Calcareous imperforate	90%
Calcareous perforate	10%

environments; energy being somewhat lower than that in the lamellibranch-sand environment (skeletal particles show less abrasional rounding), but still high enough to prevent the deposition of carbonate mud. The low rate of sedimentation, together with relatively little movement of the sediment, seems to favour incipient synsedimentary lithification (Shinn, 1969) and may have led to the high content of compound grains.

Type no. 3 Large perforate-foraminiferal sand (Figs. 3 + 4)

(a) Sedimentological description

This type of sediment consists mainly of whole tests of perforate large Foraminifera together with varying amounts of skeletal (red algal) oncoids. Because locally produced whole tests make up the majority of the particles, roundness, average grain size and sorting have little sedimentological meaning.

Mud matrix averages 1% and does not exceed 5%. Bulk insoluble residue averages 1% and does not exceed 5%.

(b) Faunal/floral elements

The perforate large Foraminifera are *Heterostegina* and/or *Amphistegina*. The oncoids are formed by melobesioid algae (several genera), which frequently coat mollusk shells or fragments.
Lamellibranch fragments, small gastropods and smaller Foraminifera form a minor constituent of the sediment.

(c) Synsedimentary diagenetic features

Grains show a varying degree of algal boring, but in general only a thin bored periphery is developed. Large foraminiferal chambers are often filled with cryptocrystalline carbonate cement and compound grains sometimes occur.

(d) Relation to environment

These large foraminiferal sands are concentrated on the tops of unrestricted offshore highs at depths of 10-20 fathoms, where they seem to live in association with coralline algae.

Type no. 4 Coral/algal sand (Figs. 5 + 6)

(a) Sedimentological description

The coral/algal sands consist mainly of rounded skeletal fragments, which are moderately to badly sorted and generally of coarse sand-size. The mud matrix averages 3% and is always less than 6%. Bulk insoluble residue averages 1% and does not exceed 6%.

(b) Faunal/floral elements

The dominant skeletal particles are fragments of corals and coralline algae. Mollusk and echinoid fragments and Foraminifera also occur. Most fragments belong to epifaunal/flora groups.

(c) Synsedimentary diagenetic features

Particles in general do not show algally bored peripheries. Compound grains are rare.

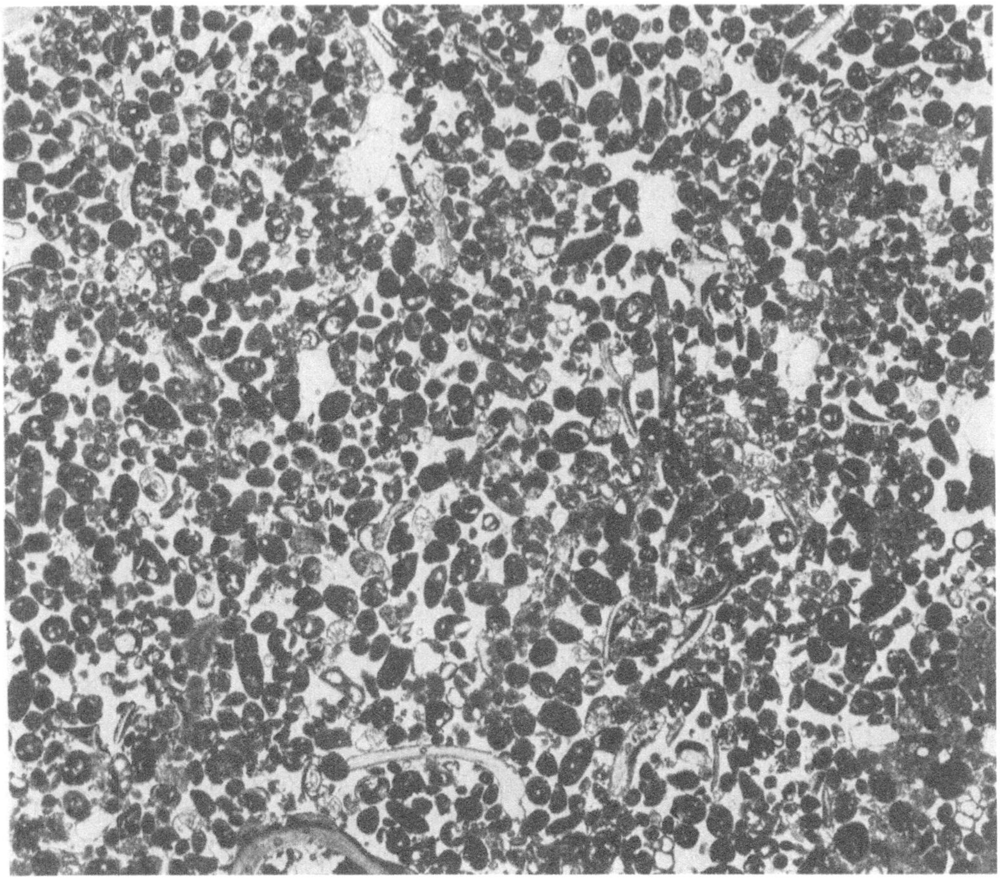

Fig. 7. Imperforate foraminiferal/pelletoid sand - Thin section of sample D 756, 12.5x

Bulk carbonate composition

Aragonite	50%
Mg calcite	30%
Calcite	20%

Sorting: moderate to good

Foraminiferal number: 1200

Grain size:

> 2 mm	1%
2-0.063 mm	92%
< 0.063 mm	7%

Bulk insoluble residue: 8.5%

Foraminiferal composition:

Calcareous imperforate	70%
Calcareous perforate	29%
Arenaceous	1%

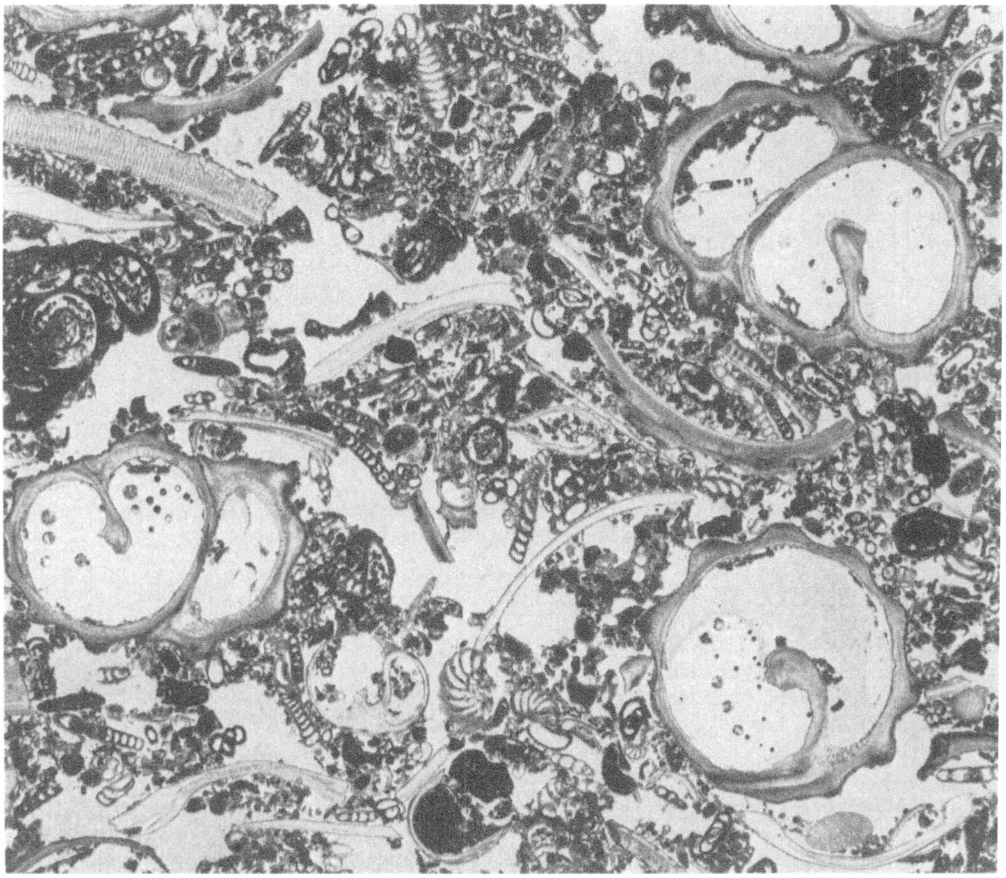

Fig. 8. Gastropod sand - Thin section of sample Q 27 (10), 10x

Note: Abundance of imperforate Foraminifera

Only faint development of micrite peripheries on certain grains.

Bulk carbonate composition:

Aragonite:	30%
Mg calcite:	70%

Grain size:

> 2 mm	11%
2-0.063 mm	81%
< 0.063 mm	8%

Sorting: poor

Bulk insoluble residue: 2.7%

Foraminiferal number: 5720

Foraminiferal composition:

Calcareous imperforate	87%
Calcareous perforate	12%
Arenaceous	1%

(d) Relation to environment

The coral/algal sands are detritus from organic reefs developed on areas of hard rock bottom, e.g. on the windward sides of islands. The sands accumulate around these source areas, particularly down wind as "sand-tails" or bars (see Purser, this volume). They may also form coastal spits where the reefs fringe the Arabian coastline (see Shinn, in this volume).

The coral/algal sands occur in relatively small patches. Deposition is in unrestricted, high-water-energy conditions where rates of sedimentation are relatively high, which probably prevents attack by boring algae.

Type no. 5 Imperforate foraminiferal/pelletoidal sand (Fig. 7)

(a) Sedimentological description

This sediment is extremely variable in character. Its most constant components are imperforate Foraminifera and algally bored skeletal grains or pelletoids.
Faecal pellets are often dominant and compound grains are also common. Traces of insoluble residue (generally quartz) are often present.

(b) Faunal/floral elements

These are dominated by miliolid and peneroplid Foraminifera with variable amounts of small perforate Foraminifera.
High-spired cerithiid gastropods are often abundant, as are diverse lamellibranchs, dasyclad algae and ostracods. Some grains exhibit algally bored peripheries.

(c) Synsedimentary diagenetic features

Gastropods and articulated lamellibranch shells are very often filled with fibrous aragonite and cryptocrystalline cements.

(d) Relation to environment

These carbonate sands characterise, but are not limited to, restricted environments. They nearly always occur in moderate to low energy settings in water depths of less than 5 m. They are typical of wide tidal flats. Virtually all grains are derived locally, often from the shallow lagoonal environments adjacent to the tidal flats.

Type no. 6 Gastropod sand (Fig. 8)

(a) Sedimentological description

These sediments consist of poorly sorted, medium to coarse particles composed mainly of high-spired gastropods, and coarsely broken lamellibranchs in subequal amounts. Accessory particles include pelletoids, large compound grains, and imperforate Foraminifera. Traces of insoluble residue (mainly quartz) are often present.

(b) Faunal/floral elements

A low-diversity fauna of gastropods and lamellibranchs (the former being more abundant) are characteristic of these sediments. Imperforate Foraminifera (peneroplids and miliolids) are often abundant.

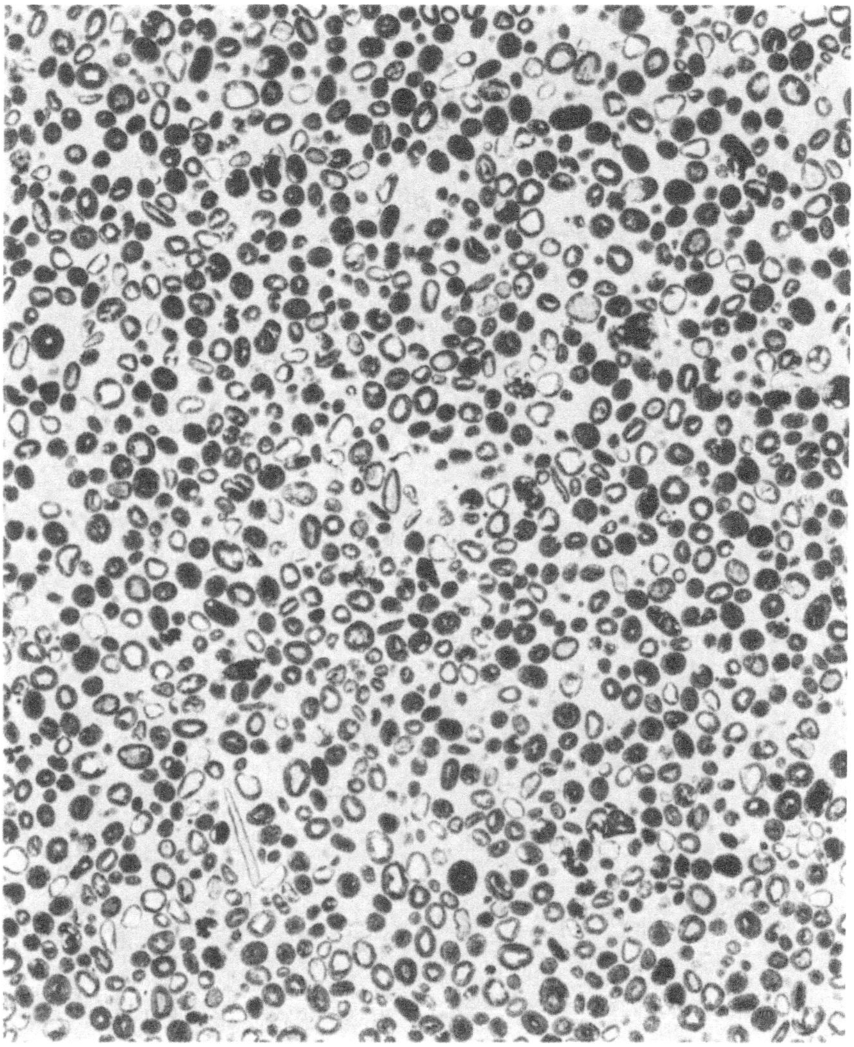

Fig. 9. Ooidal sand (oolite) - Thin section of sample D 843, 14x

Note: Good sorting and aragonitic coating of virtually all particles

Bulk carbonate composition:

Aragonite	90%
Mg calcite	10%

Grain size:

> 2 mm	-
2.0.063 mm	97%
< 0.063 mm	3%

Sorting: good

Foraminiferal number: 16

Bulk insoluble residue: 15.3%

Foraminiferal composition:

Calcareous imperforate	52%
Calcareous perforate	48%

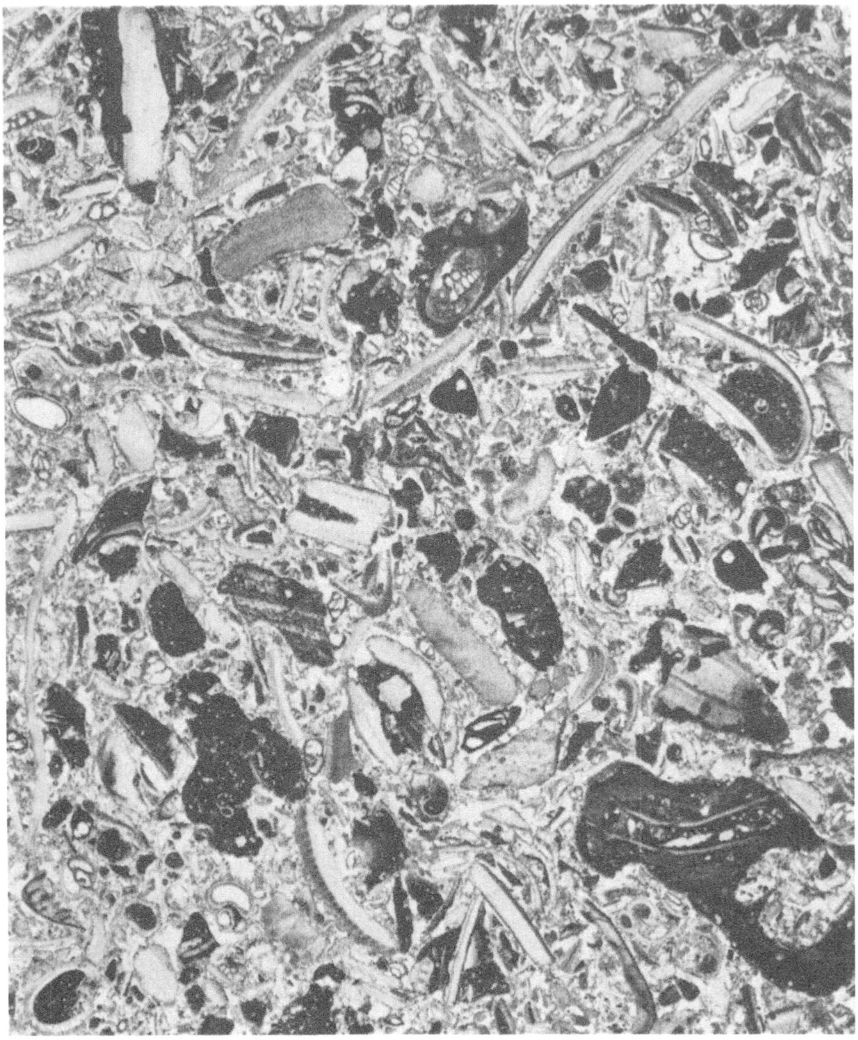

Fig. 10. Lamellibranch muddy sand - Thin section of sample T 19, 10x

Note: "Black" compound particles and certain Foraminifera and other skeletal remains. Virtual absence of micritized peripheries on skeletal grains

Bulk carbonate composition:

Aragonite	55%
Mg calcite	35%
Calcite	10%

Grain size:

> 2 mm	11%
2-0.063 mm	75%
< 0.063 mm	14%

Sorting: poor

Bulk insoluble residue: 1.3%

Foraminiferal number: 800

Foraminiferal composition:

Calcareous imperforate	55%
Calcareous perforate	15%
Arenaceous	30%

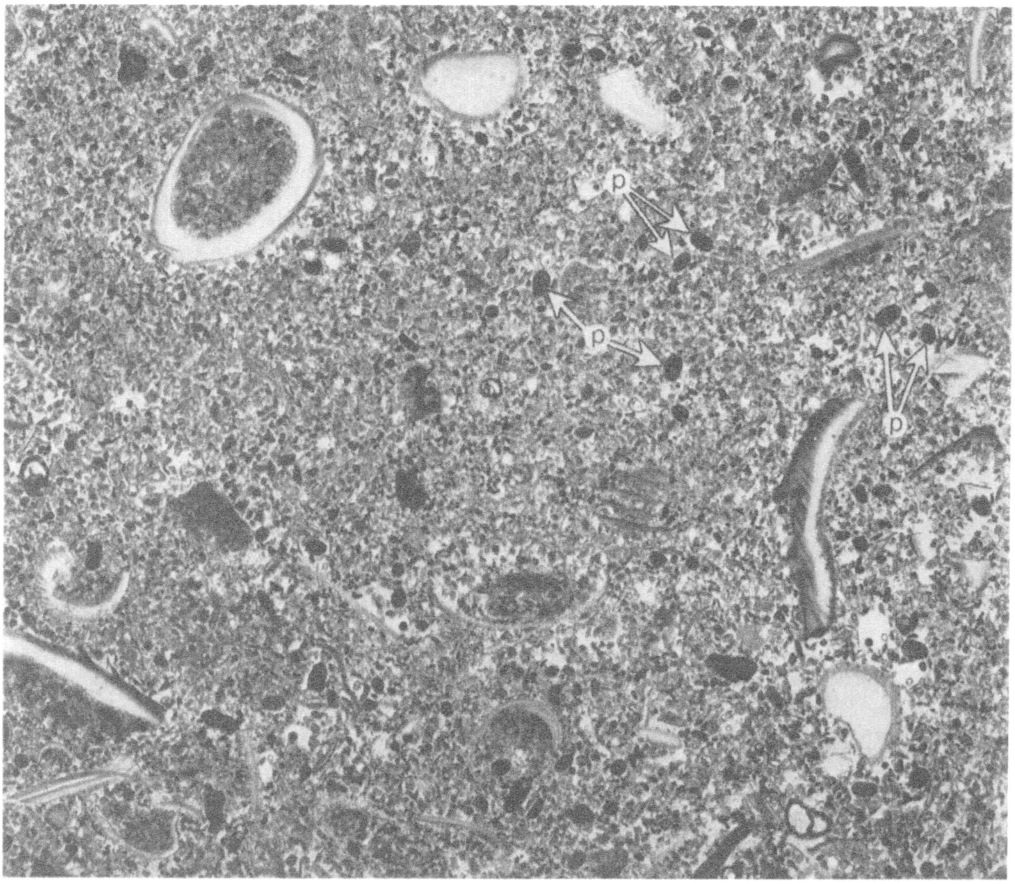

Fig. 11. Lamellibranch muddy sand - Thin section of sample T 1673, 10x

Note: Average grain size much smaller than in sample T 19 (Fig. 10)
Soft faecal pellets (p) and blackened particles.

Bulk carbonate composition:

Aragonite	55%
Mg calcite	30%
Calcite	10%
Dolomite	5%

Grain size:

> 2 mm	2%
2-0.0.63 mm	63%
< 0.063 mm	35%

Sorting: poor

Bulk insoluble residue: 3%

Foraminiferal number: 10730

Foraminiferal composition:

Calcareous imperforate	52%
Calcareous perforate	26%
Arenaceous	22%

(c) Synsedimentary diagenetic features

These include a high degree of algal micritization of many grains, and the formation of sand-sized compound grains by the precipitation of fibrous or cryptocrystalline cements.

(d) Relation to environment

These sediments are confined to depths generally less than 5 m and characterise, but are not limited to, restricted environments. As such, these sediments occur in protected coastal embayments and on the lee sides of coastal barrier complexes. The abundance of gastropods and peneroplid (large imperforate) Foraminifera is probably due to the presence of widespread rock substrates supporting abundant brown algae.

Type no. 7 Ooidal sand (oolite) (Fig. 9)

(a) Sedimentological description

These sediments consist mainly of ooids, but molluskan fragments can occur in minor amounts. The ooidal grains often have quartz nuclei.

(b) Faunal/floral elements

Scattered lamellibranchs and gastropod shell fragments are often present; most are not indigenous to the oolite-forming environment.

(c) Synsedimentary diagenetic features

The rare gastropods and articulated lamellibranchs are often infilled with fibrous or cryptocrystalline cements.

(d) Relation to environment

The oolitic sediments occur in areas of active tidal currents. These currents are often reversible in direction and maintain the grains within the area in which frequent movement can occur. The environments in which the oolites are found are always within a few kilometres of the continental shoreline, in water depths of less than 3 m and with salinities only slightly above normal. Oolites are apparently readily transported and most tend to accumulate on beaches or as aeolian dunes. This is especially common along the Trucial Coast. Scattered ooliths can occur at all depths within the Gulf, but C^{14} dating shows that all deeper occurrences are relict, having been formed earlier during the Holocene transgression. They have subsequently been intermixed with contemporary, deep-marine sediments, probably by bioturbation. The origin and distribution of oolitic sediments is discussed by Loreau and Purser, elsewhere in this volume.

MUDDY SANDS

Type no. 8 Lamellibranch muddy sand (Figs. 10 and 11)

(a) Sedimentological description

The lamellibranch muddy sands consist of angular skeletal fragments and angular to rounded black compound grains in a muddy matrix. This sediment type has a wide spectrum of grain size and variation in constituent composition. Foraminifera and other skeletal fragments often show black infills or staining; the origin of the blackening is not fully understood (see Houbolt, 1957 and Sugden, 1966).

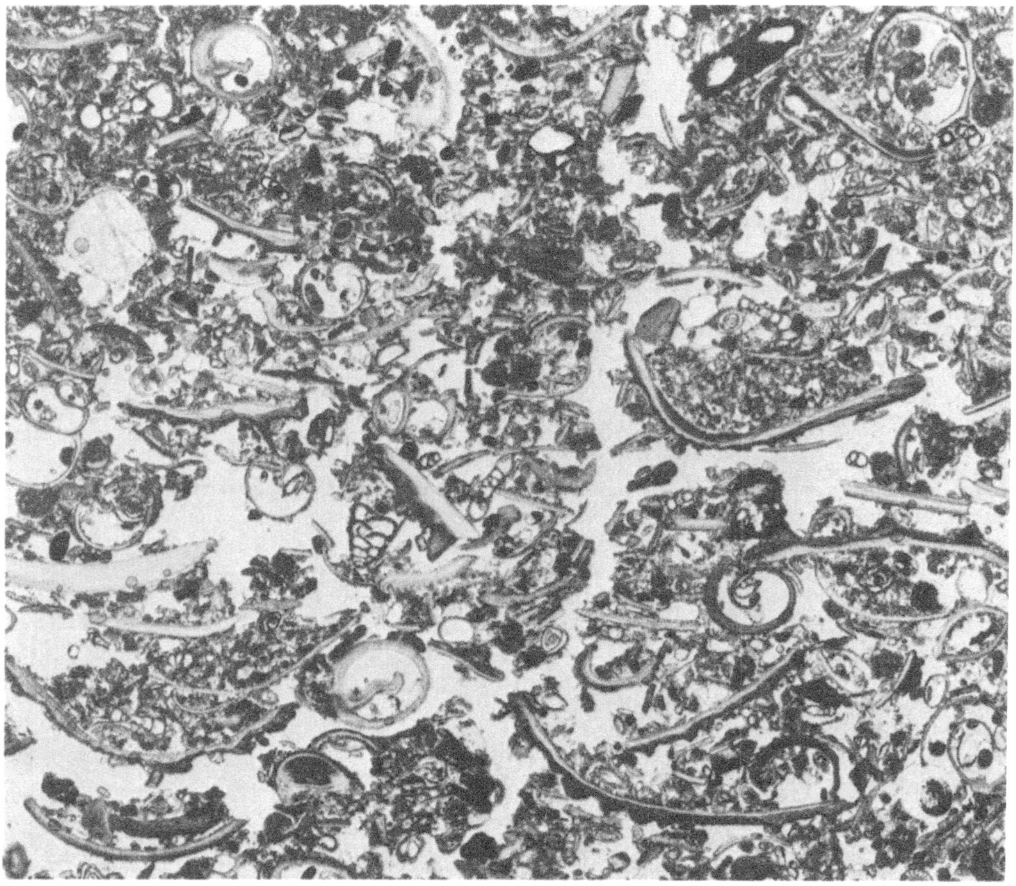

Fig. 12. Gastropod muddy sand - Thin section of sample T 1391, 14x

Bulk carbonate composition:

Aragonite	30%
Mg calcite	60%
Calcite	10%

Sorting: poor

Foraminiferal number: 3190

Grain size:

> 2 mm	4%
2-0.063 mm	54%
< 0.063 mm	42%

Bulk insoluble residue: 18.3%

Foraminiferal composition:

Calcareous imperforate	84%
Calcareous perforate	16%

The matrix averages 25% of the sediment. While bulk insoluble residue averages 3% and, in general, does not exceed 4%, some samples show an exceptionally high bulk insoluble residue (T 111 = 14.5%, T 128 = 25.5%, T 938 = 12.6%).

(b) Faunal/floral elements

Skeletal grains are mainly fragments of infaunal lamellibranchs (commonly the species Timoclea layardi), smaller Foraminifera, and echinoid fragments; the last occur particularly in the medium-sized muddy sands. In the finer sediments, soft faecal pellets may occur in varying quantities.

(c) Synsedimentary diagenetic features

Skeletal fragments in general do not show algally micritized peripheries.

(d) Relation to environment

C^{14} dating of the black particles in the sediment indicates that their age ranges from 3440 ± 170 to 4840 ± 170 years.
In any one sample they are invariably older than the non-blackened grains and may, in part, represent relict material.

The muddy lamellibranch sands occur in shallow marine, unrestricted, low-energy environments ranging in depth from 10 - 30 fathoms. They constitute widespread sheets between the highs on the Arabian homocline* and grade laterally into muds in some of the deeper depressions.

Type no. 9 Gastropod muddy sand (Fig. 12)

(a) Sedimentological description

These sediments consist of poorly sorted, medium to coarse sand and gravel-sized particles, comprising mainly gastropods, relatively unbroken lamellibranchs, imperforate Foraminifera, large compound grains (or lumps), and soft faecal pellets. Variable insoluble residues, mainly quartz, depend on proximity to aeolian sources.

(b) Faunal/floral elements

These are essentially similar to those occurring within the gastropod sands.

(c) Synsedimentary diagenetic features

Particles are generally less "micritized" than the gastropod sands; unbroken lamellibranchs may have a fresh appearance. Incipient submarine lithification is rare, but in the Gulf of Salwa it can result in the widespread formation of irregularly shaped, gravel-sized lumps of lightly cemented rock.

(d) Relation to environment

Gastropod muddy sands favour lower energy, often restricted conditions and occur in very protected embayments, exemplified by the Gulf of Salwa, at depths down to 10 m. Virtually all particles are indigenous to this environment.

---

* This term is defined and discussed in the introductory article in this volume by Purser and Seibold.

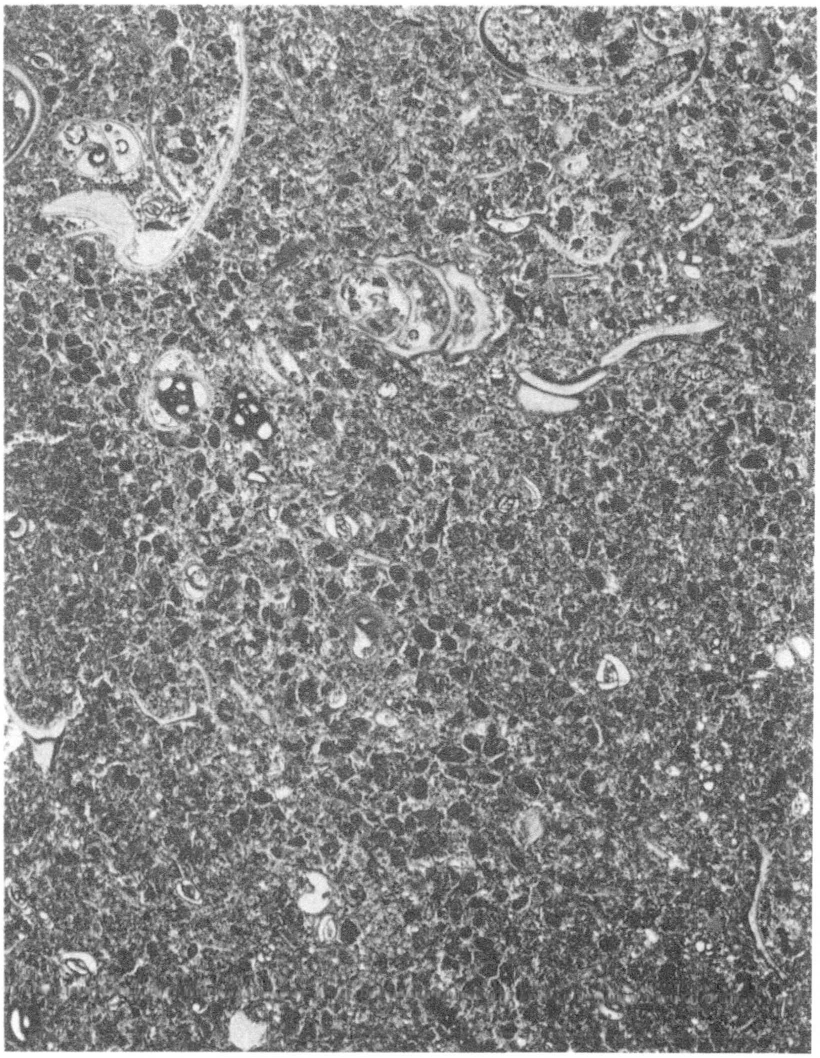

Fig. 13. Lamellibranch mud (ins. res. < 10%) - Thin section of sample T 805, 10x

Note: Abundant soft faecal pellets produced mainly by infaunal sediment feeders.

Blackening of some particles.

Bulk carbonate composition:

Aragonite	55%
Mg calcite	30%
Calcite	10%
Dolomite	5%

Sorting: moderate to poor

Foraminiferal number: 2500

Grain size:

> 2 m	2%
2-0.063 mm	25%
< 0.063 mm	73%

Bulk insoluble residue: 9%

Foraminiferal composition:

Calcareous imperforate	84%
Calcareous perforate	15%
Arenaceous	1%

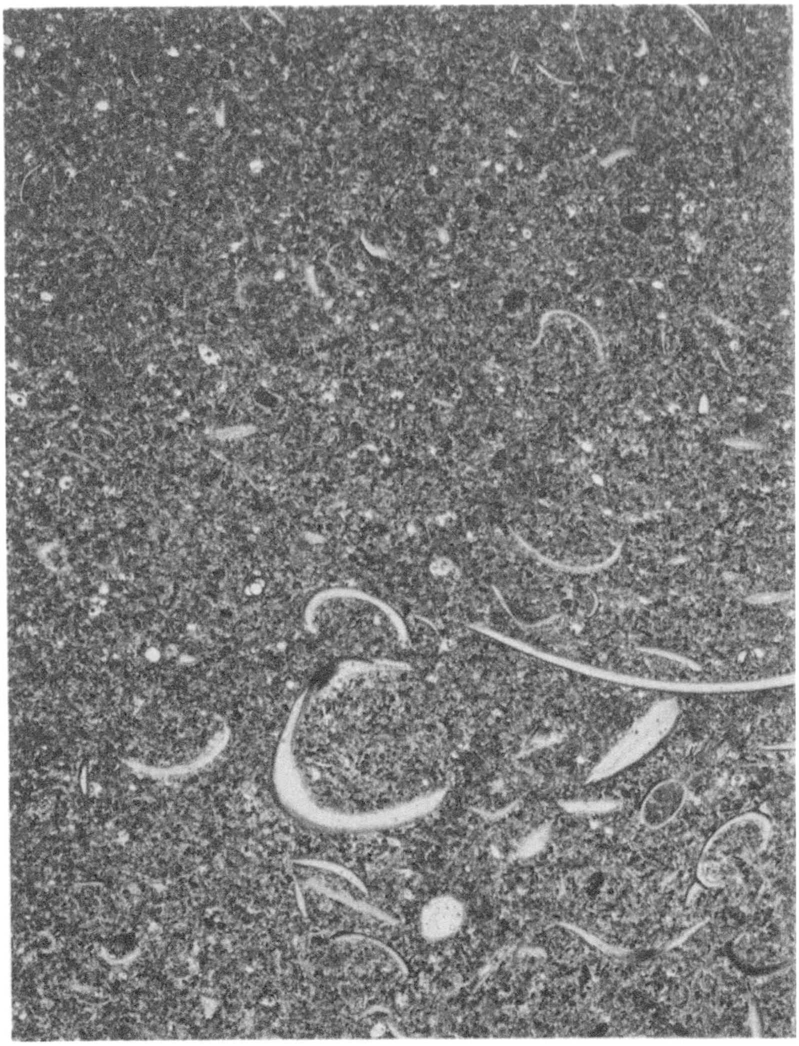

Fig. 14. Lamellibranch mud (ins. res. < 10%) - Thin section of sample T 22, 10x

Note: Overall resemblance with lamellibranch muddy sand shown in Fig. 11 (the sample is classified as a "mud" because matrix comprises more than 50%).

Bulk carbonate composition:

Aragonite	55%
Mg calcite	35%
Calcite	10%

Sorting: poor

Foraminiferal number: 10,000

Grain size:

> 2 mm	2%
2-0.063 mm	34%
< 0.063 mm	64%

Bulk insoluble residue: 4%

Foraminiferal composition:

Calcareous imperforate	69%
Calcareous perforate	17%
Arenaceous	14%

MUDS

Type no. 10 Lamellibranch mud (ins. res. < 10%) (Figs. 13 and 14)

(a) Sedimentological description

The lamellibranch muds (i.r. < 10%) contain whole or broken skeletal particles, some of which are blackened. The matrix averages 80%, and bulk insoluble residue 6% of the bulk sediment.

(b) Faunal/floral elements

Skeletal grains are mainly lamellibranch fragments and whole, small shells. Determinable forms belong mainly to the genera *Corbula*, *Nucula* and *Phacoides*. Soft faecal pellets are sometimes abundant. Scanning-electron-microscope investigations revealed numerous coccoliths (see Hughes Clarke and Keij, this volume), but of one or two species only.

(c) Synsedimentary diagenetic features

Blackening of skeletal particles.

(d) Relation to environment

The mollusks and Foraminifera occurring in this sediment are infaunal and, for the greater part, indigenous. The lamellibranch mud accumulates as extensive patches in broad depressions, (10-20 fathoms) mainly on the proximal homocline. The presence of common coccoliths (plankton) in the lamellibranch mud is interesting partly because their abundance in ancient carbonates has often been interpreted as an indication of deeper-marine environments.

Type no. 11 Argillaceous lamellibranch mud (ins.res. > 10%) Figs. 15 and 16)

(a) Sedimentological description

These sediments are distinguished from the preceding type only by their higher insoluble residue, which averages 21%.

(b) Faunal/floral elements

Skeletal grains are mainly lamellibranch fragments, whole shells being essentially the same species as in type 10 lamellibranch muds. Scanning-electron-microscope examination of the mud fraction of some samples reveals the presence of coccoliths.

(c) Synsedimentary features

Frequent blackening of grains.

(d) Relation to environment

Mollusks and Foraminifera occurring in the sediment are infaunal and indigenous to the environment of deposition. The black particles, including ooliths, give dates up to 12,200 ± 400 years BP and sometimes occur in large quantities.

Argillaceous lamellibranch muds are the most widespread sediments in the basin and occupy the deeper marine, distal homocline and axial areas. In the NW parts of the Gulf they apparently occur closer to the Arabian shore. Their distribution is probably the result of an interplay of two factors - proximity to siliciclastic source, and higher carbonate production in the shallow proximal parts of the marine homocline.

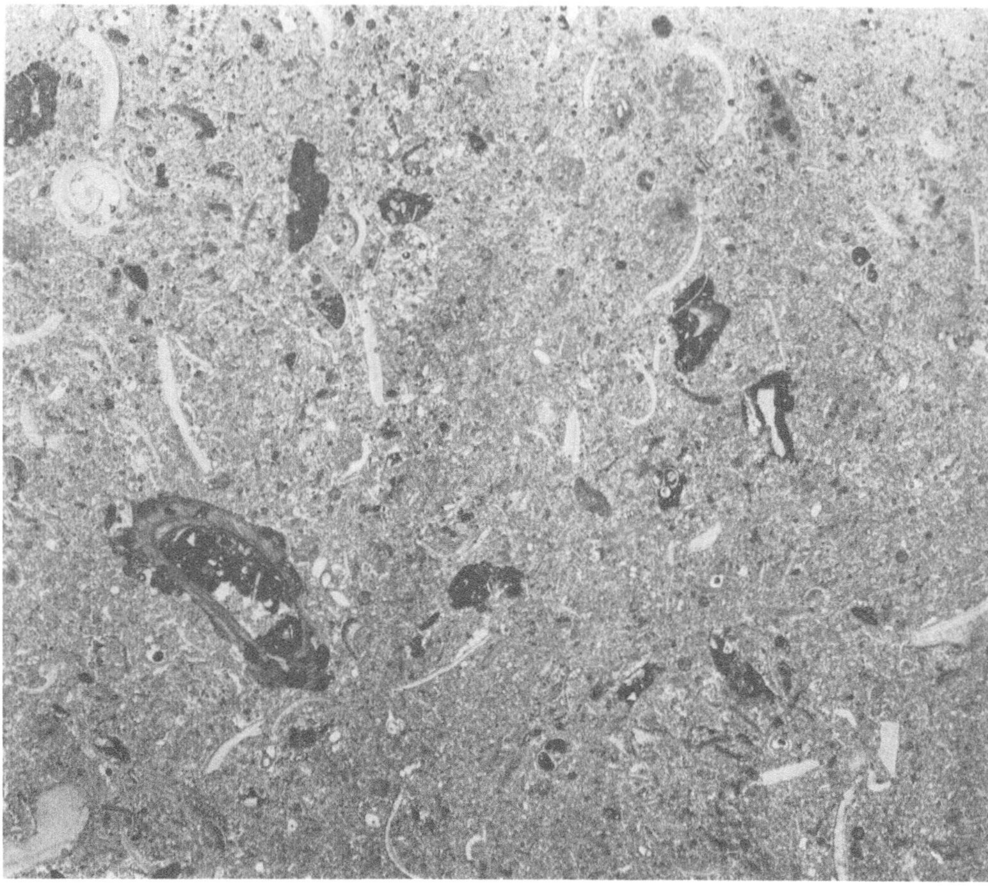

Fig. 15. Argillaceous lamellibranch mud (ins. res. > 10%)

Thin section of sample T 860, 10x

Note: Occurrence of "black" particles (relict)

Bulk carbonate composition:

Aragonite	40%
Mg calcite	35%
Calcite	15%

Sorting: poor

Foraminiferal number: 900

Grain size:

> 2 mm	2%
2-0.063 mm	22%
< 0.063 mm	76%

Bulk insoluble residue: 17%

Foraminiferal composition:

Calcareous imperforate	35%
Calcareous perforate	35%
Arenaceous	29%
Planktonic	1%

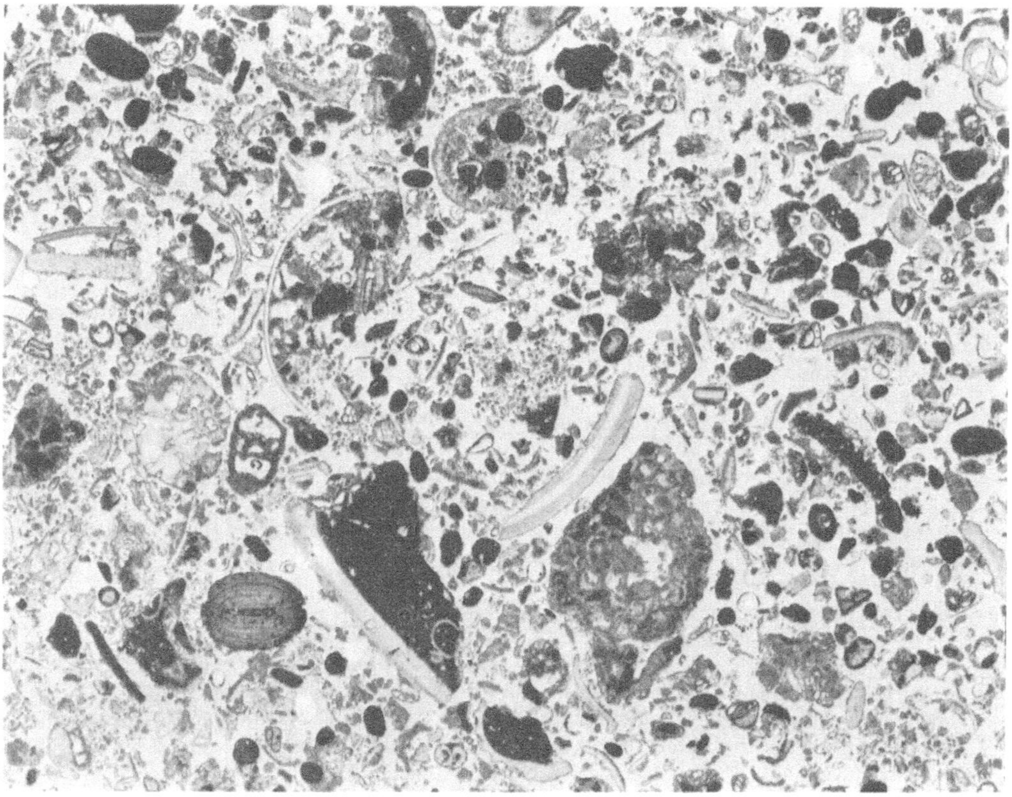

Fig. 16. Argillaceous lamellibranch mud (ins. res. > 10%)

Thin section of sample T 1297, 10x

Note: Photographed area of sample actually constitutes a muddy sand; numerous "black" particles (i.e. ooids) are relict.

Bulk carbonate composition:

Aragonite	50%
Mg calcite	30%
Calcite	15%
Dolomite	8%

Sorting: moderate

Foraminiferal number: 2900

Grain size:

> 2 mm	13%
2-0.063 mm	63%
0.063 mm	34%

Bulk insoluble residue: 7.7%

Foraminiferal composition:

Calcareous imperforate	14%
Calcareous perforate	50%
Arenaceous	24%
Planktonic	12%

Type no. 12 Imperforate foraminiferal/gastropod mud (Fig. 17)

(a) Sedimentological description

The sediment consists of light-grey carbonate mud with scattered skeletal grains dominated by imperforate Foraminifera. The mud may be extensively pelleted and when dried may appear as a pelletal sand. Artificial compaction of the soft pellets gives a "clotted" texture. Insoluble residues are less than 5% except when close to an aeolian source.

(b) Faunal/floral elements

The sediments are characterized by a uniform low-diversity fauna consisting of miliolid and peneroplid Foraminifera, often with cerithid gastropods and small lamellibranchs.

(c) Synsedimentary diagenetic features

These are generally absent. Grains are fresh in appearance and rarely have algally bored peripheries. Boring micro-algae are probably inhibited by the infaunal mode of life of most mollusks, or by the rapid burial of the skeletal grains.

(d) Relation to environment

These sediments occur in highly protected, shallow-water (< 5 m) environments within coastal embayments and on the lee sides of barrier complexes. Salinities in these environments frequently exceed 50 ppm and, together with the strongly fluctuating water temperatures, probably account for the monotony of the fauna characterising these sediments.

NON-CARBONATES

Type no. 13 Quartz sand and muddy quartz sand

(a) Sedimentological description

These sediments consist of well-sorted and rounded quartz and subsidiary carbonate grains with minor amounts of angular skeletal debris. They are generally characterized by large-scale, unidirectional cross-bedding and grade downwards into muddy (carbonate), finer quartz sands. The latter lack cross-bedding, are bioturbated, and can contain numerous angular skeletal fragments.*

(b) Faunal/floral elements

The cross-bedded quartz sands, although deposited in a marine environment, contain only rare, scattered mollusk fragments and occasional miliolid Foraminifera. The muddy quartz sands contain a richer fauna of mollusks, scattered burrowing echinoids, and miliolids.

(c) Synsedimentary diagenesis

Where these sands occur in supratidal environments, they frequently contain poikilitic gypsum and traces of Holocene dolomite.

(d) Relation to environment

In the Persian Gulf, where carbonate sedimentation is dominant, relatively

*For photographs and further discussion of these sands, see contribution by Shinn in this volume.

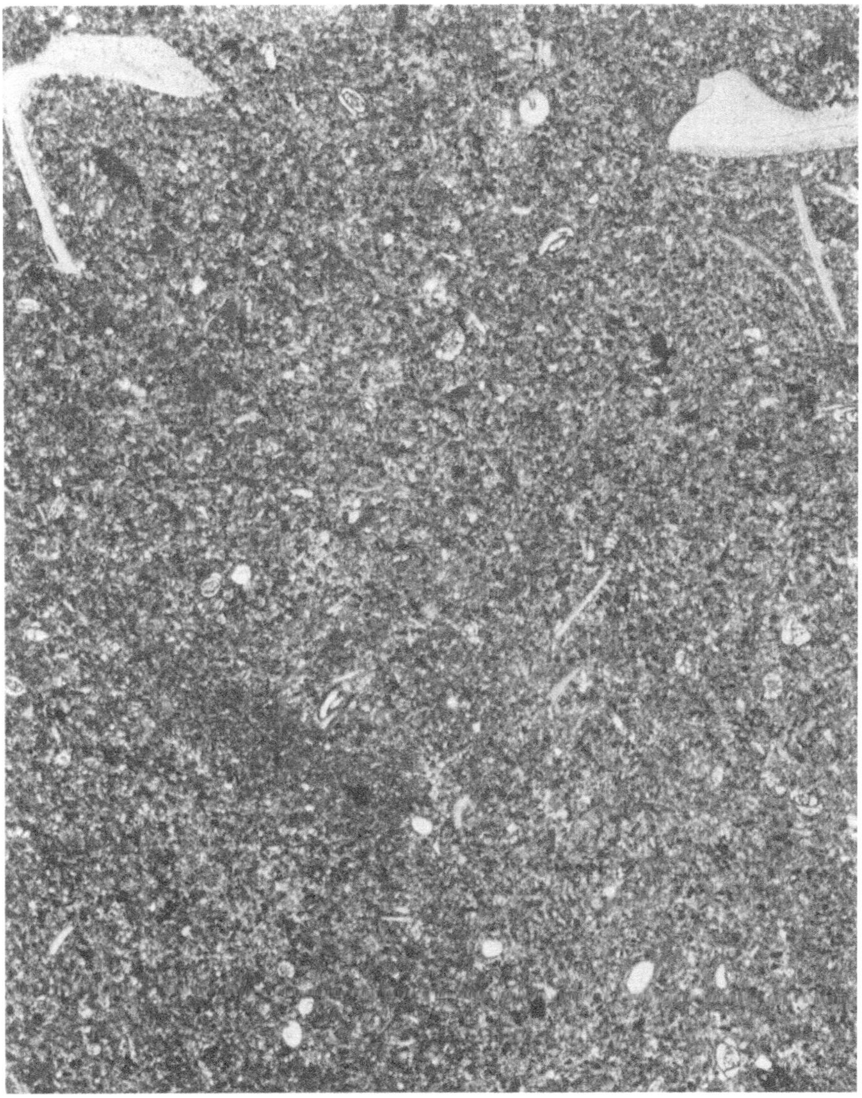

Fig. 17. Imperforate foraminiferal/gastropod mud - Thin section of sample D 719, 15x

Bulk carbonate composition:

Aragonite	20%
Mg calcite	50%
Calcite	30%

Grain size:

> 2 mm	1%
2-0.063 mm	10%
< 0.063 mm	89%

Bulk insoluble residue: 16.9%

Sorting: poor

Foraminiferal composition:

Calcareous imperforate	77%
Calcareous perforate	23%

Foraminiferal number: 3220

pure siliciclastic sands are accumulating in the marine environments only along lee coasts. This situation is typified by the SE coast of Qatar Peninsula, where quartz dune sand is blown into the adjacent inter- and shallow subtidal coastal areas. This sand grades seawards into fine muddy quartz sand, and finally into autochthonous carbonate muds (discussed by Shinn, in this volume).

Type no. 14 Sedimentary gypsum

(a) Sedimentological description

When gypsum is precipitated from standing water in highly restricted lagoons, it accumulates as thin crusts interbedded with the prevailing sediments; these are often aeolian.
Sedimentary structures have not been recognized.

(b) Faunal/floral elements

In the single area examined in any detail (Khor Odaid, SE Qatar), the highly impoverished fauna consisted only of one species of the ostracod genus *Cyprideis*.

(c) Synsedimentary diagenetic features

None are known (unless one regards the gypsum precipitate itself as a diagenetic product).

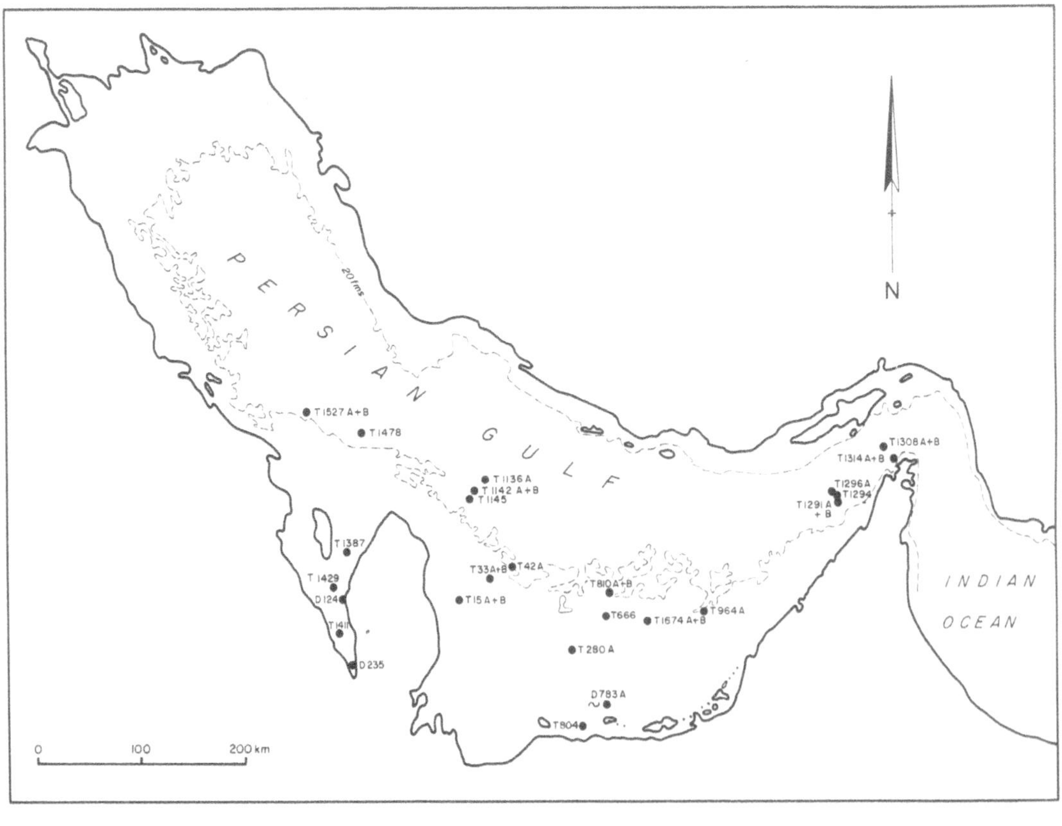

Fig. 18. Location of samples dated by carbon-14

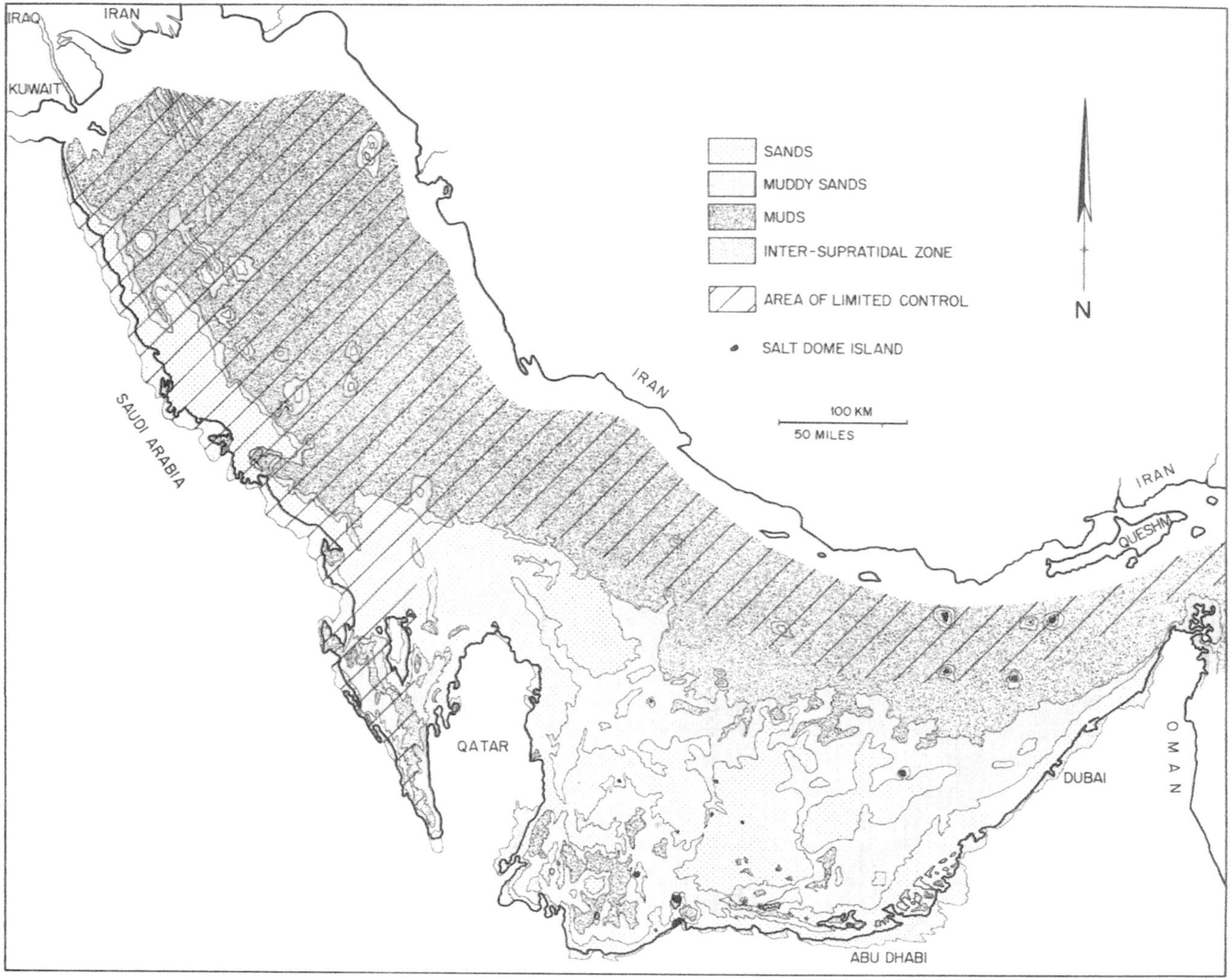

Fig. 19. Regional distribution of principal sedimentary textures

(d) Relation to environment

Primary evaporites occur in highly restricted lagoons (e.g. Khor Odaid in Qatar and Dohat Dhulum in Saudi Arabia, Bramkamp and Powers, 1955). Such super-saline lagoons occur at the ends of long restricted embayments and generally have depths of less than 2 m.

Gypsum, anhydrite and celestine also are widespread on the supratidal sabkhas in the SE Persian Gulf. These sulphates, however, are a true diagenetic product and generally replace pre-existing carbonate.

III. DISTRIBUTION OF THE SEDIMENTS

A. The map and profiles

The distribution of the described sediment types in the Southern Persian Gulf is illustrated by means of profiles and maps (appended at the back of this volume). In constructing the maps, it has been necessary to generalise the observed distribution of individual sediment types as this may vary rapidly over short distances, especially in areas of particularly irregular topography. The occurrence of the sediment types shown on the six profiles is therefore to be taken as a more accurate guide to variability in sediment type than the sediment map.

B. General trends

From the map and profiles it can be seen that the central parts of the Persian Gulf are characterized by widespread argillaceous lamellibranch muds whose textures and generally unbroken or angular skeletal grains reflect the low-energy environments that prevail in these deep-marine areas.

A one passes up the Arabian homocline towards the Arabian shoreline, there is a progressive increase in the granular fraction within the sediment (Fig. 19). This is due both to increasing energy conditions with winnowing of mud, and to increases production of skeletal carbonate. The rapid change from mud-supported sediments of the distal homocline to essentially grain-supported, lamellibranch muddy sand of the proximal homocline, generally coincides with the foot of the 20 fathom (36 m) terrace (Kassler, this volume). The relatively low-energy, lamellibranch muddy sands are the most widespread sediment type on the proximal homocline.

Where the gently sloping Arabian sea floor is situated at depths of 10 - 20 m, its sediments seem to be influenced only slightly by wave action, resulting in the winnowing of mud and silt. The lack of transport of the sand particles in these moderate-energy environments and slow rates of sedimentation seems to favour the formation of the compound grain/lamellibranch sands constituting a lag deposit.

Coastal environments are characterized by sediment patterns (discussed by Shinn, Purser and Evans, and Evans et al. in this volume). Both energy and restriction clearly influence the character of these coastal sediments and these factors, to a large extent are determined by coastal orientation with respect to the strong NW' winds, and by the presence or absence of offshore barriers.

Because most of the Arabian coast is exposed to the NW "shemal" winds, its sediments are largely high-energy, mainly (broken) lamellibranch sands. These are locally replaced by ooidal sands which are especially well developed at the seawards ends of tidal channels in Abu Dhabi, and by coral/algal fringing reefs and associated coral/algal sands in the most exposed, windward coastal settings, e.g. the N end of Qatar Peninsula.

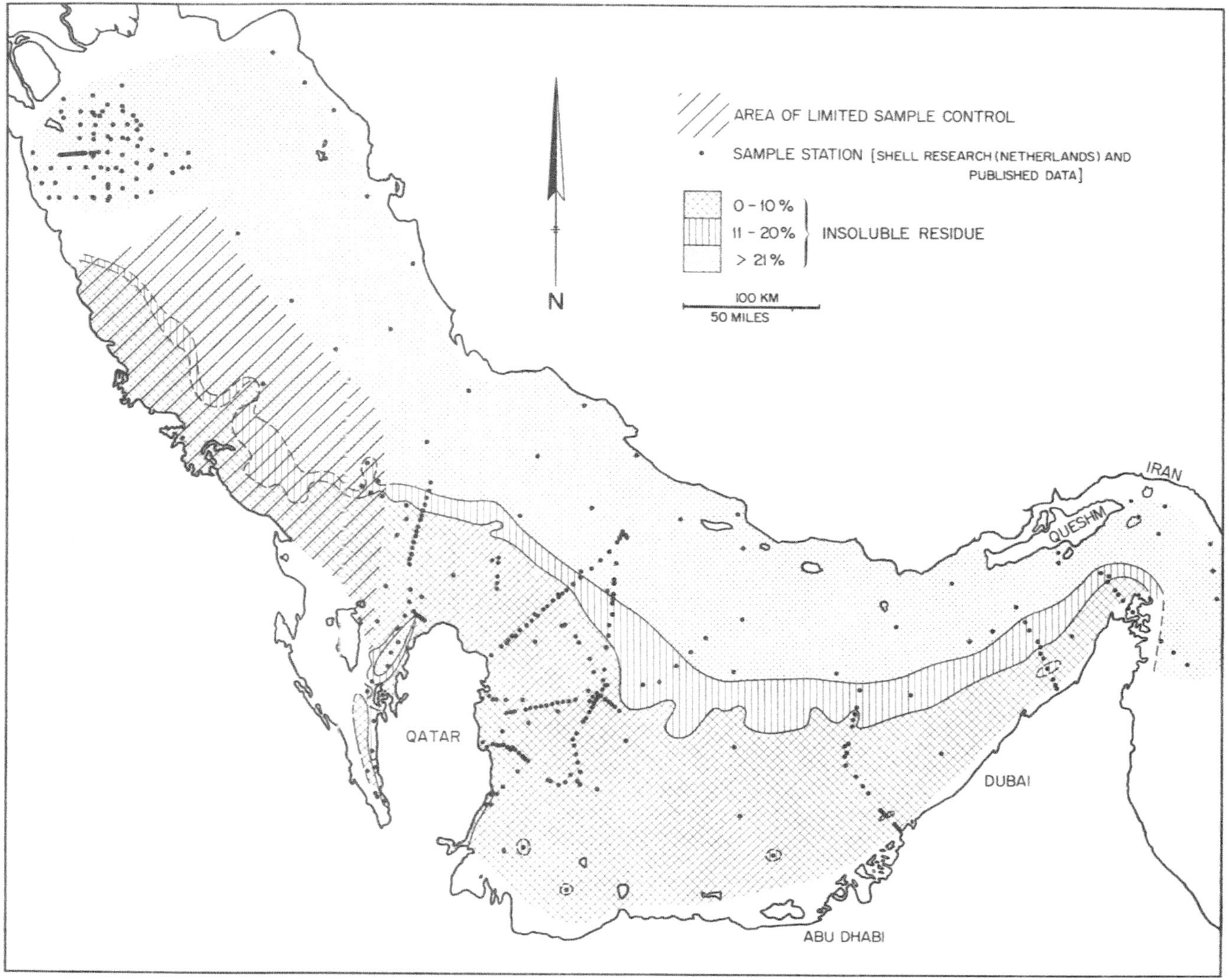

Fig. 20. Regional distribution of insoluble residues

Coastal embayments and lagoons protected by spits and offshore barriers, are dominantly low energy and often restricted areas where gastropod muddy sands and imperforate foraminiferal/gastropod muds accumulate. Where these water bodies are isolated from the main body of Persian Gulf water, they may favour the precipitation of primary gypsum.

The protected embayments and lagoons can be flanked by wide inter- and supratidal flats which are the sites of extensive stromatolitic algal mats, secondary gypsum, anhydrite, and other evaporite minerals, and dolomite (see Bush, Butler, Evans et al., elsewhere in this volume). These coastal swamps grade landwards into the rocky or dune-covered deserts of Arabia.

In sum, the sediments accumulating on the gently inclined Arabian sea floor tend to grade from impure carbonate muds near the centre of the basin to high-energy, bioclastic and ooidal sands, together with local muddy embayments on the coast. This very general trend is complicated by numerous local variations associated with an intricate system of bathymetric highs and lows superimposed upon the regional slope. Many of these highs are structurally based (Kassler, this volume), and the sediment types and patterns associated with them are discussed by Purser (this volume).

Large perforate foraminiferal sand is accumulating on, or near, the crests of highs occurring in exposed areas, often along the outer edges of the proximal homocline. In contrast, highs situated nearer the coast in shallow-marine settings are generally capped by coral/algal reefs, best developed around their windward sides, and elongate sand tails consisting mainly of coral/algal sand extending leewards from the highs.

Depressions on the proximal homocline are areas of lamellibranch-mud accumulation. The largest of these occurs landwards of the Great Pearl Bank barrier, SE of Qatar, where low-energy environments are partly the result of lateral protection from wind-driven waves by Qatar Peninsula.

Within the muddy sediments, the marked difference between the axial areas and the muds in sheltered, shallower locations lies in the non-carbonate content. This is particularly well illustrated by the distribution of insoluble residues (see Fig. 20).

C. Principal variations

The six regional profiles (appended at the back of this volume) across the Arabian homocline demonstrate how the general sedimentary trends outlined in the preceeding section may vary.
These modifications result mainly from:

(i) differing orientation of the homocline with respect to strong NW winds,

(ii) presence or absence of barriers, such as the Great Pearl Bank,

(iii) differing inclination of the homocline.

Because these essentially structural factors vary laterally along the Arabian coastline, in general sediments are <u>not</u> arranged in a geometically simple pattern of individual belts trending parallel to the regional coast. Furthermore, the complexity of sediment pattern increases towards the coastline.

TABLE I. C^{14} DATINGS

(see also Fig. 18)

Sample no.	Material examined	Water depth in metres	Age (C^{14}) (years BP)
D 124	mollusk shells from lightly lithified sediment	18	< 240
D 235	ooidally coated quartz sand	4	11,100 ± 400
D 783	compound grains	9	4840 ± 170
T 15	compound grains fresh and micritized shells	16	3440 ± 170 190 ± 120
T 33	compound grains fresh mollusk shells	24	4200 ± 170 290 ± 140
T 42	compound grains	28	2460 ± 140
T 280	compound grains	19	3540 ± 170
T 666	micritized mollusk shells and debris	16	300 ± 140
T 804	rounded bioclastic sand	6	980 ± 150
T 810	compound grains mollusk shells	23	2360 ± 140
T 964	compound grains	20	2180 ± 150
T 1136	compound grains	50	8830 ± 260
T 1142	compound grains mollusk shells	41	7820 ± 240 760 ± 140
T 1145	lithified mollusk muddy sand	37	2880 ± 170
T 1291	compound grains mollusk shells	67	14.400 ± 400 4000 ± 180
T 1294	lithified mollusk muddy sand	68	10.700 ± 300
T 1296	compound grains	76	12.200 ± 400
T 1308	compound grains mollusk shells	86	12.100 ± 300 2820 ± 150
T 1314	compound grains mollusk shells	103	12.500 ± 400 1340 ± 140
T 1387	mollusk shell fragments	6	3050 ± 180
T 1411	mollusk shell fragments	8	3810 ± 220
T 1429	mollusk shell fragments	18	5380 ± 260
T 1527	compound grains mollusk shells	36	3020 ± 160 590 ± 140
T 1478	lithified mollusk muddy sand	41	2120 ± 150
T 1674	compound grains mollusk shells	32	8340 ± 240 950 ± 130

Sedimentation around Bathymetric Highs in the Southern Persian Gulf

B. H. Purser[1]

ABSTRACT

Bathymetric highs in the Persian Gulf have been grouped into three classes based on their setting on the Arabian homocline: outer (basin centre), intermediate and inner (coastal). In general, sediments on these highs range from open marine, perforate foraminiferal or coral/algal sands on the basin centre highs to pelletoidal sands, carbonate muds with restricted faunas, and algal stromatolites and evaporites on the coastal highs.

Sediment patterns on and around highs vary according to the regional setting of the high. They tend to be concentric around basin centre highs and become progressively asymmetric towards the coast, due to the accretion of bioclastic sands on their leeward sides. These accretion tails may even link coastal highs with the adjacent mainland shore. Peninsulas are thus formed which can completely alter the system of coastal currents and the types and patterns of coastal sediment.

Sediment patterns are also related to the diameter of the high. When highs exceed 5 km in width, double sediment tails ("bulls horns") are formed by curving, elongate barrier sand ridges on the windward quarters of the high. The shelter due to these barriers creates lagoonal environments in the centre of the high where somewhat restricted, muddy sediments accumulate. Sediments on the leeward flanks of wide highs are also muddy and contrast markedly with the sand tails that accumulate on the lee sides of smaller highs.

Sediment around salt domes may contain minor amounts of exotic material brought to the surface by the diapir. This exotic fraction remains close to, and downwind of the dome. In addition, some domes are ringed by faint bathymetric depressions which may reflect rim synclines. These lows also favour the accumulation of muddy sediments. The diapiric salt itself, although close to the surface in some domes, does not outcrop and does not seem to influence contemporary sedimentation. In all other respects, sedimentation around salt dome highs is similar to that around other highs in the basin, irrespective of their origin.

Diagenetic processes are also distributed systematically, partly according to the regional setting of the high. Highs submerged below about 10 m, especially in basin centre and intermediate settings, favour active submarine lithification on their crests. Emergent highs in exposed settings favour beach rock lithification and vadose leaching, but no dolomitisation. This latter process is active on the extensive tidal flats accreting around the lee sides of wide coastal highs.

---

1 Lab. de Géologie Historique, Université de Paris Sud (Orsay).

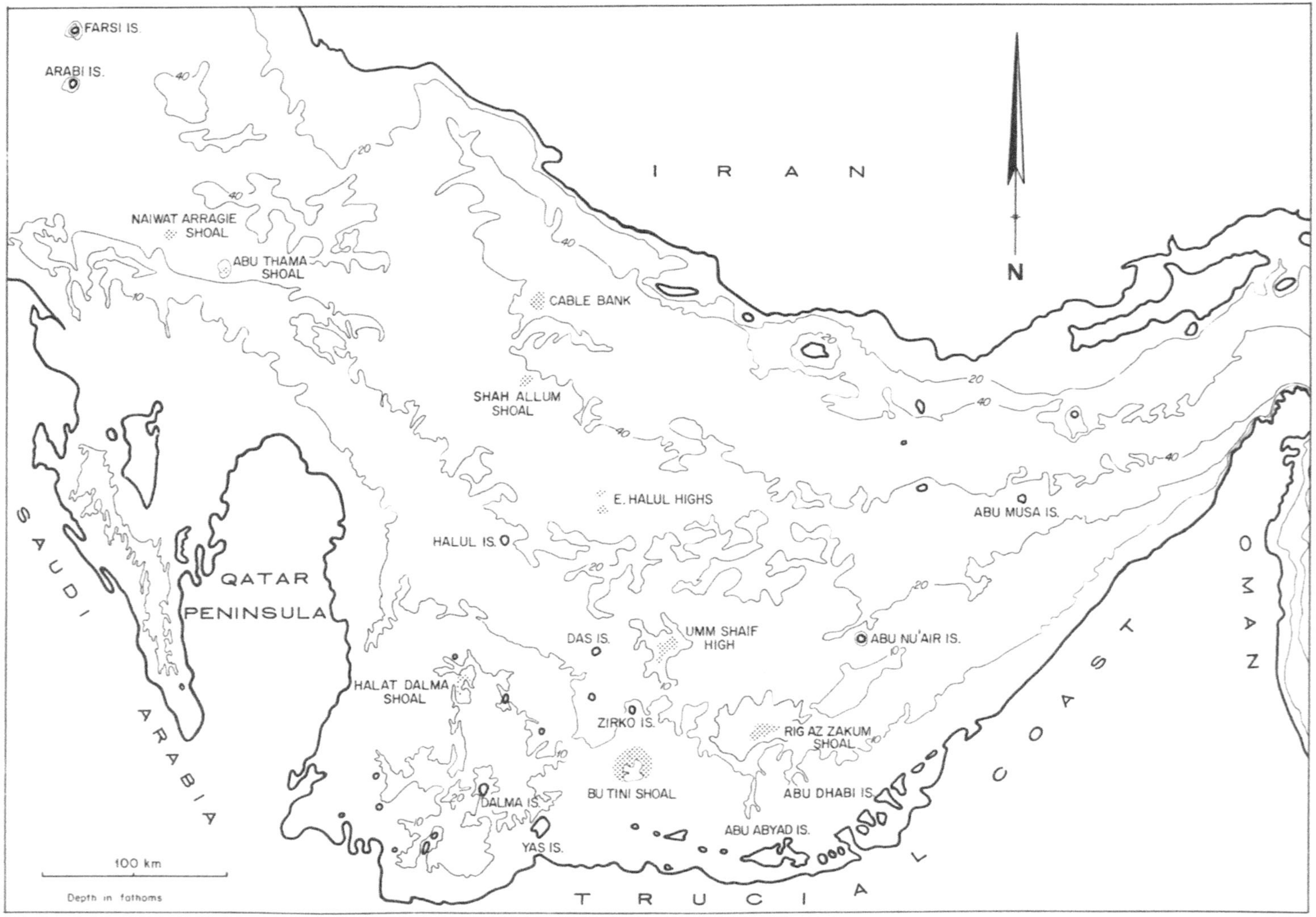

Fig. 1. Distribution of bathymetric highs discussed in this publication

I. INTRODUCTION

The sea floor of the Persian Gulf slopes gently from the Arabian coastline towards the axis of the basin, which lies relatively close to the Iranian coast. This slope, termed here the "Arabian homocline", is very irregular in detail. Bathymetric highs and lows of varying amplitude occur throughout ranging in diameter from tens of metres to tens of kilometres. Especially in the deeper, axial parts of the basin, most of these highs are submerged to depths exceeding 20 m. In shallower, coastal areas many are sufficiently shallow to favour reef growth, while others form low islands or cays. Islands up to 100 m in altitude, formed by piercement salt domes, also occur throughout the eastern half of the basin.

The irregularity of the Persian Gulf floor, with its numerous local highs and depressions, is due to several processes which are discussed by Kassler (in this volume). While most seem to be drowned mesas or tectonic domes, the morphology and setting of some suggests a sedimentary origin. In particular, the islands of Farsi and Arabi, and certain shoals such as Abu Thama and Naiwat Arragie (N of Bahrain), rise steeply from the central parts of the gulf. Their situation W of the salt-piercement province precludes that form of tectonic origin, whilst their extreme geographic isolation relative to the coastline, smallness (4 km^2) and marked vertical relief suggest they are not drowned mesas. It is more likely that these, and perhaps other similar highs, owe their relief to localized high sediment production which persisted during the post-glacial rise in sea level. Many of these basin centre highs are sites of active coral reef growth and synsedimentary lithification.

In sum, the size and vertical relief, which strongly influence the type and distribution of sediment around these highs, are due to a combination of factors, including tectonics, subaerial erosion, and sedimentation. The principal objective of this contribution is to demonstrate some of these relationships.

II. TYPES OF BATHYMETRIC HIGH AND RELATED SEDIMENT PATTERNS

Of the several interrelated factors that determine the sedimentary patterns on and around the Persian Gulf highs, the regional setting is probably the most important. Water depth, both on and around the high, strongly influences the degree of circulation and restriction, and therefore the type of sediment produced on the high. It also determines the patterns of sediment around the high. Regional setting has therefore been chosen as the basis for subdividing highs into three classes: outer homocline (basin centre), intermediate homocline, and inner homocline (coastal).

Outer homocline (basin centre) highs

Morphology : The central parts of the basin, where water depths generally exceed 50 m, are characterized by numerous highs with marked vertical relief. Many are salt diapirs and are thus of structural origin. The latter are generally symmetric in outline and range from islands, such as Bu Musa, to submerged shoals, such as Shah Allum shoal and Cable Bank (Fig. 1). All have one feature in common - they are surrounded by relatively deep water.

Sediment types and patterns: Depth of submergence determines not only the degree of water agitation and hence the texture of the sediments accumulating on the crest of the high, but also (through light penetration, oxygenation, etc.) the nature of the biota inhabiting the high (Hughes Clarke and Keij, this volume).

Houbolt (1957) has shown that, on Cable Bank and other shoals (e.g. an unnamed shoal 55 km ENE of Halul Island, Fig. 2) which are deeply submerged below wave base, the crestal sediments are large perforate-foraminiferal sands. Little coarse sediment is swept off the tops of these submerged highs, there being a very rapid transition into adjacent basinal muds around the foot of the high. These highs cannot normally be detected from the presence of exotic debris in the adjacent deep marine

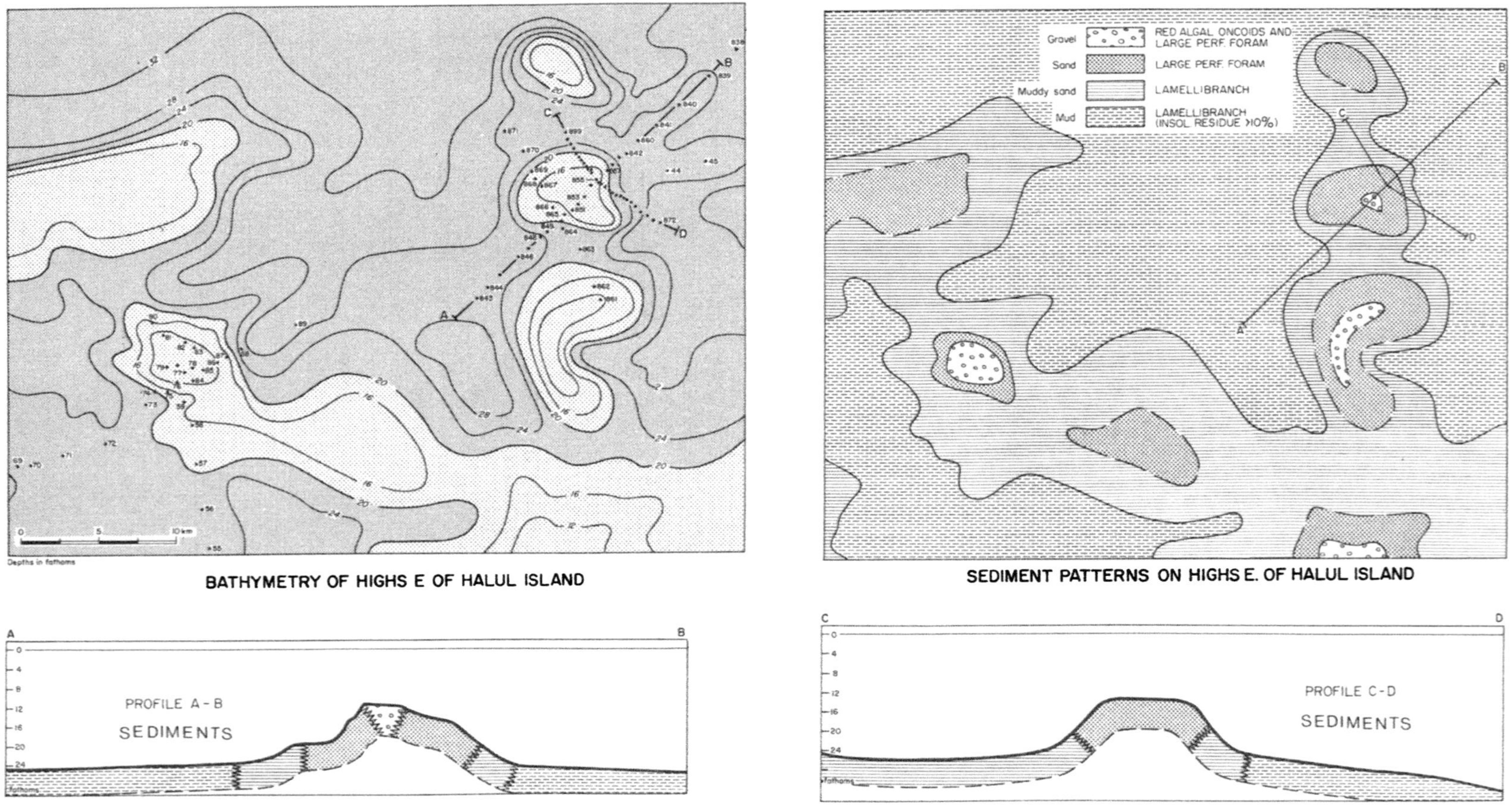

Fig. 2. Bathymetry and sediment distribution on, and around typical "outer homocline" (basin centre) highs

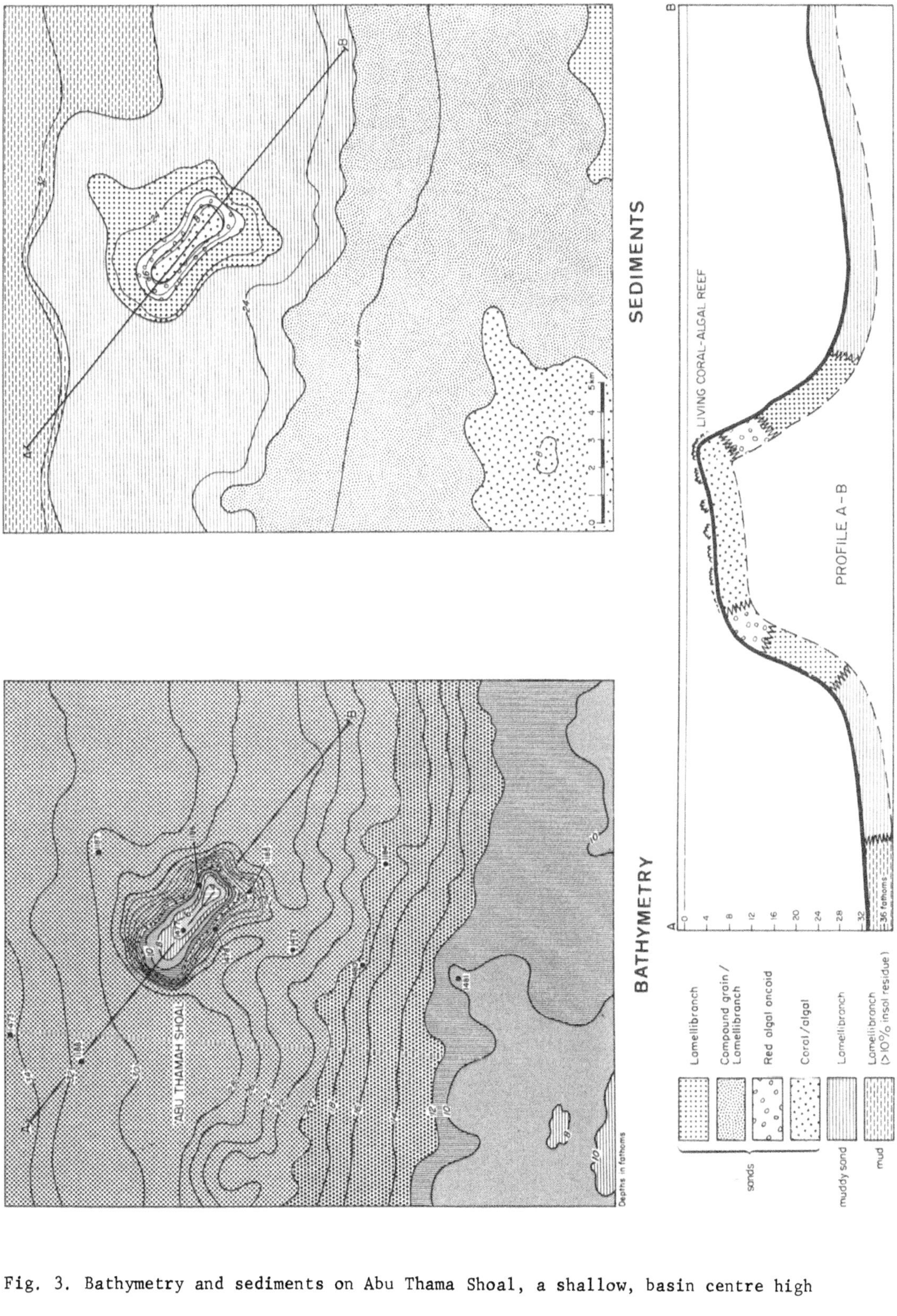

Fig. 3. Bathymetry and sediments on Abu Thama Shoal, a shallow, basin centre high

sediments, even in a down-current direction.

The crests of many of these deeply submerged highs are covered by compound grains and large lumps of lithified sediment identical in composition to the associated, nonlithified sediment forming on the high. This submarine lithification is probably favoured by the moderate energy conditions on the top of the high, which are too feeble to transport sand grains but are of sufficient strength to winnow mud. Shinn (1969) has suggested that these conditions favour submarine lithification, and it is possible that certain of these highs owe their vertical relief to this process.

Shallow submerged highs such as Shah Allum shoal (Fig. 1) are capped by reefs and coral/algal sand, and surrounded by deep marine, lamellibranch muds. Because the crests of these highs are within wave base, the bioclastic sediments formed on the high may be swept off during storms. Houbolt (1957) has shown that at Shah Allum shoal most of this sediment seems to remain close to the high because it falls into adjacent, deep, low energy environments. Fine sand and silt however, may be transported up to 15 km downcurrent. As a result, the presence of the high can be detected from the nature of the fine sediments originating on the top, but only from a downcurrent direction. This accumulation of fine debris is insufficient to create appreciable submarine relief.

The Abu Thama (Fig. 3) and Naiwat Arragie shoals N of Bahrain have coral/algal reefs on their crests. The flat top of each high slopes gently northwards and the reefs are best developed along the slightly shallower, leeward edge of the high. Topographic culmination along the leeward edge of the high may be due to sediment being maintained on the downcurrent edge due to wave refraction around the flank. This concentration of sediment on the leeward edge of small highs in exposed settings has been observed in several parts of the Persian Gulf and is a parallel to the lee-side sand cay positioning on many of the reef build-upsin the Great Barrier Reef (see Maxwell, 1968).

Emerged highs such as the salt dome island of Bu Musa, or sand cays such as the minute and isolated islands of Farsi & Arabi (Fig. 1), support coral/algal reefs. Although slightly better developed on their windward sides, free water circulation around each high favours active reef growth on all sides. In the case of Farsi and Arabi, sediments derived from fringing reefs have accumulated and lithified along the leeward fringe, forming islets of lithified coral/algal sand and gravel. This local accumulation is due mainly to sediment trapping. Large coral heads, transported across the flat top of the high, probably during storms, are prevented from falling into adjacent deeper waters by reefs fringing the lee side of the high. Large coral boulders and other reef debris are trapped in the leeward reef to form a ridge just below sea level. Sediment is piled against this ridge and lithifies as beach rock to produce an island which progrades slowly upwind. Finer carbonate sand and silt is swept back off the top of the high into adjacent deep water where it sinks below wave base and accumulates around the flanks of the high. Sediment patterns tend to be arranged concentrically around the bathymetric culmination. Fine sand and silt, however, may be more widely distributed in a down-current direction, but in quantities insufficient to create a sedimentary build-up.

Because of their exposed settings, basin centre (outer homocline) highs are surrounded by freely circulating seas and there are no protected environments, even on the lee sides of relatively large salt-dome islands. Sediment patterns tend to be concentric, and sediments derived from the crest of the high remains in its vicinity.

Intermediate homocline highs

Morphology : The sloping sea floor is characterized by a great variety of marine highs and depressions, the origins of which are discussed by Kassler (elsewhere in this volume). They vary in diameter from several hundred metres to more than 50 km

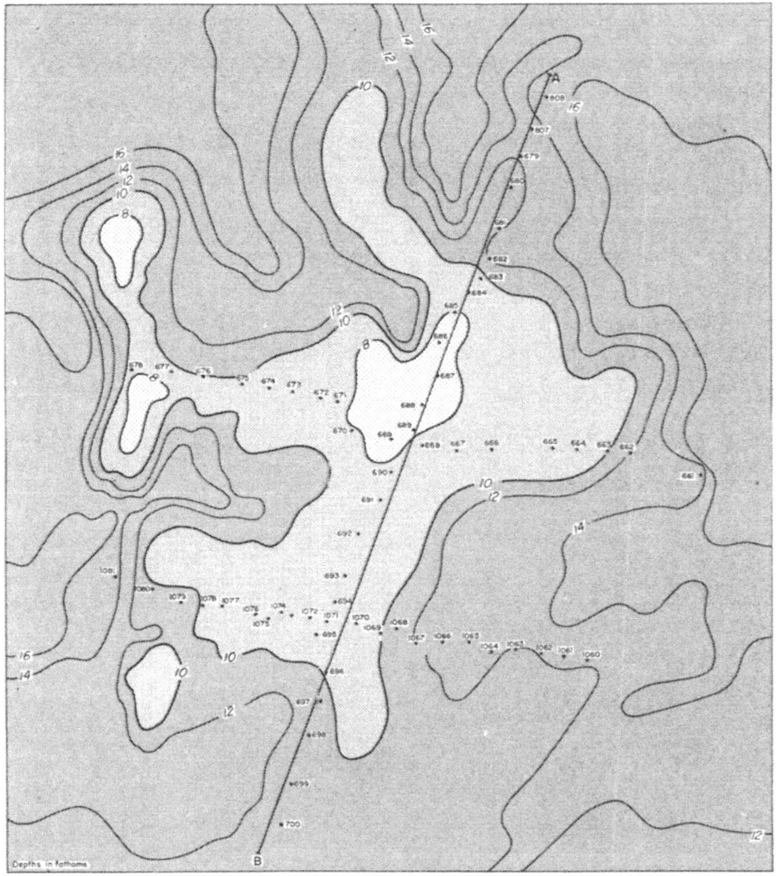

BATHYMETRY OF UMM SHAIF HIGH

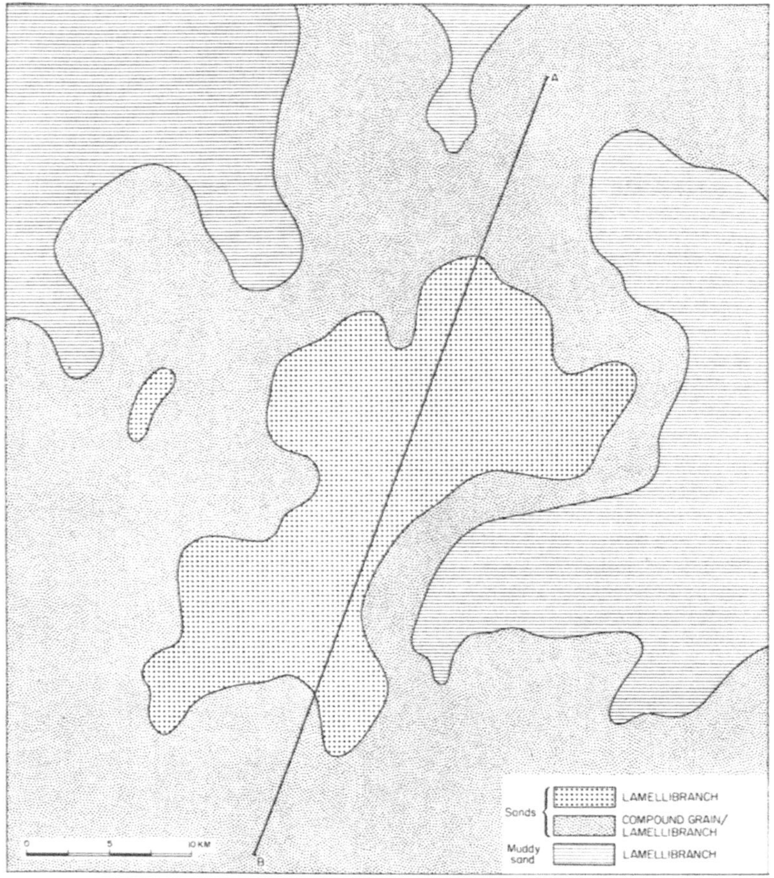

SEDIMENT PATTERNS ON UMM SHAIF HIGH

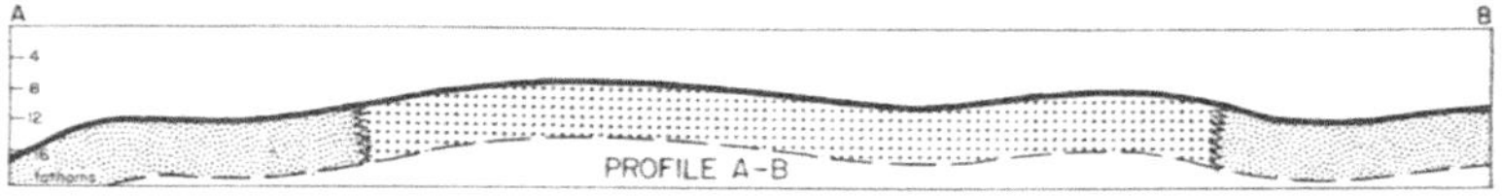

Fig. 4. Bathymetry and sediments on Umm Shaif high, a large, relatively deep, "Intermediate Homocline" high

(Umm Shaif and Rig az Zakum). Many emerge as islands due to salt diapirism in the SE parts of the gulf. Others are low sand cays whose origins are essentially sedimentary. A limited number are emergent pre-Holocene mesas. All have one feature in common: they are surrounded by relatively shallow water varying in depth from 20 fathoms (36 m) on the outer parts of the homocline, to less than 5 fathoms near the coast.

Sediment types and patterns : Because most of these highs are surrounded by relatively shallow water, there is less contrast between sediment forming on the high and that generated on the sea floor around it. In addition, sediment produced on the high is more widely distributed around its flanks because the bottom is close to, or above, wave base. Lateral transport becomes more pronounced around highs relatively close to the regional shoreline.

On highs whose tops are submerged below wave base, there is relatively little transport of sand, and hence no preferential piling of it on,or around,certain parts of the high. Umm Shaif high demonstrates this phenomenon (Fig. 4). This high, the relief of which is probably in part the product of the anticlinal doming that forms the underlying oil field, attains a minimum depth of about 14 m, much of the crest being submerged to 15-20 m. Because of the depth, the sediment forming is mainly lamellibranch sands, which grade downflank into compound grain sediments. Unconsolidated sediment constitutes only a thin veneer (5-10 cm) on limestone that frequently resembles the overlying sediment in composition. This suggests submarine lithification in these moderate energy environments on the crest of the high. Sediments surrounding the high are mainly lamellibranch muddy sands.

Carbonate sands formed on highs with crests that culminate above wave base, such as Rig az Zakum, are susceptible to transport. Rig az Zakum (Fig. 5) probably owes much of its relief to the underlying structure, which forms one of the largest oil fields in the SE parts of the Persian Gulf. The bathymetric high culminates as a sedimentary ridge along its leeward (SE) side with depths of less than 2 m. The shallowest sediments are coral/algal gravels which grade downflank into lamellibranch sands which are lithifying to form compound grains and sheets of submarine limestone. Compound grain sediments are confined mainly to the windward sides of the high at depths of 20-25 m. They occur at somewhat shallower depths (15 m) around the flanks, but are replaced by lamellibranch muddy sands and muds on the leeward side of the high. Their fluctuation in depth and absence on the leeward side of the high is probably related to decreasing water agitation.

The asymmetric nature of the Rig az Zakum high, with the sedimentary culmination along its leeward edge, is almost certainly the result of local sediment piling, related perhaps to a wave system which refracts around the leeward (SE) flank limiting the overspill of carbonate sand. In consequence, sediment patterns, although generally concentric, are somewhat asymmetric with respect to the high as a whole (Fig. 5), maximum thickness probably occurring slightly downwind of the tectonic culmination.

Virtually all emergent highs have well developed fringing reefs on their windward (N or NE) sides. Sediment produced by continued breakdown of these reefs is distributed downwind as a series of one or more sediment tails. Saltdome islands such as Halul, Das and Abu Nu'air, situated in relatively exposed settings on the outer parts of the homocline, have living reefs on all sides except the leeward (SE), from which a submerged sediment tail accretes and may be traced some 10 km downwind. Islands nearer the coast, such as Dalma, Zirko and Yas, have reefs concentrated mainly on the windward margins. Leeward sediment tails may be traced laterally for 15-20 km (Fig. 6). In general, the longest tails are associated with the highs in the shallowest settings, favouring the lateral transport of sediment (Fig. 6).

The amplitude of sediment tails, such as that associated with Dalma Island, (Fig. 12) suggests that they attain 20 m in thickness. The tails probably began to form during a lower sea level. Tails on the outer parts of the slope appear to be thickest, having had a longer, pre-Recent sedimentary history. Throughout their evolution, the orientation of the sediment tail seems to have remained relatively

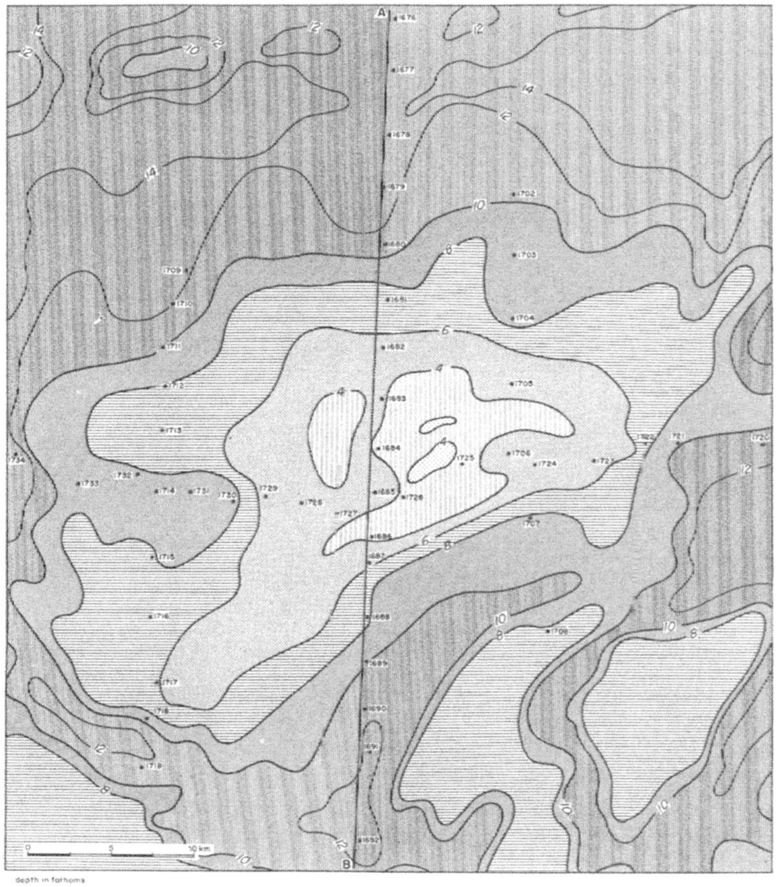

BATHYMETRY OF RIG AZ ZAKUM HIGH

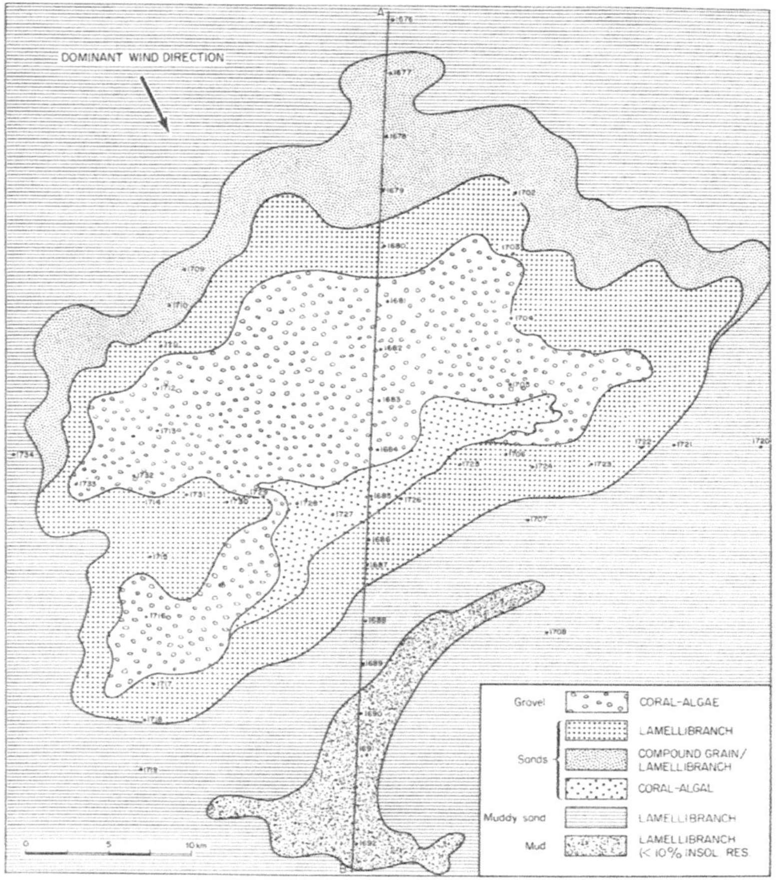

SEDIMENT PATTERNS ON RIG AZ ZAKUM HIGH

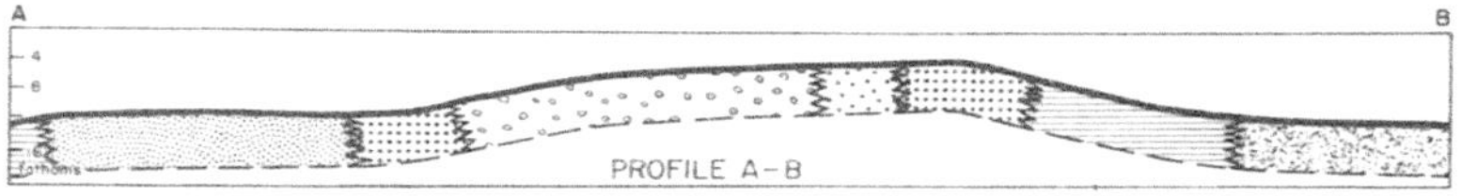

Fig. 5. Bathymetry and sediments on Rig az Zakum, a large shallow, "Intermediate Homoclire" high

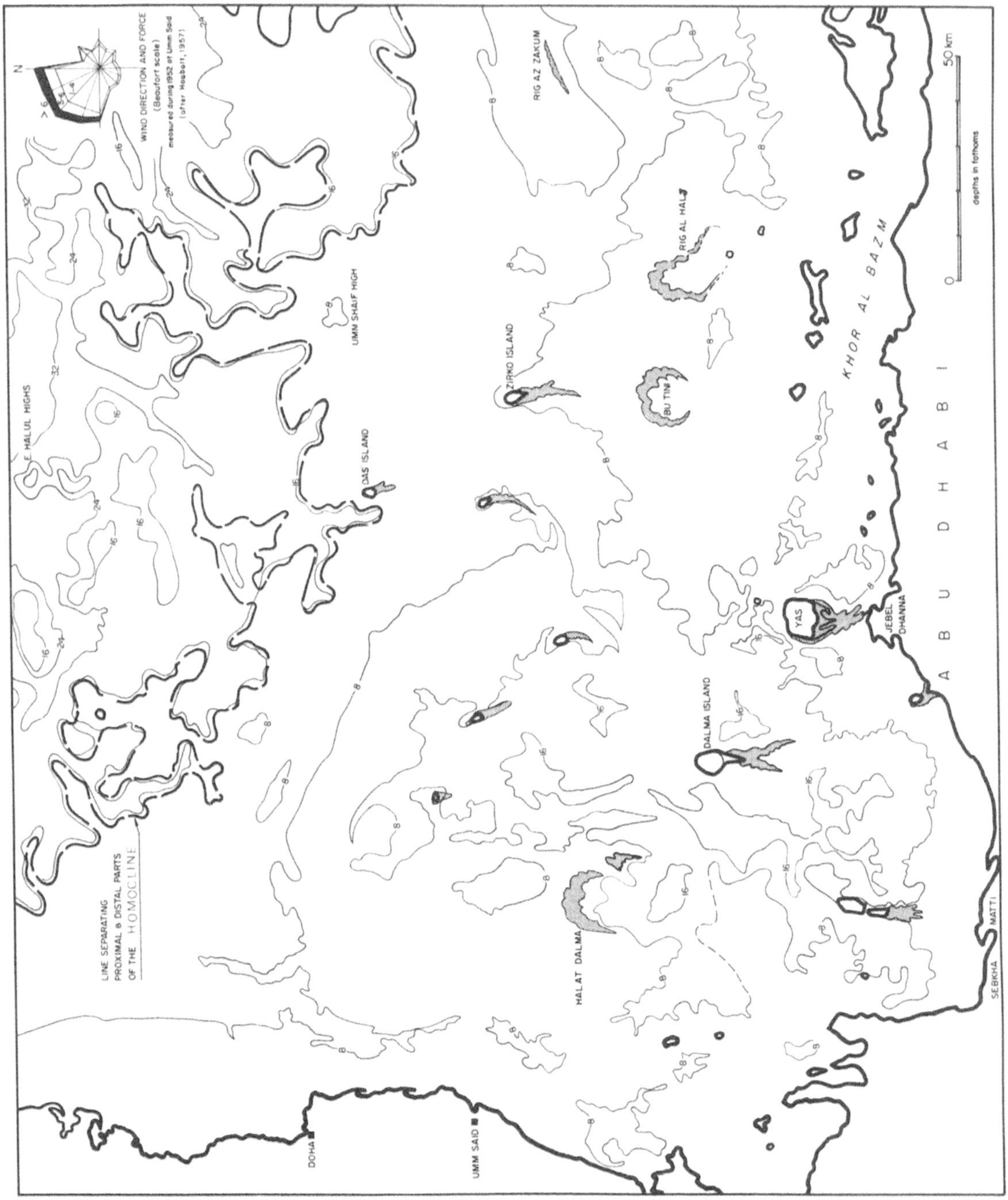

Fig. 6. Relationship between wind direction and orientation of sediment tails and barriers in the SE parts of the Persian Gulf

constant. Certain tails, including those of the Dalma and Zirko islands, tend to branch (Fig. 12), the weaker branch in each case being to the E. This may be due to a slight modification of wind direction during the evolution of the tail, or to a secondary transport direction which is still active.

Most sediment tails are lithified. The preservation of their characteristic submarine morphology may be regarded as evidence of an active submarine diagenesis. Lithification may be limited to thin, superficial crusts which encase friable carbonate sand (Shinn, 1969). These hard surfaces support small reefs on the tail S from Dalma Island. It is probable that many of the sediment tails consist of cross-bedded carbonate sands, the bedding and texture of which reflect a progressive build-up to sea level.

Because the prevailing wind direction is relatively constant throughout the Gulf, most highs are affected in a similar manner by wind-driven currents and waves. Sediment tails and other sand piles associated with these marine highs exhibit a constant S or SE orientation (Fig. 6) over most of the basin. Such constancy may also characterize series of highs in certain ancient sedimentary basins and facilitate their subsurface prediction as potential hydrocarbon reservoirs.

Influence of the width of the high on sediment pattern : The diameter of the high strongly influences both the amount of sediment transport across its crest and the degree of protection on its leeward side. On highs with crests situated close to or below wave base, there is free water movement across the high, as at Umm Shaif. Areas of shallow bottom, however, tend to slow down and reduce water movement and wave intensity, and thus sediment transport. If those parts of the high that occur above wave base are relatively narrow, the effects are minimal and sediment may be transported across the high to be deposited behind its leeward edge. However, when these shallow areas exceed some 5 km in width the movement of waves and currents across the high is seemingly reduced to the level where sand can no longer be transported.

The crestal area of Halat Dalma shoal, off SE Qatar (Figs. 7 and 8), measures some 10 km in a NW-SE direction and supports a well developed fringing reef along its windward margin. Coral/algal sands produced by the continued breakdown of this reef are transported downwind across the crest of the high. Towards the centre of the high, presumably due to decreasing energy conditions, the sands stabilize 3-5 km downwind from their source (the reef) and build an arcuate complex of bars or mega-ripples whose general orientation is E-W, normal to the dominant "shamal" wind. This sand-bar system emerges at low tide, is varyingly lithified, and thus constitutes a barrier close to the geographic centre of Halat Dalma high.

On Halat Dalma the crestal-barrier complex gives sufficient protection to the environments in its lee for carbonate muds to be deposited; the muddy sediments that occur in the deeper environments off the high therefore extend up the leeward slopes towards the crest to areas of less than 5 m depth immediately behind the crestal barrier (Fig. 8).

The presence of a crestal barrier prevents the formation of the leeward sand tail characteristic of the small highs. Instead, each extremity of the barrier complex curves downwind to form a pair of tails that extend southwards for some 10-15 km (Fig. 8). The tails and the crestal barrier that unites them constitute a broad sedimentary arc or "bull's horn". Much of this barrier is lithified, at least on its surface; its amplitude suggests a thickness of about 15 m.

A further stage in the sedimentary development of the wider highs is demonstrated by Bu Tini shoal (Fig. 9). Somewhat bigger than Halat Dalma, the flat-topped Bu Tini high is some 20 km in diameter and is mostly above wave base, average depths being less than 2 m. Fringing reefs occur on all sides, but are best developed around the windward (NW) periphery. As at Halat Dalma, coral algal sands are transported downwind from the reef system to form a barrier bar complex 2-3 km leeward of the reef. The barrier of this extensive high is thus comparatively closer to the

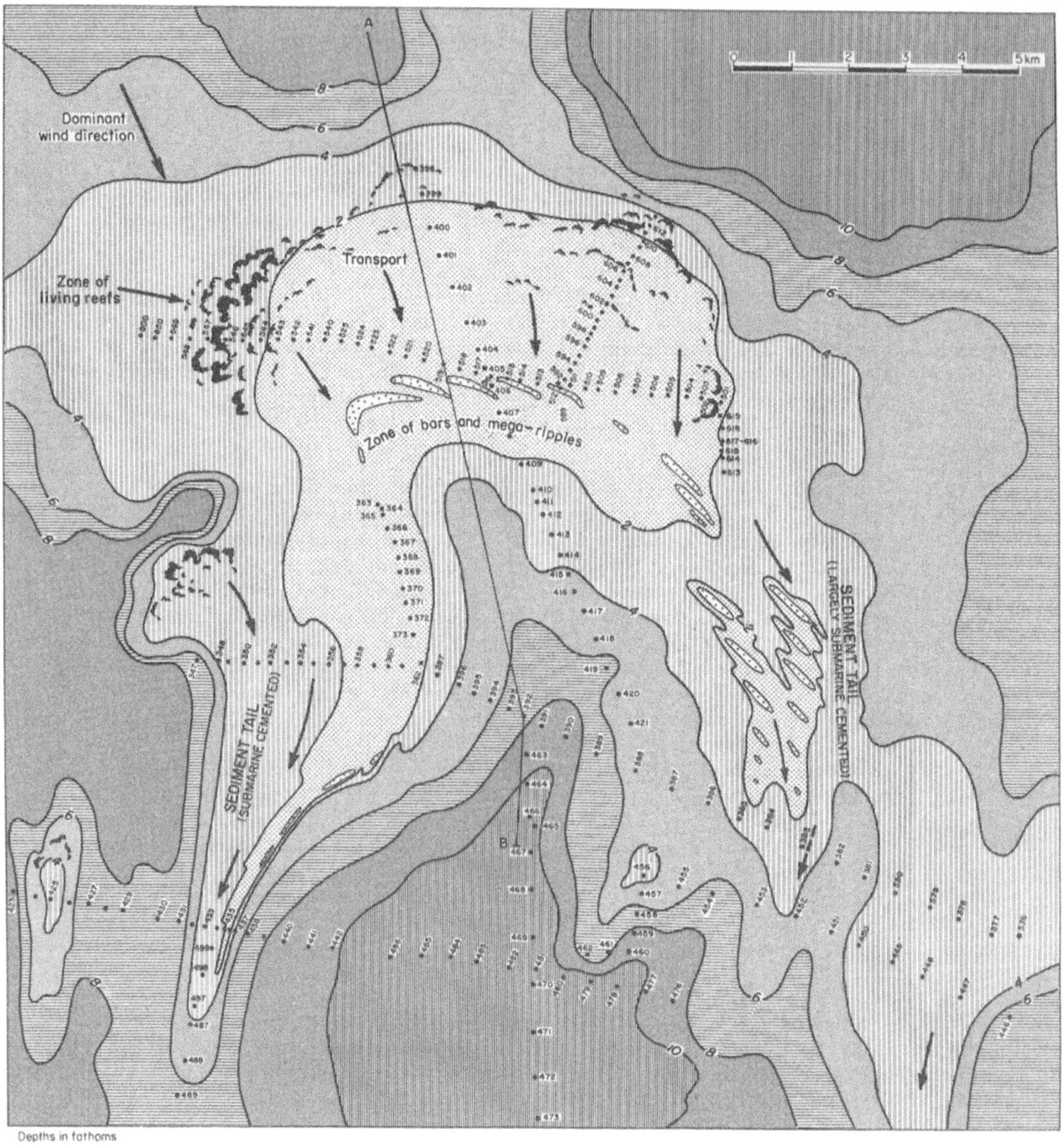

Fig. 7. Bathymetry of Halat Dalma Shoal (E. Qatar).
This relatively wide, locally emergent, "Intermediate Homocline" high is constantly swept by the NNW wind-driven waves and surface currents

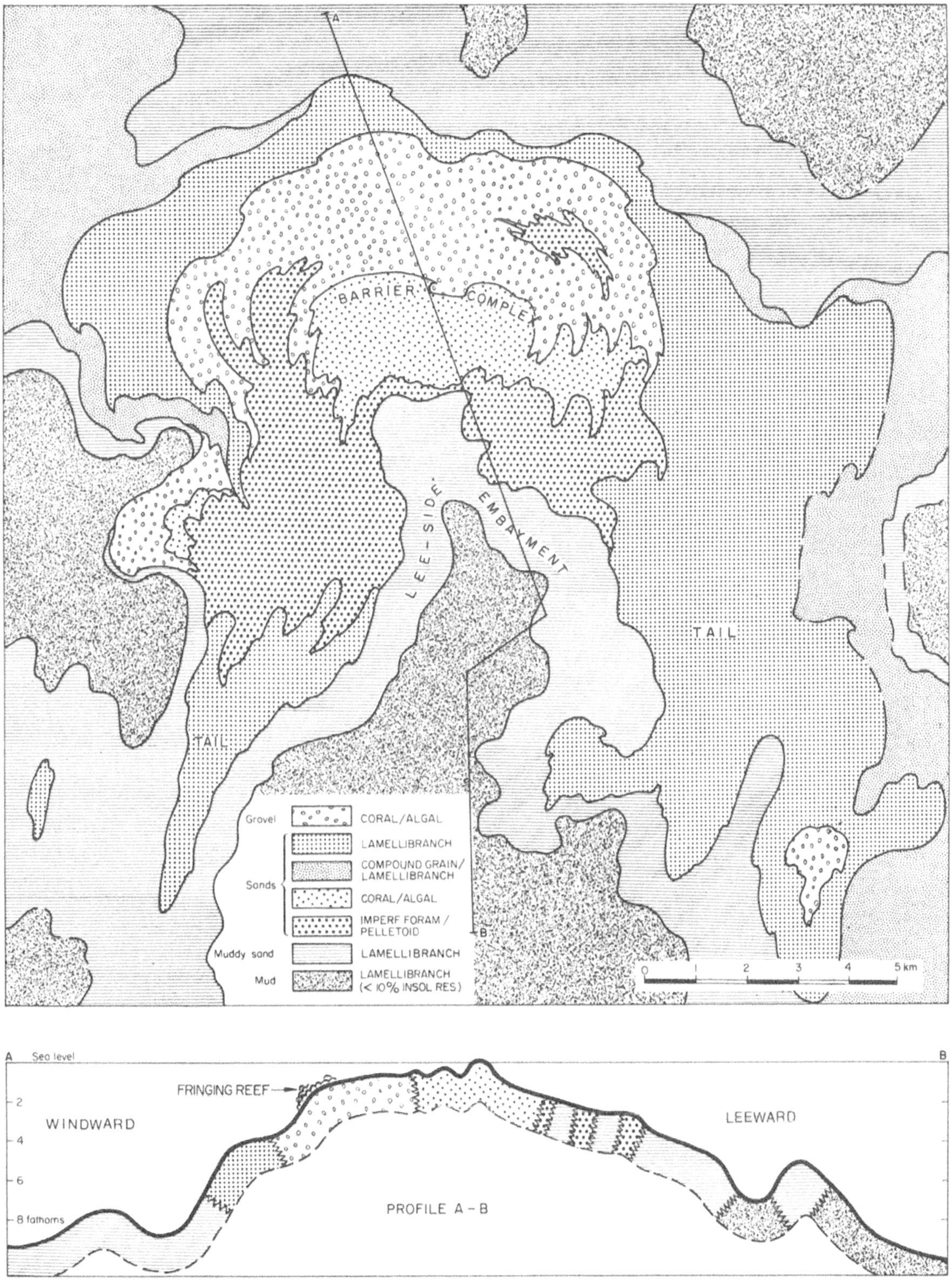

Fig. 8. Sediment distribution on Halat Dalma Shoal, showing characteristic asymmetric geometry closely related to NNW wind and wave directions

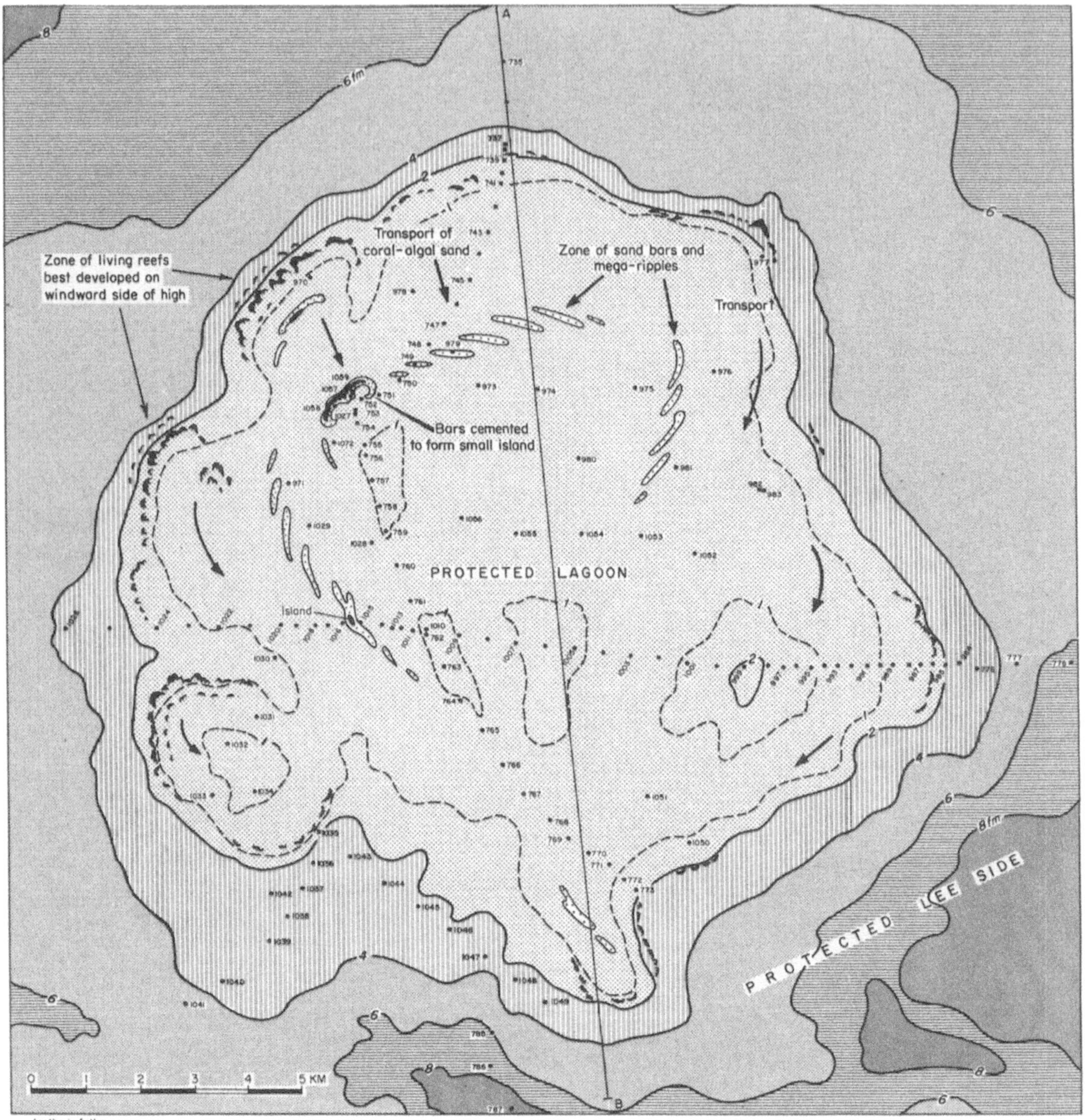

Fig. 9. Bathymetry of Bu Tini Shoal (Abu Dhabi).
This relatively wide (15-20 km) high is swept by wind-driven waves along its reef-fringed, NNW periphery. Bioclastic sand transported down-wind has created an arcuate "bull's horn" barrier which is locally emergent and stabilized both by submarine and beach rock lithification. This barrier encloses a shallow, muddy lagoon

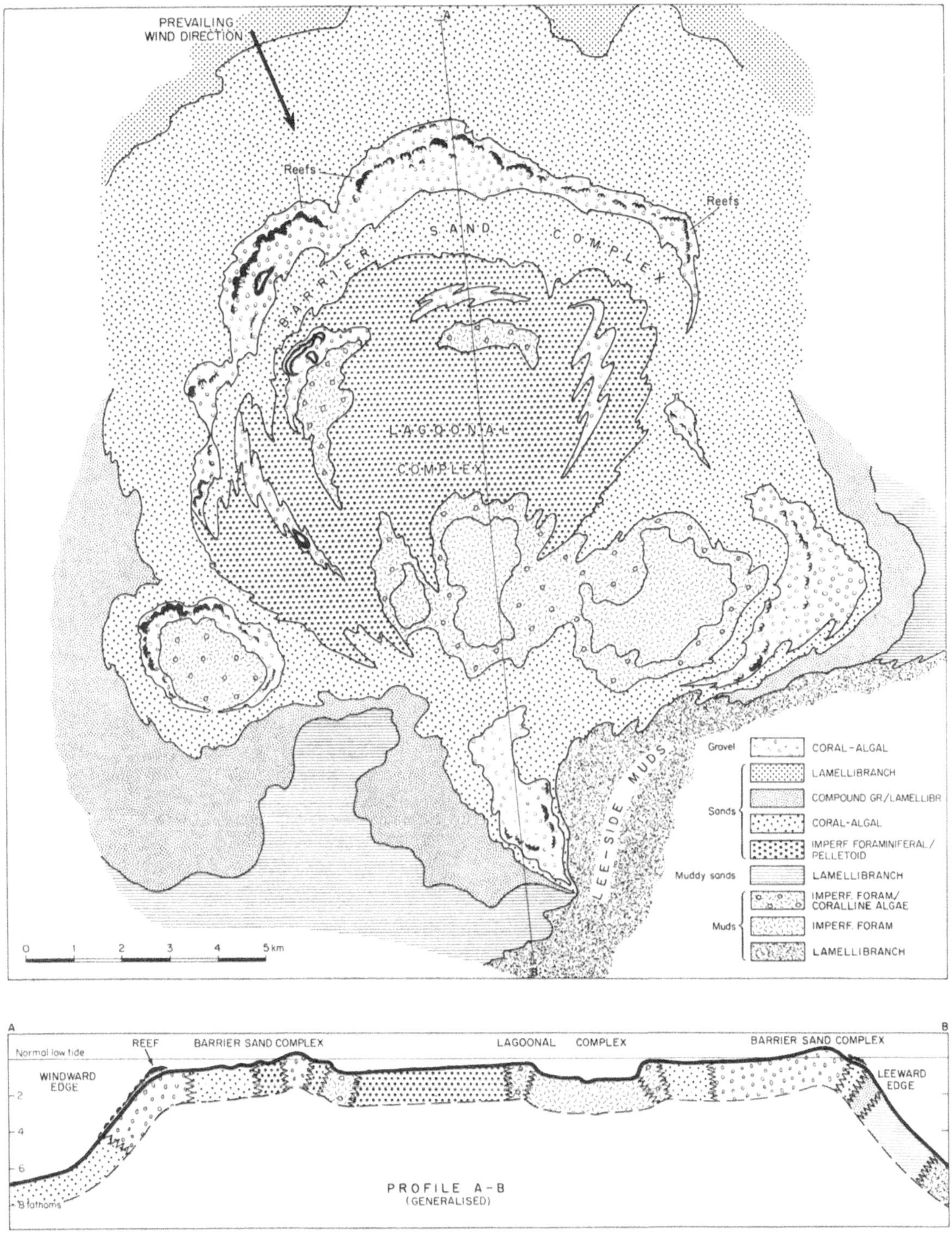

Fig. 10. Sediment types and patterns on Bu Tini Shoal

windward edge than that at Halat Dalma. Individual bars within the complex "bull's horn" barrier system are up to 1 km long and are stabilized by an active intertidal and shallow submarine lithification. Several of the larger bars combine to form permanent atoll-like islands, or cays, upon which plant life is established. These islands are increasing in size by a system of tidal-flat accretion on their leeward sides and by lateral sand spits at their extremities. The small tidal-flats support stromatolitic algal mats and other sedimentary features associated with emergence.

The barrier on Bu Tini high is more effective than that on Halat Dalma. It separates coarse coral/algal sands on its windward side from pelletoidal sands on its lee. The latter grade into carbonate muds which occupy minor depressions on the crest of the high (Fig. 10). The environments leeward of the barrier bar complex may be termed "lagoonal"; they occupy the greater part of this flat-topped high.

As at Halat Dalma, the barrier bar complex on Bu Tini extends downwind as a pair of sediment tails. Each tail is a complex of bars whose long axes are oriented subparallel to the axis of the tail (parallel to the dominant NW wind). Together with the barrier system, they enclose the crestal lagoon on three sides (Fig. 10). A poorly developed sand tail extends southwards from the high, but is seemingly inactive and probably predates the formation of the "bull's horn" barrier and lateral sand tail complexes.

Bu Tini shoal exemplifies an advanced stage of sedimentary development on the shallower highs. Its barrier system is virtually complete and has resulted in the development of lagoonal environments on the crest of the high. Carbonate muds thus extend across the topographic culmination and form tidal flats only 2-3 km leeward of the reef complex. It is not unreasonable to predict a further spread of these tidal flats leeward from the barrier bar complex and over the shallow lagoonal sediments. Dolomitization probably would be associated with the resulting "sabkha" phase.

Wide, salt-dome islands emerging above sea level, although lacking central lagoons and barrier bar systems, nevertheless have certain sedimentary features comparable with Bu Tini. Yas salt dome island (Fig. 11) is some 8 km in diameter and is characterized by a well developed fringing reef around its windward side. Detritus from this reef is transported laterally around the island as a series of spits, the longest being 2-3 km, which enclose small lagoons. In addition, a well developed submarine tail of oolitic sand, discussed by Loreau & Purser (in this volume), extends SE from the lee side of Yas, linking the island with the mainland shoreline.

The leeward side of Yas island is a coastal plain several kilometres in width which has developed from lateral accretion by a complex system of lateral spits.

As at Bu Tini, protection resulting from the excessive width of the high favours lee-side mud deposition; tidal-flats on the lee side of Yas are muddy and support a flourishing stromatolitic algal community. Other larger islands, such as Abyad, also have wide tidal-flats on their lee sides with evaporites and probably dolomite.

One may conclude that wide highs, such as Bu Tini and Yas, offer sufficient protection from the "shamal" winds to permit widespread deposition of muddy sediments in the downwind "shadow" for distances of up to 15 km. Thus, the leeward carbonate sand tails that characterize many small highs less than 5 km in diameter are replaced by downwind muds in the case of larger highs.

Inner homocline (coastal) highs

Morphology: Highs situated close to the mainland shoreline of the Trucial Coast may considerably affect coastal sedimentation. Most are situated near the axis of the Great Pearl Bank barrier (discussed by Purser & Evans in this volume), where they

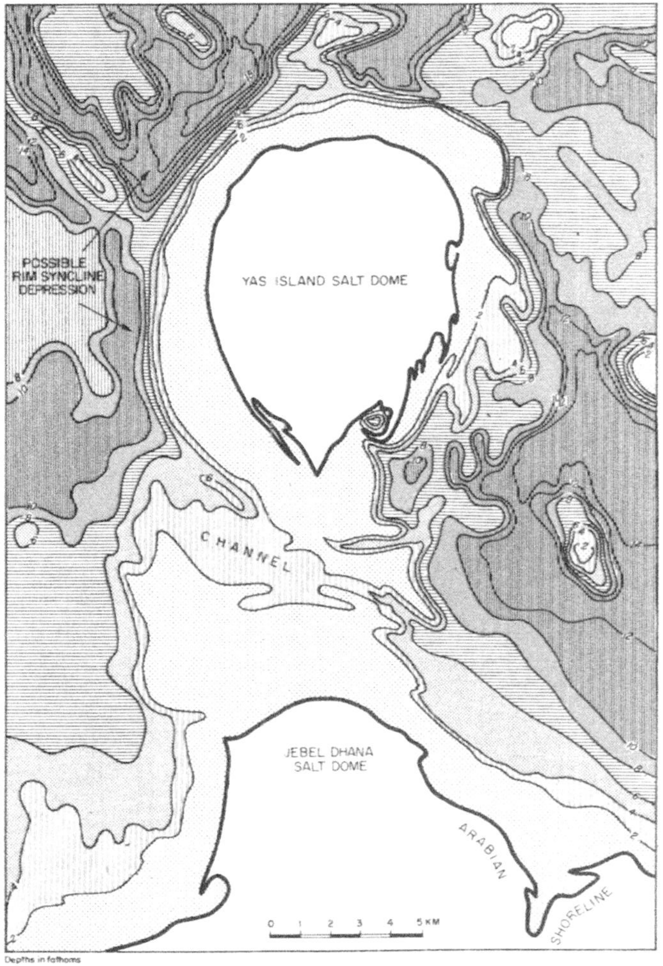

Fig. 11A. Bathymetry around Yas Island

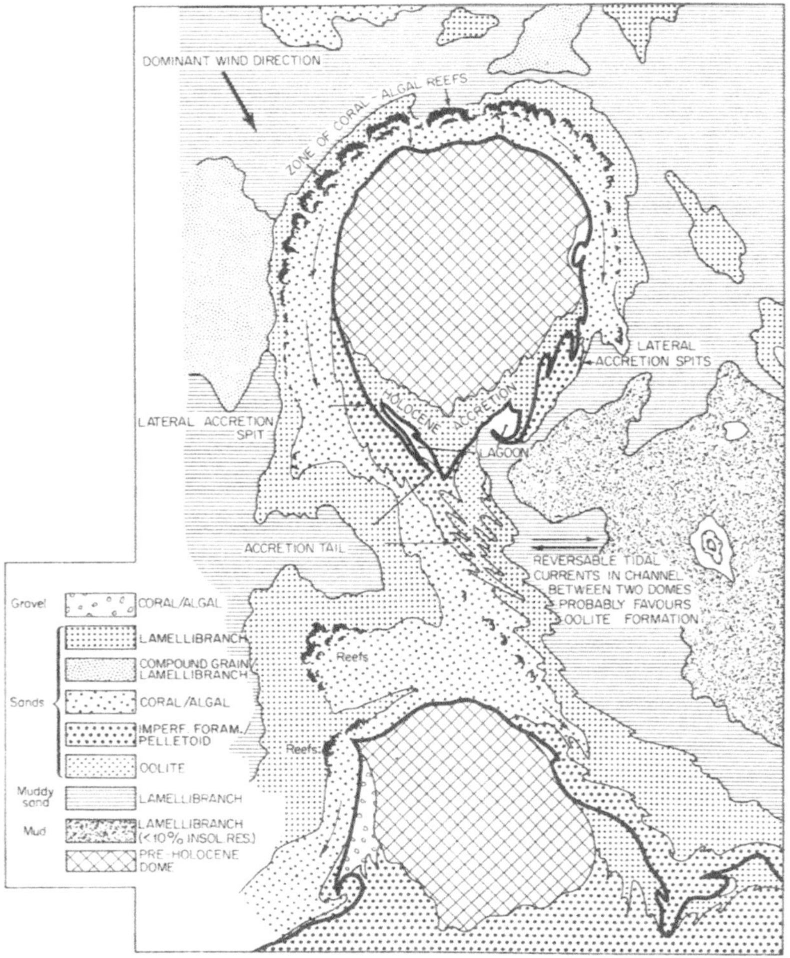

Fig. 11B. Sediment patterns around Yas Id.: leewards accretion is linking this high to the adjacent continent

constitute low islands and shoals surrounded by complex patterns of reefs and sand bars. Most highs have a pre-Holocene core of "miliolite" limestone around which sediments are accreting.

Sediment types and patterns: The shallow nature of these coastal environments results in active sediment transport around the highs to form sediment tails which extend from their leeward side (Fig. 11). This results in a progressive shallowing between the high and the adjacent mainland shore, provoking an increase in tidal current velocity and perhaps in the formation of oolitic sand bars, as between the island of Yas and coastal Abu Dhabi (Fig. 11).

Sustained sedimentation on the lee side of coastal highs in E Abu Dhabi has resulted in many being incorporated into the regional coastline as sedimentary peninsulas (or tombolas), the axes of which are oriented parallel to the prevailing NW wind direction. In consequence, many peninsulas are oriented somewhat obliquely to the regional orientation of the Trucial-Oman shoreline, Abu Dhabi "island" being a good example. The intervening depressions form lagoons with entrance channels having reversible tidal current systems which, together with suitable water chemistry, has favoured the develoment of spectacular oolite "deltas" at the seaward end of each channel. Thus, the presence of a series of bathymetric highs situated near the continental shoreline has provoked a complex sedimentary evolution, the various stages of which have successively modified the nature and distribution of coastal sediments, as discussed by Purser & Evans elsewhere in this volume.

III. SEDIMENTATION AROUND SALT DOMES

Salt diapirs constitute clearly defined highs in the E half of the Persian Gulf and the salt itself is virtually at the surface on many of the islands in the Trucial Coast Embayment. It is nowhere visible, however, and the salt apparently does not influence contemporary sedimentation.

However, the salt diapirs have brought much exotic material to the surface, including Palaeozoic dolomites and volcanic rocks, together with traces of heavy minerals including magnetite. This material is transported down the flanks of the island and mixed with indigenous Holocene carbonate sediments around its shore. Much of the exotic material accumulates close to its source as large cobbles and boulders, especially on the lee side of the island. Finer sands and silts are transported by marine currents and are incorporated into the sediment tail, together with the contemporary carbonates. The presence of the surface diapir can thus be detected only from a downwind direction. The presence of a salt diapir is not very apparent, even in the downwind sediments because the exotic material derived from the diapir constitutes only a very minor part of the sediment accumulating around the dome.

Certain salt dome islands, including Yas and Dalma, (Figs. 11 and 12) are flanked by faint bathymetric depressions which may be the surface expression of rim synclines. In common with most bathymetric depressions, these favour the accumulation of muds and muddy sands.

Apart from the presence of minor exotic sediments and the possible effects of rim synclinal depressions, salt-based highs affect contemporaneous sedimentation in the Persian Gulf in a manner comparable with any other marine high, irrespective of its origin.

IV. DIAGENETIC PROCESSES ASSOCIATED WITH MARINE HIGHS

Certain synsedimentary diagenetic processes are associated with marine highs in the Persian Gulf. These include submarine lithification, vadose lithification and leaching, and dolomitization.

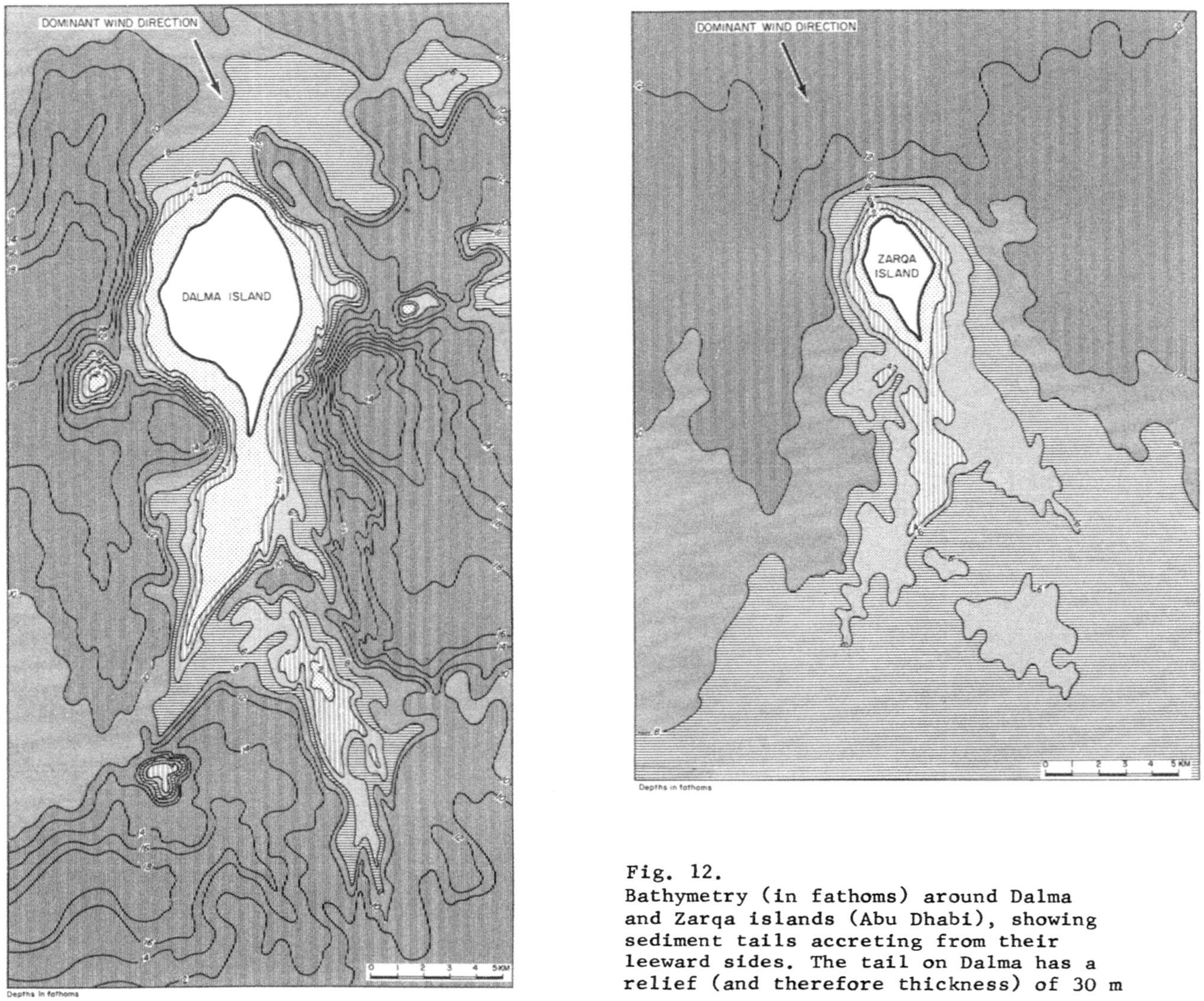

Fig. 12.
Bathymetry (in fathoms) around Dalma and Zarqa islands (Abu Dhabi), showing sediment tails accreting from their leeward sides. The tail on Dalma has a relief (and therefore thickness) of 30 m

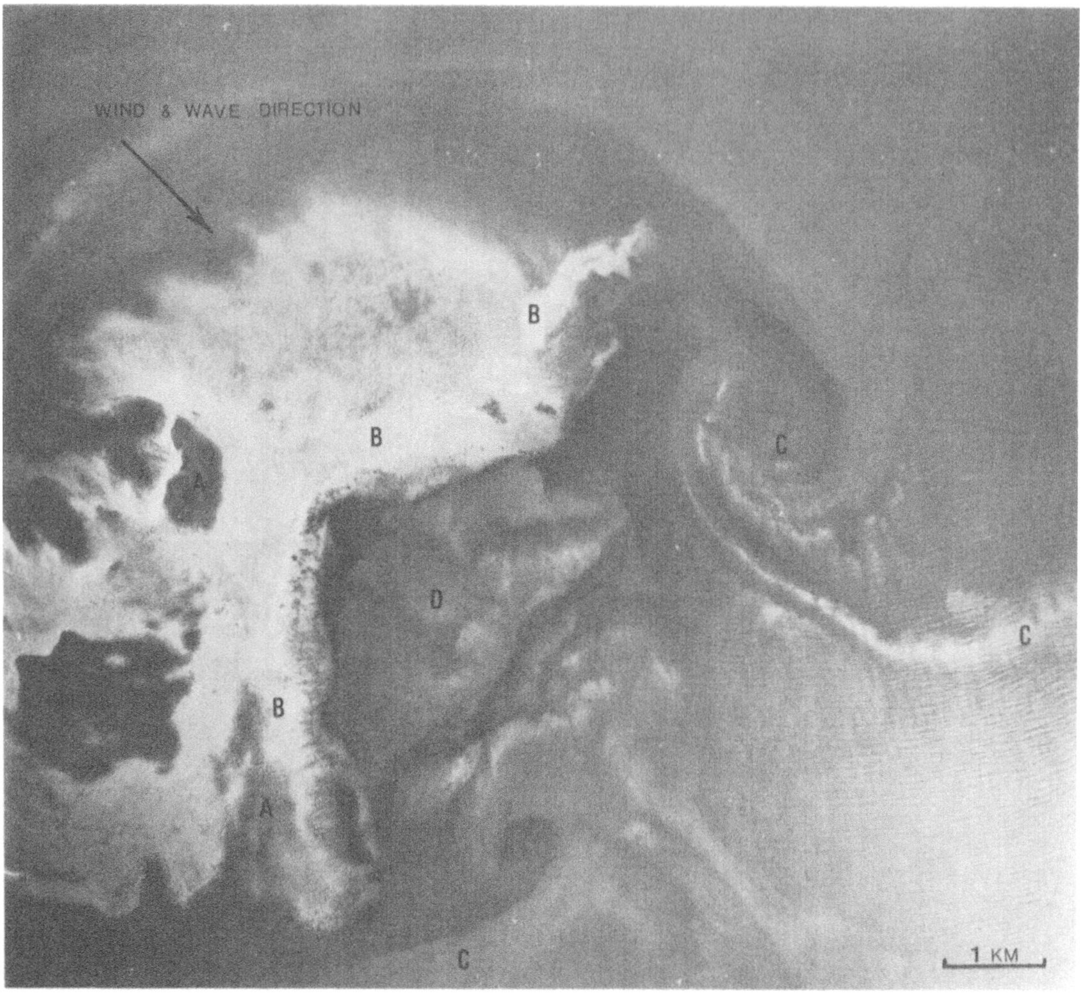

Fig. 13. Aerial photograph showing sediment patterns on, and around a relatively small high in offshore Abu Dhabi. This shallow (but completely submerged) high is swept by NW, wind-driven waves and surface currents which influence the distribution of living patch-reefs (A). Coral-algae sand is swept back to accumulate along the leeward edge of the high (B). Carbonate muds are carried in suspension (C) around the flanks of the high to accumulate on its leeward side (D)

Submarine lithification

Carbonate sediments may lithify shortly after their deposition. The process seems to be most active on sandy sea floors in moderate energy environments where adjacent grains remain in contact for relatively long periods (Shinn, 1969). It takes place close to, or at, the sediment-sea water interface and seems to cease when any given sediment surface is buried by rapid sedimentation. The crests of most highs are rocky and covered only by a thin skin of loose sediment. This is especially true of highs submerged to more than 10 m. In many cases samples have been recovered in which large cobbles of limestone were identical in composition to the loose sediment from which the sample was taken. At other localities cores showed transitions from loose sediment to dense limestone of identical sedimentary composition. In all cases the interparticle cements were fibrous or microcrystalline aragonite or magnesium calcite.

It is possible that some of the rocky highs, especially in the deeper parts of the basin, owe part of their present relief to active synsedimentary submarine lithification (verbal suggestion, E.A. Shinn). Sediments produced locally on the crests of drowned highs are fixed on the crest by the processes of lithification. They are surrounded by nonlithified muddy sediments in the adjacent deeper marine environments. On shallow or emergent highs the rate of sedimentation and active transportation discourages large scale submarine lithification. Accretion tails have thin crusts of submarine rock, especially at their proximal ends, but their flanks are generally unlithified, probably because of excessive rates of overspill sedimentation. Shinn (1969) has suggested that rapid sedimentation removes any sediment from the source of its cement, namely circulating sea water.

Vadose lithification

Where the crest of the high has emerged, either as a salt-based island or as a barrier sand complex, there is active beach rock formation. The resulting fabrics are described by Evamy elsewhere in this volume. Non-marine vadose lithification and leaching are frequently active towards the interior of the accretion plain.

Dolomitization and evaporite formation

When highs expand in size by sedimentary accretion, sabkhas on large islands such as Marrawah, Abyad and Yas are areas of active supratidal gypsum formation, and probably of dolomitization. The supratidal flats favouring these processes seem to form on the crest of the high when its diameter at sea level exceeds about 5 km. These processes occur especially on highs in coastal settings.

The types of synsedimentary diagenetic processes and fabrics associated with highs in the Persian Gulf vary according to the position of the high in the basin. Basin centre highs are often sites of active submarine lithification and rarely areas of large-scale vadose lithification or dolomitization. Highs in shallower settings may also be areas of active submarine lithification when they are submerged by more than about 10 m. Emergent highs, including salt-dome islands, may have small areas of intertidal lithification and leaching, but little dolomitization. Finally, coastal highs, because they often consist partly of wide supratidal flats, are frequently areas of active vadose diagenesis, and probably dolomitization.

Acknowledgements:

This contribution is based on research which was carried out by the author and his collegues for "Shell Research Laboratory (Rijswijk, The Netherlands). The author wishes to thank both "Shell Research" for permission to publish its data, and G.R. Varney and others who were associated with this project; many of the principles evoked in the preceeding pages are founded on discussion with these colleagues.

Carbonate Coastal Accretion in an Area of Longshore Transport, NE Qatar, Persian Gulf

E. A. Shinn[1]

ABSTRACT

In early Holocene times coastal embayments near Khor and Dakhirah on the E coast of Qatar Peninsula contained little sediment, and cliffs of Tertiary dolomite constituted the irregular coast line. After stabilization of the present sea level, fine grained,carbonate sediment and dolomite out-wash from the land began to fill and regularize the embayments. Concurrent longshore transport of Recent coarse grained carbonates from the north created "chenier" beaches with hook-shaped spits at their southern ends, which gave added protection to these pre-Recent embayments, creating an environment in which fine-grained carbonates could accumulate. Sedimentation continued within the coastal embayments until wide supratidal flats eventually formed. Contemporaneously, additional cheniers and spits were forming seawards of the older one creating quiet water embayments seawards of the older, high energy cheniers and spits. Fine grained carbonates accumulated between successive cheniers. During this five to six thousand year period, a complicated tidal channel system developed on the flats behind and between successive cheniers and spits. Progradation of this tidal-flat-channel belt and the construction of new cheniers has preserved the numerous channel deposits.

Chenier beach and spit development has straightened the coast. When the remaining embayments are eventually filled the character of deposition will probably change to that of simple beach accretion. Thus, the tidal-flat complex described in this text may, in time, be preserved as a distinct body of sediment differing notably from the sands and coral reefs at present accumulating a short distance off-shore.

The area of sedimentation described is 18 km long and up to 7 km wide. Its thickness exceeds 8 m in places. The system of coastal accretion by carbonate sediment influenced by longshore transport differs markedly from the pattern and textures of Bahamian tidal flats.

GEOLOGICAL SETTING

The NE coast of Qatar Peninsula is composed of Tertiary dolomite generally less than 15 m in altitude. Long embayments cut into these Tertiary rocks which constitute the "basement" for present-day lagoonal sedimentation. The water in these lagoons is generally less than 4 m deep, and more than 80 % of the lagoon illustrated

1 Shell Research B.V., Rijswijk, The Netherlands.
Present address: Shell Oil Company, Houston, USA.

Fig. 1. Map of SE Persian Gulf, showing area studied

in figure 2 is less than 2 m deep. Shallow probing and coring reveal sediment thickness in excess of 7 m, indicating that pre-sedimentation depths in these lagoons probably exceeded 8 m.

Salinity within the lagoons fluctuates between 41 $^{o}/oo$ and 44 $^{o}/oo$ according to Hughes Clarke (1966, personal communication). Offshore salinity averages approximately 39 $^{o}/oo$. Lagoonal water temperatures range seasonally from 5° C to slightly over 40 ° C.

NW prevailing winds, termed "shamals", produce southward moving longshore currents on both coasts of Qatar Peninsula (Fig. 1). Coarse bioclastic carbonates derived from extensive offshore coral reefs and "reef flats" at the northern end of the peninsula are transported southwards and deposited as extensive carbonate cheniers, bars and spits.

SEDIMENT CHARACTER AND DISTRIBUTION

Carbonate sediments are accumulating along the coast in three major environments:

The subtidal (in the open sea or lagoons permanently below low tide).

The intertidal (between normal high and low tides).

The supratidal (above normal high tide, but within the range of spring and storm tides; the low energy flats are termed "coastal sabkhas" by Glennie (1970).

A. Subtidal sediment

Character

Most subtidal sediment in the embayment area illustrated in figure 2 consists of mixtures of soft pelleted mud and silt-sized carbonate. These sediments, in which pellets are often squashed beyond recognition, contain varying amounts of winnowed skeletal grains. The sediments are predominantly grey in colour and the smell of hydrogen sulphide gas they give off suggests reducing conditions. They lack primary sedimentary laminations and are highly burrowed and churned (fig. 3).

The fine grained, subtidal, embayment sediments contain as much as 30 % dolomite. The dolomite crystals are less than 30μ in size thus falling in the same size range as aeolian dust deposited during strong winds; as much as 65 % dolomite is present in aeolian dust both on- and offshore Qatar. This airborne dolomite and that washed from the Tertiary rocks during sporadic rains, is thought to be the origin of the dolomite in these subtidal sediments. This is further supported by X-ray data which show that the Tertiary dolomite, the airborne dolomite, and that in subtidal sediments have a Ca/Mg mol % ratio of 50:50*. Proven Recent dolomite in this area, on the other hand, has a Ca/Mg mol % ratio between 53:47 and 55:45 (Illing, Wells & Taylor, 1965).

Subtidal sand accumulations occur as winnowed lags in channels and as a thin narrow belt of moving, rippled sand seaward of, and adjacent to, active chenier beaches (Fig. 8). In channels, these sands invariably contain grey, carbonate mud and silt, and more than half the carbonate sand grains are darkened, giving the sediment a "salt and pepper" appearance (Fig. 4). The moving sands seawards (E) of the chenier beaches, on the contrary, are clean, well sorted and lack salt and pepper grains.

---

* Analysis by K. de Groot, Shell Research B.V. (Rijswijk, The Netherlands)

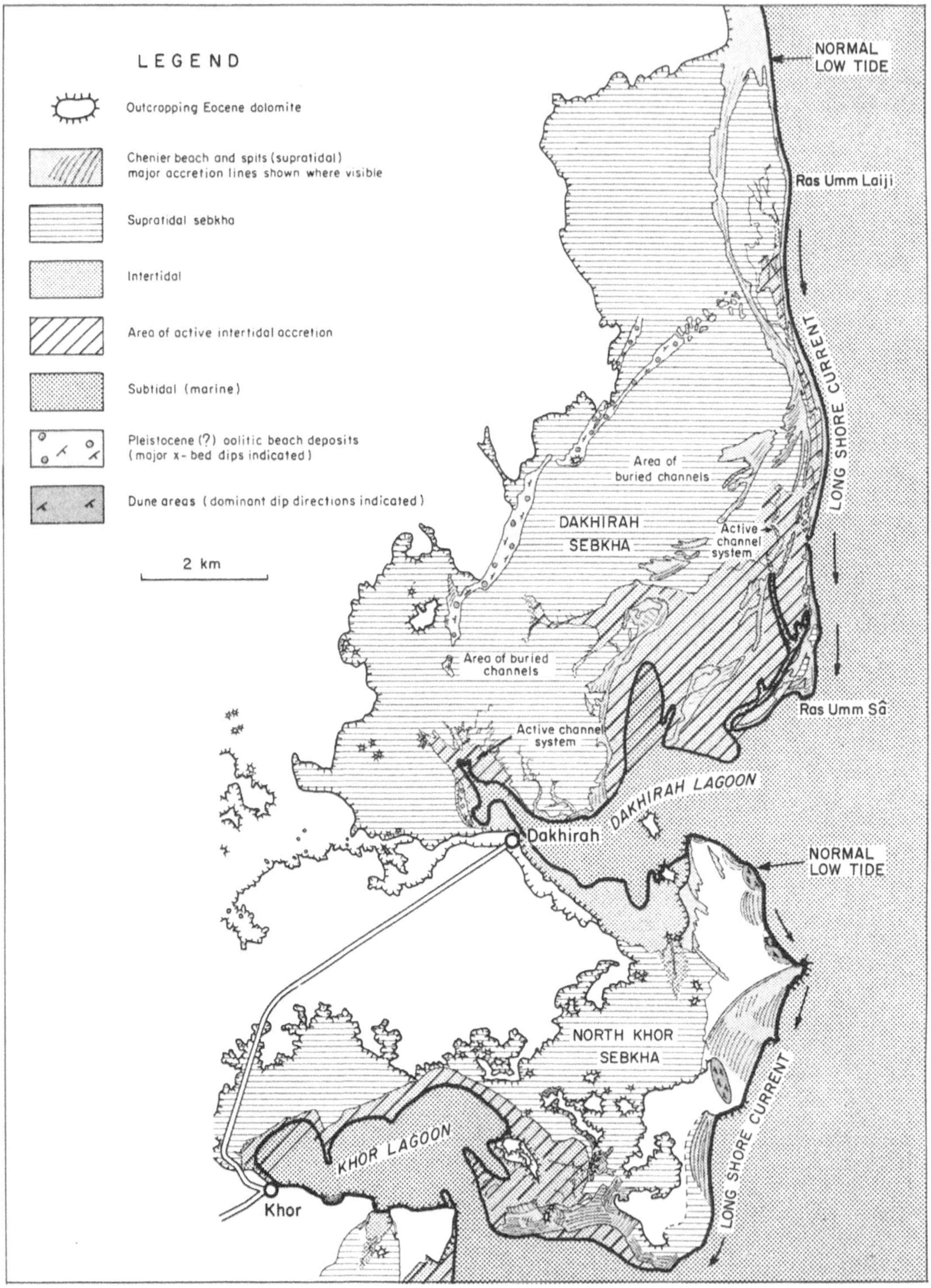

Fig. 2. Map of NE Qatar Peninsula showing the distribution of the principal sedimentary units

Distribution

Subtidal muds and silts constitute the thickest and most extensive deposits, in places more than 8 m thick. These sediments underlie all the extensive intertidal and supratidal sediments and cover the floors of all the lagoons in Qatar.

The salt and pepper sands are limited in occurrence and are preserved in meandering channel deposits beneath supratidal flats and within active channels of the present intertidal zone.

B. Intertidal sediment

Character

Intertidal sediments resemble those of the subtidal zone in being predominantly pelleted carbonate muds and silts containing varying amounts of skeletal material. They can be distinguished mainly by their light-tan (oxidized) colour, numerous iron-stained root tubes and burrows, and "birdseye" vugs. The latter, described by Shinn (1965), occur as randomly scattered, bubble-like holes in burrowed and churned intertidal sediment, and as planar vugs in well-laminated sediments (Fig. 5). Sedimentary laminations are rare, mainly as a result of disturbance by burrows and plant roots, although some stromatolitic algal laminations exist in channels in the upper parts of the intertidal zone.

Both fine and coarse grained intertidal sands are present as local deposits in channels, miniature aeolian dunes, and beaches. These accumulations are described in a subsequent section. Dolomite is present in intertidal sediments in roughly the same proportions as in the underlying subtidal sediments. Its composition suggests a detrital origin.

Distribution

Intertidal sediments constitute a belt a few metres to a kilometre wide. Where this zone is wide, it is muddy and laced with numerous, small, meandering tidal channels (Fig. 6), while narrow intertidal zones lack channels and their sediments are sandy. Wide intertidal belts predominate within the embayments where they are sinuous and in places even perpendicular to the regional coast line. Where the belt is wide and channelled, there is an unchannelled flat a few hundred metres wide in the lowest (seaward) part of the intertidal zone. In some places these lowest intertidal sediments, which are sandier than the surrounding sediments, are cemented to form a crust 2 - 10 cm thick.

Small drifts of sand accumulate on the muddy interchannel area of the intertidal zone. These dunes, or drifts, are fixed by plants and thus restricted, like the salt grasses and mangroves, to the intertidal zone. When the intertidal belt accretes seawards, the plants die and the dunes are blown away.

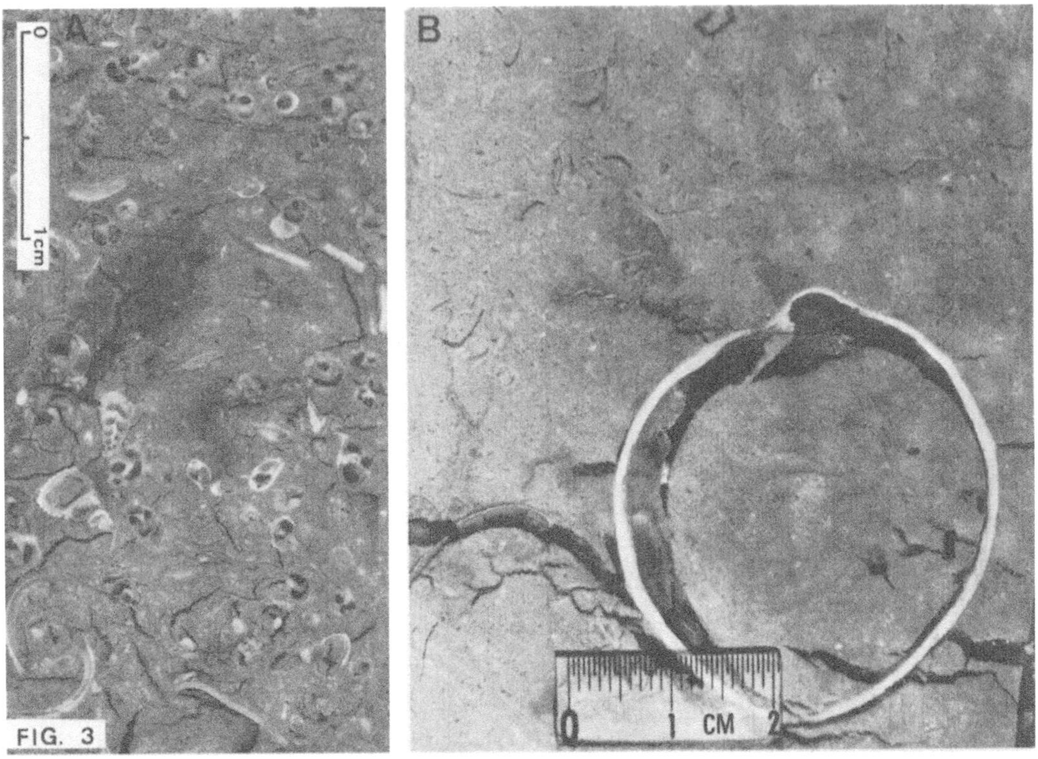

Legends to Figs. 3 and 4 on the following page

C. Supratidal sediment

Character

Though predominantly muddy, supratidal sediments are extremely variable. Lags of skeletal sand 2 - 3 cm thick deposited during flooding tides are common. Pellet rich muds are often firmer than those of intertidal sediments and pellet shapes are more easily recognized in cores. These sediments are characterized by supratidal laminations, birdseye vugs, stromatolites, mud cracks and light (oxidized) colour. However, on deflated or eroded parts of the supratidal flats, the above features, except for colour and birdseye vugs, may be absent; thus in parts of the area studied it is impossible to distinguish intertidal from supratidal sediments.

Between 10 % and 20 % dolomite was recorded in supratidal sediments by Illing, Wells & Taylor (1965). The amount in the area studied by the writer is much less and there are two types; dolomite with equal molar per cent Mg and Ca, and dolomite which is Ca enriched. Only the latter is thought to be contemporary.

Distribution

Supratidal flats, or "sabkhas" are the predominant feature in the area studied. These brown, barren flats are bounded landwards by pre-Recent outcrops and seawards by intertidal flats and chenier beaches. These flats attain 5 km in width and may be more than 10 km long. The thickness of the supratidal sediments may exceed 50 cm, but in many parts the surface has been blown away and underlying intertidal sediments are present beneath a salt or gypsum crust 2 - 3 cm thick. These flats are flooded only during the summer months when strong winds often blow from the E.

In many areas pinnacles of underlying Tertiary dolomite protrude through the supratidal flats to form steep, rubble-encrusted "mesas", the nature and possible geological significance of which are discussed by the author elsewhere in this volume.

TIDAL CHANNELS

Preserved channels:

Trenches cut across buried channels on the supratidal flats revealed channel sediments consisting of a basal, lithified, "salt and pepper" lag approximately 10 cm thick, overlaid by some 50 cm of graded, uncemented, muddy channel lag containing "salt and pepper" grains. These sediments grade upwards into light-grey muds approximately 25 cm thick. Sandy, stromatolitic sediment rests disconformably on the light grey mud and the remaining 0.5 - 1 m of sediment consists of sorted, cross-bedded, light brown, carbonate sand containing birdseye vugs. This cross-bedding is truncated by the deflated surface of the supratidal flat. The intertidal sediments on either side of the channel overlie the basal channel lag and cemented layer.

Fig. 3. Subtidal sediments (plastic impregnated cores).
Plastic-impregnated slices of core of subtidal sediment. Lack of sedimentary structures due to burrowing. Lamellibranchs (? *Lucina* sp.) and cerithid gastropods are main elements

Fig. 4. Channel sediments.
"Salt and pepper sands". A shows darkened grains in winnowed sediment from a channel, B shows similar grains in plastic-impregnated core of channel-lag sediment

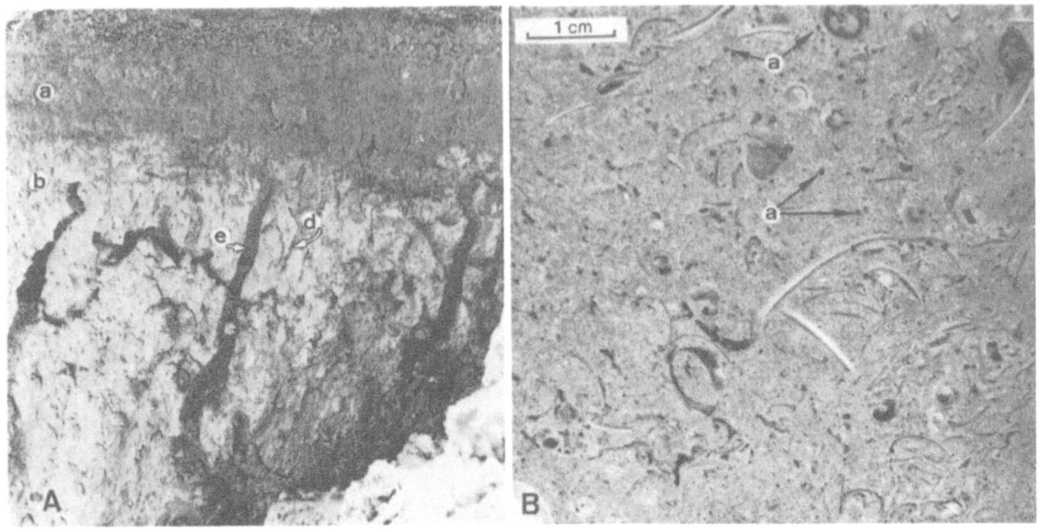

Fig. 5. Inter and supratidal sediments.
A. Contact between supratidal (a) and intertidal sediments (b). Note abundant "birdseye" vugs (c) in nonlaminated supratidal sediment, and burrows (d) and mangrove root casts (e) in intertidal (wall of pit).
B. Plastic-impregnated core of intertidal sediments showing numerous mollusks and characteristic bubble-like "birdseye" vugs (a)

Evolution of tidal channels:

Development and preservation of the tidal channels may readily be observed by following a particular channel across the intertidal belt (Fig. 7). Most active channels may be traced landwards to a point where they die out beneath supratidal flats and all intermediate stages of burial may be observed. The stages are as follows:

A cemented layer beneath a few centimetres of soft sediment has been observed locally on the broad, low intertidal flats seawards of the channel zone. This layer becomes less cemented and disappears seawards but may be traced landwards beneath the intertidal and supratidal flats. Active intertidal channels erode down to, but do not cut through this layer, and winnowed channel lags similar to the grains within the cemented layer are deposited on top of it. Most channels, therefore, are restricted to a common depth by this underlying hard layer. If a channel is traced landwards it gradually becomes shallower and channel sediments become muddier. Eventually a point is reached where the muddy channel floor is rarely flooded and stromatolitic algae flourish. This landwards part of the channel system is filled mainly by wind-blown sand. As a result, the channel fill lacks the vertical decrease in grain size characteristic of silici-clastic, water deposited, channel sediments.

CARBONATE "CHENIER" BEACHES

Beaches ridges* fifty to one hundred metres wide, two or more metres thick and 12 km long, are a characteristic sedimentary unit in the area studied. For the most part these narrow cheniers rest on intertidal, supratidal and even channel deposits, indicating that they have migrated landwards during formation. The southern ends of these features have hook-shaped spits which curve landwards and overlie subtidal sediments (Fig. 9). Cross-bedding in the inter- and supratidal part of the straight segment of chenier beaches is distinct from that of the hook-shaped spits at their southern ends.

---

* Similar ridges are termed "cheniers" on the Gulf Coast.

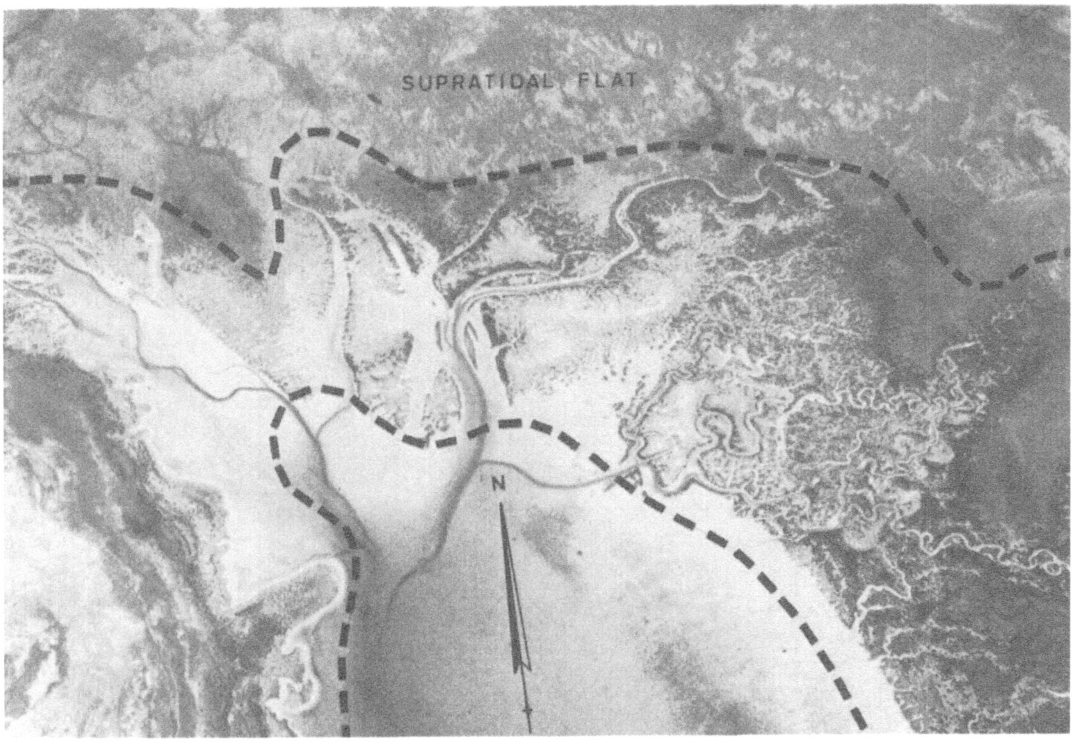

Fig. 6. Oblique aerial view of intertidal channeled belt NW of Dakhirah village (Fig. 2). Subtidal (below) and supratidal flat (above) bounds intertidal flat. Vegetation and sand drifts are restricted to channelled, intertidal belt

Cross-bedding in carbonate chenier beaches:

The chenier beaches in Figure 8 have cross-bedding which dips away from the sea. A few centimetres from the base where the deposit rests upon tidal-flat sediments (Fig. 10), there is a characteristic 0.5 m-thick set of cross-beds that incline predominantly landwards with dips as steep as 30°. Above this steeply dipping unit cross-beds dip landwards, but at angles of less than 5°. Within the gently dipping beds, coarse bioclastic debris, including corals and bored Tertiary dolomite pebbles, form 2 - 5 cm thick lenses the long axes of which are parallel to the beach. A thin layer of seaward-dipping beach beds is present as a coating on the seaward side of these chenier beaches.

Cross-bedding in spits:

Festoon cross-bedding is abundant in the hook-shaped spits at the southern, prograding ends of chenier beaches (Fig. 9). Their dips range through 360°. There is a tendency for most spit bedding to be destroyed and replaced by simpler, more permanent chenier bedding during the course of chenier evolution.

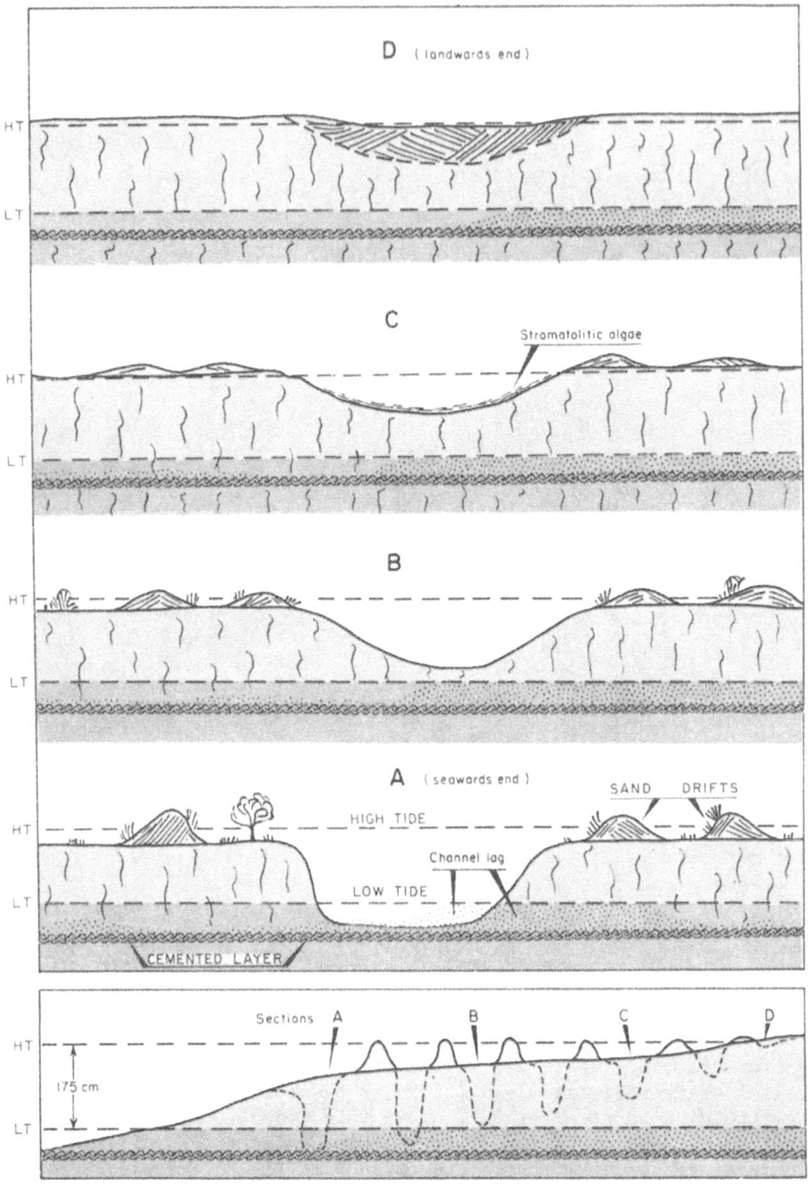

Fig. 7. Successive cross-sections along a channel showing progressive changes from active outer part (A) to inactive inner part (D). Bottom, longitudinal section shows relative locations of sections A-D. Note that channel lag is deposited on cemented layer (in section A) and that final filling of channel is aeolian sand (section D). In figure C stromatolitic algae have begun to grow on channel floor and are eventually covered by cross-bedded sand (Fig. D)

Evolution of carbonate cheniers:

Study of successive aerial photographs dating from 1947 to 1966 and periodic field observations show that most active accretion takes place on spits at the southern ends of chenier beaches. The spit complex at the southern end of the outer active chenier beach Ras Umm (Fig. 9) accreted more than 30 m in one year.

Fig. 8. Chenier beach in the process of landward shift. Left arrow shows remnants of spits which have become cemented around their intertidal borders (beach rock). Right arrow shows remnant of intertidal "spit rock" which has been left behind as chenier beach advanced landward. Destruction of these cemented spits produces slabs and cobbles which are reincorporated into the chenier

Fig. 9. Oblique aerial photo of spit at Ras Umm Sa (see Fig. 2) taken in April 1965. The entire spit has been built since 1957. Arrow indicates cemented intertidal sediment less than 20 yr old

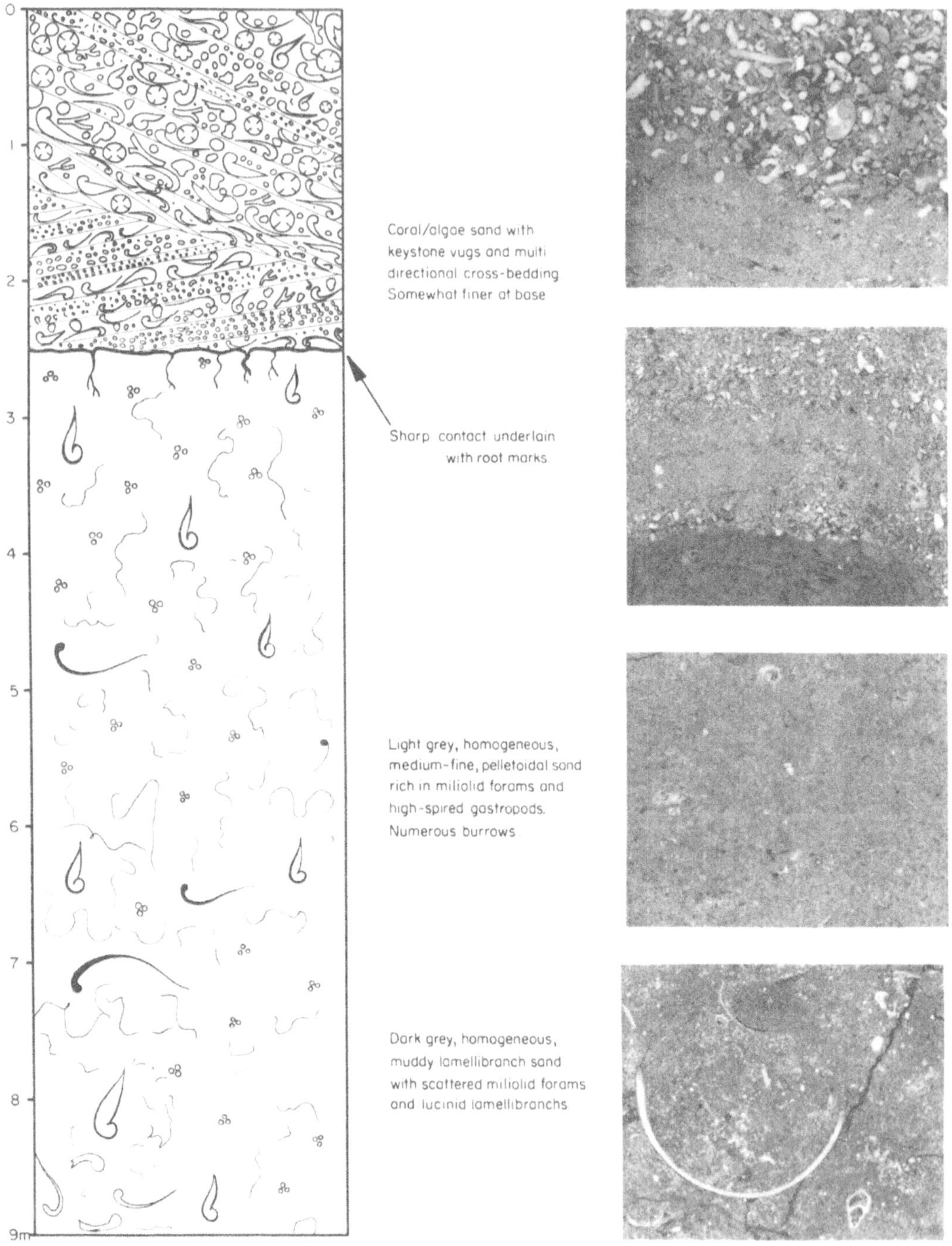

Fig. 10. Lithological sequence penetrated by well drilled on a chenier beach, Dakhirah, NE Qatar

As spits extend the cheniers southwards, that part which has been constructed becomes subject to erosion, especially during periods of strong easterly wind. Through such action the chenier is constantly straightened and pushed landwards over tidal-flat deposits, the process of spill-over producing a narrow ridge with characteristically landward-dipping cross-beds. The steeply dipping, basal cross-bed unit represents the leading edge of the landward accreting chenier. This leading edge is always steep because this portion of the beach is deposited below sea level during high water storm flooding; the overlying, gently dipping beds are constructed by waves spilling over the beach ridge.

Not all spit bedding is destroyed and replaced by chenier bedding. If a new chenier and spit complex forms seawards of an older one, it will divert longshore currents and wave action favouring the preservation of the older one. The complicated spit bedding will thus be preserved at the end of the older chenier and muddy tidal-flat sediments will accumulate around and stabilize the feature.

Trenches reveal that all the preserved older cheniers in the area have spit bedding in their southern extremities and simple chenier bedding in the remaining N, up-current part. The major part of the preserved cheniers is composed of planar, landward-dipping beds.

DISCUSSION AND GEOLOGICAL IMPLICATIONS

This study indicates that carbonate tidal-flat accumulation in areas of longshore transport may produce many parallel, linear sand bodies not formed in areas lacking longshore transport, such as W. Andros Id. in the Bahamas. The geological implication, therefore, is that not all ancient carbonate tidal flats are necessarily extensive blankets consisting essentially of lime muds, as previous studies have suggested (Shinn, Ginsburg & Lloyd, 1965).

The formation of a lower intertidal cemented layer in front of an accreting channel belt has the unique function of limiting channel erosion to a common depth. The result is that, through lateral channel migration, a sheet-like deposit of winnowed channel lag is produced which might not be recognized as a channel deposit if seen in outcrop; channel boundaries are not necessarily erosional and therefore would not be recognized. The only part of the channel that might be recognized is the landwards segment where distinct channel shapes are filled with cross-bedded sands. Since this portion of the channel deposit lacks the upward decrease in grain size characteristic of noncarbonate channel deposits, inter-channel areas could be misinterpreted as peneplaned, mud build-ups surrounded by cross-bedded sand. Indeed, few tidal-channel deposits have been recognized in ancient tidal-flat deposits.

Many of the features described have been recorded in areas of silici-clastic accumulation by Van Straaten (1956) and others. The similarities between silici-clastic and carbonate deposition are becoming increasingly apparent and suggest that many of the processes and patterns established by the study of silici-clastics may be applied also to carbonate areas. One notable exception, however, is the scale of tidal channels. The small tidal channels described here, generally less than 100 m wide, are much shallower than those on the North Sea coasts.

Recent Intertidal and Nearshore Carbonate Sedimentation around Rock Highs, E Qatar, Persian Gulf

E. A. Shinn[1]

ABSTRACT

Terrestrial morphology around the southern side of the Persian Gulf, especially in Qatar, is often characterized by low mesa-like hills of Tertiary dolomite and limestone. Many of these hills protrude through modern intertidal and supratidal sediments. They also occur in the subtidal environment elsewhere in the Persian Gulf. Most of these features probably began to form during a lower stand of sea level. It is suggested that similar erosional remnants in some ancient deposits may easily be misinterpreted as "bioherms" or localized "carbonate build-ups".

REGIONAL SETTING

Figure 1, a view across the sabkha north of Khor lagoon, on the E coast of Qatar, illustrates a typical mesa composed of massive dolomite of Eocene age. This dolomite is extremely fine grained and no fossils are visible. When freshly broken, the rock is light grey (except near the top of the unit where it is rusty brown), porous, contains numerous, sometimes chertified burrow casts, and is laminated locally. On many mesas, however, this upper portion of the unit has been removed by erosion.

Landwards of this coastal area, the Eocene dolomite forms more massive, continuous outcrops which constitute the essential morphology of the Qatar Peninsula. Between the less eroded area inland to the west and the area described in this contribution all stages of mesa formation may be observed. Further seawards are drowned mesas of similar shape, often capped by coralline sediments. Much less is known about these features. Within the present tidal-flat areas there is a complete spectrum of residual rock masses ranging from exposed to completely buried (Fig. 1), permitting a reconstruction of their Holocene history of erosion and subsequent burial in Recent carbonate sediment.

Mesas in the tidal-flat environment

The most characteristic feature of these mesas is their scree, i.e. the "flank beds", that partially cover their slopes (Fig. 2B). In most cases these "flank beds" are cemented, or partially cemented, and stand at a steep angle. They have two main components:

1 Shell Research B.V., Rijswijk, The Netherlands.

Present address: Shell Oil Company, Houston.

Fig. 1. Mesas of Tertiary dolomite in the process of burial by Holocene tidal flat sediments

A. Erosion-sculptured Eocene dolomite "mesa" **protruding** through Recent sabkha sediments on North Khor sabkha.

B. Dolomite mesa almost buried in Recent sediments. A $4\frac{1}{2}$ metre core of Recent sediment was taken to the right of the man. Close-up photos of this mesa are shown in Fig. 3

1. Angular Eocene dolomite clasts derived from the mesas (Fig. 3). These range in size from sand to automobile-sized boulders. Cracking and subsequent recracking of the angular dolomite scree is thought to be due to thermal expansion and contraction. In desert climates, such as that of Qatar, the surface rocks may experience diurnal temperature changes ranging from near freezing over 50° C. Minute expansion cracks are continuously filled with dust and with each further expansion more sediment enters and prevents the cracks from closing, thus progressively forcing the segments apart.

2. Recent mollusks, coral, and calcareous algal fragments derived from the adjacent modern intertidal and subtidal environments.

The Eocene dolomite fragments generally form the upper parts of the "flank beds" which, in the "mounds" studied, are 50 cm or more above high spring tide. Recent faunas occur only in the lower parts of the flank beds, below the level of spring and storm tides. Both types often merge and interfinger with the fine grained, sometimes stromatolitically laminated, intertidal and supratidal sediments which envelop them.

Fig. 2A. A "streamlined" mesa that faces marine conditions (windward) to the right and supratidal flats to the left (See figs. B and C.).

2B. Cemented, gastropod-rich, laminated flank beds on leeward side of mesa in Fig. A.

2C. Windward side of mesa in Fig. A showing grey dolomite mudstone core exposed by coastal erosion. Upper 50 cms. is angular dolomite scree which is continuous with bioclastic bed shown in Fig. B

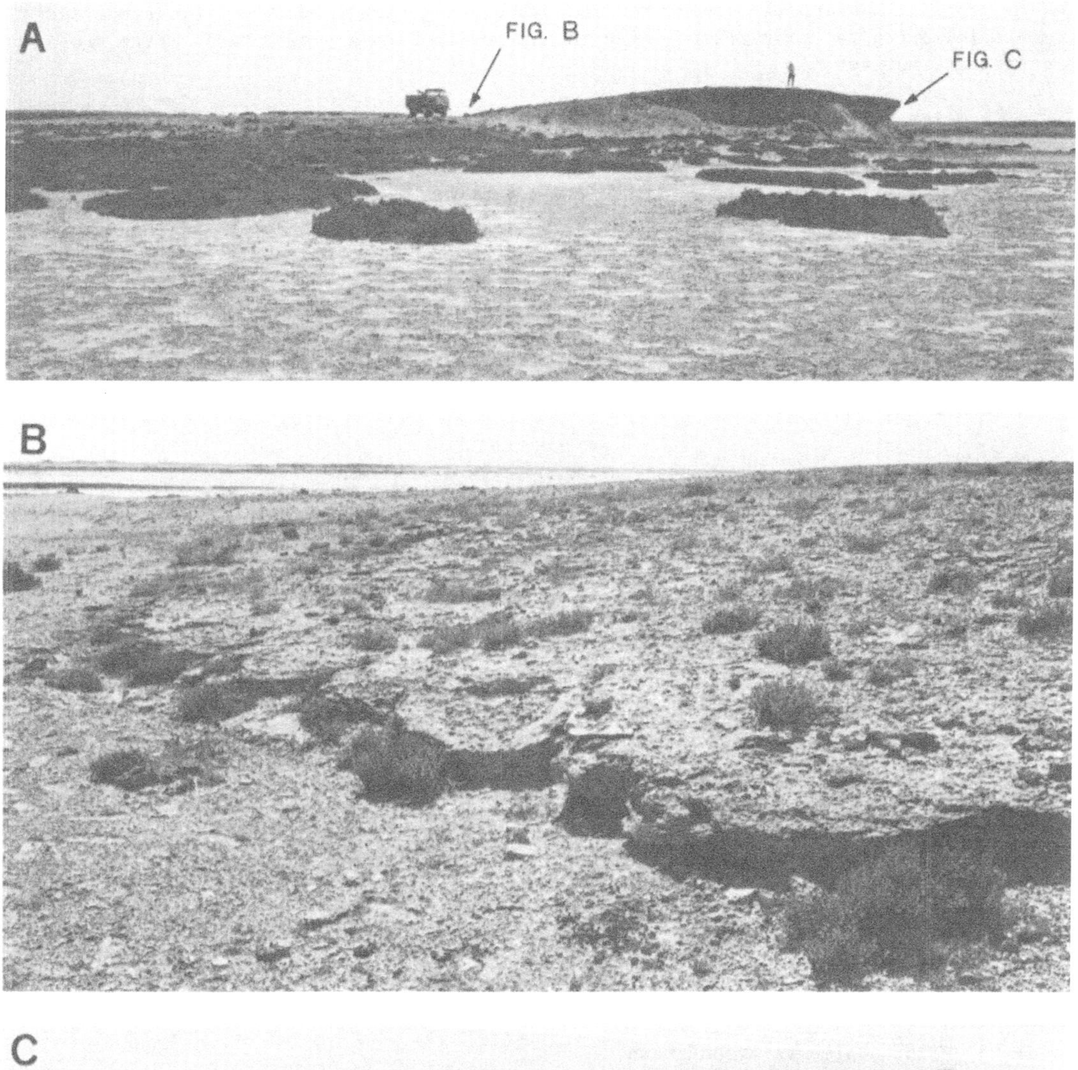

Legend to Fig. 2 A-C on the preceding page

The subaerially sculptured mesas are being slowly "drowned" by Recent sediments following the post-glacial rise in sea level. During slow burial three processes are active:

1. The mesas still exposed above the Recent tidal-flats continue to weather and produce fragments that are transported downslope.

2. The lower flanks of these exposed Eocene outliers act as beaches during storm-tide flooding, and high energy conditions at this break in slope result in the deposition of bioclastic sediments. These coarse sediments merge and interfinger with surrounding unconsolidated fine grained,lagoonal, intertidal and supratidal sediments.

3. Both the predominantly clastic Eocene dolomite and Recent bioclastic parts of flanks beds are lithifying.

Fig. 3. Flank sediments around mesa.
A. Same boulder as that to right of man in Fig. 2. Cemented angular dolomite scree often breaks into such pieces.
B. Detail of an individual piece of cemented angular dolomite which has fractured in situ. "Soil-like" dolomite sediment and carbonate cement (calcite) fill the cracks

Mesas in the subtidal environments (Doha Bay)

These features (Fig. 4) differ from those described above in that they may be capped by as much as 4 m of muddy,unconsolidated Recent sediment in which "float" whole and broken corals of two genera, Acropora, and Porites. The upper surface of this sediment, which is about 1 m below sea level at low tide, is often cemented to form a stable base for thickets of branching Acropora. Coral growth is limited to the northern, windward side of these features. The outline and spacing of these features are similar to those in nearby tidal-flats, suggesting a similar origin; although probing with a metal rod shows that they are underlaid by rock highs, it has not yet been proved that the rock is the same as that described above. Regardless of their age, however, these rock highs permit the initiation of coral growth in what would otherwise be a muddy, and therefore unsuitable, environment. Continued organic production has perpetuated the highs.

The exact processes involved in these subtidal areas are not completely known. All that can be stated is that coral, algal, and molluskan production is localized on rock highs because the surrounding sediments are mainly unfavourable muds. Continued build-up, accompanied by intermittent cementation, has produced slopes as steep as 20° where sediments rich in coralline and molluskan debris interfinger with surrounding, fine grained, carbonate muds.

Fig. 4. Aerial view of submarine, coral-capped "build-ups" in Doha Bay. Rock high (of unknown age) underlies about 12 feet of unconsolidated sediment. (Boat is about 50 metres long)

Geologic implications

The question arises as to whether similar older dolomite or limestone mesas embedded in younger tidal-flat rocks would be misinterpretèd as "organic build-ups" or "reefs" if encountered in ancient rocks. Subaerial sculpturing might explain the steep-sided "cores" and flank beds seen in some ancient "reefs" and "bioherms". Many of these are built of dolomitised cores which have often been attributed to selective secondary dolomitization, and relatively undolomitized flank beds. The Recent examples described also have dolomite cores and bioclastic limestone flanks, suggesting that the distribution of dolomite in some ancient examples may not be entirely secondary.

Some ancient "bioherms" are buried in shales. It should be pointed out that the mechanism described here is not limited to areas of carbonate sedimentation. Argillaceous muds or quartz sands (due to a regional change in source area) could bury dolomite or limestone mesas, thus making them appear to have been "killed" by an influx of terrigenous sediments.

The unconformity between "core" and flank beds might be extremely difficult to recognize in ancient equivalents, mainly because of the seemingly gradational contact produced by caliche-like weathering. Furthermore, the time break between "core" and surrounding sediment might not be of sufficient magnitude to be recognized paleontologically, or, as in the examples shown here, fossils might be absent from the core.

In sum, certain fossil accumulations that have been termed "bioherms" may therefore have been wrongly so called, simply because of the subtlety of the criteria by which true bioherms can be distinguished from the erosional forms described here.

The modern examples described in this text illustrate some of the subaerial processes that must have played a dominant role in the development of many "geologic reefs". More importantly, they demonstrate a mechanism for localizing the distribution of "reefs" and "bioherms". Although most geologists recognize the dominantly inorganic nature of many "geologic reefs", "mud mounds" and "bioherms", they often associate their distribution with local organic production related to a nearby shelf margin. The forces responsible for the distribution and initiation of the features described here are related not to organic production, but rather to subaerial sculpturing.

Sedimentary Accretion along the Leeward, SE Coast of Qatar Peninsula, Persian Gulf

E. A. Shinn[1]

ABSTRACT

In an area S of Umm Said, Qatar, NW "Shamal" winds have piled up quartz dunes which are rapidly migrating into the sea, prograding the coast to produce a quartz sand sabkha. This supratidal flat is 40 km long, 7-10 km wide and up to 30 m thick. The present sand supply is almost exhausted and within a few thousand years this type of sedimentation will probably cease. The coastal area will then almost certainly revert to carbonate coastal sedimentation similar to that in most other parts of the Persian Gulf. Thus, in a relatively short period of time, a porous, silici-clastic sand lens will be surrounded by carbonate.

INTRODUCTION

Description of the Umm Said area:

The SE part of Qatar Peninsula, near Umm Said (Fig. 1) is characterized by an aeolian, quartz dune complex which is rapidly spilling into the sea (Fig. 2). Scattered dunes, mainly of the barchan type, move under the influence of NW "shamal" winds across deflated, wind striated, Eocene dolomite in the interior of the peninsula. This Eocene rock surface, less than 5 m above sea level in this area, is separated from the sea by a quartz sand sabkha or supratidal flat 40 km long and 7 to 10 km wide. At the boundary between sabkha and Eocene outcrop there is a 20 to 100 m wide, Holocene carbonate sand beach composed principally of a pearl oyster coquina, which stands less than 2 m above present sea level. Barchan dunes migrating across the deflated Eocene dolomite surface, the old carbonate beach and the sabkha finally spill into the sea beyond, thus prograding the sabkha.

Present beach and accretion slope:

At the boundary between sea and sabkha there is a 10 m wide beach berm which is less than a metre high. Seaward of the beach is a gently dipping, partly intertidal slope or foreshore (Fig. 2) 50 to 60 m wide, which is appproximately 2 m deep at low tide*. A sharp break in slope separates the foreshore from a steeply dipping (20° to 30°) accretion slope which extends down to a depth of 10 to 12 m.

---

1 Shell Research B.V., Rijswijk, The Netherlands.

Present address: Shell Oil Company, Houston.

* Maximum fluctuation during spring tides is 2 m.

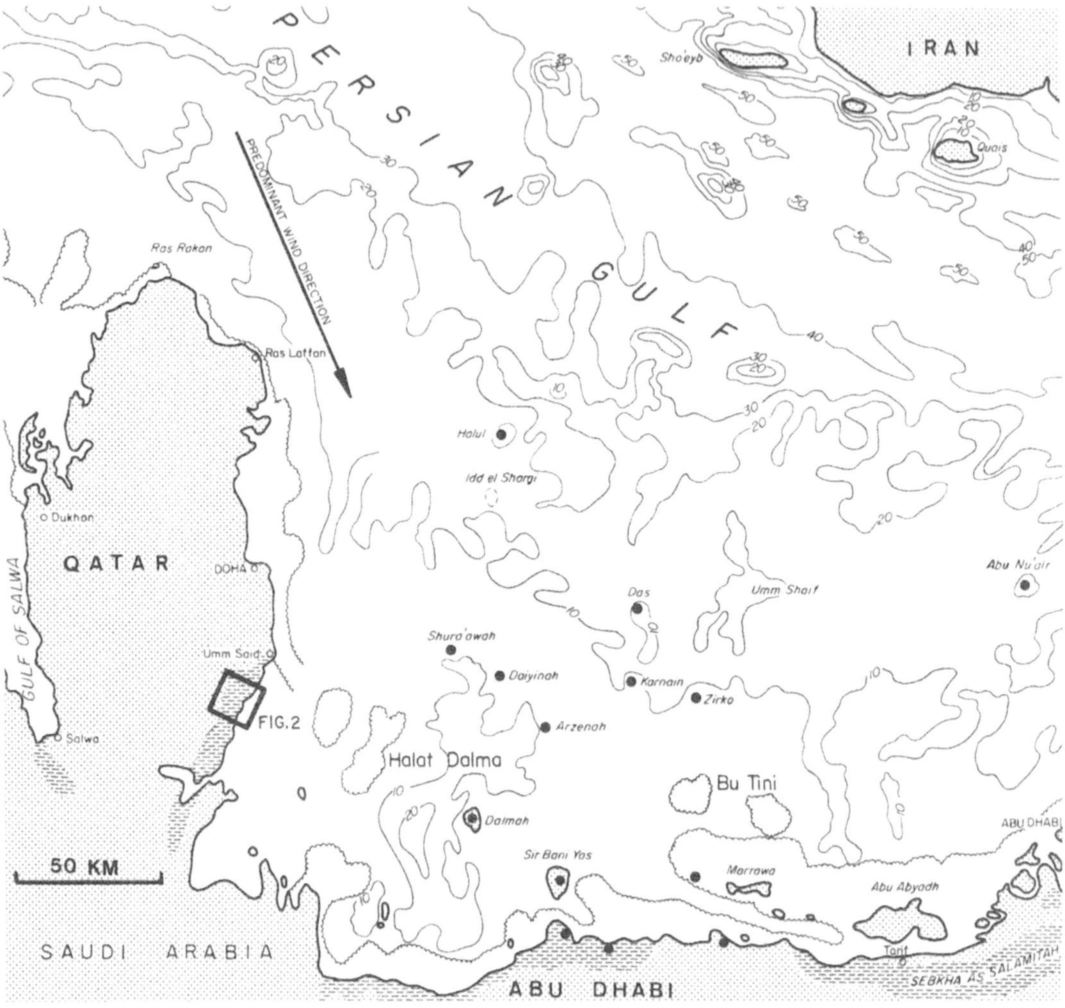

Fig. 1. Map of S.parts of the Persian Gulf, showing location of the area studied

At the foot of this latter slope the sea bottom flattens and extends seaward with a gentle dip to depths of approximately 20 m less than one kilometer offshore. The break in slope at the seaward edge of the foreshore is ornamented in many places with cusp-like features (Fig. 2) 50 m in width that are thought to be slumps resulting from overloading of the steep depositional slope during periods of rapid sedimentation. However, no evidence of slumping was noticed in cores.

Flooding, salt, and gypsum brines:

A seaward strip of sabkha, several kilometers wide, is frequently flooded during the spring and mild storm tides produced by easterly winds. Less frequently (observed only once during two years of observation), an unusually strong easterly wind combines with high spring tides to flood the sabkha as far inland as the old carbonate beach, 7 to 10 km from the sea. Flood water near the present shore drains off quickly, but further inland shallow ponds in minor depressions evaporate and, after a few weeks, produce crusts of salt, gypsum, and dried algal mat. Within a few

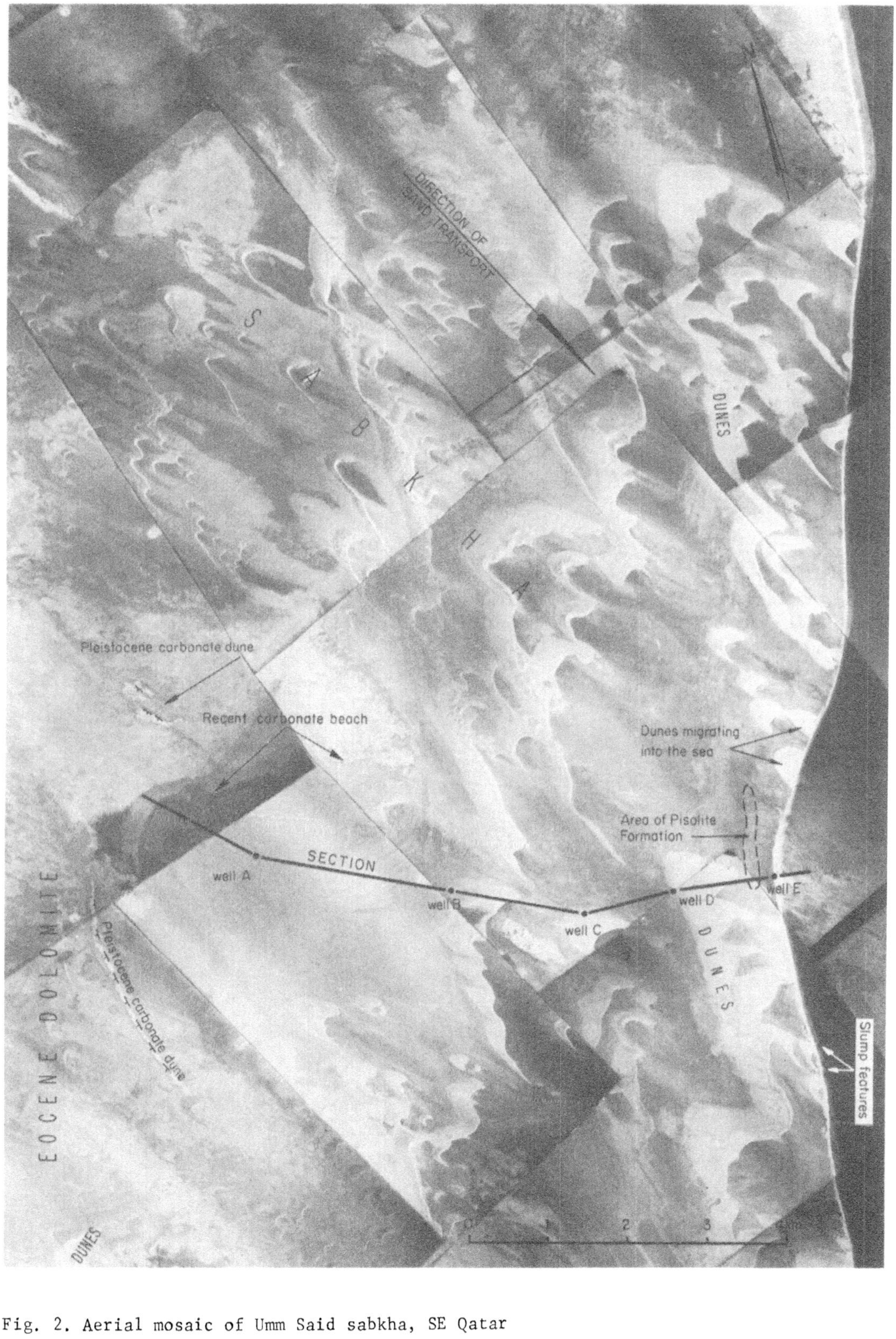

Fig. 2. Aerial mosaic of Umm Said sabkha, SE Qatar

Fig. 3. Aeolian dunes on Umm Said sabkha.
View northward along coast, from first dune north of well site E during a period of calm. Dune migration is from left to right, into the sea

Fig. 4. Same view as in fig. 3 taken during strong "shamal" wind. Sand spills directly into the sea either by gravity sliding on the forset slope, or by blowing out to sea as seen here

months, however, most of these crusts are blown away or dissolved by dew and infrequent rain*. An extensive salt crust formed in August 1965 had been removed except in the lowest part of depressions. When this study was initiated (February 1967), the water table had retreated to a point approximately 50 cm below the sabkha surface. In the inner part of the sabkha the ground water was concentrated to the point of halite precipitation; the chemical properties of sabkha brines are discussed by de Groot (in this volume).

Methods of investigation:

Five wells, A, B, C, D and E were cored along a transect in the area shown in figure 2. This transect, starting at location A near an early Holocene carbonate beach and extending out to well E at the present coast, was drilled by Shell Research NV (Rijswijk, the Netherlands) using a trailer-mounted, Midlite rotary drilling rig capable of taking oriented, 82 cm piston cores. Each well was cored, recovery generally being over 50 %, and the cores from the top 5 metres of well E were oriented. Filled core barrels were sealed and shipped to the Netherlands where they were extruded and sliced. Lacquer peels were made of one surface and the other half was impregnated with plastic and cut with a diamond saw. The methods of extracting water from these cores are discussed by de Groot (in this volume).

SEDIMENT TYPES AND PROCESSES

The principal sedimentary processes are relatively simple. Aeolian sand blows into the sea under the influence of the NW "shamal" wind, (Figs. 3 and 4). There are two important processes of aeolian transport:

- saltation across the flat sabkha in interdune areas, or from the crests of dunes which have migrated to the sea;

- slumping of sand down the foreset slope of dunes.

Most of the sand transported by either mechanism first accumulates on the beach and foreshore where it is further reworked by wave and tidal currents and distributed down the steep accretion slope slightly further seaward. Accumulation of this quartz sand therefore takes place mainly on the steep, subtidal accretion slope where it attains characteristic, seaward-dipping, accretion bedding. Sands deposited on this slope show very few traces of burrowing or mixing by organisms, probably due to the high rates of sedimentation. The surface morphology of this accretion slope (Fig. 2) suggests that slumping may locally modify the bedding. The foot of the accretion slope where the exotic quartz sands mix with indigenous carbonate muds is characterized by extensive burrowing, especially by echinoids, and by the presence of drifted seaweeds and other marine detritus.

Description of the sequence (Figs. 5-6)

The sequence is 30 m thick and composed mainly of medium to fine siliciclastic sand (quartz and feldspar) containing 5 - 15 % well-rounded, sand sized fragments of Eocene dolomite and minor traces of other nondolomitic carbonates. These sands are well rounded and frosted, and sorting ranges from moderate to extremely good. These sands are assumed to have been deposited during two separate phases of regressive sedimentation separated by a 2-3 m thick carbonate layer. The latter is probably the product of transgression. The sequence may therefore be subdivided into three units as follows:

The lower regressive quartz sand (unit 5, Fig. 5)

* Annual rainfall is less than 1 inch per year.

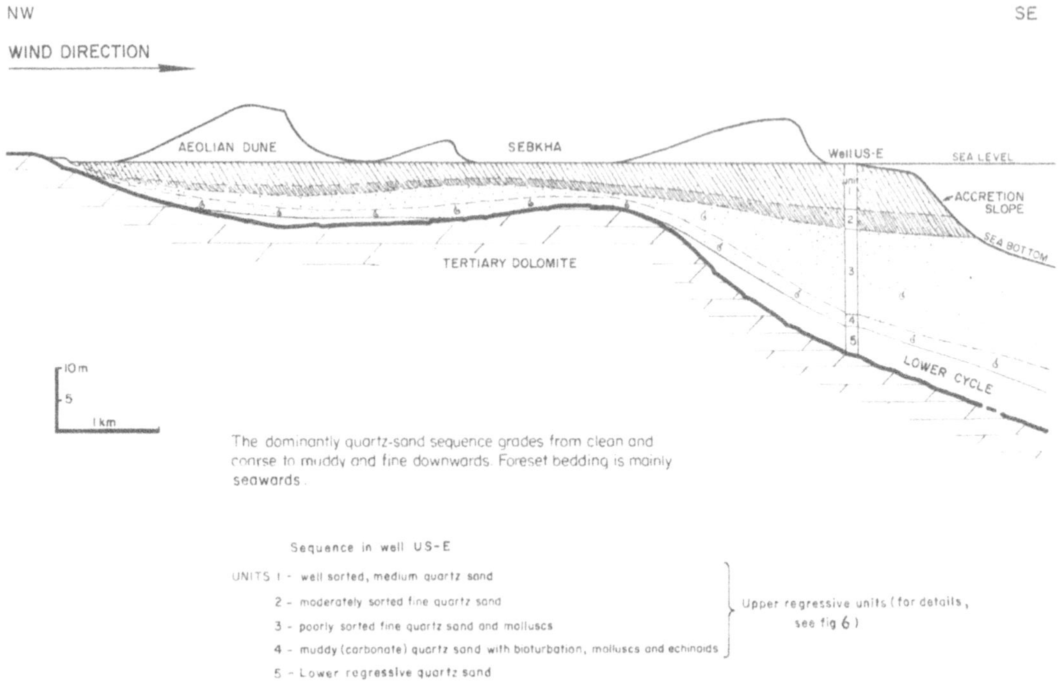

Fig. 5. Profile across Umm Said sabkha showing the lateral distribution of the principal sedimentary units

This unit consists of 3-5 m of well sorted quartz sand lying directly on the Eocene dolomite basement. The upper parts of this sand body have scattered stromatolites, mud cracks (base of well D) and oxidized, root-bearing aeolian sand(in well A). These sediments are almost certainly the product of tidal-flat sedimentation similar to that existing today at Umm Said. Because they occur some 20 m below present sea level, they are clearly the result of coastal accretion during a lower sea level.

The transgressive carbonate mud (unit 4, Fig. 5)

A 2-3 m thick carbonate"wackestone"containing typical lagoonal mollusks and associated supratidal features rises from a depth of 20 m (below sea level) in well E (drilled on the shoreline) to the surface,beyond well A, where it passes laterally into the coquina beach behind the sabkha. This carbonate layer is probably time transgressive; the sequence suggests that, during the transgressive period, sea level continued to rise until it reached approximately its present position, after which the coquinoid beach, occupying the innermost part of the sabkha, was formed. Water depth in front of the beach (at the position of well A) was probably 2-3 m when this beach was forming. The beach is similar in appearance and position to other Holocene beaches around Qatar which have been dated at between 4000 and 5000 yrs (Taylor & Illing, 1969).

The upper regressive quartz sand (units 1-3; Fig. 5)

The uppermost 25 m of the sequence are the product of present sea level, the top of the sequence being the present sabkha surface.

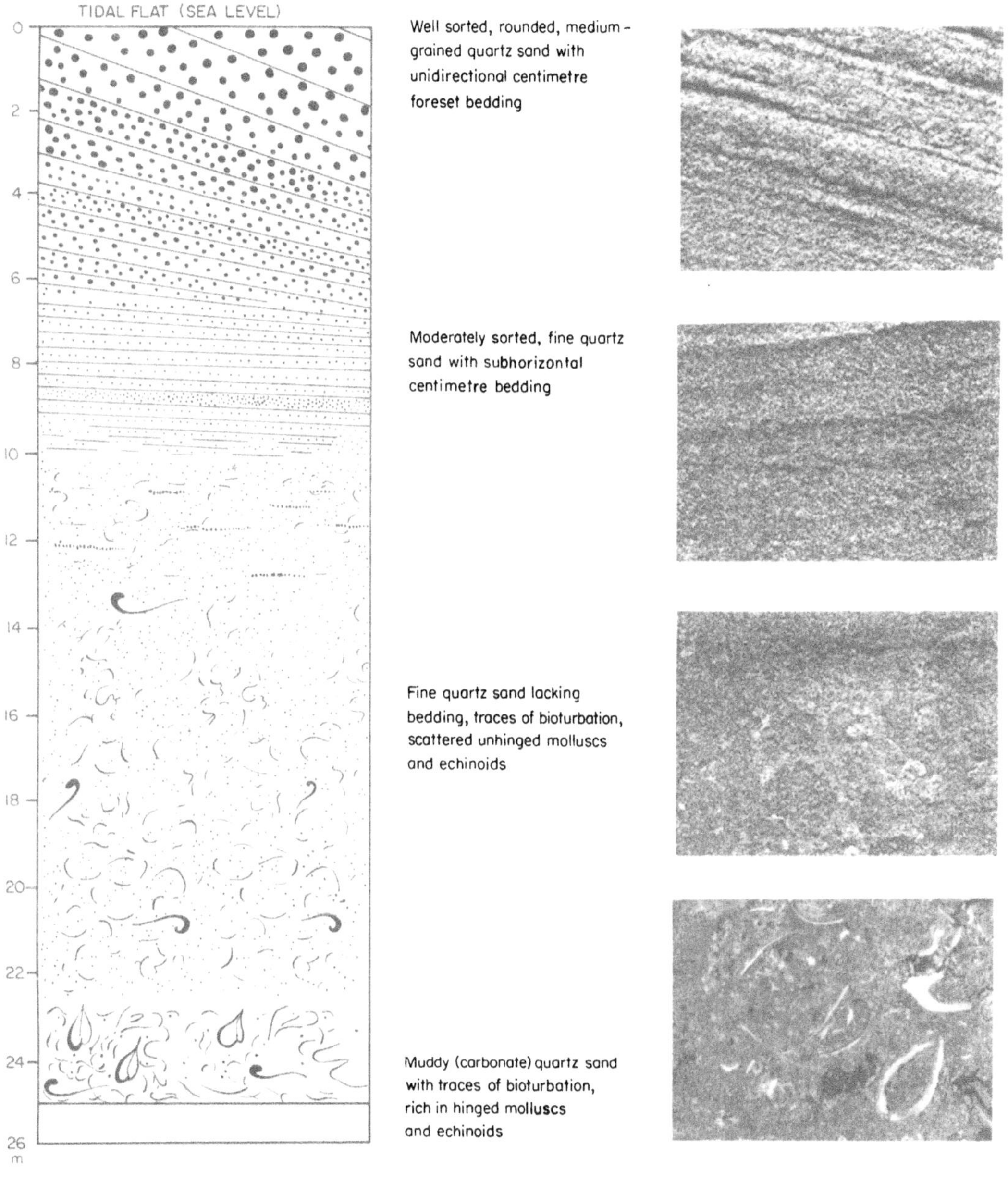

Fig. 6. Vertical sequence penetrated by Well E drilled on the seawards edge of Umm Said sabkha (see Fig. 2)

The quartz sands immediately above the transgressive carbonate unit are fine and muddy. (Fig. 7B) They were probably deposited at the foot of the beach slope. These sediments grade upwards into well sorted, medium grained quartz sands which constitute most of the upper regressive unit. (Fig. 7A) The oxidized, upper portion of this regressive sand increases in thickness in a seaward direction. Furthermore, pits excavated at, and between, all well locations revealed accretion bedding in the top meter of the sabkha. Oriented cores were not taken below the first meter, except in well E where consistent seaward dipping accretion bedding was found in the top 5 m of oriented core. It is thought that the steeply dipping (20°) beds found in the top few meters of all wells also dip toward the sea. The angle of dip gradually decreases downwards, and at a depth of 10 m in well E it apparently correlates with the present-day, almost horizontal, submarine accretion slope which is at approximately the same depth less than 200 m away. Sand grains are well sorted from the top down to the transgressive carbonate layer in well E. In all wells, grain size decreases downwards. Lamination is produced by grading of grains in each lamina.

Contact with the underlying transgressive carbonate layer ranges from gradational at wells A, B and D to sharp at wells C and E. At well C the carbonate part of the unit is locally cemented with fibrous aragonite and magnesium calcite, and is bored locally.

Source of the quartz sands

The precise source of the migrating dune sands is not known. It is probable that the dunes crossed the Gulf of Salwa, near the NW part of Qatar Peninsula, when sea level was lower. These dunes subsequently migrated across Qatar Peninsula and began forming the regressive accumulation along the leeward shore after the sea had risen to approximately its present level. Since the coqu. beach is probably 4000-5000 yrs old, the dune sands have spilled into the s and prograded the sabkha shoreline at a rate of 1-2 m/yr.

Tidal-flat pisoliths

Mollusk fragments coated with laminated aragonite, and aggregates of cemented quartz grains were found in the upper 10 cm of unconsolidated sediment 200-300 m landward from well location E on an area of the sabkha frequently flooded by high tides. These concretions, not present in any of the wells, have alternating layers of radially oriented, fibrous aragonite and concentric laminations of organic-rich, micritic aragonite (Fig. 8). They are comparable to grains described by Loreau & Purser (in this volume) from tidal-flats of W Abu Dhabi. That the pisoliths are formed in situ on the tidal flat at Umm Said is indicated by the fact that large mollusks occurring together with these pisoliths have similar coatings only on their lower surfaces, and by the absence of other obvious nearby sources. The chemistry of their formation has not been studied but it is thought that evaporating interstitial sea water provides the carbonate, which precipitates an aragonite cortex selctively on aragonitic shells or aragonitically-cemented quartz aggregates. Uncoated quartz grains do not seem to constitute a suitable nucleus. These pisoliths resemble the concretions formed in certain soils.

Fig. 7. Lacquer peels of cores from the upper and lower parts of the sequence in Well D, Umm Said sabkha

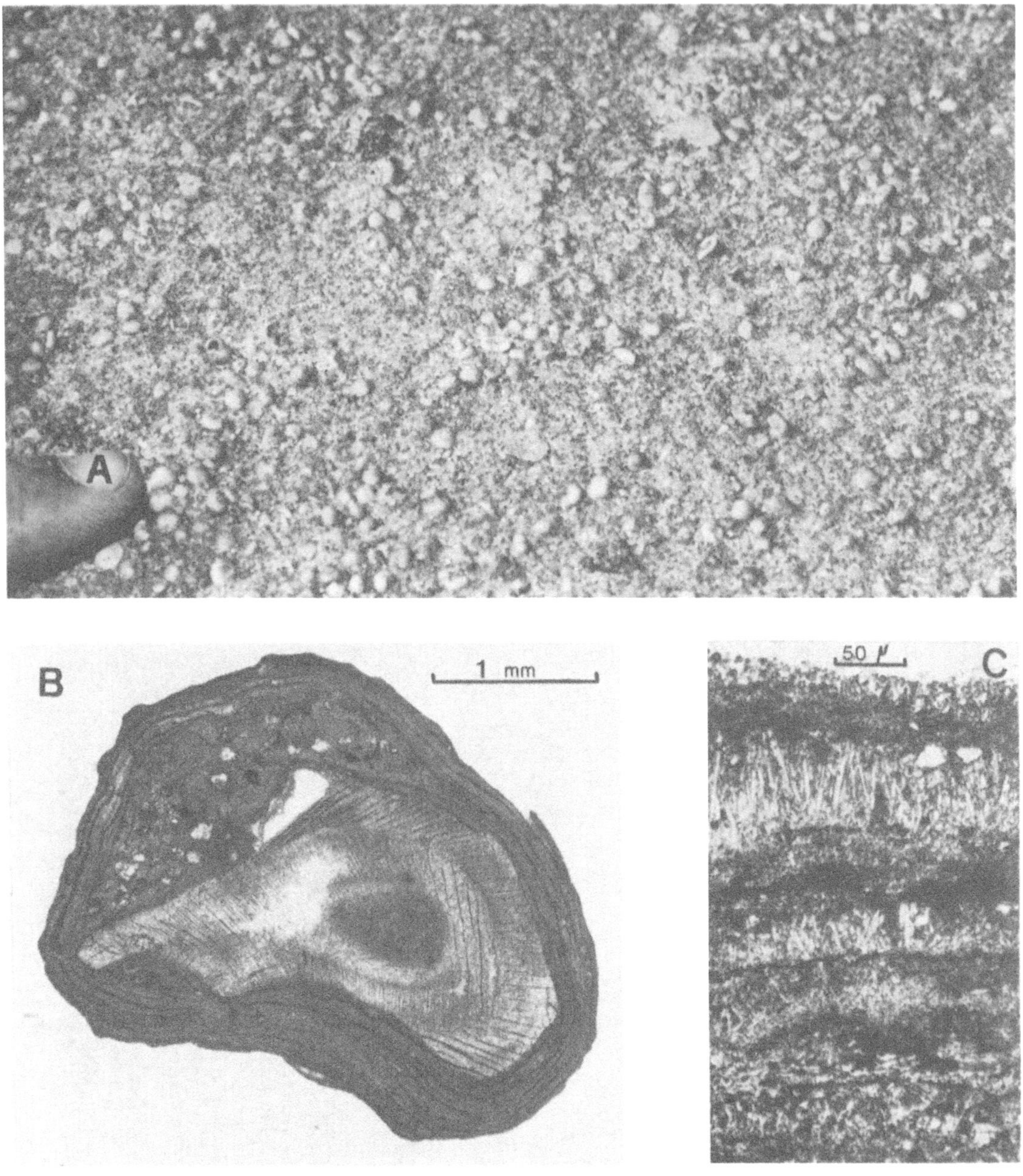

Fig. 8. Pisoliths on Umm Said sabkha:
A. Field view of pisoliths on sabkha surface
B. Section through a pisolith (thin slide)
C. Detail of thin slide showing characteristic microstructure consisting of radially oriented acicular, and micritic aragonite

DISCUSSION OF SEDIMENTARY ASPECTS

Future sedimentation

Aerial and ground observations indicate that, under the present wind regime, there is no longer a source of sand to sustain the Umm Said dune complex. If this sand continues to prograde and be stabilized at the shore, and if no major changes or wind direction occur in the next 1000 yrs, it is probable that all available sand will accumulate in the shallow, sub-tidal and sabkha deposit and thus expand and thicken the existing accumulation. Eocene outcrops landwards of this accumulation will be swept clean and, in contrast to most deltaic sedimentation, no evidence of the source of the sands will be preserved. When the supply of quartz sand is exhausted, muddy carbonate sedimentation, similar to that of the transgressive layer and the present offshore environments, will probably resume. It is likely that in time an isolated quartz sand body will be preserved in this predominantly carbonate basin. Because of limited sand supply the resulting sand wedge will tend to be elongate parallel to the coast and will thicken in a seaward direction. Similar ancient sand bodies could constitute ideal stratigraphic oil traps.

Recognition in ancient rocks

Accumulations similar to those occurring at Umm Said in the Persian Gulf may be present in ancient sequences. Where analogous quartz sand deposits are contained within ancient carbonate provinces their origin and geometry could be difficult to predict, partly because the problem is complicated by the possibility of variations in sedimentation. With an abundant sand supply, this style of sedimentation could produce a delta-shaped body in either a carbonate or a silici-clastic province which, with limited well control and insufficient sampling, might easily be misinterpreted as a fluviatile delta. It is therefore important to be able to distinguish this sabkha-type of leeward coast deposit from that of a delta or tidal-bar complex. The significant points to consider when making this distinction, none of which is sufficient evidence by itself, are listed below.

1. Absence of clay layers, which are generally present in river deltas.

2. Absence of fresh-water fauna, which is commonly associated with deltas.

3. The fauna of the adjacent marine carbonate sediments may have open-marine characteristics and not be as greatly modified as would be expected in regions of fresh water run-off around deltas.

4. Steep planar accretion dips, superficially similar to aeolian bedding, with uniform dip direction.

5. Upward increase in grain size.

6. Lack of channel deposits that show an upward decrease in grain size.

7. Possible presence of pore filling dolomite or evaporite minerals.

The accumulation described is a special type of beach and therefore has many characteristics of beach accretion. In the absence of cores or dipmeter surveys which enable one to distinguish the steep, seaward-dipping accretion bedding, it would be extremely difficult to distinguish it from a normal beach. Both have an upward increasing grain size. However, it is considered geologically more important to be able to distinguish this kind of accumulation from that of a delta or tidal-bar system. I may be possible to make this distinction by applying the criteria listed above.

Regional Sedimentation along the Trucial Coast, SE Persian Gulf

B. H. Purser[1] and G. Evans[2]

ABSTRACT

Sediment composition, surface patterns, and vertical sequences vary laterally along the Trucial Coast depending on three major factors:

- orientation of the shoreline with respect to the onshore "shamal" winds;
- proximity to Qatar Peninsula, an up-wind barrier;
- presence of the Great Pearl Bank coastal barrier.

This latter feature is paramount, its presence within the central parts of the Trucial Coast permits the subdivision of this regional shoreline into three sedimentary provinces;

The Western parts of the Trucial Coast which are protected laterally by Qatar Peninsula, the remote Great Pearl Bank having little effect. The shallow subtidal sediments are carbonate muds and the wide intertidal flats are composed of imperforate foram and pelletal sand. The protection decreases rapidly to the E where the wide Sabkha Matti embayment contains fringing reefs, oolitic and molluskan sands.

The Central parts of the Trucial Coast which are protected by the structurally-based Great Pearl Bank barrier. Because this ridge is oriented obliquely to the continental shoreline the lagoon varies in width and depth. The barrier axis rises progressively towards the E and becomes incorporated into the Arabian shore in E. Abu Dhabi. These lateral variations in the morphology of the barrier are reflected in its sediments as well as in those of the adjacent lagoon and of the mainland shoreline. The distribution of dolomite and other evaporite minerals is also related to these lateral variations.

The North Eastern parts of the Trucial Coast which are unprotected and face directly towards the entire length of the gulf. They suffer the effects of maximum wave fetch which has resulted in the development of major longshore spit systems. Although the Oman Mountain range overlooks the NE extremity of this shoreline, its detrital sediments barely contaminate the indigenous carbonates, presumably because of the arid climate. In contrast to those of the remainder of the Trucial Coast, the vertical sequence of coastal sediments in this area is transgressive, probably due to downwarping of the N end of the mountain range.

The 600 km Trucial coastline is essentially linear. Nevertheless, it is thus seen to exhibit very marked lateral variations in its sediments and their diagenetic modifications both of which are determined mainly by the proximity of up-wind barriers. An understanding of these relationships may help in the estimation of the proximity and orientation of ancient barriers and shorelines.

1 Lab. de Géologie Historique, Université de Paris Sud (Orsay)
2 Dept. of Geology, Imperial College of Science and Technology, London.

INTRODUCTION

The SE shores of the Persian Gulf, between the base of Qatar Peninsula and the N end of Masandam Peninsula, generally referred to as the Trucial Coast, is an area of relatively pure carbonate sedimentation. Although essentially linear, in detail its morphology is exceedingly complex. The complexities, the product of both pre-Holocene erosion and sedimentation, and Holocene sedimentation, result in rapid lateral variations in both the environments and sediment patterns. These contrast markedly with the relatively simple distribution of sediments in the offshore areas.

This contribution is a synthesis which emphasizes the major regional factors controlling coastal sedimentation. It is an attempt to envisage this coastal complex as a single regional shoreline exhibiting major lateral variations, similar to those which must have existed along the shores of certain ancient carbonate basins.

The research upon which this synthesis is based has been carried out by two organizations: students and staff of Imperial College, London who have worked mainly in the central areas of the Trucial Coast,and geologists of Shell Research, Rijswijk, the Netherlands who explored the W parts of this shoreline, in the area of Sabkha Matti (Fig. 1), and who undertook expeditions along the Khor al Bazm lagoon and adjacent barrier; a core drilling program was also carried out by the staff of this company in the sheikdom of Ras al Khaimah under the direction of B.D. Evamy. The authors express their gratitude to the various organizations which sponsored this research and, in particular, to Shell Research for permission to publish results.

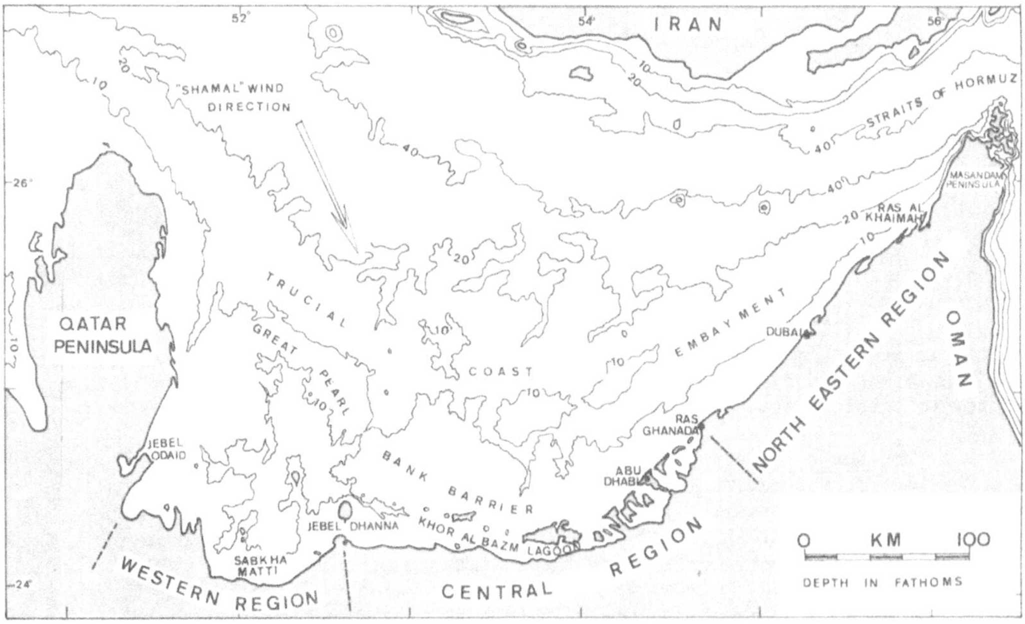

Fig. 1. Map of the Trucial Coast embayment showing morphology and major sedimentary provinces of the Trucial Coast

THE PRINCIPAL FACTORS CONTROLLING SEDIMENTATION ALONG THE TRUCIAL COAST

The floor of the Persian Gulf slopes gently from the Arabian shore towards the axis of the basin which lies close to the Iranian coast. Because waves and surface currents are generated mainly by the NW "shamal" winds the Trucial Coast, lying at the extreme SE end of the basin, is subjected to a maximum degree of water agitation. This fundamental feature of coastal sedimentation is modified along the shoreline by three factors:

- the orientation of the shoreline with respect to the onshore winds;
- the presence of a lateral "up-wind" barrier - the Qatar Peninsula;
- the presence of an offshore barrier - the Great Pearl Bank.

Of these three factors, the latter would seem to be the most important. The Great Pearl Bank Barrier (Fig. 1), in contrast to Qatar Peninsula, is oriented in an E-W direction, having an "Iranian" tectonic trend. This ridge fringes the Trucial Coast between Jebel Dhanna in the W and Ras Ghanada in the E. Its presence has a major effect on sedimentation within the central parts of the Trucial Coast

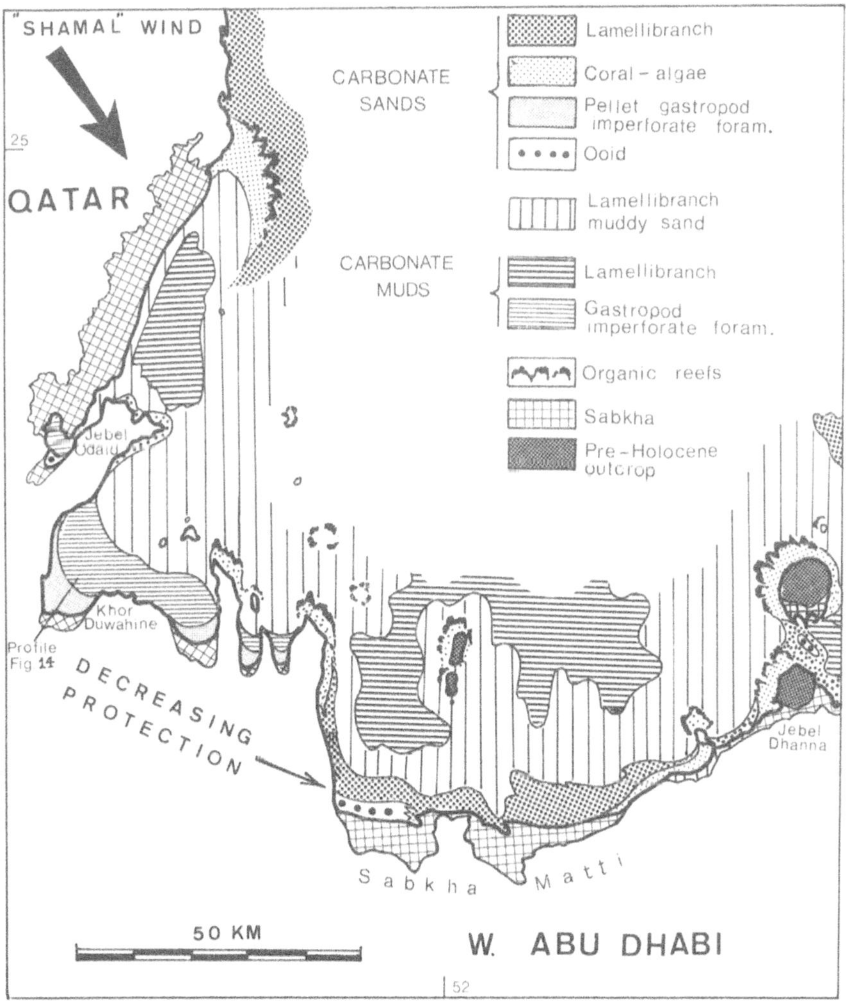

Fig. 2. Regional sediment pattern in the Western Region of the Trucial Coast

permitting the subdivision of the 600 km long Trucial Coast shoreline into three major morphological and sedimentary provinces: the Western, Central, and North Eastern Trucial Coast (Fig. 1).

SEDIMENTATION IN THE WESTERN PARTS OF THE TRUCIAL COAST

The W parts of the Trucial Coast, between the base of Qatar Peninsula near Jebel Odaid and Jebel Dhanna 150 km to the E, consists of a wide embayment which faces almost directly into the "shamal" winds. However, this area lies in the lee of Qatar Peninsula and is thus protected from the full effects of wind-generated waves and currents, this protection decreasing progressively towards the E. Little influence is exerted by the Great Pearl Bank barrier which, in this area, is too remote to have any significant effect on coastal sedimentation. This Western province may be readily subdivided into two sub-provinces:

The Western sub-province consists of a series of N-S oriented rocky peninsulas separated by wide embayments (Fig. 2). These peninsulas, probably delimited by fractures related to the "Arabian" tectonic system, support numerous fringing reefs in spite of the relatively high salinity (45-50 ‰) of this isolated corner of the basin. Individual embayments attain depths of approximately 5 m; their shores are flanked by wide, intertidal flats and narrow beaches backed by sabkhas, which may attain 3 km in width. These shallow embayments are highly protected by the close proximity of the Qatar Peninsula to the W. Their level, subtidal floors are covered with burrowed carbonate muds, which are extensively colonized by marine grasses; these sediments are extremely rich in imperforate Foraminifera and gastropods. When traced landwards the bottom steepens suddenly to form a clearly defined sedimentary scarp (1-3 m in height) with its upper edge coinciding with the lower limits of the intertidal zone (Fig. 14). This minor scarp coincides with a change from the carbonate muds of the subtidal zone to the carbonate sands of the, 1 km wide, intertidal flats above. The sands of the latter area are rich in miliolid and peneropolid Foraminifera, fragments of dasycladacian algae (Acetabularia sp.), ovoid pellets and small ooids. The surfaces of these intertidal flats are extensively rippled; they are burrowed by numerous crabs in their outer parts, and covered by a carpet of dark green filamentous algae on their inner parts. The surface sediments of these intertidal flats are frequently lithified into thin crusts of beach rock which are sometimes broken to form extensive sedimentary breccias.

These embayments in the extreme W end of the Trucial Coast, of which Khor Duwahine is typical, are areas of active sedimentary accretion. Skeletal carbonate, consisting mainly of imperforate Foraminifera, cerithid gastropods and dasycladacian algae, is produced dominantly within the shallow subtidal environments of the embayment. The clearly defined scarp, which marks the lower limit of the intertidal zone, is almost certainly an accretion surface, with poorly defined seawards-inclined bedding. Although the exact mechanisms of its seawards, progradation have not been determined, its morphology clearly suggests seawards or oblique longshore transport, somewhat analogous to that described by Davies (1970) in Shark Bay, W Australia. The progressive lateral accretion of this shallow subtidal scarp leads to a characteristic sedimentary sequence illustrated in figure 14B.

The Eastern sub-province is an open embayment, the landward part of which is called Sabkha Matti, and whose coastline is fringed by a high (1-3 m) storm beach which extends 55 km westwards to Jebel Baraka. This ridge is composed essentially of molluskan shell sand and gravel except at its lateral extremities: at its W end, near Sila, these skeletal carbonates are replaced by oolitic sands which are forming on the adjacent tidal flats (discussed by Loreau and Purser, in this volume), while at its E end the sediments are composed essentially of red algal debris derived from the adjacent fringing reefs. These marked lateral changes in composition along the beach do not reflect any major variation in the environment of beach deposition but are the consequence of changes in the adjacent offshore environments of sediment formation; they are typical of coastal environments exposed to dominantly onshore wind and wave action.

Coastal accretion in the Sabkha Matti embayment differs from that associated with the coastal spit systems of NE Qatar, described by Shinn elsewhere in this volume. Whereas the storm beach at Sabkha Matti is prograding mainly in a seawards direction to form a sheet of carbonate sand, the coastal spit system of NE Qatar is more strongly influenced by longshore transport: these spits migrate laterally across muddy intertidal flats, developed on their leewards sides, to overlie these tidal flat sediments as isolated ridges or "cheniers" (Price, 1965). Because the storm beach in the Sabkha Matti embayment is prograding seawards its basal contact tends to be gradational: coarse carbonate beach sands with low angle inclined bedding, oriented mainly in a seaward direction, grade downwards into finer, intertidal and subtidal sands with complicated cross-stratification. The resulting sequence is summarised in Fig. 14B. Coastal spits may have a similar composition and texture but their internal bedding is usually steeper and oriented mainly in a landwards direction. Furthermore, because these spits have been driven landwards during storm conditions, their sharp basal contacts are underlaid by lagoonal or tidal flat muds (this sequence is demonstrated by Shinn's Fig. 10 elsewhere in this volume).

Fig. 3. Morphology and sample (Shell Research) distribution in the W half of the Central Region, showing the Khor al Bazm Lagoon and adjacent Great Pearl Bank Barrier

SEDIMENTATION IN THE CENTRAL PARTS OF THE TRUCIAL COAST

Between Jebel Dhanna and Ras Ghanada, some 250 km to E, the Trucial Coast is strongly influenced by the Great Pearl Bank barrier system. This barrier is oriented obliquely to the Arabian coastline into which it is progressively incorporated. Its increasing proximity to the shoreline as it is traced eastwards results in a complex coastal morphology which includes both barriers and protected lagoons, (Figs. 3 and 8).

The barrier is markedly asymmetric. North of Jebel Dhanna the axis is submerged to depths of 5-10 m but it rises progressively to the E to produce a broad ridge surmounted by a series of carbonate sand shoals and small islands, all of which become more numerous and progressively larger towards the E. This progressive shallowing along the axis results in progressively higher energy conditions on the barrier and increasing restriction within the adjacent Khor al Bazm lagoon. The barrier also exhibits a cross-sectional asymmetry, having a steep lagoonal slope and an irregular, but nevertheless more gently sloping, seaward slope.

The Great Pearl Bank barrier is parallel to the structural axis of the Persian Gulf (Kassler,elsewhere in this volume) and may therefore be considered as part of the late Tertiary "Iranian" tectonic system. The marked asymmetry across the barrier axis may be due partly to spill-over sedimentation along its lagoonal side (Fig. 6). However, the exposure of pre-Holocene limestone islands concentrated along the inner side of the barrier suggests that modern sedimentation alone has not developed this asymmetry. The most plausible explanation is that it is mainly the result of pre-Holocene erosion; as the structural dip is northwards, towards the axis of the Persian Gulf, the development of subsequent drainage, parallel to the strike, would have created an asymmetric,cuesta-type topography resembling the present day Tuwaiq escarpment in central Saudi Arabia. Subsequent drowning and sedimentation have modified this essentially structural feature to produce the morphology seen today. The principal features of obliquity to coastline, axial plunge and asymmetric profile seem to be the main factors determining sediment composition on and around the barrier, within the adjacent lagoon,and along the Arabian shoreline.

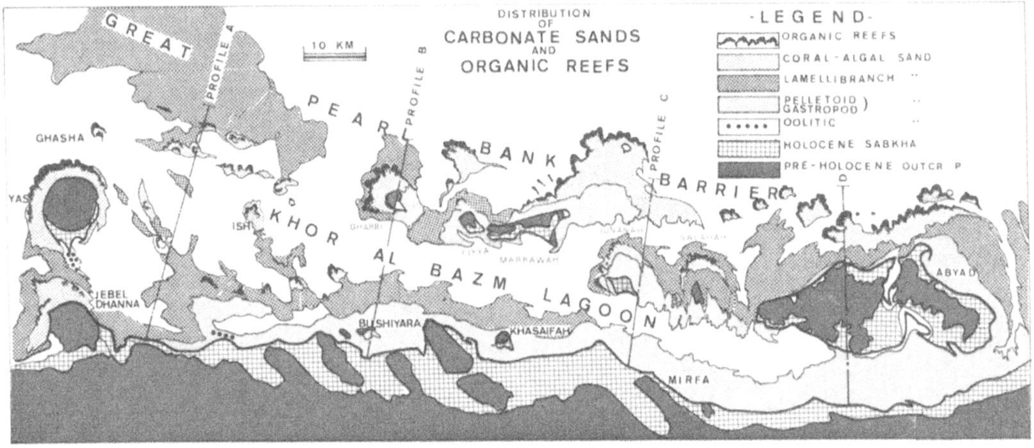

Fig. 4. The distribution of carbonate sands and organic reefs within the W part of the Central Region, Trucial Coast

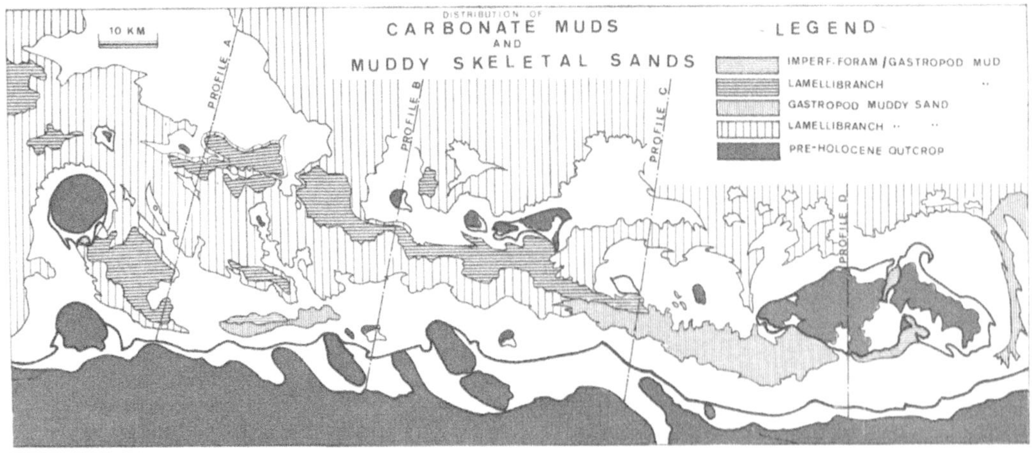

Fig. 5. The distribution of carbonate muds and muddy skeletal sands within the W part of the Central Region, Trucial Coast

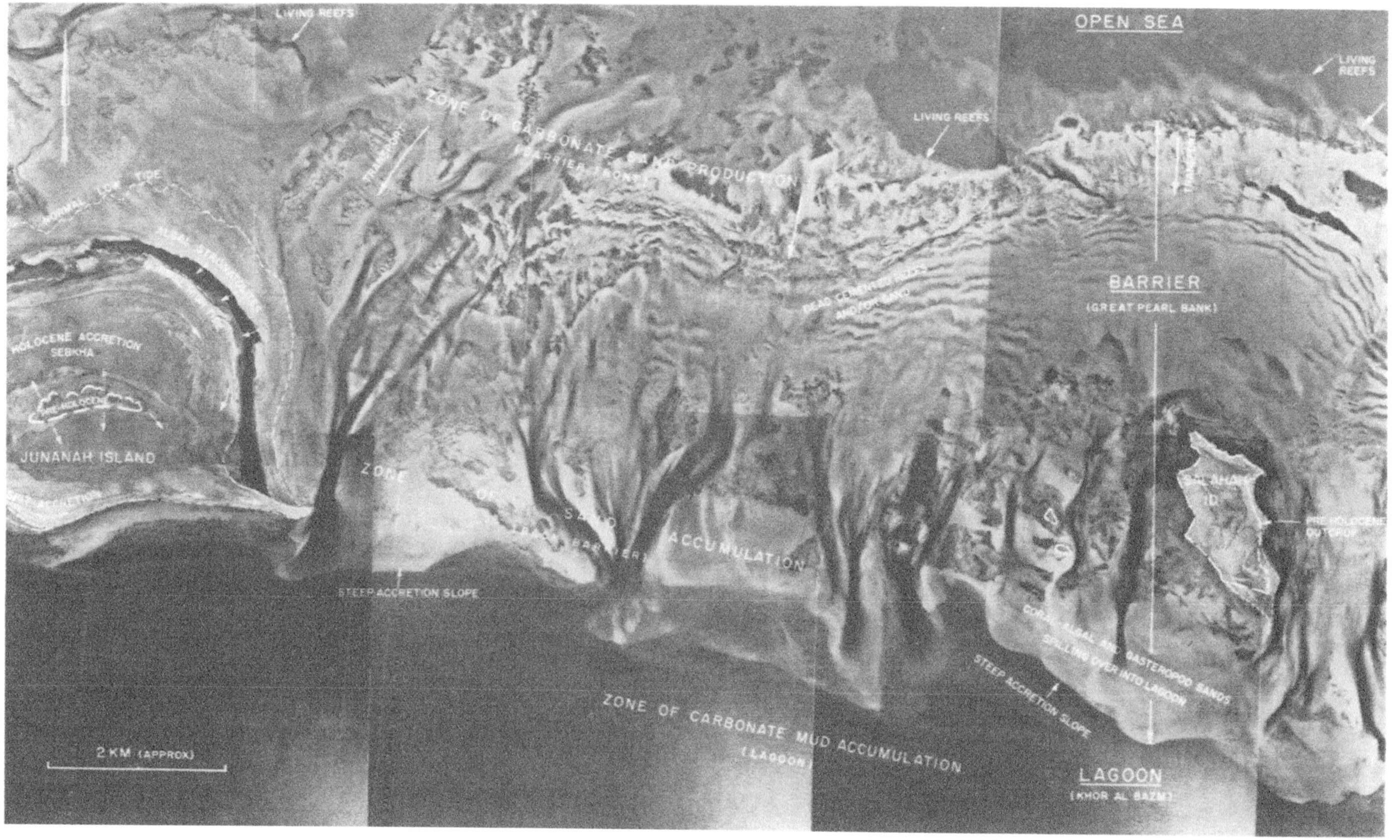

Fig. 6. Aerial photo-mosaic showing the morphology and sediment patterns on the central parts of the Great Pearl Bank Barrier, Trucial Coast

Sedimentation on the barrier

At its W end the water seems to be too deep (5-10 m) for prolific reef growth and the barrier is covered mainly by lamellibranch sand (Fig. 4). Seawards these sands grade into muddy lamellibranch sand and mud, a similar but more rapid change taking place shorewards towards the deeper waters of the adjacent lagoon. Eastwards along the barrier the water shallows progressively and there is a gradual increase in the amount of coral and calcareos algae which make numerous small patch reefs. These are concentrated mainly along the inner, shallow edge, of this asymmetric ridge. Carbonate sand shoals, composed essentially of skeletal debris become emergent at low tide, mainly along the inner edge of the barrier. These sands have lithified to form small islands such as Gharbi these islands having fringing reefs developed along their windward sides; they are, however, not sufficiently large to allow the accumulation of intertidal muds and the development of evaporites.

Eastwards from the island of Gharbi (Fig. 3) the barrier axis develops into a flat platform, submerged to depths less than 2 m, and widens to approximately 20 km. This increase in width is accompanied by an increase in the size and number of sand shoals and islands which thus create an increasingly effective barrier and a more protected lagoon. The increasing width also results in the coral-algal reefs being restricted to the seaward side of the barrier and they thus become progressively more remote from the adjacent lagoon.

The increasing width of the barrier and its associated islands culminates in the large islands of Marrawah and Abu Abyad (Fig. 3). Both have cores of Pleistocene limestone (termed "miliolite"), and other pre-Holocene rocks which have acted as nuclei around which Holocene carbonate sediment has accreted. Extensive fringing reefs have developed on their seaward (windward) sides to produce sheets of dead biostromal boundstone which attain several km in width (Fig. 6). Detritus from these reefs have been swept back across the barrier to accumulate as beaches around the ever-enlarging islands situated along the inner edge of the barrier. When islands attain diameters greater than 2 km they offer sufficient protection on their lagoonal sides for the accumulation of intertidal carbonate muds and pellet sands, often accompanied by algal mats and associated dolomite and gypsum. Thus, sedimentation is actively creating environments favourable for the development of early diagenetic minerals; these sites are concentrated along the inner edge of this asymmetric barrier, relatively close to the lagoon.

A complex system of bars of skeletal sand are strung along the inner edge of the barrier while channels, floored with coarse skeletal debris, locally traverse the barrier. Sediments are swept across the barrier to spill into the lagoon as steep accretion slopes or small deltas at the ends of channels (Fig. 6). Lagoonwards transport of sediment is thus producing a vertical sequence which probably consists of lagoonal muds sharply overlain by skeletal sands with inclined bedding oriented mainly towards the lagoon (Fig. 14). Continued accretion may result in these sands being overlain by pelletal sand and carbonate mud with stromatolites and evaporitic minerals, including dolomite, as the islands on the lee side of the barrier accrete across the lagoon.

East of Abu Abyad Island the barrier and associated lagoon of Khor al Bazm lose their identity and are replaced by a complex of islands and peninsulas which characterize the coastline of E. Abu Dhabi (Fig. 8). This change in coastal morphology may be regarded as the result of the merging of the Great Pearl Bank barrier with the Arabian shore, blurring the relatively simple morphological units so clearly distinguished further to the W.

With progressive lateral emergence of the barrier towards the E, the islands of Pleistocene limestone become more frequent, being concentrated mainly in an elongate zone which appears to be the prolongation of the Great Pearl Bank ridge. Each island is growing by accretion around a Pleistocene rock core. However, in contrast to the accretion around the islands further to the W, this accretion

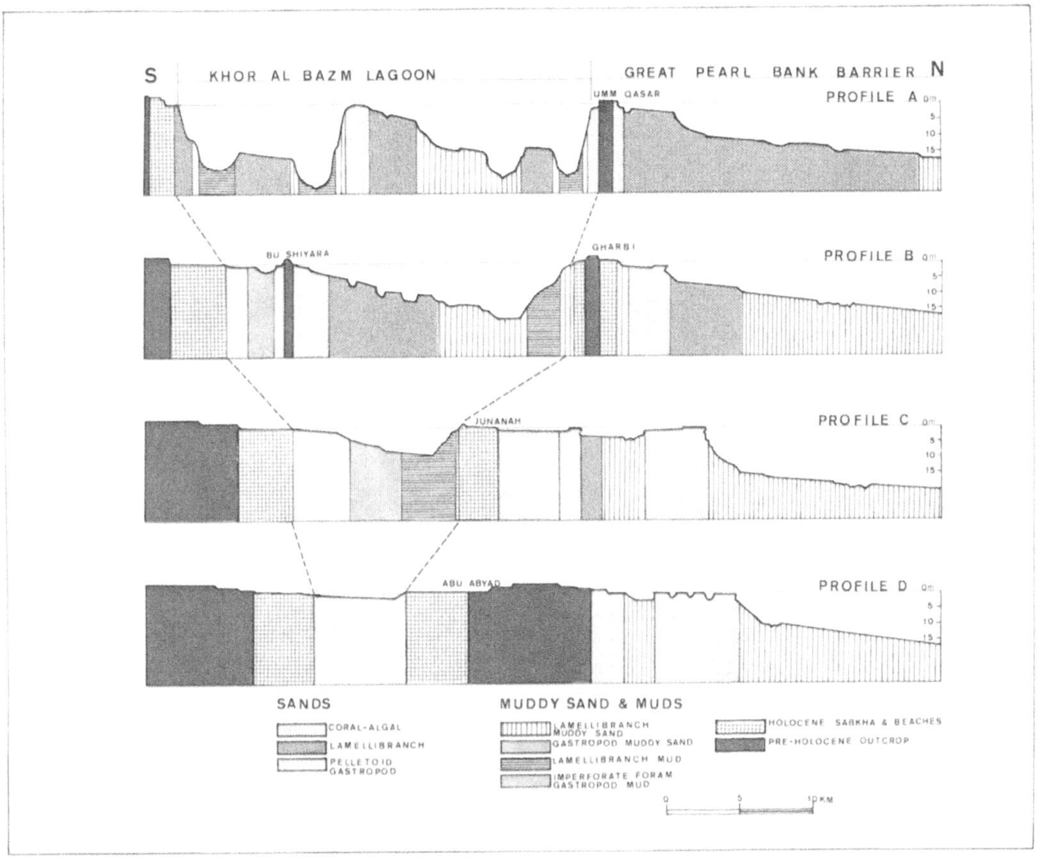

Fig. 7. Profiles across Khor al Bazm Lagoon (W half of the Central Region) showing major lateral changes in morphology and sediment types (see Figs. 4 and 5)

occurs mainly as "tails" of sediment oriented normally to the barrier axis. Because the barrier has approached much closer to the mainland, the sediment tails have locally accreted to this shore (Fig. 8) to form tombolas. These morphologies are well developed between Abu Abyad and Abu Dhabi islands; accretion lines clearly visible on aerial photographs suggest that the 15 km long Ras al Khaf Peninsula has formed in this manner. It would thus appear that sediment tails have been able to accrete across the shallow lagoon. These tails are the lateral equivalence of the spill-over sheets which occur further to the W where the barrier is bordered by the deeper Khor al Bazm lagoon. They are reminiscent of the "wantijen" which have formed on the leeward side of barrier islands in the N. Netherlands (Evans, 1970), where they are less well-developed because of the more oblique orientation of onshore waves. Van Veen (1950), however, had predicted that accentuated morphologies comparable to those found in Abu Dhabi would develop under onshore wind conditions.

Concurrently with the growth of landward tails, there is also a striking lateral spit accretion parallel to the axis of the barrier, which extends both to the E and the W from the Pleistocene rock cores (Figs. 8, 9). West of Abu Dhabi Island this lateral accretion is symmetrical about the Pleistocene nuclei and is less active than tail accretion; E of this island, however, spit accretion becomes progressively dominant at the expense of tail accretion. Furthermore, the easterly component of this lateral spit system progressively exceeds that to the W as one proceeds eastwards and finally, E of Sadiyat Island, accretion occurs only on the E side of the Pleistocene core (Fig. 8). This progressive change from an essentially down-wind "tail accretion" in the W to a unidirectional lateral spit accretion in the E is probably the consequence of an increasing longshore current component which is related to the increasing fetch towards the exposed E end of the gulf.

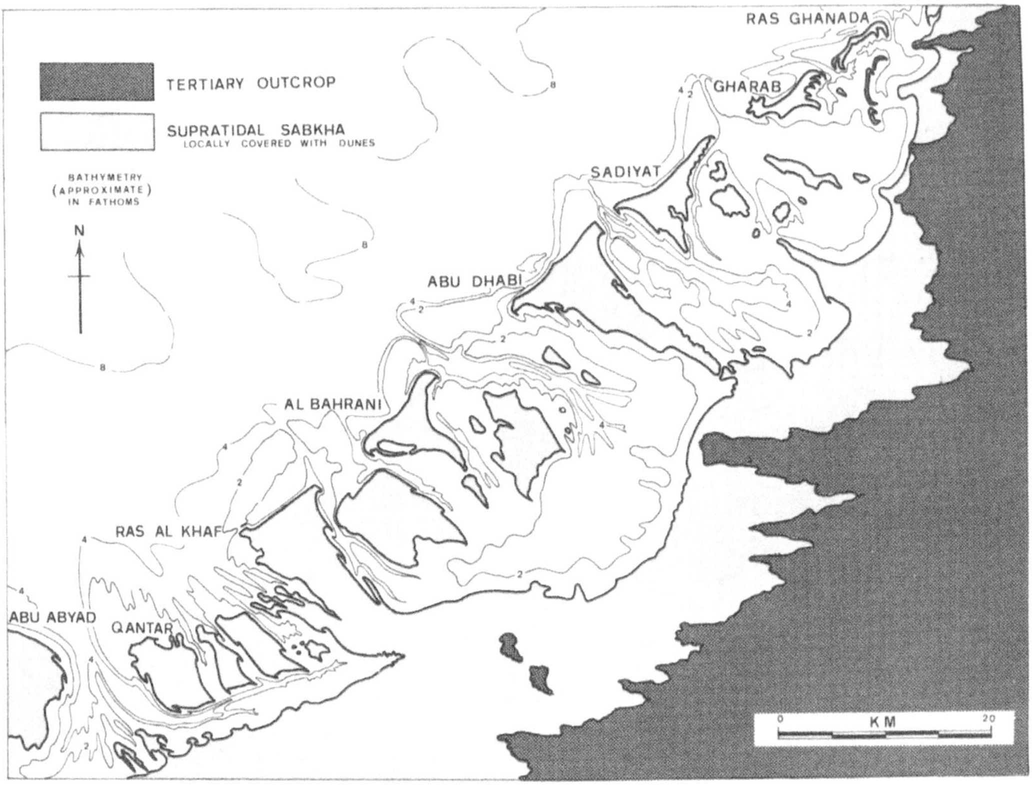

Fig. 8. Coastal morphology of the E half of the Central Region (E. Abu Dhabi), Trucial Coast

In the emergent extremity of the barrier system where down-wind "tail accretion" is a major feature, large channels up to 10 m deep and several km in width, have developed. Most of these would appear to have envolved mainly as elongate, inter-tail depressions, and as such are not primarily the product of tidal movements, although tidal scour has certainly deepened them. Their oblique orientation with respect to the regional shoreline (Fig. 10) is essentially parallel to the adjacent accretion tails, whose orientation is determined mainly by the predominant NW "shamal" wind. Sediments within these channels include skeletal and oolitic sands and gravels near their seawards ends, and pellet sand and carbonate muds near their lagoonal extremities.

The seawards ends of the inter-island channels terminate in spectacular oolite deltas discussed by Loreau and Purser, and by Evans et al., elsewhere in this volume. These deltas are oriented mainly seawards their geometry being clearly related to ebb-flow within the channels, although small deltas occur locally within the channels and are oriented towards the lagoon e.g. between the islands of Sadiyat and Gharab. The Abu Dhabi tidal deltas are genetically related to the tidal channels and thus post-date channel formation. Because the channels are the direct result of tail and lateral spit accretion, these deltas represent a relatively late phase in the evolution of the Abu Dhabi barrier complex. They probably became an important physiographic element of the region once barrier island growth was sufficiently advanced to restrict the tidal flow to fairly narrow outlets and thus produce the conditions necessary to their formation. Such oolitic tidal deltas are seen to be the characteristic element of a particular type of barrier which limits shallow lagoons, both in the SE parts of the Persian Gulf and between the Exuma islands along the E margins of the Bahamian platform. These features are also analogous to deltas composed of silici-clastic sediment at the entrances of channels traversing the barrier island complex of the N. Netherlands (Evans, 1970).

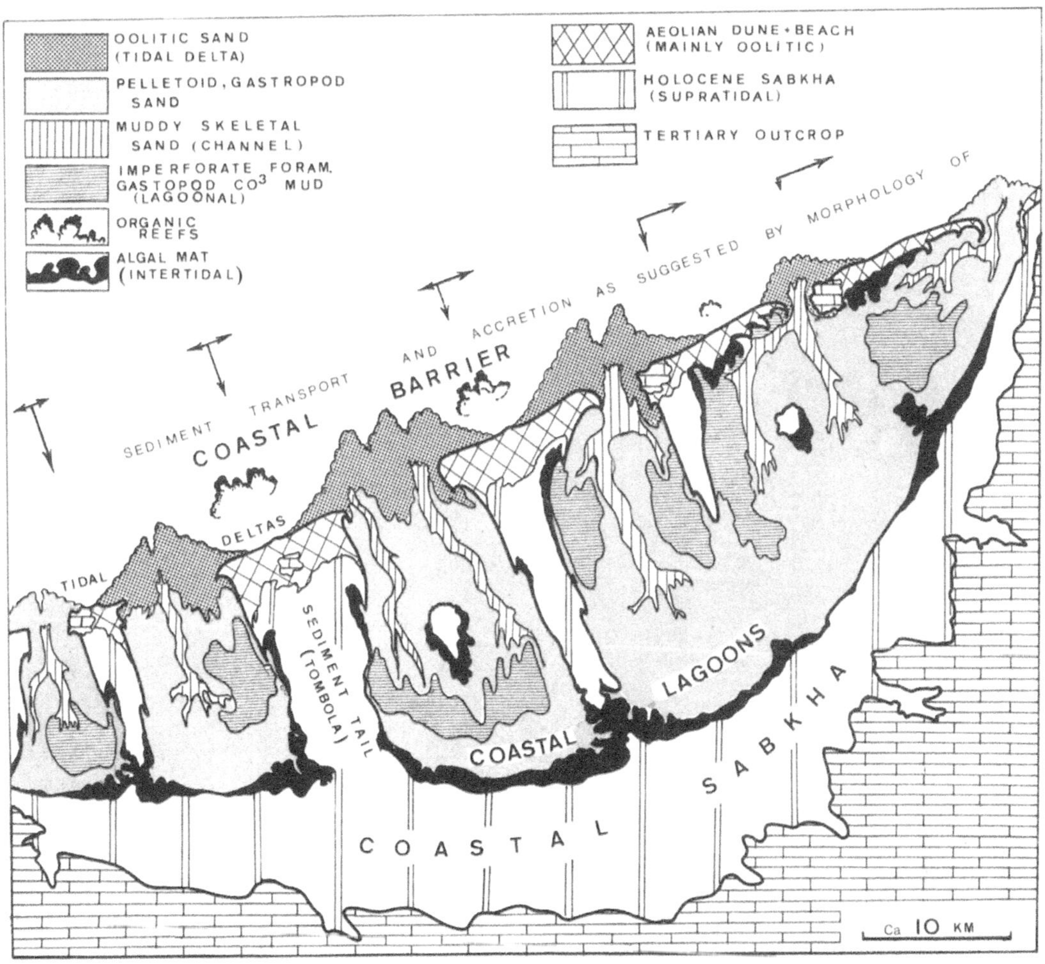

Fig. 9. Schematic sketch-map of the E half of the Central Region (E. Abu Dhabi) showing the geometry of the principal sedimentary units

Large scale production of oolitic sand on the tidal deltas has provided considerable volumes of sediment, much of which has been driven onto the barrier islands by onshore waves and which has contributed effectively to the lateral accretion of the barrier complex. These sands have also been piled up on the barrier to form spectacular aeolian dunes.

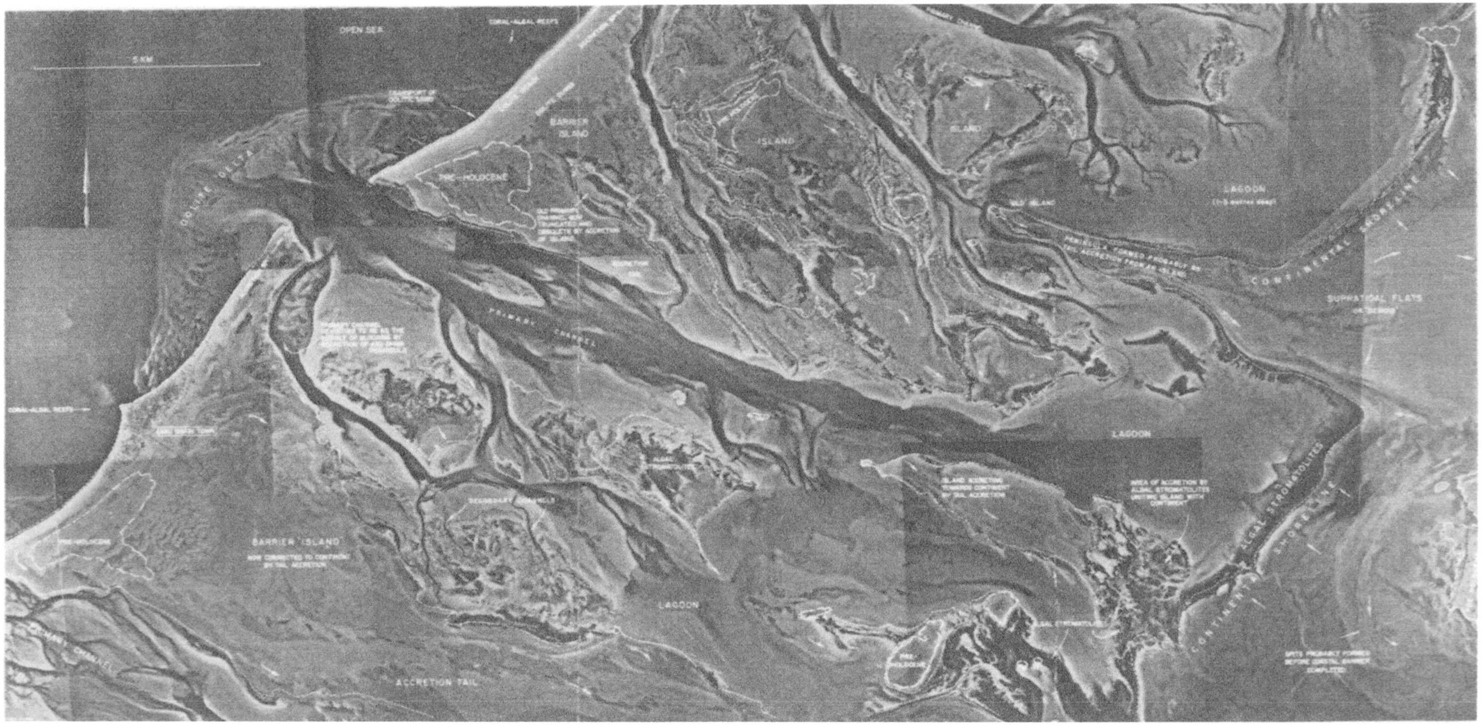

Fig. 10. Aerial photo-mosaic showing the morphology of the coastal barrier and lagoons, E. Abu Dhabi, Trucial Coast

Small fringing and patch reefs are present on this E part of the barrier and are best developed immediately seawards of the larger barrier islands between the tidal deltas (Fig. 9). Small lagoons have developed locally between these reefs and the barrier to form a double lagoonal system which is similar to, but smaller than, the windward lagoons of the Bahamian platforms and SE Florida.

Sedimentation within the lagoon

The Great Pearl Bank barrier encloses a lagoonal complex whose considerable variations in width and depth are closely related to the morphology and the position of the adjacent barrier with respect to the regional shoreline. At its W end, immediately N of Sabkha Matti, the lagoon attains 140 km in width and 40 m in depth (Fig. 1), and is an open marine environment. However, N of Yas Island (Fig. 3) the barrier is situated only 40 km from the shores of Central Abu Dhabi and encloses the clearly defined Khor al Bazm lagoon. This latter feature becomes progressively narrower and shallower until it loses its entity E of Abu Abyad Island; from here eastwards it is dissected into a series of small secondary lagoons separated by the tails of sediment which have accreted across the lagoon from the barrier islands.

Decreasing width and depth together with the fragmentation of the lagoon, eastwards along its axis, leads to a relative increase in its isolation from the open waters of the adjacent Persian Gulf. The increasing restriction is reflected in the salinity, which ranges from 40 ‰ near the W end of the Khor al Bazm to 50 ‰ near its shallower E extremity and attain values of 60 ‰ within the secondary lagoons in E Abu Dhabi. Water temperatures show similar variations (Evans et al., in this volume). These lateral changes within the lagoon clearly influence the nature and distribution of sediments (Fig. 7). Those within the deeper, unrestricted,W part of the Khor al Bazm are comparable with sediments existing throughout much of open Persian Gulf: well developed coral-algal reefs cap the numerous local highs within this part of the lagoon and border its adjacent mainland shores (Figs. 4, 7). There is a marked concentration of carbonate muds with lamellibranchs immediately behind the barrier due to the protection afforded both by the deeper water close to the barrier and to lateral protection by the barrier itself. Towards the E, as the lagoon becomes more restricted, there is a marked reduction in faunal diversity and consequently a change in sediment character. Corals, Melobesioid algae and echinoiderms become rare whilst both gastropods and imperforate Foraminifera, together with their debris,become extremely abundant (Murray, 1966).

The very shallow (2-5 m),highly protected area immediately S of the large island of Abu Abyad includes a narrow zone of carbonate mud rich in imperforate Foraminifera (Fig. 5). The remainder of the lagoon in this area, in spite of its highly protected nature, has very little carbonate mud and the sediment consists mainly of pelletal and compound-grain carbonate sand. This sand forms only a thin veneer over a rock pavement of similar petrographic and mineralogical composition, suggesting that carbonate sands are being lithified within this rather restricted lagoonal environment. This submarine limestone exhibits large polygonal fracture patterns up to 400 m in diameter, clearly visible on aerial photographs (Kendall and Skipwith 1969; Shinn 1969).

In E. Abu Dhabi, where the lagoonal system is discontinuous (Fig. 8), carbonate muds rich in imperforate Foraminifera and gastropods occur in the lagoonal ends of the channels and in intertidal and subtidal areas in the lee of the barrier islands. However, the dominant sediment is a hard pellet sand rich in imperforate Foraminifera and gastropods, which covers most of the shallow lagoon floors and the broad intertidal flats bordering them (Evans and Bush, 1969, and Evans et al., in this volume). These carbonate sands are very thin and overlie a cemented crust which is in the process of lithification. The progressive faunal and sedimentary variations found along the axis of the Khor al Bazm lagoon further to the west are here compressed into distances approximately one tenth of those in the latter area, and have their greatest variation normal to the regional shoreline, and not parallel to it as in the latter area.

Table I. The relationship between the degree of emergence of the Great Pearl Bank barrier and the lateral variation in sediment composition, and distribution of diagenetic minerals, within the adjacent lagoon and along the continental shoreline

	W		E
BARRIER	ESSENTIALLY SUBMERGED	PARTIALLY EMERGED	ESSENTIALLY EMERGED
	Between Ghasha and Gharbi islands	Between Gharbi & E end of Abyad islands	Between Abyad Id. and Ras Ghanada
	Average depths 5-10 m; Locally emergent as small (1 km diam.) islands along lagoonal margin.	Average depth on axis 2-5 m; Large islands along lagoonal edge with intertidal flats;	Depth in channels 2-10 m; 75 % of barrier emergent as barrier islands with down-wind tails and/or lateral spits due to longshore transport
	Lamellibranch sand and small patch reefs along lagoonal margin	Well developed reefs on seaward flank, lamellibranch and coral-algal sand on axis and as spill-over sheets along lagoonal edge; mud and evaporites on intertidal flats	Oolitic sand; very small reefs in channels and in front of barrier islands. Wide intertidal flats with algal mats and evaporites along lagoonal margins
LAGOON	Average depth 20 m	Average depth 2-5 m	Lagoon dissected by accretion tails into secondary lagoons (depth 2-3 m)
	Salinity ca. 40‰	Salinity ca. 40-50‰	Salinity 50-60‰
	Lamellibranch muds in lee of barrier; coral-algal sands on lagoonal highs, molluskan sand and coral-algal debris nearer shore	Limited areas of imperforate foram. mud along lee of islands; pelletal and compound grain sands widespread; frequent submarine lithification	Imperforate foram. muds in very sheltered areas behind barrier islands; hard pelletal sands dominant; lithified submarine crusts frequent
CONTINENTAL SHORELINE	Rocky headlands with narrow tidal flats in embayments	Regular morphology consisting mainly of 1-2 km wide sabkhas	Maximum development of sabkha plain (2-5 km) accreting over adjacent lagoon
	Molluskan sand, fringing reefs, local pellet sand in embayments; abundant beach rock, little evaporite.	Pellet and compound grain sands; abundant imperforate forams and cerithid gastropods; algal mats and evaporites.	Mainly pelletal and imperforate foram. sand; local cerithid gastropod ridges, well devel. algal mats; maximum devel. of dolomite and evaporite minerals.

Coastal sedimentation along the mainland (continental) shoreline

The composition of the coastal sediments is determined mainly by the nature of the adjacent lagoon and barrier. In the W, near Jebel Dhanna (Fig. 3), the lagoon is relatively wide and the barrier offers little protection; the shoreline sediments therefore consist mainly of skeletal sand and have fringing reefs. Intertidal lithification is a characteristic feature and sedimentary breccias consisting of penecontemporaneously broken beach rock are common. Eastwards along the shores of the Khor al Bazm there is a progressive change in character of the sediments of the mainland shoreline which is related to the progressive shallowing and narrowing of the adjacent lagoon, and to the increasing protection offered by the frequently emergent Great Pearl Bank barrier. The disappearance of small fringing reefs in the nearshore waters of the adjacent lagoon results in the absence of coral debris in the sediments of the shoreline. Skeletal carbonate sands of the shores of the western lagoon are replaced by pelletal and compound grain carbonate sands towards the more protected E parts (Figs. 4, 7) (Kendall and Skipwith, 1969).

The Arabian shoreline of the more protected eastern end of Khor al Bazm lagoon and those around the smaller secondary lagoons in E. Abu Dhabi, have wide intertidal flats with well developed blue-green algal mats (Kendall and Skipwith, 1964). Here, as in the adjacent lagoons, sediments consist mainly of pelletal sands with abundant cerithid gastropods. Locally, small beach ridges composed almost entirely of gastropod sand and gravels fringe the shoreline. In other areas small swamps, colonized by the black mangrove Avicennia marina and other halophytic plants have developed. These latter areas favour the deposition of carbonate muds and soft pellet carbonate muds which are intensely bioturbated by crabs. The shallowness of the lagoons, together with the very effective protection provided by the coastal barrier complex, has led to very active intertidal flat accretion which has produced the wide sabkha plains characteristic of E. Abu Dhabi (Evans et al., 1969, Butler, 1969). This coastal sabkha consists of a wedge of Holocene marine carbonate sediment which passes downwards into pre-Holocene, aeolian, quartzose sands. The character of the Holocene sequence is closely related to the protected nature of the adjacent lagoons with their abundant production of carbonate sediment and its shorewards transport leading to the progradation of the coastal sabkha. An accretion plain, approximately 5 km wide, has developed southwest of Abu Dhabi Island in the last 4000 years (Evans et al., 1969). Landwards of this wedge of Holocene sediment the pre-Holocene quartzose dune sands have been deflated to form a flat surface which extends the Holocene sabkha towards the rocky plains and dunes of the N. Rub-al-Khali.

The high salinity and temperature of both the lagoon and the interstitial waters of the sabkha, together with the extreme aridity and high air temperature, have led to the extensive development of dolomite and other evaporite minerals, of which gypsum and anhydrite are the most common within the supratidal sediments of this sabkha plain. The genesis and distribution of these diagenetic minerals are discussed by Butler et al., and Bush, elsewhere in this volume. The progradation of the shoreline with its various carbonate sediments and evaporite minerals has created a characteristic regressive "sabkha sequence" illustrated in Fig. 14B.

In the extreme NE of this Central Province of the Trucial Coast the highly protected lagoons appear to have been completely filled by these accretionary processes. This has permitted coastal dunes to migrate across the sabkha to cap the regressive sequence with a layer of skeletal and oolitic sand with inclined bedding which is oriented mainly to landwards.

The variation of sediment type and the distribution of evaporite minerals laterally along the regional shoreline of the Central Trucial Coast is thus clearly the consequence of variations of both the morphology of the adjacent lagoon and its barrier. These relationships are summarized in Table I.

Evolution of sedimentation within the Central Province of the Trucial Coast. (Fig. 11)

The present coastal morphology, together with the distribution of pre-Holocene limestones within the coastal areas of Central and E. Abu Dhabi, indicates that a shallow, westerly-plunging regional depression, now occupied by Khor al Bazm lagoon, was a major element of pre-Holocene morphology. This depression was flanked by a rock ridge which also plunged to the W and which was oriented obliquely to the regional coastline (Fig. 11A). The orientation of this ridge and associated depression appears to be related to the regional structural strike, its relief having been accentuated by subsequent pre-Holocene, subaerial erosion. This Great Pearl Bank barrier has expanded seawards by the growth of fringing reefs on its seawards side. Bioclastic sediments on its crest have been piled around Pleistocene limestone cores to form accretionary islands of carbonate sand. This Holocene accretion has increased the efficiency of the ridge as a protective barrier and has led to the deposition of carbonate muds and pellet sands in the adjacent lagoon and also by lowering the degree of water agitation, to the build-out of intertidal flats. The shallowing of the lagoons probably facilitated the accretion of sediment tails from the lee side of

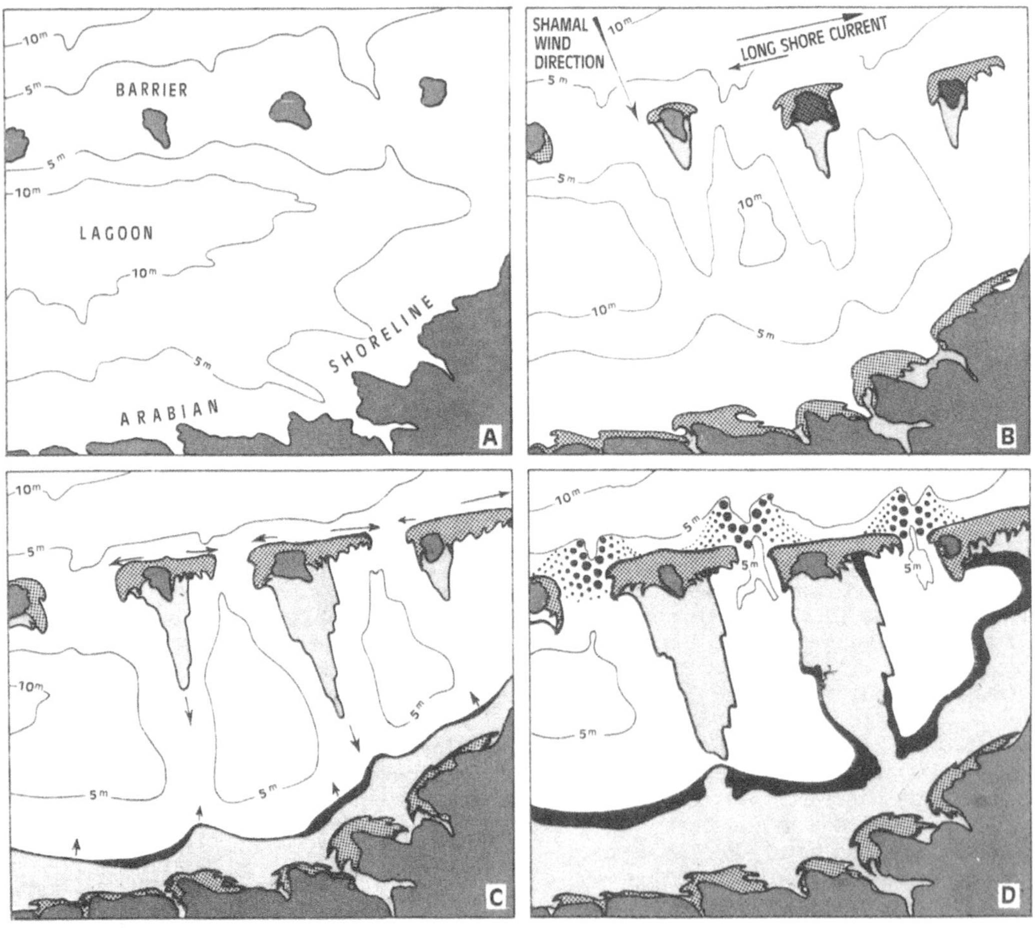

Fig. 11. Sketch-maps showing interpreted evolution of coastal barrier and lagoonal systems in E half of the Central Region (E. Abu Dhabi)

barrier islands (Fig. 11B). Where the lagoon was very shallow these tails accreted across it to the mainland shore, subdividing the the eastern part of the initial depression into a series of isolated secondary lagoons which are now a characteristic feature of E. Abu Dhabi. In addition to "tail" (downwind) accretion, the efficiency of the barrier was progressively increased by the accretion of lateral spits which extended the islands parallel to the shoreline and led to the progressive narrowing and deflection of the inter-island channels. The progressive narrowing of these inlets ultimately led to the formation of oolitic tidal deltas. All of these processes have led to the increasing isolation of the coastal lagoonal area from the open waters of the Persian Gulf and were accompanied by a progressive change in the sediments and the faunas as the environment evolved: beaches composed of skeletal carbonate sands have been replaced by wide intertidal flats composed of pelletal carbonate sands and muds with well developed algal mats, particularly on the mainland shore. The high salinity and temperature of the increasingly restricted lagoonal waters, together with the high air temperature, have led to striking diagenetic changes in the intertidal and progradational coastal plain sediments and to the extensive development of dolomite, gypsum, anhydrite and other evaporitic minerals (Shearman, 1963; Curtis et al., 1963; Kinsman, 1966, 1967; Butler, 1969).

Continuation of the present sedimentary regime may result in the closing of channels between the barrier islands, both by continued lateral growth of spits and by sedimentation within the lagoons. This will result in the development of highly restricted lagoons possibly leading to the precipitation of primary gypsum comparable to that forming within the present lagoons along the coast of Saudi Arabia (Bramkamp and Powers, 1955). As the lagoons become completely filled, aeolian coastal sands may transgress over them. Coversely, the barrier may migrate over the infilled lagoons as cheniers; there is some evidence of this process near Ras Ghanada in the extreme NE extremity of this barrier complex. This system of accretion may be expected to extend laterally down the lagoonal axis towards the W, and it is probable that the Khor al Bazm, having become filled with lagoonal sediment, will be dissected into a series of secondary lagoons by sediment tails accreting down-wind from the adjacent barrier. Present beach ridges of skeletal sands, and fringing reefs along the mainland shore, will be replaced progressively by intertidal flats of pelletal sand or carbonate mud supporting carpets of algal mats (stromatolites), and the development of dolomite and associated evaporite minerals.

SEDIMENTATION IN THE EASTERN PARTS OF THE TRUCIAL COAST

The Trucial Oman Coast between Ras Ghanada and Masandam Peninsula, 180 km to the E (Figs. 1, 12), is essentially linear. It consists of storm beaches backed by coastal dunes composed essentially of skeletal carbonate sands. In the extreme NE near Sharjah, Umm al Qaiwain, Hamra and Ras al Khaimah, a series of sub-parallel spits has prograded the coastline some 5-10 km to seaward. The resulting sediment patterns are comparable to those of N.E. Qatar, described by Shinn elsewhere in this volume. Individual spits composed of skeletal carbonate sand enclose lagoons in which carbonate muds presumably dominate. As in NE Qatar, channels dissect the spits and these terminate seaward in small tidal deltas composed of pelletal sand. The resultant barrier complexes extend laterally near Umm al Qaiwain (Fig. 12) for more than 15 km. Inland of these coastal features the extensive quartzose dune fields of the Arabian desert extend inland to the alluvial fans which skirt the Oman Mountains (see Glennie, 1970; for details of the continental facies).

The regional coastline faces obliquely into the WNW "shamal" winds and associated waves and surface currents. It is a particularly exposed coastline, facing the entire length of the Persian Gulf, and thus suffers the effects of maximum fetch. Because of the absence of inherited offshore barriers, deep water impinges directly onto the shore. These factors combine to make this Eastern province an area of maximum water agitation and effective longshore transport.

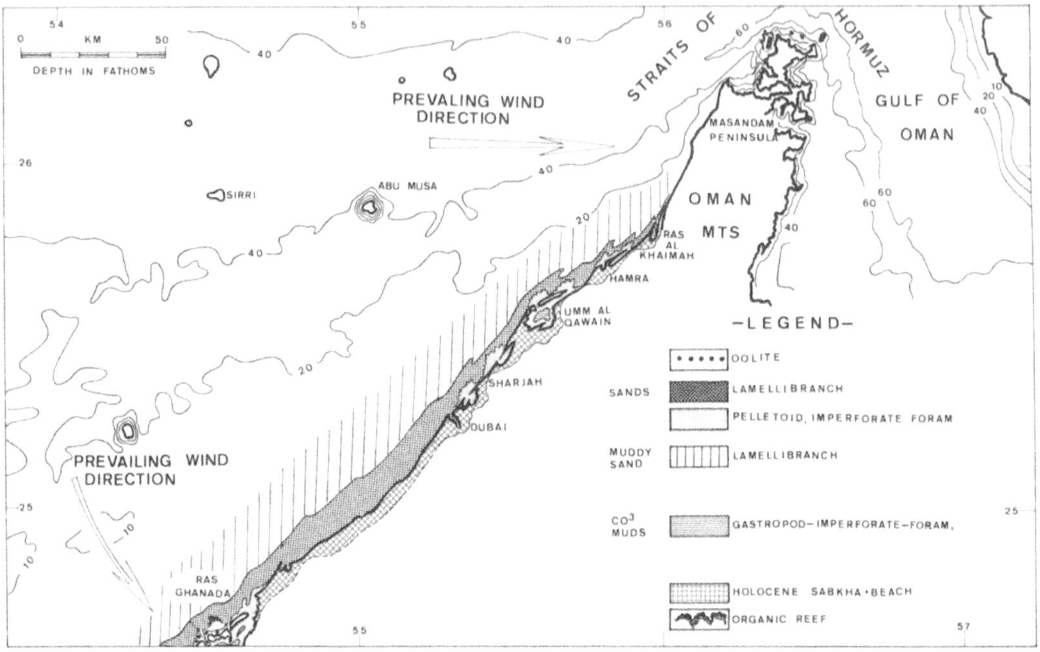

Fig. 12. Morphology and distribution of the major sediment units in the NE Region of the Trucial Coast

Whether or not the linear character of this coast is due mainly to its exposed position is not known. Henson (1951) has suggested that it is a faultbounded coast. Its regularity is probably due in part to its close proximity to the adjacent Oman Mountains; detritus derived from these mountains must have spread over the adjacent area during periods of lower sea level and more extensive rainfall, and in so doing would have buried any topographical features that existed to produce a relatively smooth surface which was later drowned by the rising sea. This additional event may have removed the topographical features which have such an important effect on Holocene sedimentation in Central Abu Dhabi.

The evolution of the extreme NE part of the Trucial Coast has been revealed by seven cored holes drilled across this coastal plain to an average depth of 50 m in the vicinity of Ras al Khaimah (Fig. 13) (Evamy and de Groot, unpublished Shell Research manuscript). These wells passed through a Quaternary sequence ranging from wadi gravels and aeolian quartz sands at the base, to marine skeletal sands and carbonate muds at the surface (Fig. 13). The marine part of the sequence, approximately 10 m in thickness, is probably the only part which is Holocene in age. The sequence is essentially a transgressive one and is most probably the result of the post-glacial rise in sea level; however, the presence of northerly-tilted terraces on the flanks of the nearby Oman Mountains suggests that this transgressive sequence could be related partly to tectonic movements.

The NE section ot the Trucial Coast is terminated by the cliffs and rocky shorelines of the Masandam Peninsula which forms the N extremity of the Oman Mountains. These are composed essentially of limestone and are drained by a series of deeply incised wadis which terminate in spectacular alluvial fans, some of which reach the coast (Glennie, 1970). A similar morphology in temperate latitudes would result in a flood of detritus from the adjacent mountains, especially where these were composed of silicate rocks. However, in this arid climate, this is not the case and the presence of the mountain system is barely evident in the Holocene marine sediments accumulating at its foot. Although the coastal sands contain scattered limestone pebbles, the nearby offshore sediments are composed of relatively pure molluskan sands.

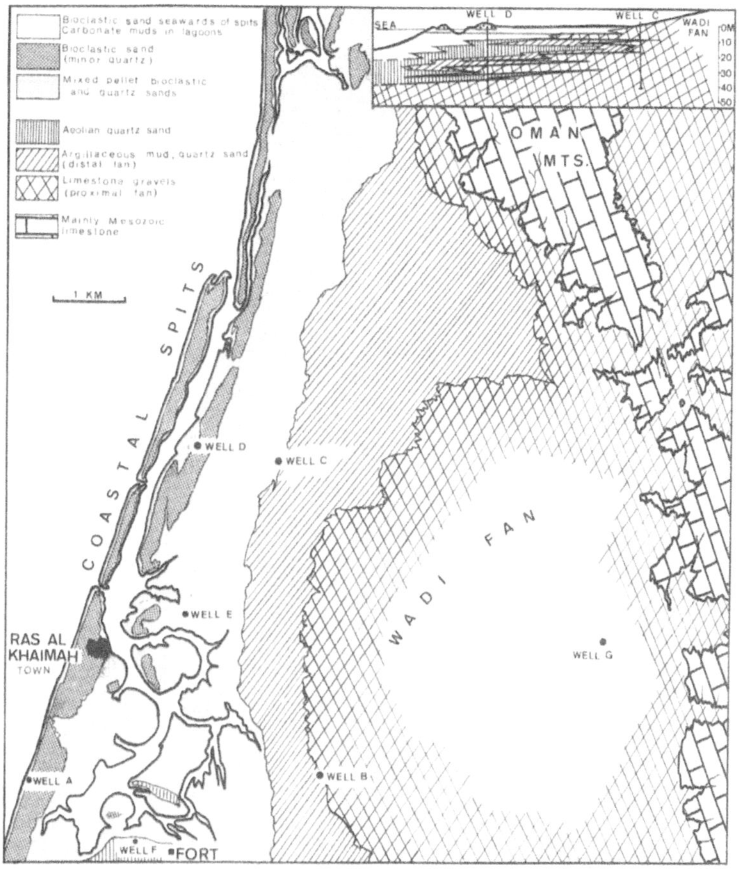

Fig. 13. Coastal morphology and geometry of the principal sediment units, NE extremity of the Trucial Coast (Ras al Khaimah); based on unpublished maps prepared by B.D. Evamy

CONCLUSIONS

Very little terrigenous material is supplied to the Trucial Coast from the adjacent Arabian landmass, mainly because of its arid climate. Most sediments are indigenous carbonates whose composition and geometry are determined by the inter-related effects of coastal morphology and wind-controlled waves and surface currents. Because these "shamal" winds are directed towards the shore the presence, or absence, of coastal barriers is clearly expressed in the nature and geometry of the shoreline sediments and their diagenetic minerals which include dolomite, gypsum and anhydrite. Shoreline and lagoonal sediments also reflect the relative proximity, width and degree of emergence of the adjacent barrier. Geologists concerned with the prediction of sedimentary and diagenetic patterns related to offshore barriers and shorelines, may improve their prediction by understanding the inter-relationships seen in this Trucial Coast model.

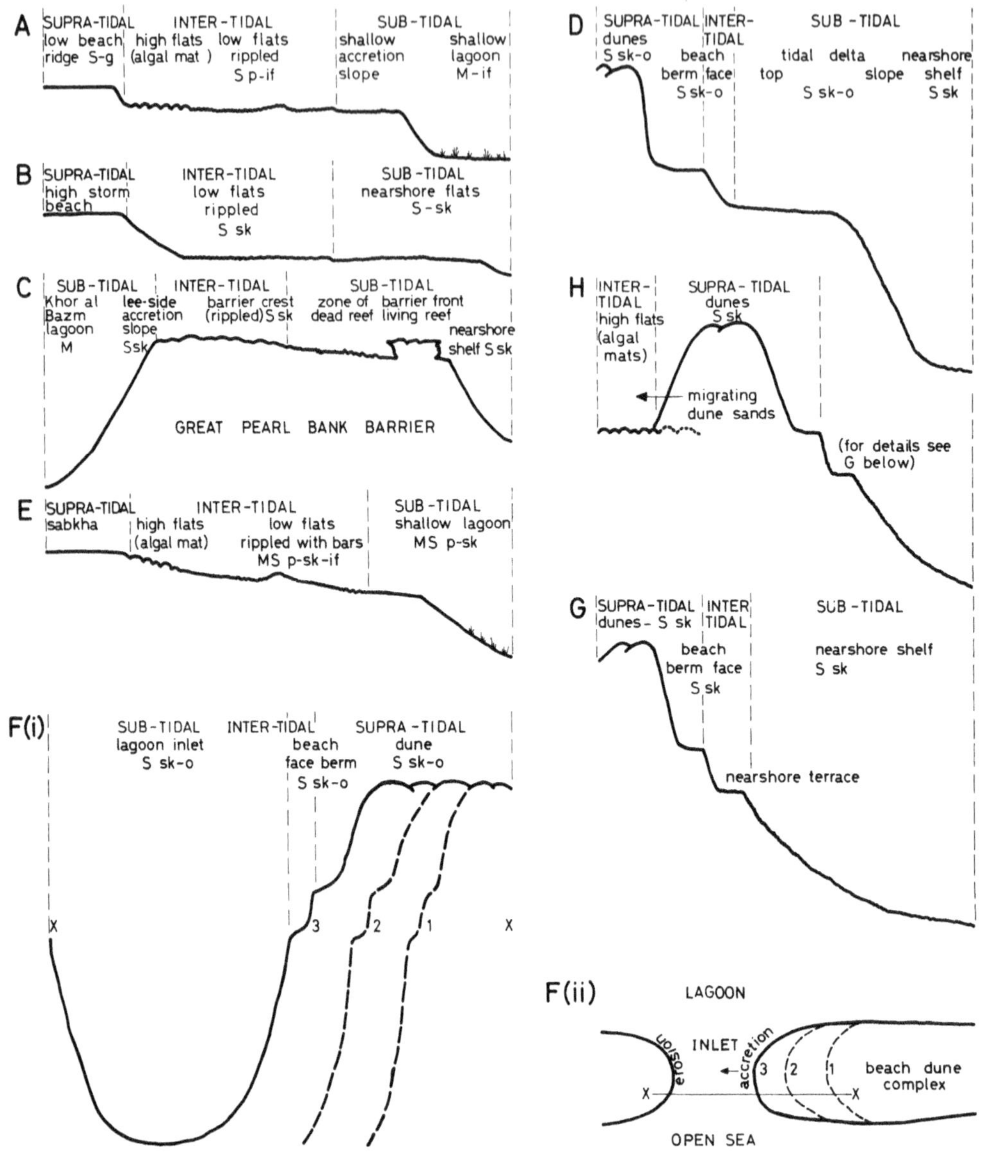

Fig. 14A. Typical sedimentary profiles of the Trucial Coast.

A. Intertidal flats bordering protected embayment, Khor Duwahine, W Region, (extreme W part of Abu Dhabi, - see Fig. 2).

B. Storm beach flanking exposed embayment; Sabkha Matti, W Region, (Abu Dhabi, see Fig. 2).

C. Great Pearl Bank Barrier, W part of Central Region (see Fig. 6).

D. Prograding barrier island-tidal delta complex, Abu Dhabi Island, E Abu Dhabi

Legend continued on Fig. 14B

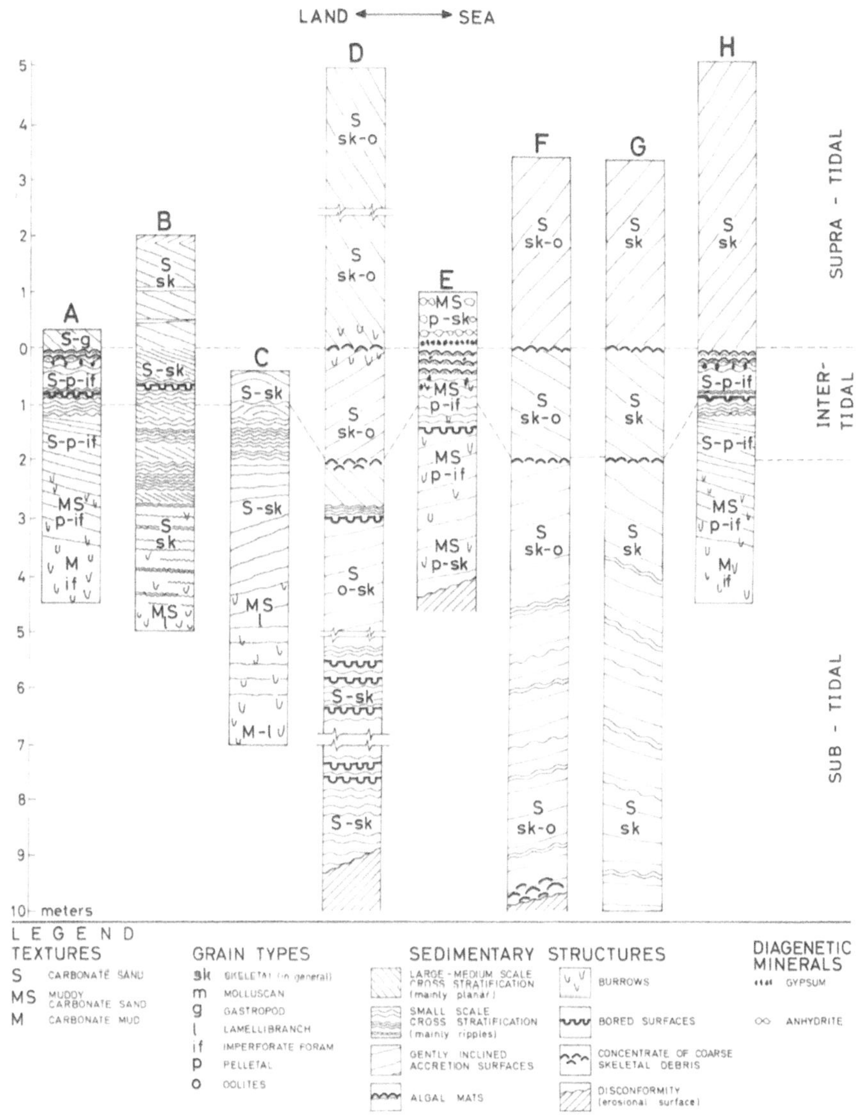

Fig. 14B. Vertical sequences produced by sedimentary accretion on the profiles illustrated on Fig. 14A.
--- legend continued:
E. Prograding landward margin of protected lagoon, Central Abu Dhabi.
F. Laterally accreting beach-dune complex; the sediments between 2-10 m (Fig. 14B) probably will show through cross-stratification oriented both seawards and lagoonwards; Central Abu Dhabi.
G. Prograding beach-dune complex between Ras Ghanada (E. Abu Dhabi) and Dubai (see Fig. 12).
H. Landwards migrating dune sands capping infilled lagoon, Ras Ghanada, E. Abu Dhabi (see Fig. 12)

The Oceanography, Ecology, Sedimentology and Geomorphology of Parts of the Trucial Coast Barrier Island Complex, Persian Gulf

G. Evans[1], J. W. Murray[2], H. E. J. Biggs[3], R. Bate[3], and P. R. Bush[1]

ABSTRACT

The coastal barrier and related lagoonal and intertidal flat systems in NE Abu Dhabi are composed of relatively pure carbonate sediments. Although this complex includes a great variety of micro-environments and sediment types, these may be grouped into seven principal units: nearshore shelf, frontal reef, tidal deltas, frontal (barrier) beaches and dunes, lagoonal channels, lagoonal terraces, and intertidal flats. The principal faunal and floral components of each unit are discussed and listed in distribution tables, careful distinction being made between living and dead assemblages.

The character and sedimentary composition of this coastal complex has evolved mainly as the consequence of rapid carbonate production and onshore transport of these sediments. Cementation in the shallow subtidal and intertidal environments has produced extensive, diachronous crusts of Holocene limestone, while the hot, arid climate has stimulated the formation of dolomite and other evaporite minerals within the coastal sabkha. Most of these sedimentary and diagenetic features would permit the recognition of similar environments in ancient rocks.

INTRODUCTION

A coastal complex of barrier islands and lagoons borders the SW shore of the Persian Gulf. This complex shows lateral variations in physiography, oceanography and its resulting faunas, floras and sediments, as well as in its geological evolution (Purser and Evans this volume). The region was comparatively unknown until the early 1960s, apart from brief references in papers by Emery (1956) and Sugden (1963). Since that date extensive, although still incomplete, studies have been made by members of the Geology Department of Imperial College London (assisted by members of the staff of the British Museum of Natural History and of Mobil Research and Development, Dallas USA., and by various members of other British Universities) as well as by Shell Research Laboratory, Rijswijk, The Netherlands. These studies have been further pursued by members of the initial Imperial College team who have since worked from Princeton University, University of Southern California and Esso Research Laboratory, United States.

1 Dept. of Geology, Imperial College, London.

2 Dept. of Geology, University of Bristol, Bristol.

3 British Museum (Natural History), London.

The general physiography of the area and its associated sediments, fauna and geological development have been discussed in numerous publications, some of which are printed in this symposium (Curtis R. et al., 1963; Shearman D.J., 1963; Evans G., Kinsman D.J.J. and Shearman D.J., 1964; Kinsman D.J.J., 1964; Evans G. Kendall, C.G. St.C. and Skipwith P.A., D'E; 1964; Kinsman D.J.J., 1964; Evans G. and Shearman D.J., 1964; Shearman D.J. and Skipwith P.A., D'E., 1965; Murray J.W., 1965 (a) and (b); Murray J.W., 1966 (a), (b), (c); Kinsman D.J.J., 1966; Shearman D.J., 1966; Kendall C.G.St.C. and Skipwith P.A., D'E., 1968; Butler G.D., 1969; Evans G., Nelson H., Schmidt V. and Bush P., 1969; Evans G. and Bush P., 1969; Kinsman D.J.J., 1969; Kendall C.G.St.C. and Skipwith P.A.,D'E., 1969 (a) and (b); Evans G., 1970; Murray J.W., 1970 (a) and (b); Shearman D.J., Twyman J. and Karimi M.Z., 1970; Bate R.H., 1971).(In addition, considerable data are available in the unpublished theses of the University of London; Kinsman 1964; Butler 1966; Skipwith 1966; Kendall 1966; Twyman 1969; Bush 1972).

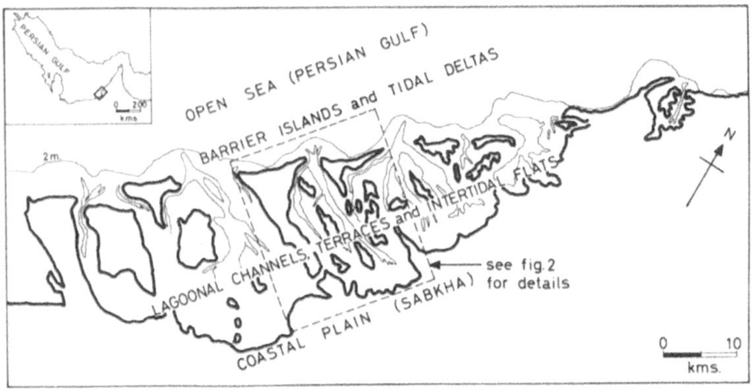

Fig. 1. Map of the Trucial Coast showing the location of the area described in this paper

The barrier island - lagoon complex is particularly well developed in the Sheikdom of the Abu Dhabi (Fig. 1). A description of the physiography, oceanography, ecology and sedimentology of one of these lagoons and associated islands will serve to emphasize the main features of this coastal region. It is characterized by a very arid climate, and there is no fresh water reaching the coast, and it therefore lacks a detrital sediment supply except for that brought by wind. Consequently, it is a region dominated by calcium carbonate deposition accompanied by the development of evaporites on and within the sediments of the adjacent coastal plain.

The Sadiyat - Abu Dhabi lagoon is bordered to the NE and SW by the islands bearing those names, landward by the low coastal plains - the sabkha - of the Arabian mainland, and to seaward it passes through a narrow inlet into the open waters of the southern Persian Gulf (Fig. 2).

The islands and the adjacent coastal plain are composed essentially of unconsolidated, with some consolidated, Quaternary calcareous sediments. Some Tertiary sediments occur in the area; they form a series of low bluffs at the landward margin of the adjacent coastal plain and farther W form isolated hills (jabals) on the plain; however, apart from one small jabal, which forms an island in the landward parts of the Abu Dhabi-Sadiyat lagoon, these rocks have not been observed elsewhere within the barrier-lagoon complex. Rocks of similar lithologies have, however, been encountered in boreholes on the seaward parts of the islands at depths of 17 to 20 m below approximately low water spring tide, and at 25 m below that level approximately 18 km offshore.

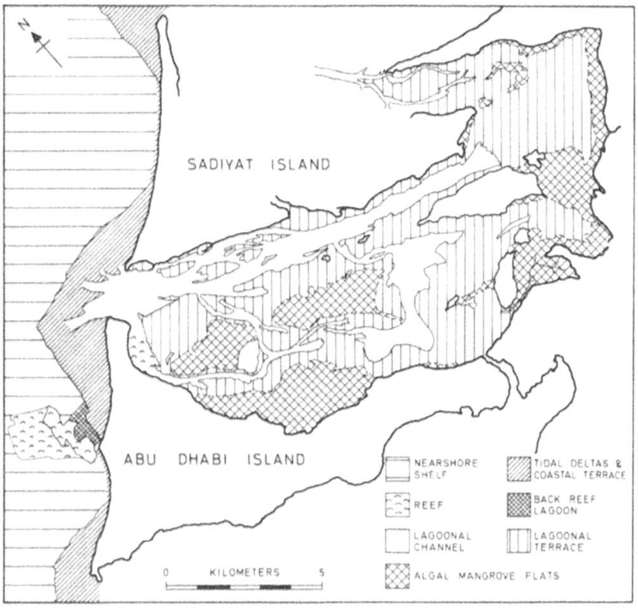

Fig. 2. Sub-environments of the Abu Dhabi - Sadiyat barrier island complex

Small isolated masses of a quartzose calcarenitic limestone ("miliolite" of earlier workers) occur on the islands and scattered throughout the lagoons as small islets, and rocks of a similar lithology crop out extensively in the lagoon channels. This limestone shows well developed trough cross-stratification and in some areas contains extensive branching structures which appear to be fossil remains of plant roots. This deposit has been interpreted as having an aeolian origin, and the orientation of its cross-stratification suggests that it was formed by winds having the same orientation as the modern "Shamal" (i.e; NNW). Numerous boreholes and excavations on the mainland of Abu Dhabi have shown that a brown, unconsolidated, cross-bedded quartzose carbonate sand underlies much of the region. This is similar in texture and composition (table 1) to the calcarenitic limestone and both are thought to be one and the same deposit, the former merely being cemented patches of the latter.

TABLE I. Comparison of the composition of the uncemented brown sand, and the lithified "miliolite"

	"Miliolite"	Brown Sand
LOW Mg CALCITE	61%	56%
QUARTZ	24%	26%
DOLOMITE	6%	9%
FELDSPAR	9%	9%

The outcrops of calcarenite ("miliolite") are frequently capped by a limestone of variable facies, sometimes containing corals and elsewhere consisting essentially of a gastropod coquina. This upper limestone lies on a sharp plane of erosion, cut in the underlying cross-bedded calcarenite, and often contains fragments of this in its basal unit. Also, in some places, mollusks can be seen in borings extending down from the erosion surface into the underlying "miliolite" calcarenite. This capping limestone has obviously resulted from a transgression when marine waters eroded the aeolian limestone ("miliolite"). The exact age of this event is not known but it is thought to have probably taken place during the high sea levels of the last interglacial. This level need not have been very different from present sea level as the erosional plane separating the two deposits is rarely more than 1 to 3 metres above present high water mark. Where the transgressive limestone is absent a caliche-like crust caps outcrops of calcarenite.

The cross-stratified calcarenite ("miliolite") and its uncemented equivalent (the brown quartzose carbonate sand), are thought to form the core of the islands, and to have acted as nuclei around which the unconsolidated Holocene carbonate sediments accumulated during and subsequent to the post glacial rise of the sea to its present level (Purser and Evans, this volume).

OCEANOGRAPHY

The area is one of high air temperatures and very low rainfall. The dominant winds blow from between 315° and 015°. The strong "Shamal" wind strikes the coast approximately at right angles. A diurnal on-off shore wind system also exists, light winds blow offshore in the mornings and these develop into strong on-shore winds in the afternoons and evenings. The "Shamal" winds are the dominant wave producers, the tidal deltas are areas of the expenditure of large amounts of wave energy. These winds and the daily onshore waves also cause considerable turbulence in the lagoons, and the sediments on the terraces are vigorously stirred by waves except in the sheltered areas on the lee sides of certain islands and banks.

The tides of the region have a large diurnal inequality and at neaps may be entirely diurnal. They range from a maximum of $2\frac{1}{2}$ m in front of the islands to approximately 1 m at the back of the lagoons. However, during periods of prolonged onshore winds the level of the coastal waters may be raised several metres above normal high water level and lead to widespread flooding of the coastal plain. Little detailed information is available on the strength of the tidal currents. However, three stations occupied for part of a tidal cycle in 1965 show that currents outside the tidal deltas attain speeds of approximately 0.25 m/sec on the surface and 0.15 m/sec on the sea floor. They appear to flow approximately parallel to the shoreline. A station in the inlets between the islands showed surface velocities of up to 0.65 m/sec and bottom velocities of up to 0.40 m/sec. A further station situated at the innermost end of the main tidal channel at the back of the lagoon gave surface velocities of up to 0.25 m/sec and bottom velocities reaching 0.20 m/sec.

The temperature of the surface water varies from 23 - 24° C in the nearshore areas to 22 - 36° C in the inner lagoon. No detailed information is available on the diurnal range of water temperatures, but it is thought to be fairly wide in the shallow waters of the lagoons and slight in the open sea, nearshore areas.

The nearshore waters and those of the lagoons have a particularly high salinity, ranging from 42.7 - 44.5 ‰ in the former, and from 53.6 - 66.9 ‰ in the inner lagoons. Even higher values (77.4 ‰) are found in some pools on the algal flats. Generally the salinities within the lagoons are highest in summer. The distribution of surface water temperatures and salinities are shown in Figs. 3, 4 and 5. Additional salinity readings from the inner lagoon in March 1969 gave values of 48.76 - 53.06 ‰ (i.e. slightly lower than the values found before the area was modified by engineering works).

The ratio of most of the elements to salinity remains constant in all the lagoonal and nearshore waters. Calcium, however, does appear to be lost, particularly from the inner lagoon waters in summer, where losses of up to 40 mg/litre and lowering of the specific alkalinity have been detected. No clear pattern of loss has been

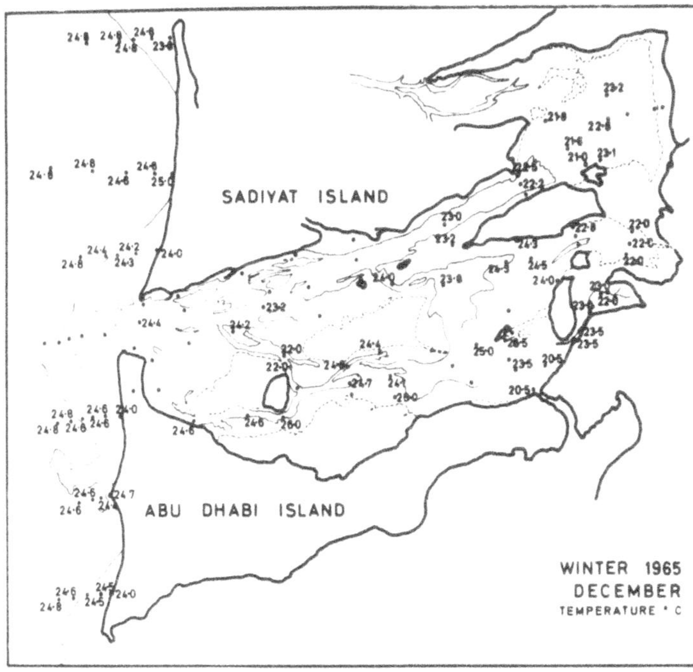

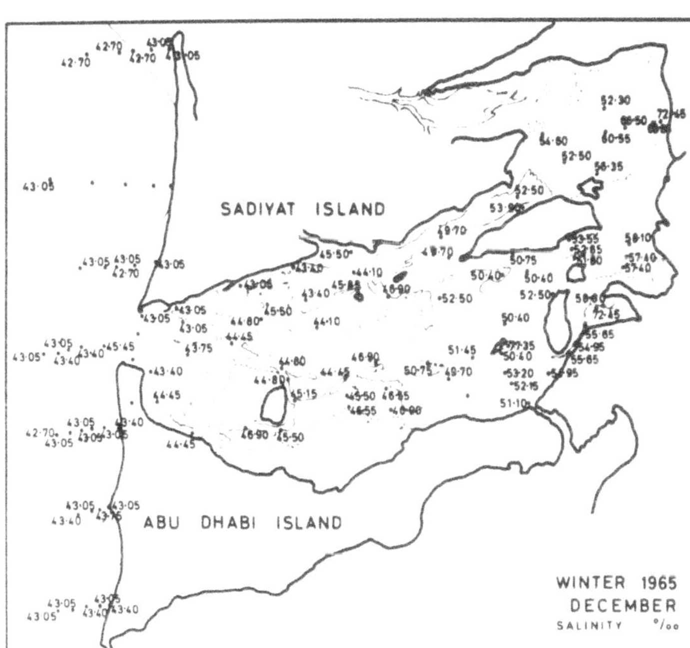

Fig. 3. Distribution of salinity and temperature of surface waters in winter

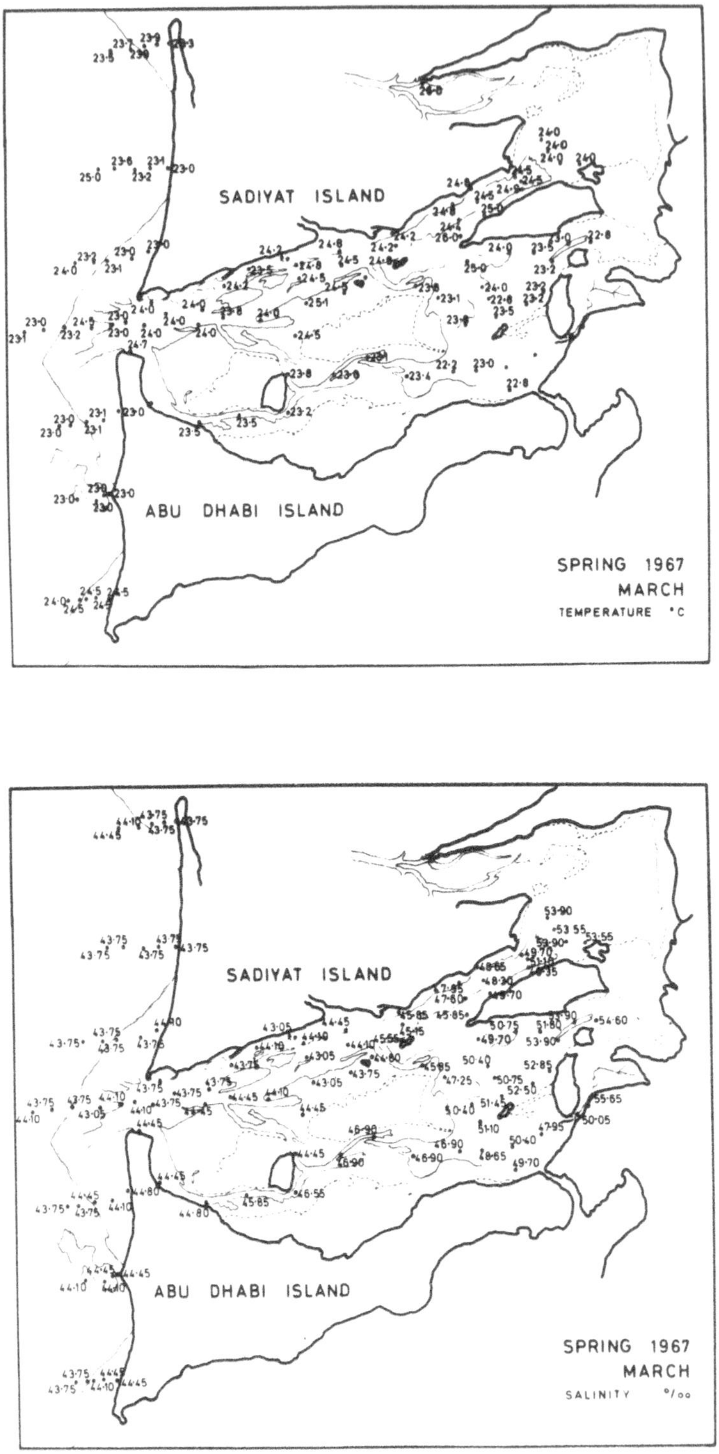

Fig. 4. Distribution of salinity and temperature of surface waters in spring

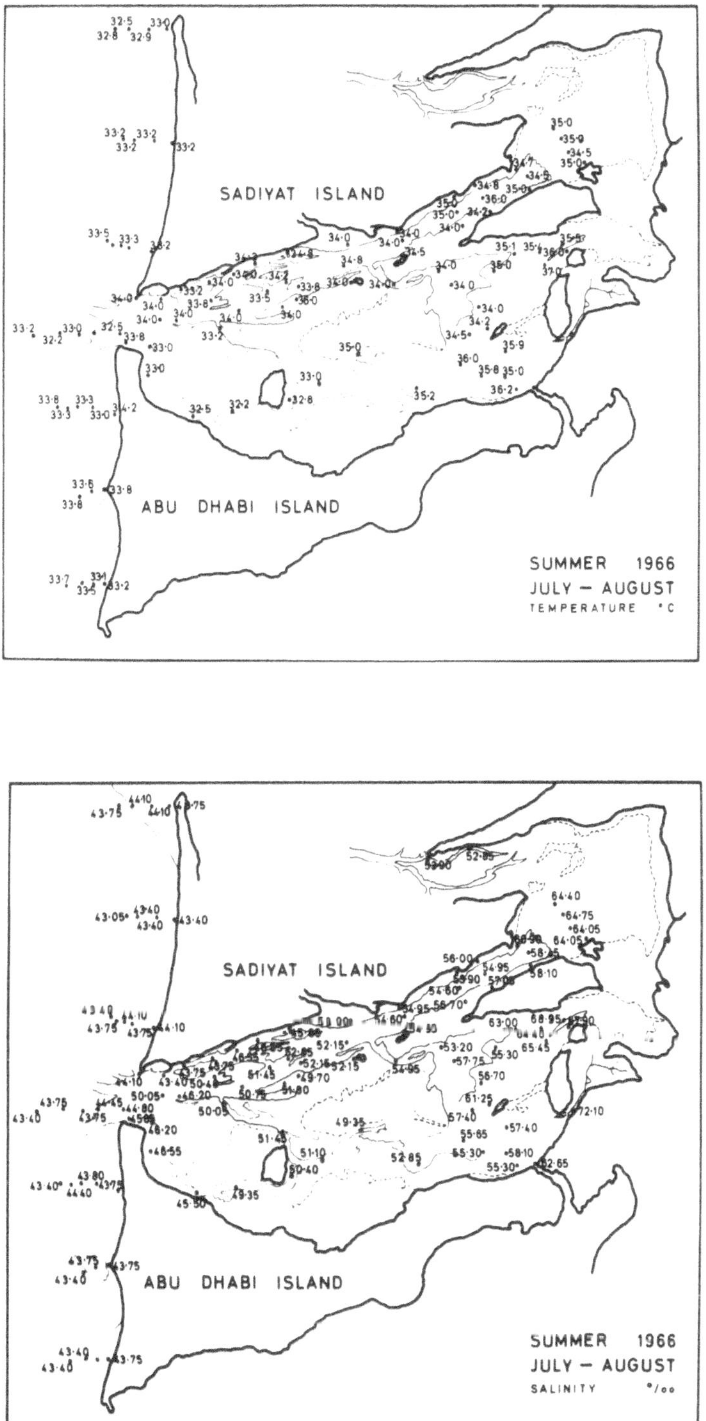

Fig. 5. Distribution of salinity and temperature of surface waters in summer

established and the mechanism of $CaCO_3$ extraction is obscure (see Kinsman, 1969).

Some reconnaissance measurements of the water nutrients show that the coastal and lagoonal waters are generally low in phosphate and nitrate (Table 2). The waters of the open sea in the nearshore zone and the waters of the outer and inner lagoons show no significant trend in phosphate content during summer or winter. Similarly, there is no observable trend in the nitrate content of the waters during winter, although it does appear to decrease when traced from the open sea to the back of the lagoon during the summer. The silicate content is higher in the waters of the inner lagoon than in those of the open sea both in winter and summer.

When the measurements for the various seasons are compared, the waters of the nearshore zone, outer lagoon and inner lagoon all show higher phosphate content in winter than in summer; the nitrate content is very variable and shows no preferential enrichment in either season, while the silicate is noticeably higher in summer. The comparatively high silicate content found in the waters at the back of the lagoon may indicate seepage of water from the adjacent rocks or from the coastal plain, as the ground waters of this plain are unusually rich in this component.

FLORA AND FAUNA

Generally, the Persian Gulf contains an impoverished Indo-Pacific fauna and flora. Many of the species common in the open waters of the adjacent Indian Ocean disappear in this area due to the adverse environmental conditions (Sugden, 1963 ; Den Hartog, 1970). This depleted fauna shows an even lower diversity in the bordering lagoons where conditions are even more extreme. As with lagoonal faunas and floras elsewhere in the world, those in the Abu Dhabi lagoons contain some open marine forms and also some which are entirely restricted to the lagoon. The distribution of these forms is controlled to a large extent by their tolerance of the variable physical and chemical environmental conditions. Ecologists recognize five critical limits of tolerance for any limiting environmental parameter: maximum limit for survival; maximum limit for successful reproduction; optimum; minimum limit for successful reproduction; and, finally, minimum limit for survival. The significance of these limits is that a species can survive, for short periods at least, under conditions which are unfavourable for reproduction.

The Abu Dhabi lagoons represent an extreme environment, with high salinities and high temperatures: foraminiferids, corals, mollusks, crustaceans (decapoda and ostracoda), echinoderms and other groups as well as the various floral groups (seagrasses, seaweeds, mangroves and other plants) all show restricted distributions in the inner lagoon when compared with those of the adjacent open sea. A study of their distribution, at any one time, gives a guide to the limiting parameters; and a study over a longer period should show, in addition, how small environmental changes may cause the tolerance limits to be exceeded and thus cause "local extinction" of species. Unfortunately it has been impossible to make such a study. However, the repeated visits to the area in the 1960s followed by a later visit in 1969 did show some major differences. These changes were induced by engineering works around the entrance of the lagoon which removed part of the tidal deltas and allowed easier ingress of open marine water from the adjacent sea; and by the dredging of an initially shallow channel into an open waterway at the landward end of the lagoon to produce a much greater connection with the lagoon lying SW of Abu Dhabi island. Also, the initiation of the dumping of sewage from the island of Abu Dhabi with its increasing population, produced important changes.

The study of the seagrasses and seaweed distribution in 1965 showed that both of these were restricted to the near shore and outer lagoon areas where salinities did not exceed 50‰ (see Fig. 6). They occurred from the level of low tide to a depth of approximately 8 m. They never occurred in the intertidal zone as intense heat kills them if they become exposed to the atmosphere. The seagrass, in 1965, was sparse in its distribution, commonly brown and looked unhealthy; it was clearly close

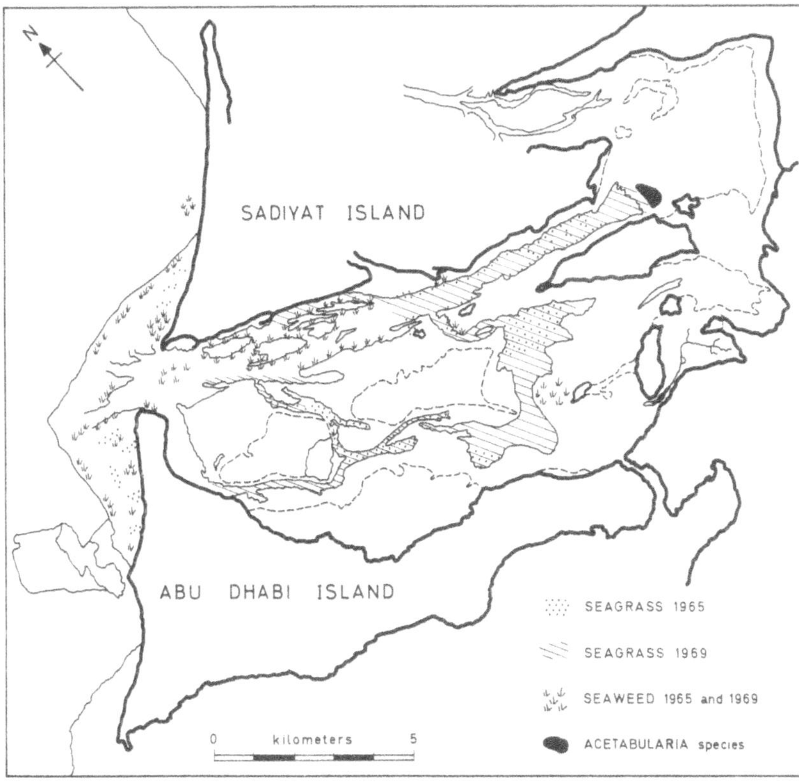

Fig. 6. Distribution of seaweed and seagrass

to extinction. In 1969, after the various changes had occurred, the seagrass was seen to be much healthier and formed a denser cover than previously; also it covered approximately twice its previous area within the lagoon (13 % of area). It appears that the greatly improved circulation,due to the dredging works; and the initiation of sewage input, had affected the environment by preventing the salinities from rising as high as those previously recorded and by increasing the nutrient supply.

In addition to changes in the seagrass community the foraminiferids showed striking changes in abundance and distribution between 1965 and 1969. As most of the foraminiferids live in association with seaweed and seagrass and their distribution is controlled by the distribution of the latter, this is not surprising. However, their increased abundance is also likely to be due to the greater supply of nutrients from sewage.

The coral reefs which consist of a rather impoverished suite of species (Kinsman, 1964; Evans and Bush,1969), also showed distinct differences over the same period. Although the reef seaward of Abu Dhabi Island (Fig. 9) was fairly healthy in 1965, the small patch reefs in the outer lagoons contained conspicuous amounts of dead coral. In contrast, in 1969, the patch and fringing reefs of the lagoon were flourishing. This again can probably be attributed to the improved water circulation; in Halat al Bahrani lagoon to the SW of Abu Dhabi Island, which has a better circulation, corals penetrate much further landwards than in the Abu Dhabi lagoon.

These artificially induced changes, which are producing more open and less extreme conditions in the lagoon, are interesting in that they are causing the lagoonal area to revert to conditions comparable to those of the earlier part of the

Holocene. At this time the barrier islands were smaller and the tidal deltas had not completely developed, so that they formed a less effective barrier. Consequently, the water circulation was much more active and environmental conditions were less extreme, probably explaining the coral and other remains encountered in artificial exposures in the inner lagoon. With continued sediment production and barrier accretion the lagoons became increasingly restricted until the reversion of this trend caused by the man-made changes in the late 1960s.

Two important points emerge from the observations: firstly, at any one time the distribution of the flora and fauna is controlled by their tolerance limits; if only one limiting factor exceeds the tolerance limit for a given species, that species cannot survive; secondly, short term changes may make the environment unfavourable (or more favourable) and this may greatly alter the distribution patterns. In a carbonate environment, where much of the sediment is of biological origin, the concept of tolerance is especially important.

The Flora

The submarine parts of the area are colonized by various seagrasses and brown seaweeds. Whereas the former colonize soft sediment surfaces in almost all subaqueuos environments, being particularly concentrated in some areas (see Fig. 6) the brown seaweeds are found where a rocky or hard substrate is present. They are thus found in areas where older limestone (mainly "miliolite") crops out and also where Holocene cemented crusts cover the sea floor.

Three species of seagrass occur: Halophila stipulacea has been found on banks in the lagoon; Halophila ovalis on the nearshore shelf, back reef lagoon and on the tidal delta and Diplanthera (Halodule) urinervis has been found in the nearshore shelf, tidal deltas, back-reef lagoon and in the channels and banks of the lagoons. It is possible that Thallassia sp. also occurs, particularly in the deeper subtidal areas (D. Kinsman personal communication to W.D. Gill 1966), although this would seem unlikely from the distributional figures of Den Hartog (1970).

The black mangrove Avicennia marina forms thick groves around the lagoon margins of some of the islands. It is often accompanied by Anthrocneumum glaucum which grows on the algal covered intertidal flats and also on the sandy surface of the islands.

Land plants form only a thin cover of vegetation and most of the islands have bare, sandy, wind scoured surfaces. However, scattered specimens of Cornulacea sp., Halopyrum mucronatum, Dipterygium glaucum, Heliotropum sp. and Halopeptis perfoliata form a thin cover of vegetation on the wind-scoured sand flat between the beach and the dune ridge, on the front of the islands, around the scattered dunes bordering the lagoon shoreline and within the dunes and wind-scoured interiors of the islands. At one or two locations small clumps of palms occur, but these may have been introduced by man.

Blue-green algal mats are widely developed on the lagoonal margins of some of the islands, and in places they form broad areas, fringed by protective beach ridges. Elsewhere, they pass seawards into bare intertidal flats, the latter type being particularly well developed along the mainland shore. In the subtidal, or only very intermittently exposed intertidal areas, a thin surface film of algae covers much of the floor of the inner lagoon terrace, and an important function in binding the sediment.

On the terraces of the inner parts of the lagoon the delicate alga Acetabularia sp. is found, although only in small numbers. This lightly calcified form may be significant in the production of carbonate mud but its infrequent occurrence make this doubtful.

Although little detailed work has been done on the group, diatoms have been noticed in sediment samples, particularly in those collected adjacent to reefs both inside and outside the lagoons. Rarely, coccoliths have been found in the muddy sediments of the nearshore and lagoonal areas (W. Diver personal communication).

Lithothammnion sp., *Lithophyllum* sp., *Goniolithon* sp. and other red algae coat the coral on the reefal areas and *Jania* sp. is commonly found attached to weed in the nearshore tidal delta and lagoon terrace areas.

TABLE 2. The nutrient content of the nearshore and lagoonal waters

NUTRIENTS[1]

	Summer			Winter		
a) PHOSPHOROUS AS PHOSPATE						
	Mean	Max.	Min.	Mean.	Max.	Min.
Near Shore	0.14	0.32 -	0.00	0.31	0.67 -	0.04
Outer Lagoon	0.04	0.19 -	0.00	0.34	0.71 -	0.17
Inner Lagoon	0.16	0.49 -	0.00	0.36	0.56 -	0.11
b) NITROGEN AS NITRATE						
	Mean	Max.	Min.	Mean.	Max.	Min.
Near Shore	1.45	2.41 -	0.34	0.88	1.61 -	0.33
Outer Lagoon	0.51	1.00 -	0.00	0.78	1.79 -	0.10
Inner Lagoon	0.58	1.03 -	0.25	0.97	1.71 -	0.40
SILICON AS SILICATE						
	Mean	Max.	Min.	Mean.	Max.	Min.
Near Shore	2.63	10.67 -	1.26	0.99	2.73 -	0.17
Outer Lagoon	5.51	9.72 -	1.42	1.78	3.38 -	0.89
Inner Lagoon	13.24	22.58 -	4.69	2.68	6.16 -	0.98

[1]All results expressed as micro-gram atoms/litre.

The Fauna

The main faunal groups living in the area are discussed in the following pages and their broad distribution is shown in table 5. Accounts of some of these groups have already been published and the results of the studies on other groups will be published more fully in the near future.

The details of the foraminiferids in the sediments of the area have been discussed in several papers (Murray, 1965 (a) and (b); 1966 (a), (b) and (c), 1970 (a) and (b).

There is great variation in Quinqueloculina and Triloculina. This produces considerable problems when an attempt is made to speciate these forms and it was therefore decided to sub-divide them into the morphological groups shown in Fig. 7. The use of the Fisher diversity index α is based on Fisher, Corbett and Williams (1943).

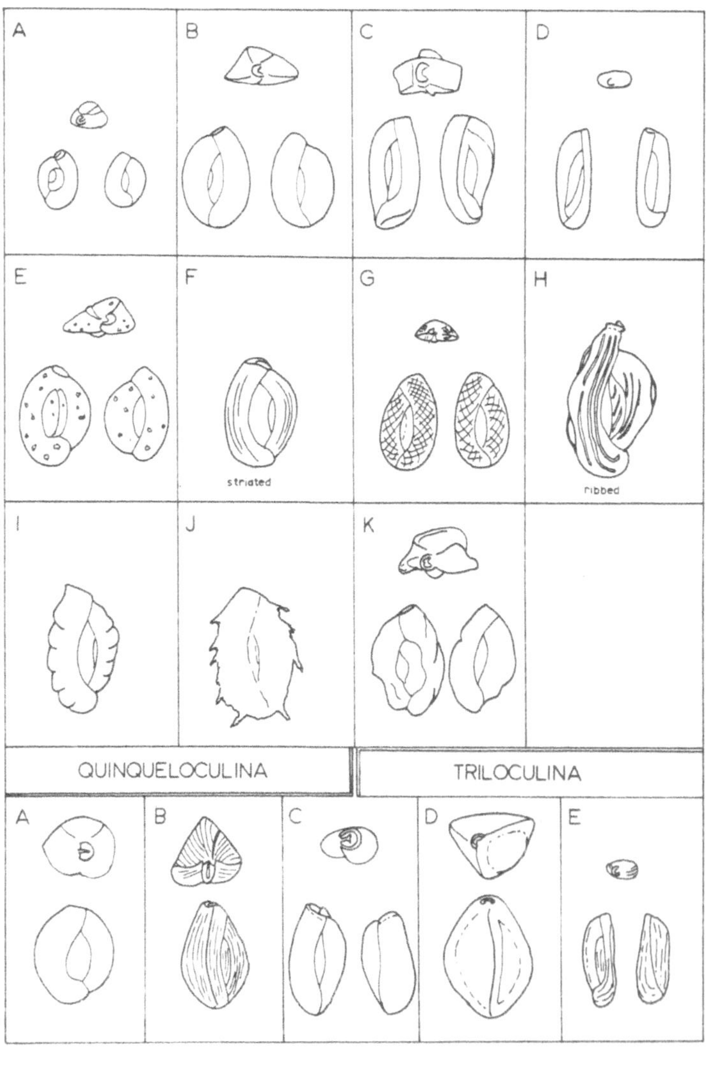

Fig. 7. Morphological groups of Quinqueloculina and Triloculina

The foraminiferids live mainly in association with seaweeds and seagrass and their distribution is thus "clumped" . On a small scale the inhabitants of one plant are geographically isolated from those on another. However, when the assemblages from different plants are compared they show a reasonable degree of similarity, not only in the species present, but also in their abundance: thus seaweed assemblage pairs show a peak similarity of 60 - 70 %; for seagrass assemblages the values are 40 - 75 %. Generally, the sediment surface in the Abu Dhabi region is an unfavourable substrate for foraminiferids.

The dead assemblages found in the sediments have all been transported to some degree. Their diversity is commonly slightly higher than that of the living assemblages. The same species are present (see table 3) but the dominant dead species may not be the same as in the living assemblage. There were marked differences in abundance in living forms in the collections made in 1965 and 1969 (see Murray, 1970, for details) and therefore it appears that the dead assemblages may give the best idea of the "average" assemblage. However, when the three sub-orders are used as end members on a triangular plot, both living and dead assemblages occupy the same distincitve field; and this is one that generally characterized hypersaline environments. It is, namely, a high Miliolina and rare Textulariina assemblage (Fig. 8).

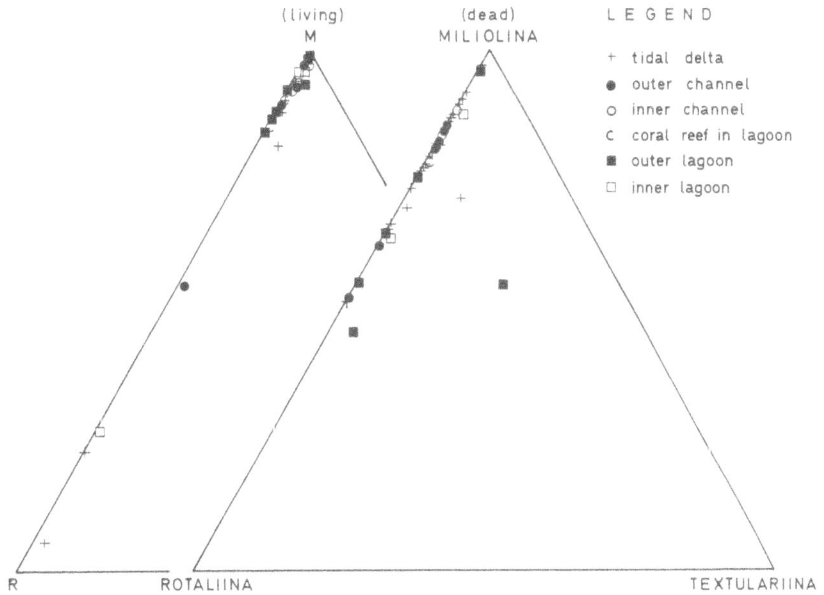

Fig. 8. Typical hypersaline foraminiferid assemblages

Sponges are common, encrusting and growing on rock, coral and weed throughout the area.

Corals are found forming reefs to the seaward of Abu Dhabi Island, and as a patch reef at one locality on the edge of the tidal delta to the SW of Abu Dhabi Island.

Coral is also found in the outer lagoon, forming small fringing reefs, and in one locality as narrow strips crossing a channel floor, immediately NE of Abu Dhabi Island (Fig. 9). Nowhere does coral extend more than 3.5 km into the lagoon from the entrance. The forms found in the area have been listed in table 5. The collections represent an impoverished Indian Ocean assemblage, presumably because of the adverse salinity and temperature conditions of the Persian Gulf. However, Kinsman (1964) has shown that some species seem to tolerate more adverse conditions than was originally thought. The collections made around Abu Dhabi Island are very similar to those made on the edge of the reef flats in the Seychelles - a region where very adverse conditions are encountered (B. Rosen, personal communication). The dead coral is often encrusted with calcareous algae and at one locality in the outer lagoon a large number of diatoms were found in the associated sediments.

Polyzoa are found in the sediments throughout the area. They coat rock, shell and weed, and are often transported onto the beaches with their host weeds during storms.

The area supports a rich molluskan fauna. Approximately 120 species have been recognized. Some are common to all sub-environments whereas others have a rather restricted distribution. Over forty species are found in the offshore and nearshore areas, seventy two in the outer lagoon and only six in the inner lagoon.

Ostracods form an abundant part of the microfauna in the sediments of the area, although in some sub-environments (tidal deltas and outer lagoon) their numbers are drastically reduced. The least diverse population occur in the hypersaline, shallow intertidal pools; in these localities the number of species is small although the abundance of individuals is considerable. The nearshore shelf, back-reef lagoon and lagoon channel sub-environments are areas where the numbers of both species and individuals are greatest. The characteristic assemblages are discussed under the various sub-environments (see also table 5 and Bate (1971) for further details). The distribution of both living, dead, and moulted ostracods is comparable. Little displacement of ostracods appears to have taken place and only two species (Cyprideis sp. and Genus A sp.) appear to have been affected in this way.

Two species of barnacles have been found living on rocky areas and on the pneumatophores of mangroves within the lagoon, and a third form has been found on floating debris.

Many species of crab (decapods) have been found in the area. They are very active burrowers in various environments. Although their activity on the beaches and intertidal flats is very conspicuous they probably are equally active in many of the sub-aqueous environments. In many places they have completely destroyed the stratification and have formed distinctive sedimentary structures. As with other groups, they show a noticeably lower diversity in the inner lagoon when compared with the fauna in the more seaward sub-environments.

The echinoids in the area appear to be restricted to the open waters of the Persian Gulf, except for Echinometra malthei which occurs in the outer lagoon associated with reefs. A whole specimen has never been found, dead or living, in the sediments collected from the inner lagoon, although skeletal fragments are sometimes found in the sandy fraction of the sediments from this sub-environment. Similarly, the asteroids appear only in the open Gulf areas except for one species that extends into the outer lagoon where it is often associated with coral reefs.

The ophiuroids are found in the open Gulf and also in the outer lagoon where they are associated mainly with coral reefs, but none have been found in the inner lagoon. Holothurians are found in various sub-environments.

In addition to the invertebrates, large numbers of fish live in the nearshore and lagoonal areas, and large shoals are often found at considerable distances from the lagoon entrance. Small sharks are occasionally seen in the lagoons and near shore areas. Rays are common, and are particularly noticeable in the inner lagoons when they are probably responsible for the breakdown of significant quantities of skeletal debris and thus are important sediment producers; they also leave very characteristic resting marks, feeding marks and trails on the surface of the sediment. Turtles are fairly common, but not abundant, in the lagoons and remains of their skeletons occur on some beaches and marshes. Dolphins are commonly seen in the nearshore areas and sometimes move into the channels of the outer-lagoon. Sea snakes are sometimes seen around reefs.

Numerous holes and burrows are common on the floor of the lagoon, the tidal delta and the nearshore areas. The animals responsible for these are elusive and difficult to trap; they are probably mainly crustaceans, fish and possibly "mud-skippers" but no definite results are available. A study of some of these burrowers and their effects would probably be most rewarding (for example see Shinn, 1968).

The area has a sparse land fauna, although the dunes and intertidal flats often have tracks and burrows left by desert foxes, gazelle, lizards, snakes and small arthropods.

TABLE 3. Comparison of some features of the foraminiferid assemblages

	Nearshore Shelf		Tidal deltas		Channels		Lagoon terrace Outer		Lagoon terrace Inner		Coral banks	
	D.	L.	D.	L.	D.	L.	D.	L.	D.	L.	D.	L.
Ammonia beccari	X		X		X		d		X	d	X	
Elphidium aff. _advena_									X	X		
Elphidium crispum	X		X									
Eponides murrayii	X		X								X	
Miliolinella spp.				X				X	X		X	d
Peneroplis pertusus	X		X	X		X	X	X	X			
Peneroplis planatus			X	d	X	d	d	d	X	d		
Quinqueloculina spp.	d	X	d	X	d	X	d	X	X		X	d
Rosalina adhaerens	X	d	X	d	X	d						X
Rotaliammina mayori		X		X								X
Triloculina spp.	X		X	X	d	X	d	d	X		X	d
diversity index α	3-/½	-	3-6½	2½-5	4½-7	4½-5	2-5½	2½-6	2-3½	1-2	6	3-4
standing crop per 30 cm^2	-	-	-	-	-	43	-	1-114	-	1-141	-	-
biomass mm^3 per 30 cm^2	-	-	-	--	-	1.26	-	0.64-1.74	-	0.02-12.02	-	-
broken specimens %	4.20		10-28	-	7-22	-	4-13	-	10-20	-	14	-

D = dead; L = Living; d = dominant; X = present

THE SUB-ENVIRONMENTS OF DEPOSITION

Several clearly defined sub-environments of deposition can be distinguished in the area under discussion. The general character of these sub-environments, their fauna, flora and the characteristics of their sediments are discussed in the following pages and illustrated by table 4, and figures 9-18. (Several plates illustrating various features of the sediments can be seen in Bathurst (1972)).

Various aspects of the environments and their faunas, floras and sediments have already been published in theses and papers already mentioned, and a more detailed discussion of the petrography, texture, and chemical and mineralogical composition of the sediments will be published in the future.

The Nearshore Shelf

The nearshore shelf nowhere exceeds depths of 10 m within the area under discussion. It is a reasonably flat surface, deepening gradually offshore. Landwards

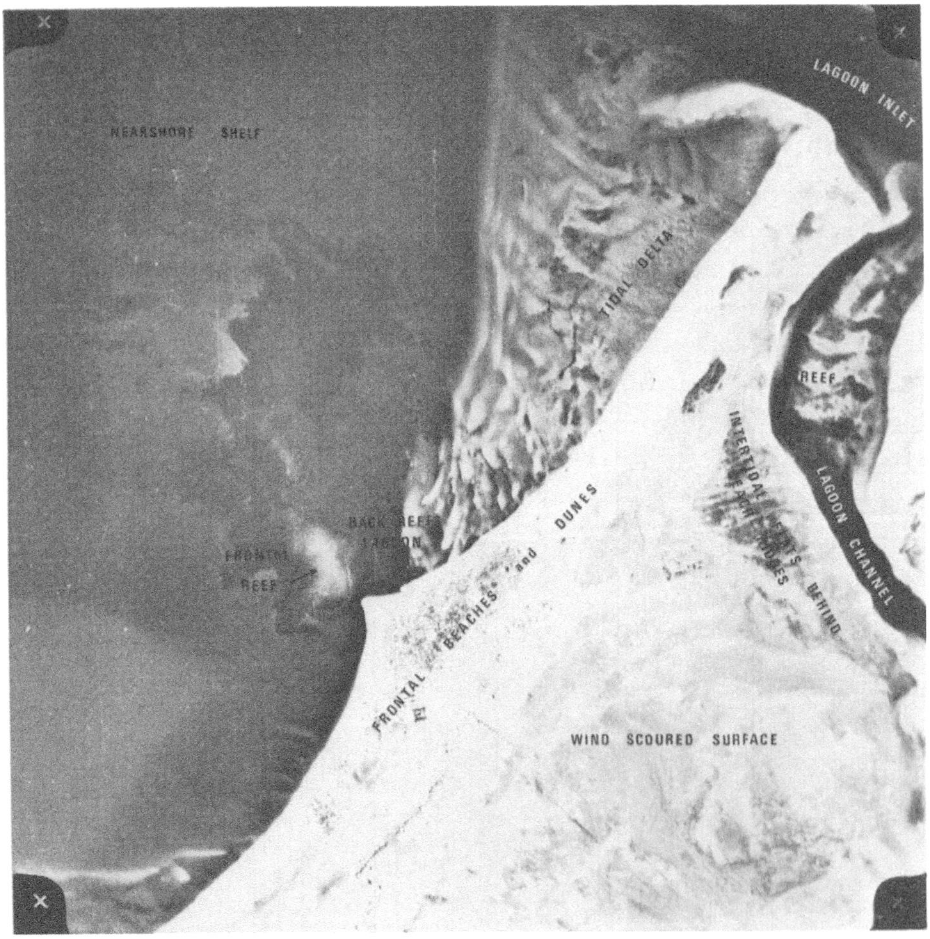

Fig. 9. Aerial photograph of the nearshore shelf, frontal coral reef, and tidal delta, Abu Dhabi Island

it passes into the relatively steep frontal slope of the tidal deltas or the slope leading up to the shallow coastal terrace which fronts the islands in the intervening areas (Fig. 9). This area is covered with grey speckled sands containing minor amounts of silt and clay and variable amounts of coarser material (mainly shell debris). These sands are composed of skeletal fragments mixed with variable amounts of pellets, some composite grains and various insolubles (mainly quartz). The skeletal component consists mainly of molluskan debris (including scaphopods) with some ostracod and polyzoan tests and foraminiferal, echinoid, algal and serpulid debris. Some oolites, which are usually small, and consist of cores with very thin envelopes, are found in the sediments adjacent to the foot of the tidal deltas, but these are rare, and the boundary between the two sediment types is often very sharp. Generally these sediments contain more insolubles (mainly quartz, but with feldspar, chert and heavy minerals including haematite are present) and more low magnesium calcite, less aragonite and lower organic carbon contents, than the sediments of the various other sub-environments (their general properties are given in Table 4, and Figures 15 and 16).

Seagrass (Halophila ovalis and Diplanthera urinervis) is found growing in the sediment, and in several places where rock is exposed seaweed is found. The floor has been observed only on rare occasions, and it is seen to be rippled often with well developed symmetrical wave-induced ripple marks.

The dead foraminiferid assemblages in the sediments are dominated by Quinqueloculina spp. Other species include Triloculina spp., Peneroplis pertusus, Ammonia beccarii, Eponides murrayi and Elphidium spp. (especially E. crispum). Locally the clinging species Rosalina adhaerens is abundant. The diversity indices are α 3-7½ with most values in the range 4-7.

The living assemblages are sparse and normally only a few individuals occur in each sample: Rosalina adhaerens, Rotaliammina mayori, Spirillina vivipara, Quinqueloculina spp. and Elphidium spp. The first of these species live clinging to shell fragments on the sediment surface. It can be seen that there is a substantial difference between the living and dead assemblages. Many of the dead specimens have a slight oolitic coating which gives the test a glazed appearance. In some cases the size of the formainiferal tests is less than would be expected in comparison with those of the adjacent tidal delta samples. Broken specimens form 4-20 % of the total dead assemblage. It seems likely that the nearshore shelf represents a burial ground for foraminiferids derived from areas of production on the nearby tidal deltas as indicated by the slight oolitic coating - a feature unlikely to be formed beneath the deeper, nearshore shelf waters.

Ostracods are present in the sediment, but their distribution appears to be patchy; they are most numerous in the area adjacent to the tidal deltas where the main lagoonal channels empty into the open waters of the Persian Gulf. This phenomenon occurs in other areas and is possibly related to the discharge of organic debris, locally enriching the nearshore shelf. Elsewhere the scanty ostracod fauna may be due to the fact that the shelf sediments contain less organic matter than those of the other sub-environments. The ostracods Cyprideis sp. and Genus A sp. recorded in this fauna, are considered to be lagoonal species which have been transported into the area.

The main forms present are: Paijenborchellina sp., Xestoleberis rhomboidea, X. rotunda, X multiporosa, X sp. A.B. & C, Pontocypris sp. A, Pontocypris sp. B, Neonesidea schulzi; Paranesidea sp. Loxoconcha ornatovalvae; L. sp. A & sp. B., Genus B. sp., Moosella sp., Cytherelloidea sp. A., Cytherella cf. punctata; Paracytheridea sp., Aglaiocypris triebeli, Thalmannia sp., Genus C. sp.; Carinocythereis cf. hamata, Caudites sp. A & B., Munseyella sp., Hemicytherura videns aegyptica, Cytherura sp., Bosquetina sp.; Sclerochilus sp., Paradoxostoma longum; Cytherois sp., Cytheroma dimorpha, Paradoxostoma sp. A.; Tanella cf. gracilis; and Hulingsina sp.

The Frontal Reef

Seaward of Abu Dhabi Island a small barrier coral reef has developed (Fig. 9), probably on an erosional remnant of "miliolite". Wave refraction around this reef has produced a small cuspate foreland on the seaward face of the island. The reef has a gently sloping seaward face and a steeply sloping landward face. Between the reef and the beach there is a small sheltered lagoonal area in which muddy carbonate sand is accumulating (see Fig. 16B).

The reef consists mainly of Acropora sp. and Porites sp. with Platygyra lamellina, Cyphastrea sp. and Stylophora pistillata being common. Much of the coral is dead and is coated with the calcareous algae Lithothammnion sp., Lithophyllum sp. Goniolithon sp. and other algae.

Many mollusks can be seen living either within the coral or in the patches of ill-sorted sediment between individual heads. Large gaping Pinna sp. have been seen living at various locations. The echinoid - Echinometra malthei - is very abundant, together with various ophiuroids and crabs. Fish are abundant around the reef and sea snakes are not uncommon.

The sheltered area between the reef and the seaward face of Abu Dhabi barrier island is the site of accumulating sediment that is finer grained than that found seaward of the islands elsewhere (see Fig. 16B). Generally these sediments have a similar mineralogy to those of adjacent areas, except that they usually contain higher percentages of magnesium calcite and are usually richer in organic carbon. These sediments usually support a dense growth of sea grass.

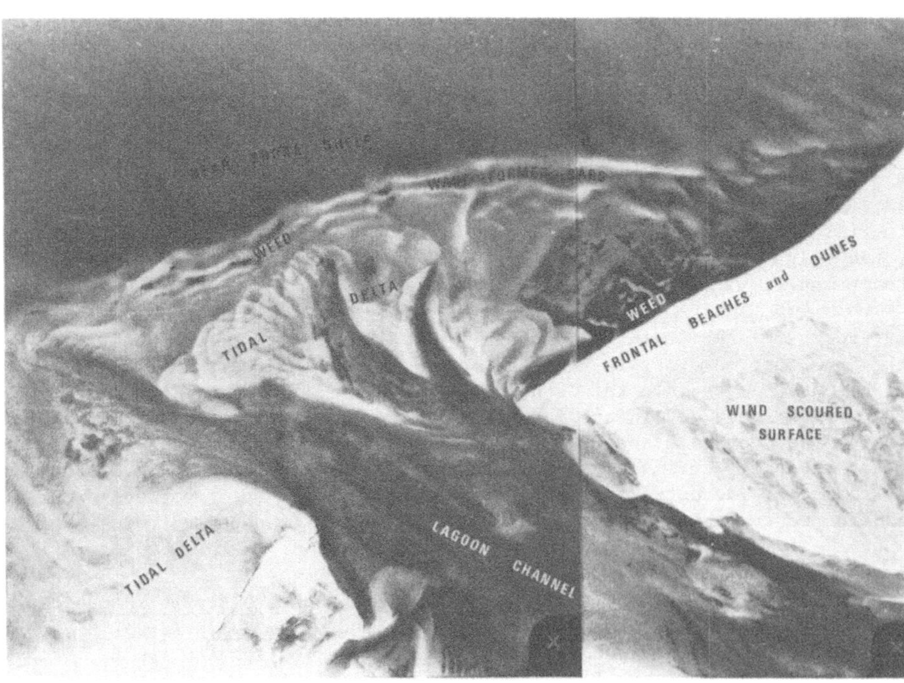

Fig. 10. Aerial photograph of the tidal delta between Abu Dhabi and Sadiyat islands

A fairly varied ostracod fauna has been found in these fine grained sediments which are accumulating in the shelter of the reef. The main forms are: Cytherura sp., Paradoxostoma sp. C, Tanella cf. gracilis, Cytherella cf. punctata, Paracytheridea sp., Thalmannia sp., Genus C sp., Paijenborchellina sp., Genus B sp., Pontocypris sp. A, Bosquetina sp., Loxoconcha sp. B., Loxoconcha ornatovalvae; Xestoleberis rotunda, X. rhomboidea, Cytherelloidea sp. B., Alocopocythere reticulata, Aglaiocypris triebeli, Cytherois sp.; Paradoxostoma longum, Hemicytherura videns aegyptica, and Cytheroma dimorpha.

One species - Alocopocythere reticulata - is found only in the sediments of this sub-environment (in the samples available); however, it is of interest to note that this form also occurs in bottom sediments from depths greater than 9 m in the Persian Gulf and in the sediments of the shallower parts of the lagoon, but never in the sediments of the nearshore shelf.

The Tidal Deltas

Where the tidal channels draining the lagoons enter the open Gulf from between the islands, broad shallow tidal deltas are developed (Figs. 9, 10) whose outer margins are defined approximately by the 2 m contour. They form a zone where a large amount of wave energy is expended; waves generated in the open waters to seaward, break along the delta edge and have raised a series of large, breaker-point bars. Further inshore other bars are found and raised ridges (levées) border the tidal channels which cross the deltas. The surfaces of these deltas are covered with many sets of megaripples and elsewhere by wave induced asymmetrical and symmetrical (mainly the latter) bedforms of smaller magnitude. In other areas large patches of bare rock are colonized by brown seaweeds (Fig. 10).

The sediments of the deltas are whitish sands mainly of medium or fine grade with only small amounts of finer material. Coarser material is sometimes present as concentrates of shell and cemented rock fragments. These sands are composed mainly of oolites mixed with skeletal debris, pellets and composite grains described in detail by Loreau and Purser, in this volume. Their textural mineralogical and chemical composition is given in Table 4 and Figures 15B and 16C. They contain less high magnesium calcite than the offshore sands and the sediments of the other sub-environments, and also to have a higher Ca/Mg ratio than these sediments.

The dominant species of dead foraminiferids in this environment are: Quinqueloculina spp. with Triloculina spp. and Peneroplis pertusus, Ammonia beccarii, Elphidium crispum and Eponides murrayi are also commonly present. Peneroplis planatus is locally abundant and on the seaward slope of the delta Rosalina adhaerens forms a significant percentage of the dead assemblage. The diversity values are α 3-6½.

In the collections made in 1961-1962 living foraminiferids were rare with the clinging forms Rosalina adhaerens and Rotaliammina mayori being the principal forms present. These cling to shell fragments and seem to be able to tolerate movement of the substrate. In the collection made in 1965, however, a sample of seaweed living on the rocky outcrops of the delta supported numerous living forms: Quinqueloculina spp., Rosalina adhaerens and Elphidium reticulosum with Peneroplis pertusus and Peneroplis planatus being common locally. These nestled in the protection of epiphytes such as polyzoans and calcareous algae (Jania sp.).

Elsewhere a sparse assemblage of living foraminiferids including Miliolinella sp., Quinqueloculina spp., Buliminella elegantissima compressa and Rosalina adhaerens were found living on seagrass colonizing the soft sediment. Further sediment samples collected in 1969 yielded living Rosalina adhaerens and Peneroplis planatus was the dominant form found living on the seaweed. Other common forms included Miliolinella sp., Peneroplis pertusus, Quinquelocolina spp., and Triloculina sp. The diversity values were α 2½ - 5.

Generally, the dead assemblages showed a greater variety of species and a slightly higher diversity index than the living assemblages, and it is very noticeable that there are significantly fewer dead foraminiferid tests in the oolitic sediments of the tidal deltas than in the sediments of other sub-environments. The broken foraminiferids constitute as much as 10-28% of the dead assemblage on the delta top and close to shore. In contrast, the foraminiferids in the sediments on the slope at the front of the delta contain less than 10% broken forms. It would appear that dead foraminiferids must be dispersed from the productive areas to adjacent areas.

Some fresh tests of ostracods have been found in this environment, which, together with the foraminiferids, are thought to be associated with the seaweed. Whether or not ostracods form a persistent population in the area is doubtful as most of the forms found are rolled and badly worn. The main forms are: Alocopocythere reticulata (badly worn), Aglaiocypris triebeli (poorly preserved), Genus C sp. (worn and coated), Mutilus sp. (worn specimens), Xestoleberis rotunda (some fresh), Carinocythereis cf. hamata (poorly preserved); Loxoconcha sp. A (some fresh); and Tanella cf. gracilis (fresh).

The Frontal Beaches and Dunes

The islands have gently curving seaward faces (although these have now been greatly interfered with by engineering works) (Figs. 9, 10).

Islands are fronted by beaches which sometimes have a narrow low tide terrace, and always have well developed beach faces and berms (the latter often have beach cusps). The beach sediments are composed mainly of skeletal and oolitic sands with skeletal debris and other components such as pellets and composite grains sometimes abundant (Fig. 17A). They contain large accumulations of coarse, transported shell debris which is concentrated either on the berm or at the toe of the beach face. Masses of brown algae are thrown onto their surface (particularly during storms) and later become buried in the sediment.

Burrowing crabs, mainly Ocypoda aegyptica, are sometimes very abundant, particularly on the berms, and these tends to destroy the regular lamination of the sediment. Landward of the berms is a wide wind scoured flat with low hummocky dunes and scattered bushes. Its surface is littered with shells, loose fragments of recently cemented limestone and other flotsam and jetsam, most of which is derived from the nearshore areas.

Dunes, in places 12 m high, form a series of transverse ridges with steep landward faces. Behind the main dune ridge, in the SW of Abu Dhabi Island, are a series of barchan dunes which extend inland for over 4 km. Over the remainder of Abu Dhabi Island the dune ridge is low and has been considerably modified by man, even before the recent explosion of the island's population. In the NE corner of the island the dune ridge has been severely deflated and only a low ridge remains with some elevated dune cores of loosely cemented oolitic sand (a shell from these gave an age of 1378±360 years BP).

High transverse dunes occur adjacent to the tidal deltas on the NE and SW ends of Sadiyat Island, whereas in the central part of the island only low dunes occur. These pass inland to deflational areas with well developed parabolic dunes. Locally the dune sand is cemented, particularly in hollows and around plant roots (cf. Evamy and Glennie, 1965). These dunes are composed essentially of oolitic sands with various admixtures of skeletal debris, pellets and composite grains (Fig. 17B). The roundness of individual oolites presumably facilitates their preferential transport by the wind and explains the purity of these sands in some areas. Elsewhere, small molluskan shells and a variety of skeletal debris form patches of coarse detritus which is sometimes fashioned into megaripples.

Foraminiferids are also transported landwards and the dune sands on Sadiyat contain Quinqueloculina spp., Triloculina spp., Peneroplis pertusus and a few rarer forms. The foraminiferids decrease in abundance away from the sea and the specimens become progressively more rounded. Nevertheless it would be difficult to recognize this sub-aerial mode of accumulation from the abundance and preservation of the foraminiferids. In fact, the skeletal debris in the dune sands may, in some cases, be better preserved than that accumulating in the sub-aqueous areas (see M. Hughes Clarke oral communication to K. Glennie (Glennie,1970)). However, sub-aerial diagenesis may have a very profound effect on these grains later in their history (Shearman, Twyman and Karimi, 1970).

The Lagoon Channels

Large channels between the barrier islands, lead from the tidal deltas into the lagoon (Figs. 10, 11 and 12). They are up to 7 m deep between the barrier islands and have steep rock sides exposing Pleistocene "miliolite"; they become less well defined towards the landward parts of the lagoon. Thick groves of seaweed grow on the rocky exposures along the channel margins. In the inner lagoon seagrass is abundant and binds the loose sediment. Elsewhere, particularly in the outer parts, the sandy sediment has been piled into megaripples in the central parts of the channels and along the margins of the islands and banks. Small fringing reefs occur in the outer parts of the lagoons along the main channel,invariably within 3.5 km of the barrier face. A series of patches of coral also floor most of a small channel immediately NE of Abu Dhabi Island in the outer parts of the lagoon (Fig. 9). In all these localities Acropora sp. and Porites sp. are the dominant forms, and are associated with abundant mollusks, echinoids, ophiuroids, etc.

The sediments of the lagoonal channels are mainly grey-speckled sands with variable amounts of coarse shelly gravel (Figs. 15C and 17C), lithoclasts of "miliolite" and some fragments of loosely cemented younger rock. In their most landward portions the channels appear to be filling with carbonate sand which often contains up to 50% carbonate mud.

The sand fraction is composed of skeletal debris mixed with pellets and considerable amounts of composite grains. Oolites are abundant, especially at their seaward ends in the proximity of the tidal deltas, and also near the intra-channel banks and terraces bordering the outer lagoon, on which they appear to be forming. Locally, fragments of calcareous algae are abundant, especially around the coral reefs. The general chemical and mineralogical composition of the sediments is given in Table 4. The Ca/Mg ratio of these sediments in the seaward parts of the lagoon is higher than that of the nearshore shelf sands, but lower than that of the sediments of the tidal deltas. The channel sediments in the more landward parts of the lagoon have the lowest Ca/Mg ratio of all sediments found in the coastal complex but contain more organic matter than those of either the nearshore shelf or tidal delta.

The dead foraminiferid assemblages from the main channel consist mainly of: Quinqueloculina spp., Triloculina spp., and Ammonia beccarii. Rosalina adhaerens is abundant in the sediments at the seaward end of the channels. The sediments of the subsidiary channels contain assemblages dominated by Triloculina spp. together with Quinqueloculina spp. Peneroplis planatus and Ammonia beccarii with Spirolina acicularis being locally abundant. The diversity index is α $4\frac{1}{2}$ -7.

Living foraminiferids were rare in the sediments collected in 1961-1962 except for one station which yielded an assemblage dominated by Rosalina adhaerens. In 1969 the foraminiferids were found to be living on seaweed and the assemblages were dominated by Peneroplis planatus with other common species including Peneroplis pertusus, Rosalina adhaerens and Triloculina sp. The standing crop was 43/30 cm^2 biomass 1.26 mm^3/30 cm^2. The diversity was α $4\frac{1}{2}$-5. In contrast to the seaweed, the seagrass appeared to support very few living foraminiferids. The dead assemblages contained 7-22% broken tests and it appears that these foraminiferids, which live on the seaweed, and to a lesser extent on the seagrass, are released to the sediment and transported by the strong tidal currents.

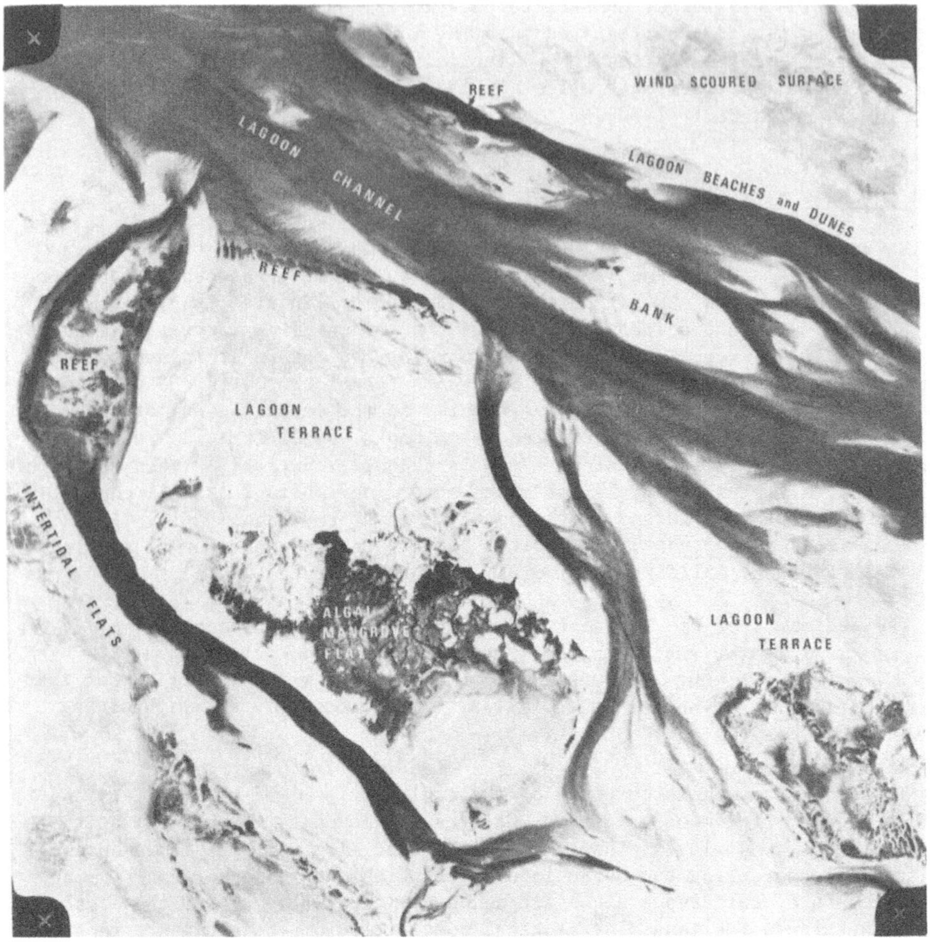

Fig. 11. Aerial photograph of the outer lagoon, NE of Abu Dhabi Island

Foraminiferids live on dead coral and its algal encrustations. The dominant live species found are Miliolinella spp. (16-18%) Triloculina sp. (34-40%) and Quinqueloculina spp. (23-24%). The diversity index is α 3-4. The dead foraminiferids in the sediment between coral heads have a diversity index of α 6. This assemblage differs from that of the sediments of the lagoon terrace by the almost complete absence of Peneroplis pertusus and Peneroplis planatus.

Ostracods seem to be more abundant in this than in any other sub-environments, with the exception of the nearshore shelf sub-environment. In addition, there is a great similarity with the assemblage found on the nearshore shelf. The main forms present are:
Carinocythereis cf. hamata, Xestoleberis rotunda, X. multiporosa, X. sp. C., X. rhomboidea, Cytherura sp., Tanella cf. gracilis, Caudites sp. A., Hulingsina sp., Paradoxostoma sp. B., Cytheroma dimorpha, Loxoconcha ornatovalvae, Loxoconcha sp. A. and sp. B., Cytherelloidea sp. A., Paracytheridea sp., Moosella sp., Cytherella cf. punctata, Genus B sp., Aglaiocypris triebeli, Alocopocythere reticulata, Neonesidea schulzi, Thalmannia sp., Mutilus sp., Paijenborchellina sp., Bosquetina sp.; and Pontocypris sp. A.

Cyprideis sp. and Genus A sp. are more characteristic of the bordering lagoon terraces and banks and appear to have been swept into the channel environment. Also, the form Genus C sp. - of which only one specimen has been found - appears to have been derived from the nearshore shelf.

The Lagoon Terrace

The main islands are bordered by a terrace within the lagoon (Figs. 11, 12 and 13). The depth of water over these features rarely exceeds 2 m and wide areas often become dry at low tide. In the landward parts of the lagoon the terraces merge to form a wide shallow terrace bordering the mainland (Figs. 13 and 14). Also, numerous banks or shoals occur in the channels and have the same characteristics as the terraces.

In many places bare rock, or rock having only a thin cover of sediment, is found on the terraces. This may be eroded "miliolite" but in many other places it is underlain by soft sediment and is obviously the result of early "in situ" cementation.

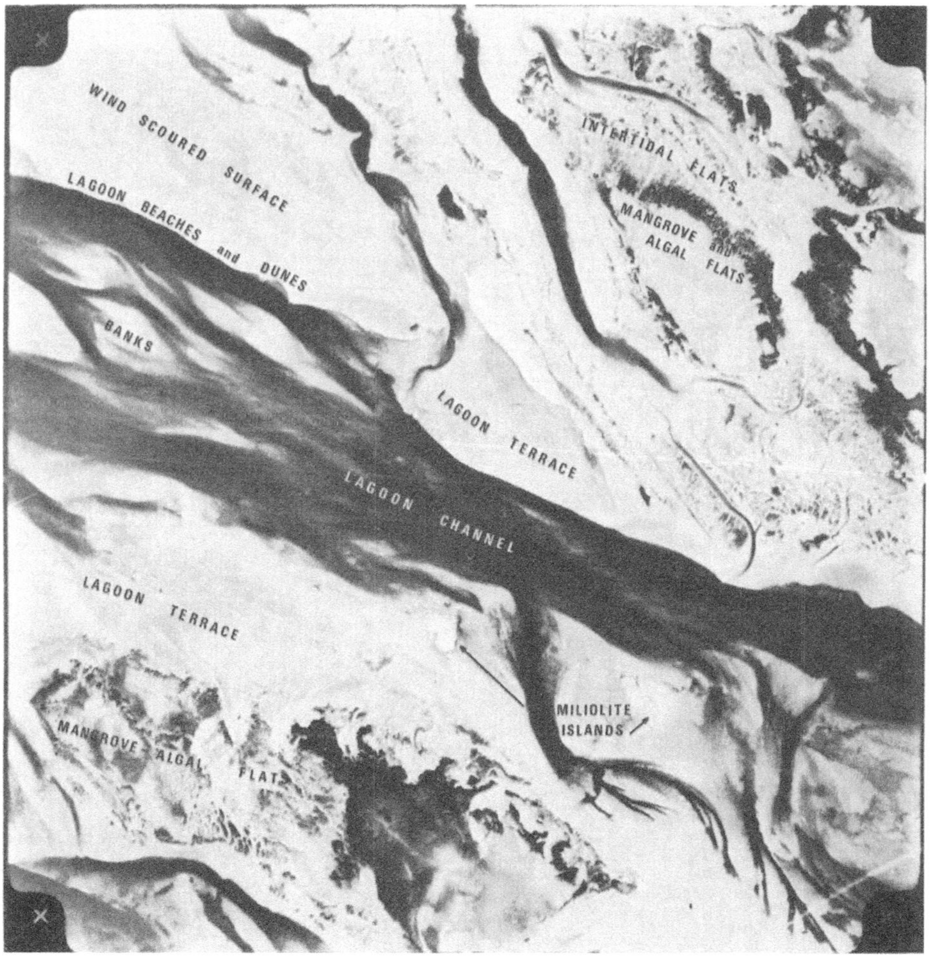

Fig. 12. Aerial photograph of the middle lagoon, NE of Abu Dhabi Island

The seaweed commonly found in the other sub-environments is usually absent from the terraces, except for a few scattered plants growing on rocky crusts. A thin layer of filamentous algae often covers the sediment and exposed rock and the lightly calcified alga Acetabularia sp. has been found growing, particularly at one locality.

The sediments of the terrace and banks are mainly whitish sands, though they may sometimes be grey or speckled with local concentrations of coarser material (usually composed of broken shell). Fine-grained material(<0.0625 mm) may be present in amounts in excess of 20% in some more sheltered areas (see Table 4, and Figures 15 and 18 for details).

The sands are composed mainly of pellets, with variable amounts of skeletal debris which may become locally dominant. Generally the skeletal debris appears to be a more major component on the terraces of the outer than those of the inner lagoon .

Foraminiferids and molluskan fragments (mainly of gastropods) are the most common skeletal constituents, together with calcareous algae, rare fragments of echinoids (whole tests have been found only around the reefs in the outer lagoon) polyzoa and other organisms.

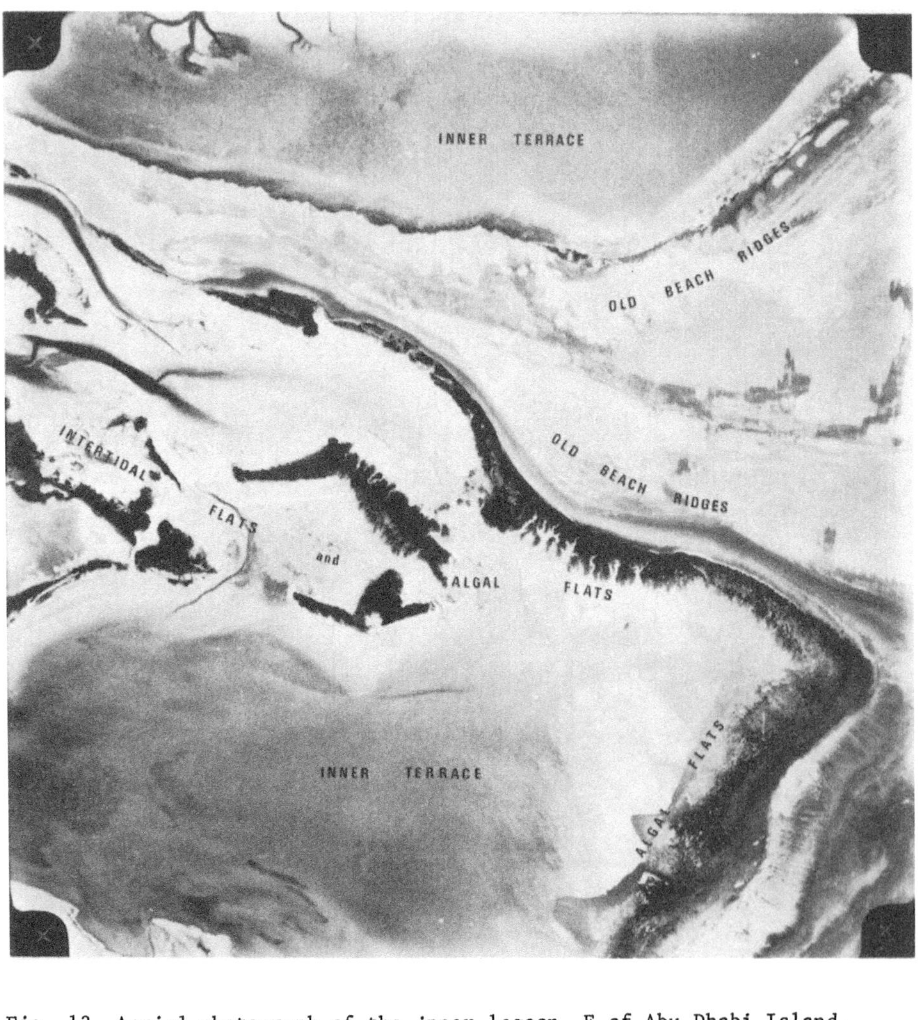

Fig. 13. Aerial photograph of the inner lagoon, E of Abu Dhabi Island

Oolites are common on the edges of the terraces and also on the narrow levées bordering the banks of the outer lagoon, but they do not appear to extend more than 5 km into the lagoon. Composite grains, which are usually accretionary, occur in this sub-environment and appear to be particularly common in the inner lagoon. The mineralogical and chemical composition of the sediments is given in Table 4. These sediments have Ca/Mg ratios which are normally higher than those of the sediments of the lagoon channels and nearshore shelf but lower than those of the tidal deltas. Generally the lagoon terrace sediments have higher organic carbon contents than those of other sub-environments and both the inner lagoon channels and the inner lagoon terrace have the highest organic content of all sediments in the area.

The dead assemblage of foraminiferids on the terraces of the outer lagoon is dominated by:
Quinqueloculina spp.; Triloculina spp.; Peneroplis pertusus. Peneroplis planatus and Ammonia beccarii. The diversity is α 2-5½. Between 4-13% of the tests in the sediments are broken. Generally living foraminiferids are rare within the sediments but samples of seaweed yielded a fairly rich fauna including:
Miliolinella spp., Quinqueloculina spp., Trioculina spp., Peneroplis pertusus, Peneroplis planatus, Spirolina acicularis, Vertebralina striata, Elphidium reticulosum and Eponides murrayi.

Subsequent re-sampling of seaweed produced an assemblage of living foraminiferids dominated by: Triloculina sp. together with Milionlinella sp., Peneroplis planatus and Quinqueloculina sp. The standing crop was 14/30 cm^2, with a biomass of 1.066 mm^3. The assemblages found associated with seagrass were generally more restricted, though a sample collected from one patch included an assemblage dominated by Peneroplis planatus (7.71%) with Peneroplis pertusus, Miliolinella sp. and Triloculina sp. The diversity was α 2½-6. The standing crop was 1-14/30 cm^2 and the biomass 0.064-1.739 mm^3/30 cm^2.

The dead foraminiferid assemblage of the sediments of the terraces of the inner lagoons include: Quinqueloculina spp. Triloculina spp., Ammonia beccarii, Elphidium aff. advena and Elphidium sp. Generally Peneroplis decrease in abundance in the inner lagoon and are absent from the innermost parts. The living foraminiferids are rare, Ammonia beccarii and Elphidium aff. advena being the most widely distributed forms (the latter seemed to be a distinctive feature of this environment). A single sample in an area with seaweed yielded a live assemblage composed of 90% Peneroplis planatus; and another with associated Acetabularia sp. yielded Elphidium advena (70%) and Peneroplis planatus (27%). The latter station had a high standing crop and biomass (132/30 cm^2 and 8.222 mm^3/30 cm^2 respectively). The diversity was α 1. Elsewhere Peneroplis planatus was the dominant species; the standing crop was 1-141/30 cm^2, biomass 0.023-12.018 mm^3/30 cm^2. The diversity was α 2. There appears to be a considerable similarity between the living and dead assemblages and this, together with the small percentage of broken specimens, indicates that most of the dead forms have not suffered much transportation.

The ostracod assemblage in the sediments of the terraces of the outer lagoon is relatively poorly diversified and is not abundant in comparison to that of other sub-environments. It contains Cyprideis sp.; Algaiocypris triebeli; Xestoleberis rotunda; X. rhomboidea; X. multiporosa; X. sp. A. and sp. C.; Loxoconcha ornatovalvae and L. sp. C. This fauna is typified by the dominance of species of Xestoleberis, although both Algaiocypris and Cyprideis are common. Loxoconcha sp. C., a finely punctate form, is also fairly common in the lagoon terrace environment.

Further landwards the ostracod fauna shows an increase in number and species in comparison to those of the outer lagoon terraces. They consist of:
Cytheroma dimorpha, Cytherella cf. punctata, Thalmannia sp., Cyprideis sp. Pontocypris sp. A., Genus A. sp., Genus B sp., Alocopocythere reticulata Loxoconcha ornatovalvae L sp. A. and C., Moosella sp., Paijenborchellina sp. Paradoxostoma longum, Xestoleberis rotunda, X. rhomboidea, X. sp. A. and C. Cytherois ? sp., Xestoleberis rhomboidea; X. rotunda.

The Intertidal Flats, Beaches etc.

The littoral complexes surrounding the islands and on the inner edges of the lagoonal terraces vary widely in type (Figs. 11-14). In many places the shallow lagoonal terrace passes landwards into a zone of intertidal flats which are usually composed of skeletal and pelletal carbonate sands with minor amounts of coarse (mainly skeletal debris) and finer carbonate mud. In all cases the grain and the mineralogical composition of the sediments clearly reflect those of the adjacent terrace sediments (Figs. 15 and 18). The intertidal flats are often extensively rippled and may have low bars, with their steep faces inclined shorewards. Sediments are often cemented to form thin crusts, and these are broken in places forming patches of sedimentary breccia.

The intertidal flats are colonized by burrowing worms, crabs and scattered bivalves. Abundant gastropods, particularly Cerithium scabridum, Nodolittorina subnodosa, Mitrella blanda, Cerithidea cingulatus and Pirenella conica are found crawling on their lower parts, and these often pass landwards into a zone characterized by abundant crabs, particularly Scopimera sp. which have covered the sediment surface with vast spreads of faecal or feeding pellets.

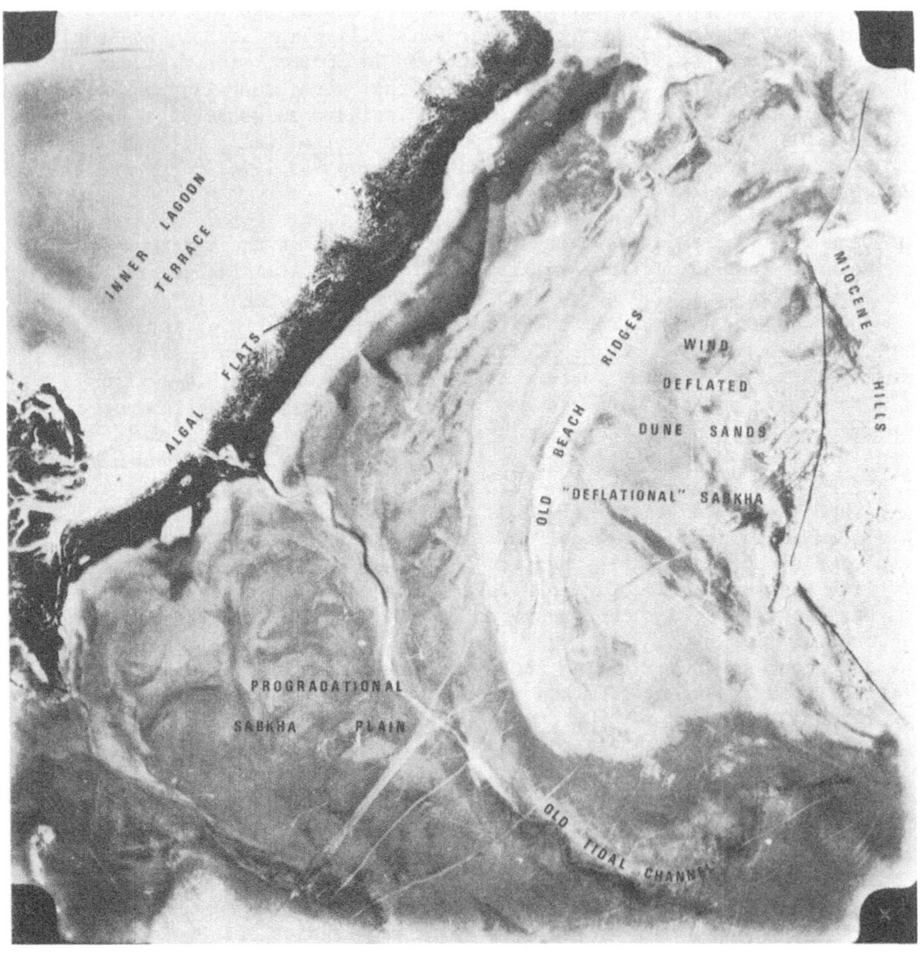

Fig. 14. Aerial photograph of the inner lagoon and sabkha, E of Abu Dhabi Island

In places, where rocks and crusts are exposed, the mollusk Brachidontes variablis occurs in clumps very reminiscent of the Mytilus edulis (common mussels), found on intertidal-flats of temperate latitudes.

Landward of the intertidal flats are narrow beach ridges often capped by dunes. These are composed of skeletal and/or pelletal carbonate sands or gravels. Many beach ridges are composed essentially of coarse concentrates of gastropod shells. These ridges are well laminated, but the beach sediments are often completely churned over by burrows of the crab Ocypoda aegyptica.

Sometimes it can be seen that the beaches have been driven shorewards; they exhibit excellent cross-stratification, often quite steep and directed shorewards, and have intertidal flat sediments exposed on their channel or open water side.

The lagoonal intertidal flats are of various types: large areas of the surface are often covered with blue-green algae on a thin cemented crust (these are mainly cemented by aragonite but have gypsum growing in them) -which is often intensely brecciated; elsewhere thick groves of the mangrove Avicennia marina occur, floored with pelletal carbonate mud which is intensely bioturbated by crabs which have piled this mud into a series of mounds of ropey and balled-up sediment. These mounds are the size and shape of cauliflower heads and have a central canal or burrow. Several species of crab have been found associated with these structures but the forms Metopograpsus messor and Cleistomata near dotilliforme are usually the most abundant and are probably the main burrow and mound producers. These swamps are cut by a complex of drainage creeks very similar to those described by Shinn et al., (1969) from Andros Id; also, bare intertidal flats, with only thin algal coatings, occur, particularly behind Abu Dhabi Island. These are composed mainly of oolitic material derived from the frontal dunes mixed with locally produced pelletal and skeletal carbonate sand and some carbonate mud and have been colonized by various burrowing crabs and surface crawling mollusks. The various types of intertidal zone often occur in close proximity to one another and many intermediate types are found.

Intertidal flats along the mainland shore of the lagoon are covered with pelletal carbonate sand sometimes with high contents of carbonate mud. They are often extensively rippled and have large populations of gastropods (particularly Cerithium rugosum, Cerithidea cingulatus, Nodolittorina subnodosa and Mitrella blanda), some crabs, worms, etc.. Landward, these sand flats pass into broad blue-green algal flats which are more extensively developed than anywhere else in the area under discussion. The sediment enclosed between the algal layers consists essentially of pelletal and skeletal (very rich in gastropods) carbonate sand and mud.

Abundant gastropods live on the surface and occasional crab burrows are found. The algal flats, which have a wide variety of growth forms, exhibit striking dessication cracks (see Kendall and Skipwith, 1968). These algal flats pass landwards into the bare, salty surface of the sabkha (see Evans, Nelson, Schmidt and Bush, 1969, for details). Small areas of rocky shoreline occur in various parts of the lagoons, where "miliolite" masses outcrop. They usually show erosion nicks with prominant overhangs as described elsewhere in tropical seas. Wide erosional surfaces exhibiting beautifully developed festoon bedding are not uncommon. They are usually covered with only a thin sheet of skeletal-pelletal carbonate sand with more insoluble residue and derived composite grains than are found elsewhere. The surfaces of small rocky "miliolite" islands and parts of the surface of other larger islands are often covered with a fine loess-like dust. This has presumably been winnowed from sediments of the seawards areas of the islands and is also derived from the desert areas inland. It may percolate into the solution hollows and cavities in the underlying limestones to produce a silt-size fill (cf. Dunham, 1969).

No living foraminiferids have been recorded on the intertidal flats although transported tests are present, especially where vegetation has been swept to the shore.

Very few ostracods are capable of surviving on the intertidal flats but two species: Cyprideis sp. and Xestoleberis rhomboidea are present in considerable numbers. The ostracod fauna contains: Genus A. sp. and Aglaiocypris treibeli: Most of these species, apart from the first two forms, have been washed in either as live specimens or as dead or moulted instars.

THE PRODUCTION OF LOCAL CRUSTS OR HARDGROUNDS

Syn-sedimentary cementation of carbonate sediments has been reported for a considerable time. However, not until the last decade has the widespread nature of submarine cementation been realized. Fundamental studies by Shinn (1969), Taylor and Illing (1969) and De Groot (1969) have now made "submarine cementation" in the Persian Gulf a well-known phenomenon. Generally, it appears to be associated with very slow sedimentation or non-deposition i.e. a diastem.

In the area examined, as in other regions of carbonate sedimentation, there is widespread intertidal and supratidal cementation. Beach rock occurs on the barrier beaches and also around the lagoon. Locally, dune sands have also been cemented, particularly in the depressions, but also locally on their seaward faces. The algal flats on islands in the lagoons are often cemented and a thin crust has developed which, in places, has disrupted during the cementation - exactly as described by Shinn et al. (1965, 1969), and by Evamy (in this volume) to produce an "edge-wise" breccia.

Subaqueous cementation has not been demonstrated in the nearshore zone as elegantly as Shinn(1969) demonstrated the phenomenon in the areas around Qatar; however, engineering reports of various boreholes offshore from Abu Dhabi make it obvious that this process is taking place.

The surface of the tidal deltas is composed of rock in several areas (usually overgrown by weed) and this is underlain by soft sediment. Cementation therefore appears to be taking place in this sub-environment under subtidal, or locally, under intertidal conditions.

Finally, the floor of the inner part of the lagoon has a well developed crust underlain by soft sediment (further crusts are sometimes found within the sediment). This area is probably sub-aqueous but is partly intertidal and, possibly during some tidal conditions, the whole of this area may be exposed. This lagoonal crust can be traced for many kilometres landwards under the prograding sabkha plain where it is overlain by prograding intertidal and supratidal sediments. (Evans et. al. 1969).

It would appear that in different sedimentary sub-environments, crusts form at different depths of water, but in each case they develop where the accumulation of sediment has slowed down (i.e. a diastem exists); this cementation may be due either to changing chemical-oceanographical conditions or merely because of changing physical conditions. Thus, on the tidal deltas, although sediment production is continuing, waves are constantly moving the oolite grains shorewards to the beach and dunes. There is also transport seawards to the outer delta slope. A condition of equilibrium, or quasi-equilibrium, on the delta top results in material immediately beneath the sediment surface being in a stable position and thus cementation takes

---

Fig. 15. Photomicrographs of sediments from the principal sub-environments.

A: Offshore, skeletal sand
B: Tidal delta, oolitic sand
C: Outer lagoon channel, skeltal-pelletal sand
D: Outer lagoon terrace, skeletal-pelletal sand
E: Inner lagoon terrace, skeletal-pelletal sand
F: Inner lagoon beach, pelletal sand

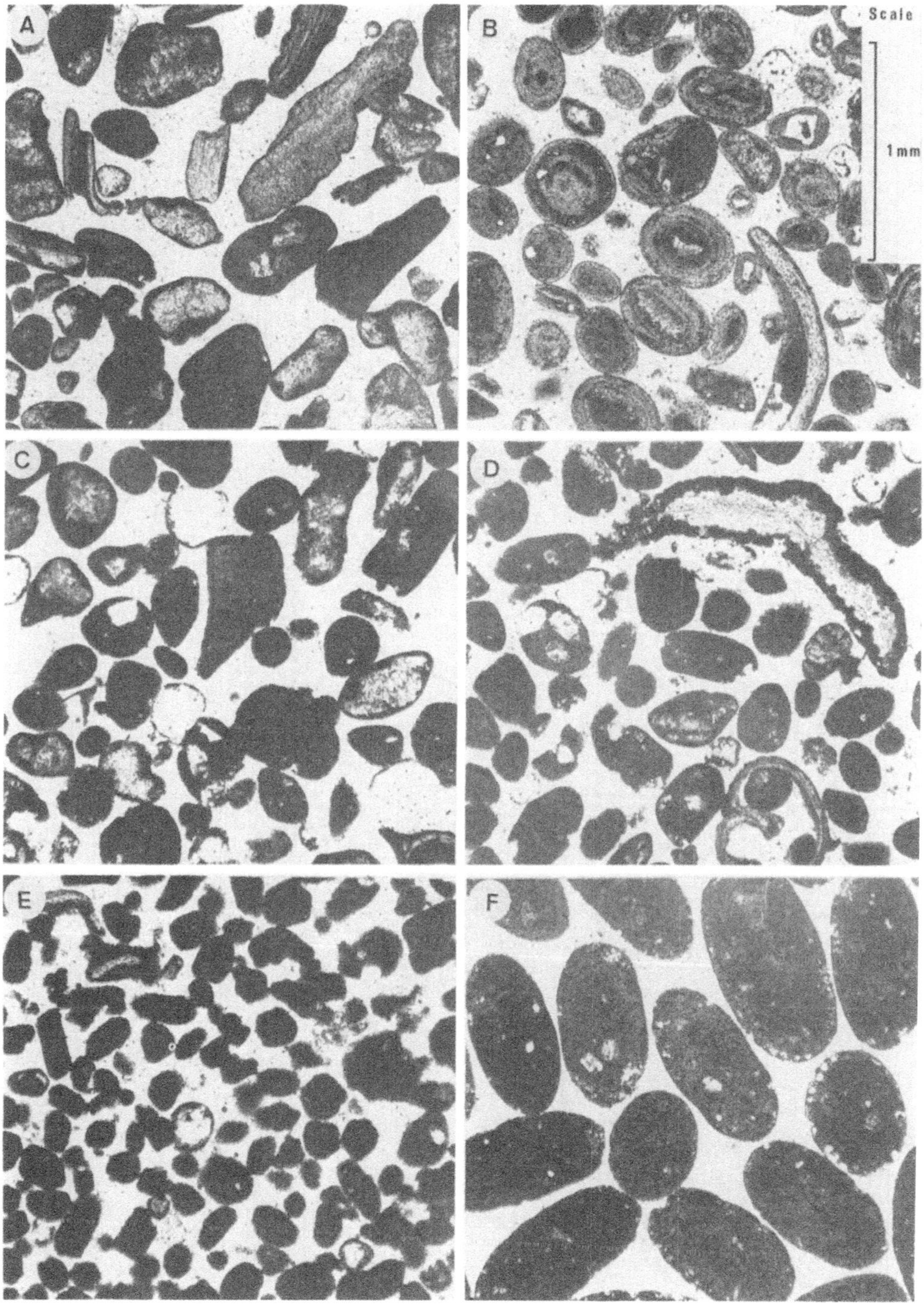

Legend to Fig. 15 A-F on the preceding page

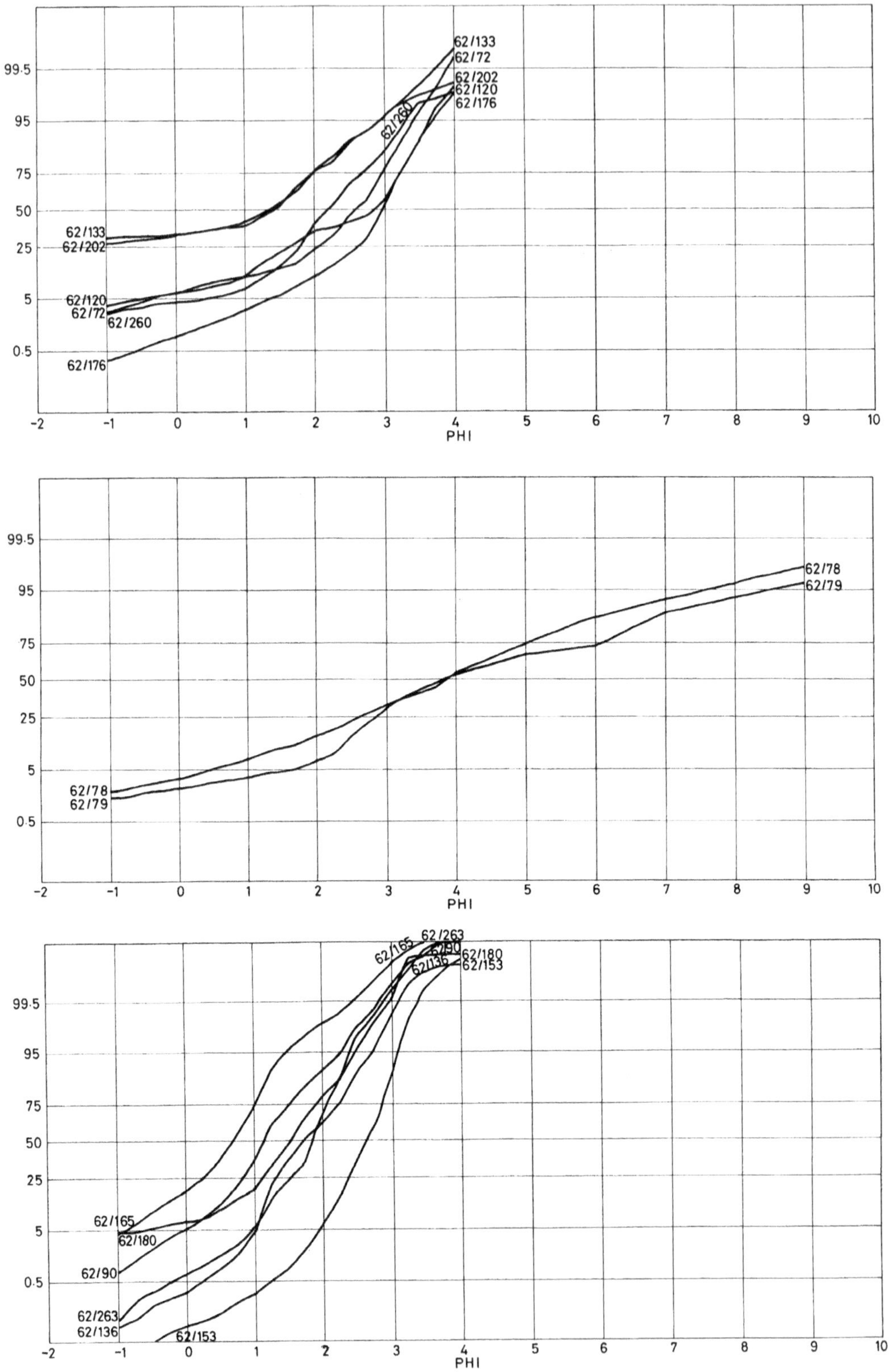

Figure 16: GRAIN-SIZE DISTRIBUTIONS
A. Nearshore shelf sands
B. Frontal lagoon (behind small coral reef) muddy sands
C. Tidal delta sands

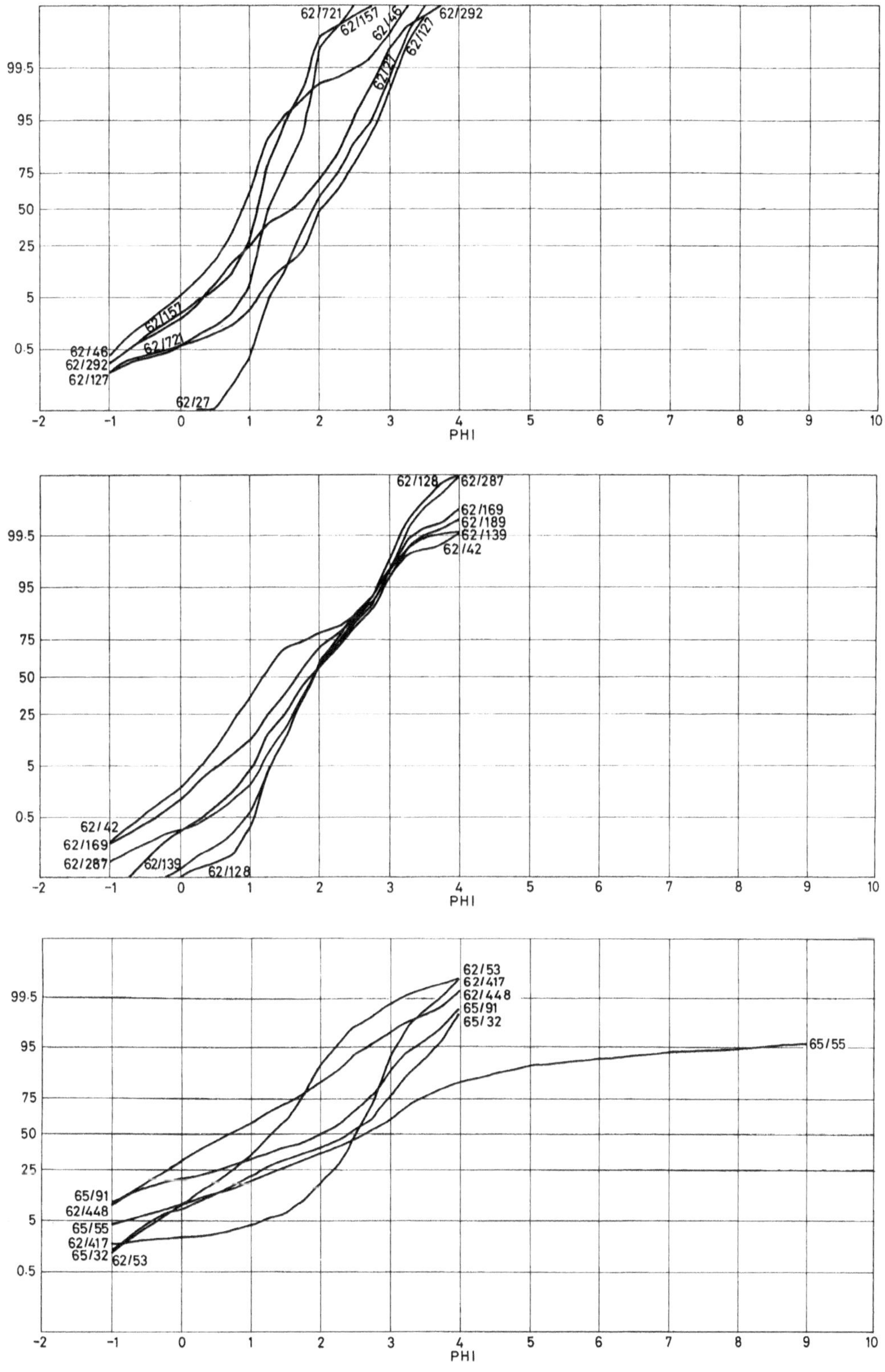

Figure 17: GRAIN SIZE DISTRIBUTION
A. Beach face sands
B. Frontal dune sands
C. Lagoon channel sands and muddy sands

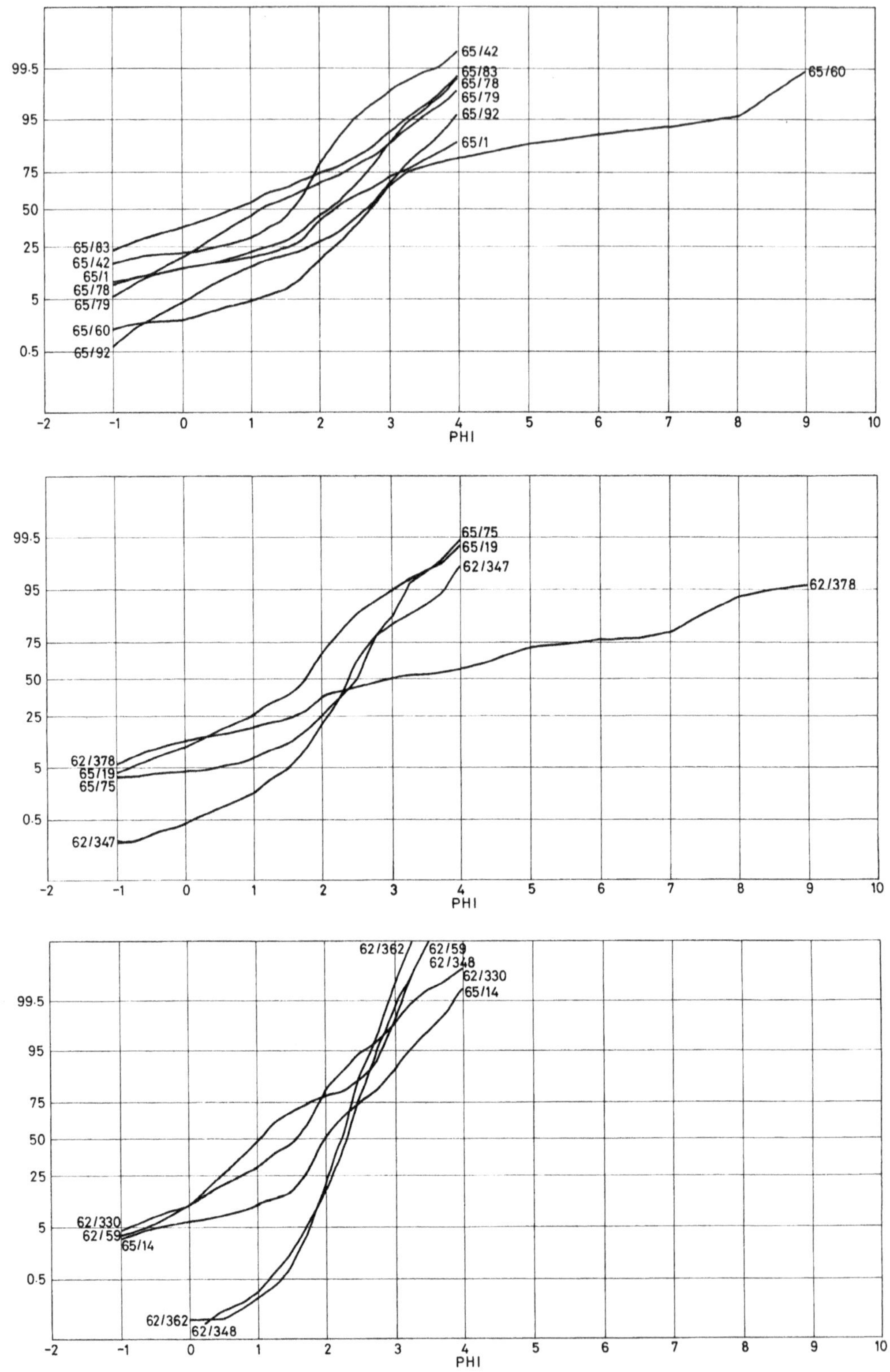

Figure 18: GRAIN SIZE DISTRIBUTIONS
A. Lagoon terrace sands and muddy sands
B. Lagoon intertidal-flat sands and muddy sands
C. Lagoon beach sands

place, sometimes preferentially around burrows. Storms occasionally break up this lithified crust, and loose fragments and linings of burrows ranging from a few grains to fragments 0.3 metres in diameter are cast on to the adjacent shore. Similarly, in the lagoon, waves and currents prevent the skeletal and pelletal material that is being produced from accumulating. This sediment is transported either landwards to accumulate on the prograding intertidal flats, or seawards into the inner ends of the lagoon channels. Again, the sediment immediately beneath the surface of the lagoon floor is stable (sometimes bound by a thin algal carpet) and cementation ensues; in this environment the depth of water varies from 1 m to intertidal, i.e. it is shallower than that on the tidal deltas. Thus, in three different environments three diastems ("hard grounds") are forming, all differing in extent, and each representing differing time durations (Fig. 19). However, all have one feature in common - they represent local sedimentary equilibrium. Each diastem will be capped by sediment of an adjacent sub-facies: prograding coastal barrier and tidal delta sediments in the case of the nearshore crusts, beach and dune sediments in the case of the tidal delta crusts, and intertidal and supratidal sabkha margin and sabkha plain sediments in the case of the lagoon crusts. Because of their widely differing extent and considerable variation in the amount of time they represent they will be of variable value for stratigraphical analysis of comparable ancient rocks.

These crusts provide a hard substrate for the attachment of seaweeds (which in turn support many epiphytes and animals), and for corals and epifaunal bivalves. During periods when crusts are not forming the fauna and flora would be different, in turn influencing the type of sediment formed (see Shinn, 1969).

TABLE 4. Mineralogical and chemical composition of the nearshore and lagoonal sediments

	aragonite	calcite	high magnesium calcite	dolomite	quartz	feldspar	Ca/Mg	organic carbon	Insoluble residue
Near-Shore Shelf	81.1%	8.0%	3.9%	1.6%	3.8%	1.6%	47.1	0.24%	13.39%
Tidal Delta	89.4%	4.1%	2.8%	1.4%	2.2%	0.3%	85.4	0.27%	3.31%
Outer Lagoon Channel	86.4%	4.3%	4.8%	0.8%	3.8%	0.0%	66.5	0.31%	7.28%
Outer Lagoon Terrace	89.9%	3.4%	3.1%	1.4%	2.2%	0.2%	75.0	0.32%	6.29%
Inner Lagoon Channel	82.1%	5.2%	7.1%	1.1%	3.9%	0.7%	28.8	0.54%	7.22%
Inner Lagoon Terrace	83.7%	5.1%	1.9%	2.4%	4.8%	2.4%	68.2	0.55%	7.40%

PRODUCTION AND DISPERSAL OF SEDIMENT

The greater part of the sediment in the area is produced locally by extraction of calcium carbonate by biological and chemical processes. Material blown to the coastal lagoons and nearshore waters by offshore winds must add some finer material to this sediment, in addition to fine dust added from the Iranian

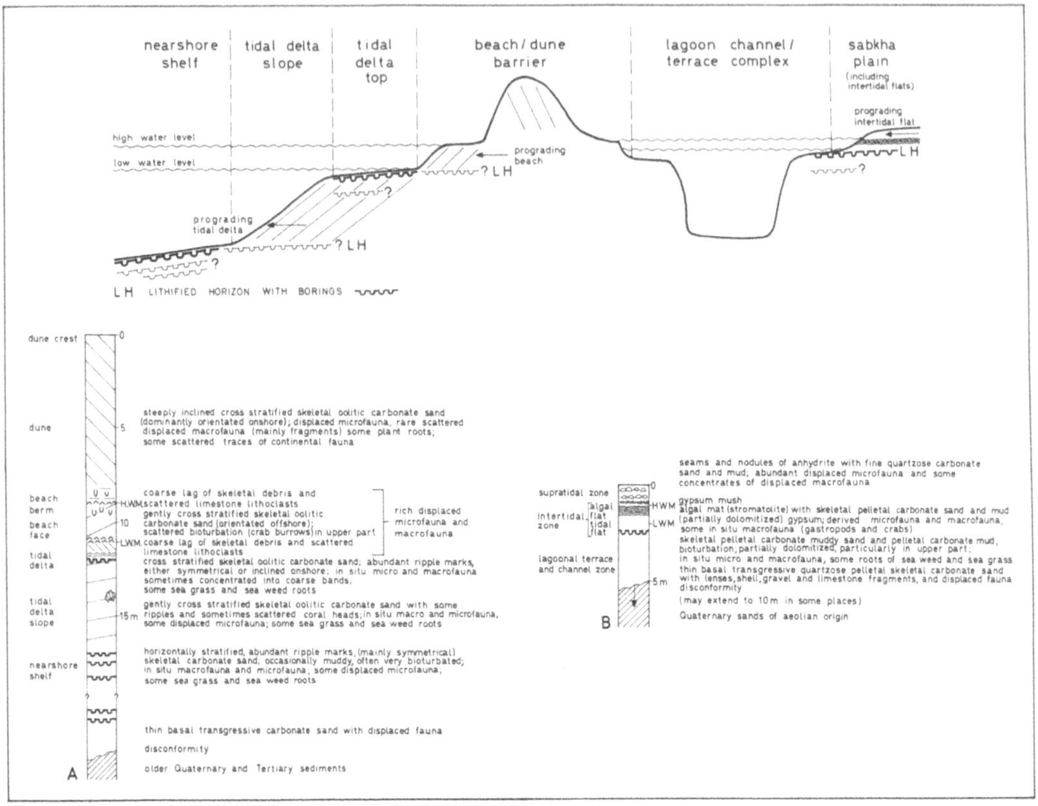

Fig. 19. Distribution of subtidal (and possibly intertidal) lithified surfaces in the nearshore and lagoonal areas of the Trucial Coast and sequences likely to be encountered in the stratigraphic column.

A. Seawards progradation of the barrier islands, and

B. Lagoonwards progradation of the intertidal flat and sabkha

side of the Persian Gulf by onshore winds. The nearshore sand is produced by "in situ" breakdown of bottom living organisms, mainly mollusks, both by organic activity and by wave and current action to produce a skeletal and pelletal sand. Mixed with this are fragments of early cemented material and clasts reworked from older, underlying deposits, and possibly some wind-transport dust.

Oolite production appears to be active on the tidal deltas although not on all of them. The junction between the oolitic sands and the nearshore skeletal and pellet sands is sharp, but some very small grains with thin oolitic coatings, and a few larger grains, are found in the near shore shelf sediments, at the foot of the delta slope. Also lighter foraminiferids, with thin oolitic coatings, are found in the nearshore shelf sediments together with ostracods characteristic of the tidal delta and outer lagoon. The oolites are not being dispersed far to seaward although the deltas as a whole are probably building slowly outwards by addition of oolites to their seaward faces. In a borehole at the seaward edge of the tidal delta approximately 7 m of oolitic sand has been penetrated. A considerable amount of oolitic sand has obviously moved shoreward, as demonstrated by the presence of high, oolitic sand dunes where deltas front the shoreline, contrasting with the low dunes where only a narrow coastal terrace and steep shore front the islands. A similar relationship, in the area of silici-clastic deposition, between offshore topography and the coastal dune pattern can be seen in Baja California (see Figures in Phleger, 1965, and personal discussions F.B. Phleger, 1967). The situation has been complicated at the NE end of Abu Dhabi Island where the dunes which are normally present have been deflated leaving only cemented dune cores. The dunes have extended up to 4 km inland at the southwest end of Abu Dhabi Island. Elsewhere, oolites have been carried lagoonwards and mixed with the lagoonal-beach-intertidal flat complexes.

The seaward face of Abu Dhabi Island is more complicated because of the presence of the adjacent reef and this has probably produced the littoral drift which has resulted in the cuspate point at the central part of its seaward face. The reef also protects part of the delta at the NE end of Abu Dhabi Island from the waves from the NW, and sand appears to be moving parallel with the delta's edge (Fig. 9) towards the shelter provided by this reef, rather than at right angles to this edge, as is the case elsewhere. This may partly be the reason for the starving of the NE corner of Abu Dhabi Island of sand, and the consequent deflation of the dunes at this end of the island.

The lagoon channels are areas of high water velocity and, in most parts places, of non-deposition. However, on their inner parts, they appear to be sheltered settling basins which are silting up.

The lagoonal terraces are locations for the production of skeletal debris, composite grains and faecal pellets. The main pellet producer is not known although there are worms living within the sediment and gastropods, crabs, rays and other animals which ingest and break down the sediment to produce considerable volumes of faecal material, most of which appears to become broken down even further. The sediments, on these terraces, are subject to stirring both by tidal currents and by wave action. This action appears to keep much of the area free from appreciable fine material, except in very sheltered areas, and waves distribute sand and coarser material to form beach ridges around the lagoonal margins. In places, small dunes have built up on top of these ridges. The fine material winnowed from the terraces appears to settle either in the inner ends of some of the channels or on intertidal flats and mangrove swamps directly behind the sheltering barrier ridges. Reaction between waves and strong tidal currents have produced levées and bars on the edges of the banks and terraces in the lagoons. Tidal currents have also been observed to transport shells and weed over the floor of the inner lagoon terrace. However, erosion of sediment is reduced in many areas by the presence of a thin layer of blue-green algae which binds the surface sediment. The importance of chemical precipitation of calcium carbonate, as against the production of carbonate mud by skeletal breakdown; is not known. Kinsman (1966) has suggested that the former is a most important process in this area. It is possible that the breakdown of the lightly calcified alga Acetabularia sp. of possibly of the species Jania sp. may be important.

Examination of these muds, under the electron microscope has shown them to consist largely of aragonite needles (W. Diver personal communication).

The considerable production of carbonate sediment in this area is indicated by the extensive progradation of both the mainland shores and the shores of barrier islands. There appears to be little evidence of any transport of coastal sediment to any great distance offshore, but the tidal deltas and barrier island beaches are probably prograding slowly seawards.

COASTAL DEVELOPMENT

An early transgression reaching a slightly higher level than that of present sea level appears to have taken place as shown by the cliffed "miliolite" and the capping of marine limestone of variable facies, common along this coast. Although sufficient evidence is not yet available, it seems likely that this represents a last interglacial sea level. The present coastal complex seems to have developed subsequently to the drowning of the outer part of the Arabian Desert during sea level rise following the last glaciation.

When the sea inundated the desert landscape the slightly cemented cores of the dunes (i.e. the aeolian "miliolite" calcarenite) and the elevated masses of Tertiary rock probably acted as nuclei for the subsequent growth of reefs, and accretion of beaches and dunes. As the gaps between barrier islands closed due to lateral accretion of the barrier, the tidal waters flooding from and ebbing into shallow waters of the open Gulf produced tidal deltas by the formation of oolitic sands. These deltas have probably prograded seaward.

Onshore movement of sand has led to large volumes of dune sand spreading inland from the coast and the individual barrier islands have developed striking down-wind tails which subdivide the lagoonal areas into separate basins. Evidence from borings in the adjacent sabkha plain indicate that the lagoons were previously more open and had a better circulation (Evans et. al., 1969). Beach ridges, composed largely of gastropods, have been found halfway along the Abu Dhabi Island; these are now separated from the lagoon by intertidal flats and small sabkhas. Radio carbon dates of sediments from these ridges gave ages of: 3465 $\pm$ 173; 3948 $\pm$ 185; 4191 $\pm$ 193 (all B.P.)

Behind the coastal barriers the lagoons became the site of deposition of pelletal and skeletal carbonate sands with some carbonate muds. The sediment was driven shorewards to accrete around the islands and along the mainland shoreline to form fringing beaches capped by small dunes, and intertidal flats. In places algal flats and mangrove swamps developed. Because of the abundance of carbonate sediment the landward parts of the lagoon have filled and the intertidal flats, colonized by by blue-green algae, have prograded seawards over the in-filled lagoons forming a coastal plain - the sabkha (Evans, Schmidt, Nelson and Bush, 1969, and Bush, Butler et al., Hsu and Schneider, in this volume). The base of this algal mat, where it is growing on the shoreline, and further inland where it is buried in the sediments under the sabkha plain, does not change in level more than 0.6 m over approximately 4.5 km, and indicates that sea level has been stable during the last 4,000 years. However, one can not be certain that there has not been a change in sea level over this period as changes in the adjacent barrier complex has reduced circulation of tidal waters and produced various geomorphological changes. Wave action has certainly been reduced in some areas over the last few thousand years.

The algal mats and other tidal flat deposits, particularly along the mainland shoreline, have been covered by aeolian sediment derived from the adjacent desert, and material carried landward by occasional storm tides. Precipitation of calcium sulphate, as gypsum and anhydrite, and reaction between the coastal plain sediments and waters, partly derived from the adjacent lagoons, have caused dolomitization and the production of other evaporitic minerals (Shearman 1963; Curtis,

Evans, Kinsman and Shearman, 1963; Evans and Shearman, 1964; Butler et. al., 1965; Kinsman, 1966; Shearman, 1966; and Butler, Bush, Hsü and Schneider, in this volume).

Inland of the maximum extension of the sea, the immediate desert dune area has been deflated down to the water table forming a low erosional plain, which continues the almost flat surface of the Holocene accretional plain landward. These accretional and erosional sections together constitute the coastal plain or "sabkha" which is so well developed in the Sheikdom of Abu Dhabi.

Conclusions

The Abu Dhabi coastal complex displays a wide variety of micro-geomorphological units, each of which is characterized by particular combinations of sediments, fauna and flora. In spite of this considerable array of detail a rather simple pattern emerges: a nearshore shelf with skeletal carbonate sands and an open marine fauna and flora; a barrier with oolitic carbonate sands and coral bioherm with an open marine fauna (usually displaced); a lagoon with pelletal-skeletal carbonate sands and carbonate muds with a restricted fauna; a coastal plain with pelletal-skeletal carbonate sands and muds, stromatolites, dolomites, gypsum and anhydrite (plus other evaporites), with a displaced restricted fauna; and, finally, a coastal desert with quartzose dune sands and some displaced marine as well as an indigenous continental fauna. These sedimentary associations are sufficiently characteristic to allow detailed interpretation of certain ancient shallow water limestone-evaporite sequences.

Acknowledgements

These studies have been supported mainly by the Natural Environment Research Council, Great Britain, but substantial financial aid has been provided by Shell Research, Rijswijk, The Netherlands; further assistance has been provided by Mobil Research and Development, Dallas, U.S.A., and minor travelling and subsistence expenses were provided by the Royal Society of London, Great Britain, and Bristol University. The project received considerable help from the Hydrographic Office of the Royal Navy during the initial stages and also local help from the Abu Dhabi Marine Areas and Abu Dhabi Petroleum Co. The writers would like to record their thanks to all these organizations as well as to A.H. Clarke, J.P. Harding, R.W. Ingle, E. White and M. Hilcote of the British Museum of Natural History, who identified the echinodermata, barnacles, crabs, corals and plants respectively. The authors wish also to acknowledge help provided by C.G. St. C. Kendall, Sir Patrick d'E. Skipwith, V. Schmidt, H. Nelson and R. Selley in the field; to the technical staff of Imperial College for assisting in the analysis of the sediments; to Miss M.E. Pugh and Mr. S. Phethean for assistance in preparing this paper and to Mrs. R. Evans for drafting the figures.

TABLE 5. The distribution of the fauna in the area. (Offshore refers to the "open shelf" of the Persian Gulf; nearshore to the nearshore shelf, tidal deltas, reef and back reef lagoon and seaward intertidal flats, beaches and dunes; the outer and inner lagoon are self-explanatory)

Legend

------- = dead

——— = alive.

COELENTERA	Offshore	Nearshore	Outer Lagoon	Inner Lagoon
Gorgonacea	———			
Siderastrea savignyana	- - - - -			
Favites halicora	- - - - -	- - - - -		
Paracyathus sp.	- - - - -	- - - - -		
Porites sp.	———	- - - - -		
Cyphastrea sp.	- - - - -	———		
Acropora sp.	———	———		
Porites cf. lutea	- - - - -	- - - - -		
Acropora cf. pharaonis		- - - - -		
Coscinaria monile		- - - - -		
Favia sp.		- - - - -		
Favia sp.		- - - - -		
Pleisastrea sp.		- - - - -		
Psammocora (Stephanaria) planipora		- - - - -		
Siderastrea liliacea		- - - - -		
Turbinaria sp.		- - - - -		
Platygyra (Coeloria) lamellina		———		
Cyphastrea microphalma		- - - - -		
Stylophora pistillata		- - - - -		
MOLLUSKA				
(1)LORICATA				
Isnochiton yerburyi			———	
(2)GASTROPODA				
Fusinus townsendi	- - - - -			
Finella scabra	- - - - -			
Finella pupoides	- - - - -		- - - - -	
Mitra (Scabricula) boyei	- - - - -			
Terebellum (Terebellum) terebellum	- - - - -			

TABLE 5

GASTROPODA (cont.)	Offshore	Nearshore	Outer Lagoon	Inner Lagoon
Iravadia trochlearis	- - - - -			
Calyptrea pellucida	- - - - -			
Scaliola bella	- - - - -			
Triphora acuta	- - - - -			
Xenophora caperata	- - - - -			
Diodora imbricata	———			
Cypraea (Errones) lentiginosa	———			
Cypraea (Erosaria) turdus	———	- - - - -		
Trochus (Infundibulops) erythraeus	- - - - -	- - - - -		
Strombus decorus	- - - - -	- - - - -		
Ancilla castanea	———	———		
Bullaria ampulla	- - - - -	- - - - -	- - - - -	
Murex custerianus	———	———	———	
Scaliola arenosa	- - - - -		- - - - -	
Diodora funiculata		- - - - -		
Murex scolopax		- - - - -		
Oliva bulbosa		- - - - -		
Oliva caerulea		- - - - -		
Terebralia (Terebralia) palustris		- - - - -		
Mucronalia lepida		- - - - -		
Persicula subflava		- - - - -		
Mitrella misera		———		
Aglaja cf. nigra		———		
Cerithium scabridum		———	———	
Cerithium petrosum		- - - - -	- - - - -	
Thais tissoti		- - - - -	- - - - -	
Nassarius pullus		———	———	
Nassarius stigmarius		———	———	
Turbo coronatus		———	———	
Cerithium rugosum		- - - - -	———	- - - - -
Mitrella (Mitrella) blanda		———	———	———
Ancilla eburnea		———	———	
Drupa margariticola		———	———	
Pirenella conica			———	

TABLE 5

GASTROPODA (cont.)	Offshore	Nearshore	Outer Lagoon	Inner Lagoon
Phenacolepas evansi			- - - - -	
Phenacolepas omanensis			———	
Ancilla cinnamomea			———	
Euchelus asper			———	
Laemodonta rapax			- - - - -	
Melampus lividus			- - - - -	
Melampus sp.			- - - - -	
Phasianella nivosa			———	
Planaxis sulcatus			———	
Nodilittorina subnodosa			———	
Rissoina distans			- - - - -	
Siphonaria rosea			———	
Thais carinifera			- - - - -	
Thais pseudohippocastanum			———	
Cerithium caeruleum			———	
Emerginula planulata			———	
Turritella fascialis			———	
Oncidium peronii			———	
Natica lineata			- - - - -	
Natica spp.			- - - - -	
Cerithidea cingulatus			———	- - - - -
(3) SCAPHOPODA				
Dentalium sp.		- - - - -	- - - - -	
(4) BIVALVIA				
Arca (Anadara) ehrenbergi	- - - - -			
Chama brassica	- - - - -			
Cultellus cultellus	- - - - -			
Cuna coxi	- - - - -			
Isognomon ephippium	- - - - -			
Tellina (Acropagia) isseli	- - - - -			
Tellina (Pinguitellina) robusta	- - - - -			

TABLE 5

BIVALVIA (cont.)	Offshore	Nearshore	Outer Lagoon	Inner Lagoon
Tellydora pellyana	- - - - -			
Arca (Andara) antiquata	———			
Arca (Scapharca) tricenicosta	———			
Arca (Scapharca) vellicata	———			
Cardita ffinchi	———			
Circe (Circe) intermedia	———			
Glycymeris (Pectunculus) pectunculus	———			
Katelysia marmorata	———			
Septifer bilocularis	———			
Tellina (Acroparginula) inflata	———			
Trachycardium maculosum	———			
Arca (Anadara) uropigmelana	———	- - - - -	- - - - -	
Chlamys ruschenbergerii	———	- - - - -	- - - - -	
Trachycardium lacunosum	———	- - - - -	- - - - -	
Spondylus exilis	———	- - - - -	———	
Lima tenuis	———		- - - - -	
Arca (Barbatia) lacerata	———		———	
Pinctada radiata	———		———	
Circe (Circe) scripta	———	- - - - -	———	———
Brachidontes (Hormomya) variabilis	———	———	———	———
Circe (Parmulophora) corrugata		- - - - -		
Divaricella cumingi		- - - - -		
Glycymeris lividus		- - - - -		
Pinna bicolor		———	———	
Pitar (Amiantis) erycina		- - - - -		
Tivela ponderosa		- - - - -		
Meretrix (Meretrix) zonaria		———		
Glycymeris striatularis		———	- - - - -	
Mactra olorina		- - - - -	- - - - -	
Cardita antiquata		- - - - -	- - - - -	

TABLE 5

BIVALVIA (cont.)	Offshore	Nearshore	Outer Lagoon	Inner Lagoon
Ctena divergens		- - -	- - -	
Dosinia alta		- - -	- - -	
Sunetta spp.		- - -	- - -	
Meretrix (Meretrix) meretrix		———	———	
Gastrochaena cuneiformis		———	———	
Gafrarium arabicum			———	
Cardium (Parvicardium) sueziensis			- - -	
Diplodonta raveyensis			- - -	
Dosinia spp.			- - -	
Pinna atropurpurea			- - -	
Plicatula plicata			- - -	
Arca (Acar) plicata			———	
Asaphis (Asaphis) deflorata			———	
Codakia (Jagonia) fischeriana			———	
Cuna majeeda			———	
Gafrarium dispar			———	
Isognomon dentifer			———	
Isognomon legumen			———	
Lithophaga lithophaga			———	
Malleus regula			———	
Musculus spp.			———	
Crassostrea cucullata			———	
Macoma arsinoensis				———
ECHINODERMATA				
Brissopsis persica	———			
Clypeaster reticulatus	———			
Clypeaster humilis	———			
Lovenia sp.	———			
Metalia townshendi	———			
Echinodiscus auritus	———			
Echinodiscus bisperforatus		———→		

TABLE 5

ECHINODERMATA (cont.)	Offshore	Nearshore	Outer Lagoon	Inner Lagoon
Paracrocnida persica		—		
Laganum depressum		—		
Echinometra malthaei		—	—	
Astropecten pugnax	—			
Astropecten indicus	—	—		
Astropecten phragmorus	—	—		
Asterina burtoni iranica	—	—	—	
Ophiopeza fallax	—	—		
Ophiothrix aff. *exigua*		—		
Ophionereis dubia		—		
Amphioblus hastatus		—		
Ophiactis savignyi			—	
Ophiothrix comata			—	
Ophiothrix savignyi			—	

CRUSTACEA (1) DECAPODA	Offshore	Nearshore	Outer Lagoon	Inner Lagoon
Halimede sp.	—			
Myra fugnax	—			
Cryptopodia fornicata	—	—		
Hyastenus dicanthus	—	—		
Leucosia sp.	—	—		
Micippa philyra	—	—		
Trachypenaeus sp.	—	—		
Ogyrides		—		
Pachycheles sp.		—		
Tetralia glaberrima		—		
Thalamita sp.		—		
Actaea savignyi		—	—	
Chloridius niger		—	—	
Gonodactylus demani demani		—	—	
Ocypoda aegyptica		—	—	
Scopimera sp.		—	—	—
Alpheus sp.			—	

TABLE 5

CRUSTACEA (1) DECAPODA (cont.)	Offshore	Nearshore	Outer Lagoon	Inner Lagoon
Cleistostomata nr. dotilliforme			—	
Cyphocarcinus minutus			—	
Gonodactylus chiragra chiragra			—	
Menaethius monoceros			—	
Myomenippe sp.			—	
Pagurus sp.			—	
Petrolistes sp.			—	
Metopograpsus messor			—	—
(2) CIRRIPEDIA				
Balanus amphitrite amphitrite			—	—
Chthamalus cf. challengeri			—	—
Balanus amaryllis		attached to drifting debris		
(3) OSTRACODA				
Xestoleberis sp. B.		—		
Pontocypris sp. B.		—		
Paranesidea sp.		—		
Cytherelloidea sp. B.		—		
Caudites sp. B.		—		
Munseyella sp.		—		
Hemicytherura videns aegyptica		—		
Sclerochilus sp.		—		
Paradoxostoma sp. A.		—		
Paradoxostoma sp. C.		—		
Cytherois sp.		—		- - - - - -
Xestoleberis multiporosa		—	—	
Xestoleberis sp. A.		—	—	
Xestoleberis sp. C.		—	—	
Pontocypris sp. A.		—	—	
Neonesidea schulzi		—	—	
Loxoconcha sp. B.		—	—	

TABLE 5

CRUSTACEA (3) OSTRACODA (cont.)	Offshore	Nearshore	Outer Lagoon	Inner Lagoon
Genus _B_. sp.		——	——	
Moosella sp.		——	——	
Cytherelloidea sp. A.		——	——	
Cytherella cf. _punctata_		——	——	
Paracytheridea sp.		——	——	
Thalmannia sp.		——	——	
Genus _C_. sp.		- - -	- - -	
Carinocythereis cf. _hamata_		——	——	
Caudites sp. A.		——	——	
Cytherura sp.		——	——	
Bosquetina sp.		——	——	
Paradoxostoma longum		——	——	
Cytheroma dimorpha		——	——	
Tanella cf. _gracilis_		- - -	- - -	
Hulingsina sp.		——	——	
Aglaiocypris triebeli		——	——	——
Xestoleberis rhomboidea		——	——	——
Xestoleberis rotunda		——	——	——
Paijenborchellina sp.		- - -	——	- - -
Loxoconcha sp. A.		——	——	——
Alocopocythere reticulata		- - -	- - -	- - -
Paradoxostoma sp. B.			——	
Mutilus sp.			- - -	
Cyprideis sp.			——	
Genus _A_. sp.			——	——
Loxoconcha sp. C.			——	——
Callistocythere sp.				——

Distribution and Ultrastructure of Holocene Ooids in the Persian Gulf

J.-P. Loreau[1] and B. H. Purser[2]

ABSTRACT

Ooids occur in many different environments in the Persian Gulf; blackened relict ooids are common at depths of 100 m in the center of the basin while others have been transported to desert environments up to 40 km from the shore. Most Holocene ooids occur in coastal areas in the S parts of the Persian Gulf in which they appear to be forming in a variety of settings. The most spectacular accumulations constitute tidal deltas associated with coastal barrier system in E Abu Dhabi. Other agitated environments include tidal bars situated in wide channels between islands (Yas and Bahrain) and the adjacent Arabian shoreline, and on open tidal flats and beaches in exposed embayments (Sabkha Matti). Although most ooids are forming in these agitated environments, significant quantities are forming within lagoons (Khor Odaid and Jebel Dhanna), and on the SE lee coast of Qatar Peninsula (Umm Said).

The structure of the cortex of ooids collected in these various environments has been studied by light microcope, SEM, micro-probe and other analyses, to evaluate the relationship between ooid structure and environment. Aragonitic nanograins seem to be associated with organic material in the cortex and occur in most ooids irrespective of environment. Elongate rods (1-2 microns) of aragonite exhibit various orientations. In agitated environments on the crests of bars,or delta levées, these have a statistically tangential orientation with respect to the nucleus, and are often tightly packed or coalescent. In ooids collected from more protected depressions between bars, aragonite rods within the outermost layer of the cortex have haphazard or radial orientation and the fabric is loose. In protected lagoonal settings this radial orientation is well developed and the ooids are often unusually big and irregular in shape.

The relationship between the structure of the external layer of the cortex and the environment in which the ooid was collected indicates that the aragonitic components grow on the nucleus initially with a haphazard or radial orientation, creating a loose fabric. This primary orientation, developed mainly while the ooid is in a relatively protected micro-environment, is subsequently modified to a secondary tangential orientation on the crest of adjacent bars or beaches where crystals are physically compacted to create a dense fabric. Lagoonal ooids tend to retain their primary radial structure and may attain considerable size in the absense of physical abrasion in these protected environments.

---

1 Lab. de Géologie, Muséum National d'Histoire Naturelle (Paris)

2 Lab. de Géologie Historique, Université de Paris Sud (Orsay)

Physical conditions favourable to ooid formation would seem to include a bar and protected channel morphology associated with reversible tidal or longshore currents which retain ooids within this favourable system. Evaporitic tidal flats seem to favour the formation of large pisooids whose formation is not related to mechanical movements.

Most Holocene ooids of the Persian Gulf have much in common with the classical Bahamian ooids. They differ mainly in their regional distribution, most forming along the continental shoreline remote from any platform edge, many accumulating in desert environments. Others form in quiet water tidal flats and have much in common with cave pearls. Although the areas of ooid formation at any given moment are small, these grains are readily dispersed. In time a sheet of oolitic sand may develop whose geometry and magnitude will not reflect that of the ooid-forming environment.

I. INTRODUCTION

Oolitic sands are accumulating along the Arabian shores of the Persian Gulf. Their presence seems to have been noted firstly by Sugden (1963) who reported scattered coated grains from the shores of Qatar Peninsula, and subsequently by Evans, Kinsman and Shearman (1964) along the shores of the Trucial Coast, where pure oolitic sands constitute a series of spectacular tidal deltas. These deltas,and the adjacent coastal barriers,are interpolated within the regional coastal sedimentary system by Purser and Evans elsewhere in this volume.

Although the tidal deltas of the Trucial Coast constitute the most spectacular oolitic accumulations within the Persian Gulf,they are far from being the only style of oolitic sedimentation. Extensive areas of oolitic sand occur along the present shoreline of Abu Dhabi in settings which are markedly different from the classical deltas, both in agitated and in protected lagoonal environments. Scattered relict ooids also occur both within the deepest, central parts of the basin, and in desert sands up to 40 km from the present shoreline. In sum, unconsolidated oolitic sediments occur in many contrasting environments in the eastern parts of the Persian Gulf. Many (perhaps most) are not in equilibrium with their present environment and their anomalous occurrence can readily be explained in terms of post-glacial transgression or lateral transport. In ancient rocks, however, these relationships would naturally be more subtile.

Oolitic sediments have been sampled with precision in a considerable variety of environments ranging from infratidal bars and levées, inter-bar depressions, beaches, lagoons, tidal flats, and aeolian dunes at varying distances from the coast. This diversity has permitted an evaluation of the relationship between ooid ultrastructure and environment which has led to a better understanding of ooid formation and subsequent,syn-sedimentary modification.

The authors express their thanks to Professor R. Laffitte(Muséum National d'Histoire Naturelle, Paris) for sustained interest in the project and to Abu Dhabi Petroleum Company (A.D.P.C.) for practical help in the field. Certain samples and carbon 14 datations have been provided by Shell Research Laboratory, Rijswijk, The Netherlands (for whom part of this research was originally carried out) to whom the authors offer their sincere thanks. The authors also wish to thank Professor J. Fabriès for facilitating mineralogical analyses in the Laboratoire de Minéralogie (Muséum National d'Histoire Naturelle) and to Professors J. Wyart, W.L. Brown, and Drs. A. Rimsky, C. Willaime and C. Desnoyers for fruitful collaboration within their Laboratoire de Minéralogie et Cristallographie (Faculté des Sciences, Université de Paris VI).

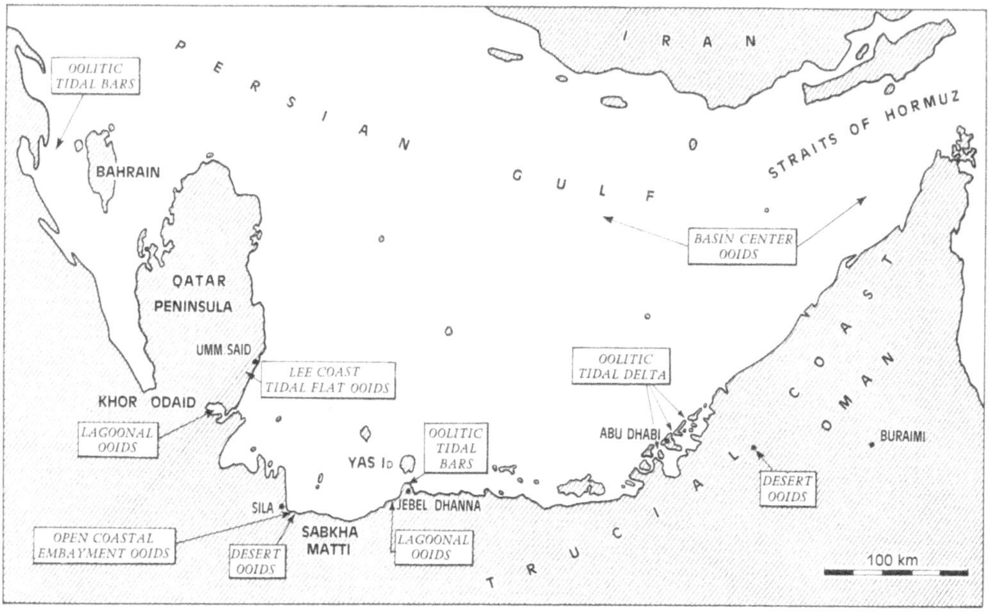

FIG. 1. Map of the southern Persian Gulf showing distribution of Holocene ooids discussed in this contribution

II. THE ENVIRONMENTS AND DISTRIBUTION OF HOLOCENE OOIDS

Holocene ooids occur as loose grains on the sediment surface in three major environments (Fig. 1) :

- basin center areas down to depths of approximately 100 m,
- coastal areas as bars, tidal deltas, beaches, lagoons and on tidal flats,
- desert environments.

While this contribution deals mainly with coastal environments, which include the bulk of Holocene ooids, the more anomalous basin center and desert ooids are also included partly to demonstrate the widespread distribution of ooids within the surface sediments of the Persian Gulf.

BASIN CENTER OOIDS

Ooids are common within the argillaceous carbonate muds of the axial parts of the Persian Gulf between depths of ca. 80-120 m. They were recorded initially by Houbolt (1957) who described them as pellets, and subsequently by Sugden (1966). Structurally, these ooids closely resemble the oolitic grains currently forming in various coastal environments. Most have a thick, finely laminated cortex and have average diameters of 1 mm. They differ from coastal ooids mainly in colour and only slightly in composition; they range from black to light beige, the darker colours predominating, due seemingly to traces of an amorphous iron sulphide, hydrotroilite (Sugden, 1966). Bulk X-ray analyses, however, indicate that these darkened grains consist mainly of aragonite.

Carbon 14 datings (by Shell Research) of blackened ooids from depths of ca. 80 m vary in age from 12,000-15,000 years confirming the suspicion that they are relict and, in this respect, comparable with the deep water ooids of the Yucatan Shelf (Logan et al, 1969). They are particularly common within the axis of the gulf

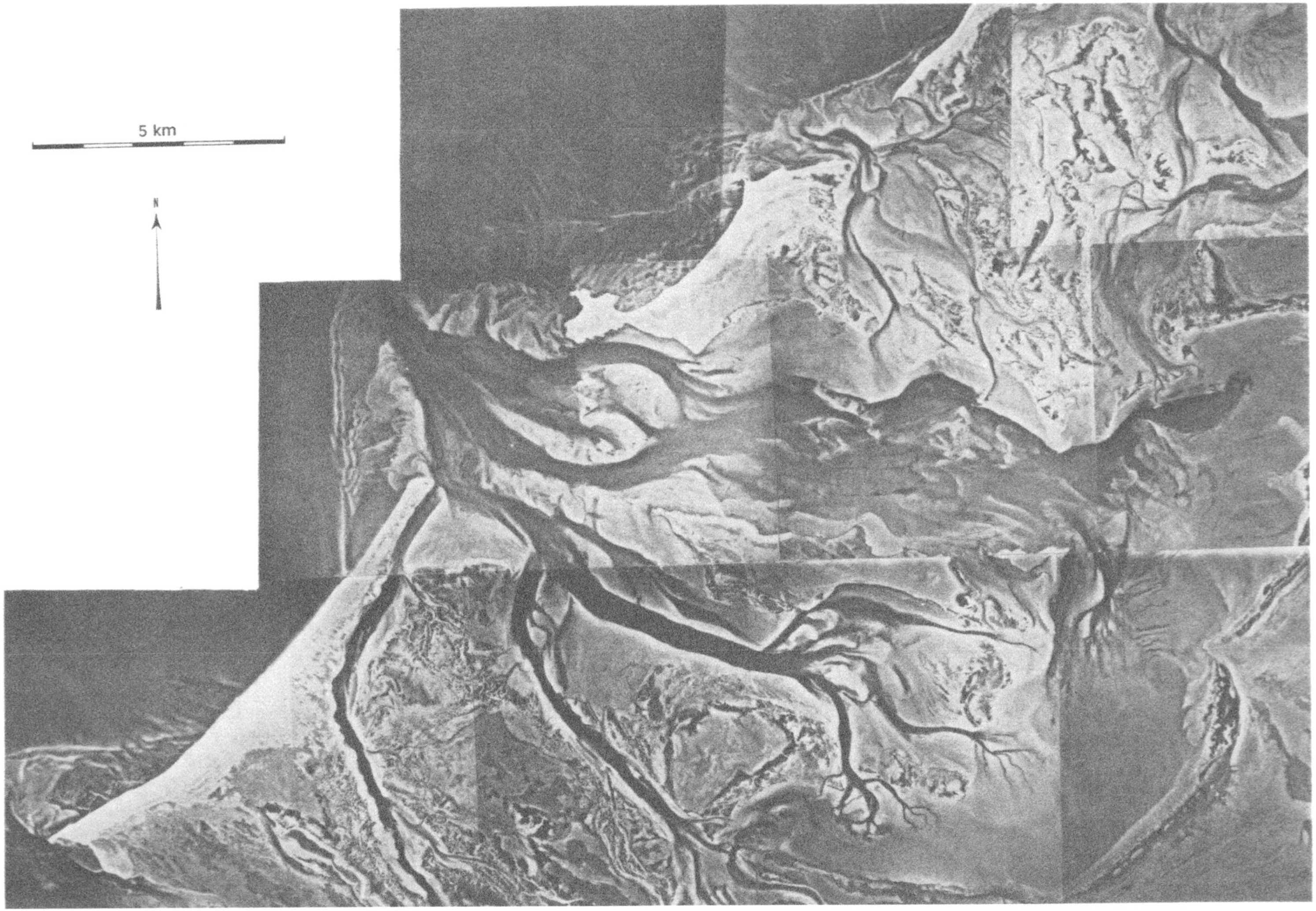

FIG. 2. Aerial photo-mosaic of oolitic tidal delta system and coastal lagoon between Sadyat and Gharab islands, E. Abu Dhabi

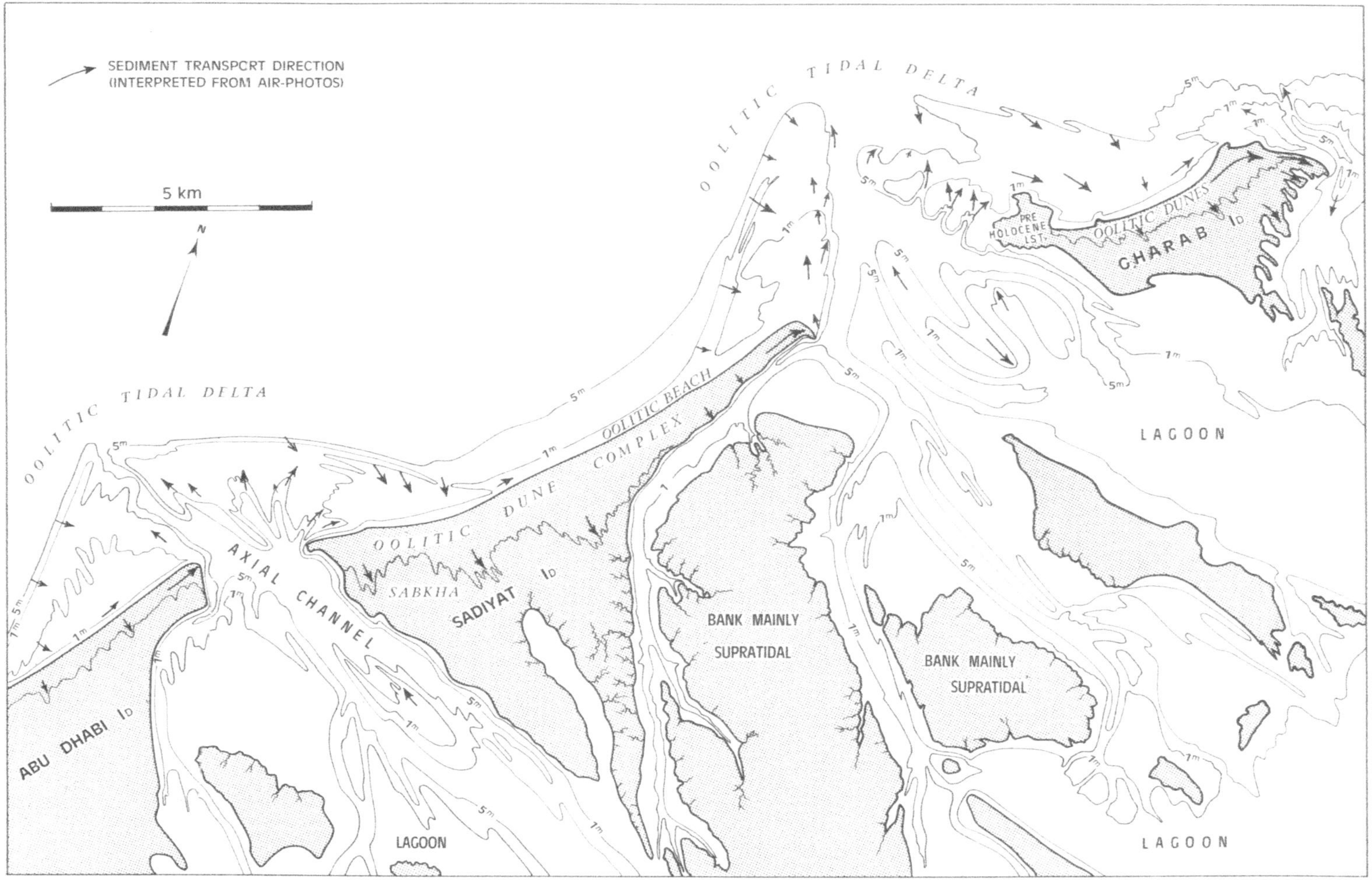

FIG. 3. Map of coastal barrier and oolitic tidal delta system showing morphology and direction of sediment transport

near the Straits of Hormuz where they may be sufficiently abundant to constitute a grain-supported texture. They are intimately mixed with other grains, including planktonic Foraminifera. These deep water sediments therefore include polygenetic grains of divers age; the relict ooids, generated during an early Holocene,lower sea level, have been mixed with modern elements including planktonic Foraminifera, presumably by the combined effects of bioturbation and slow sedimentation.

COASTAL OOIDS

Most Holocene ooids of the Persian Gulf occur within 5 km of the Arabian shore as a series of discontinuous patches. The milieux in which these ooids are forming and the geometry of their accumulations are variable. Four major coastal settings are defined :

- tidal deltas and adjacent barrier island beaches and dunes,
- tidal bars within channels,
- open coastal embayments and adjacent beaches,
- coastal lagoons and lee coasts.

The first three environments are essentially ones of considerable water agitation while the lagoons and lee coasts, on the contrary, are generally very calm.

1. Tidal deltas and adjacent barrier beaches and aeolian dunes

In E Abu Dhabi (Fig. 1) a series of barrier islands are separated by channels several km in width and up to 10 m in depth, each of which is flanked by a "delta" of oolitic sand. This series of oolitic deltas are the principal sites of modern ooid formation within the Persian Gulf. A typical tidal delta between the coastal barrier islands of Sadyat and Gharab (Figs. 2,3) has been studied by the authors in some detail. Its morphology, in common with deltas occuring between most barrier islands in E Abu Dhabi, is clearly related to the inter-island channel which links the open waters of the Persian Gulf with the extensive coastal lagoons leewards of the barrier. Although by far the greater part of the delta is oriented seawards, small deltaic lobes are also oriented towards the lagoon (Fig. 2).

The delta itself is a complex system of banks and channels whose surfaces are modified by ripples of varying dimensions. The geometry of the delta as a whole clearly indicates the dominance of seawards-flowing currents whose local effects are most strongly expressed along the border of the axial channel, mainly as "spill-over" lobes. They are due presumably to ebb-tidal currents possibly augmented by occasional offshore winds. These effects diminish away from the axial channel and are reversed locally near the periphery of the delta where bars measuring 100 m in width and 1-2 m in amplitude have somewhat steeper slopes facing landwards. The morphology of these peripheral bars is probably determined mainly by onshore waves and surface currents. A third, longshore transport direction, mainly towards the NE, is clearly indicated by the accretion of barrier islands. In sum, the morphology of the tidal delta clearly indicates that its oolitic sands are influenced by a complex system of multi-directional currents (Fig. 3) which tend to retain these sands, at least temporarily, within the delta system.

Sediment character and distribution within the delta is illustrated in figures 4 and 5. Two suites of sediment have been defined :

- a suite of pure oolitic sand (> 90% ooids within any given sample (Fig. 5 C) localized within the delta itself. Ooids comprising this suite attain maximum size (2 mm) along the edges of the axial channel whose levées of white sand are exposed at low tide. Ooids decrease in size towards the outer fringes of the delta (Fig. 5B) and at depths of ca. 2 m there is a rapid transition into a mixed sediment suite;
- a suite of mixed ooid-pellet-bioclastic grains (Fig.5D-F) containing 10-90% ooids is very widespread. These sediments occur both on the outer edges of the delta at

depths of 2-5 m, within the axial channel and, especially, as beaches and aeolian dunes on the adjacent coastal barrier island complex. The ooid fraction varies, being relatively high within the beaches and dunes, but is strongly diluted by coarse bioclastic debris within the channels (Fig. 5D). Ooids are diluted progressively with pelletoidal grains around the fringes of the delta (Fig. 5A).

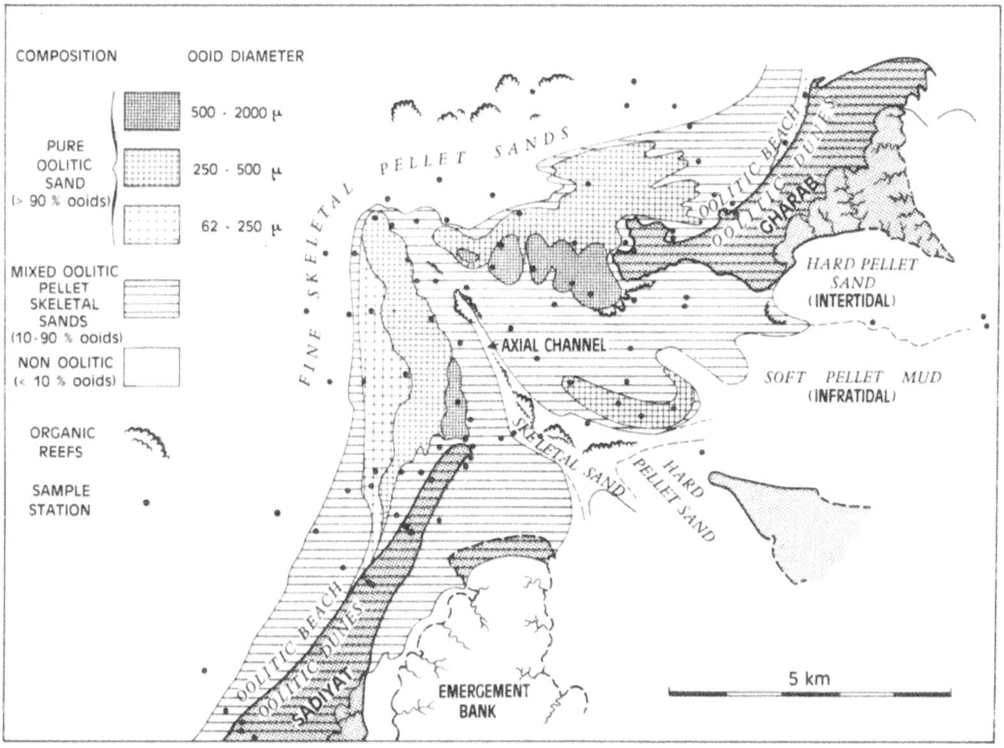

FIG. 4. Map of oolitic tidal delta between Sadiyat and Gharab islands showing distribution of principal oolitic facies and associated sediments

Dispersion of ooids :

Although ooids are a predominant grain type within the coastal barrier complex of NE Abu Dhabi, it is highly probable that their widespread occurrence is mainly the consequence of effective dispersion. The environment most favourable to their formation is believed by Kinsman (1964) and others to be the tidal delta. Grain size distribution and composition seems to confirm this hypothesis; ooids increase both in size and abundance towards the levées bordering the axial channel. Ooid size must be related partly to the degree of water agitation, the highly agitated levée environments preventing the stabilization of relatively small grains. However, the dominance of oolitic grains within the delta cannot be explained only in terms of water agitation; the adjacent barrier beaches suffer a comparable degree of agitation but generally contain less than 90% ooids and are frequently rich in bioclastic grains. Elsewhere along the coast, where tidal channels are not developed, ooids are rare in spite of the shallow, frequently agitated nature of their environments.

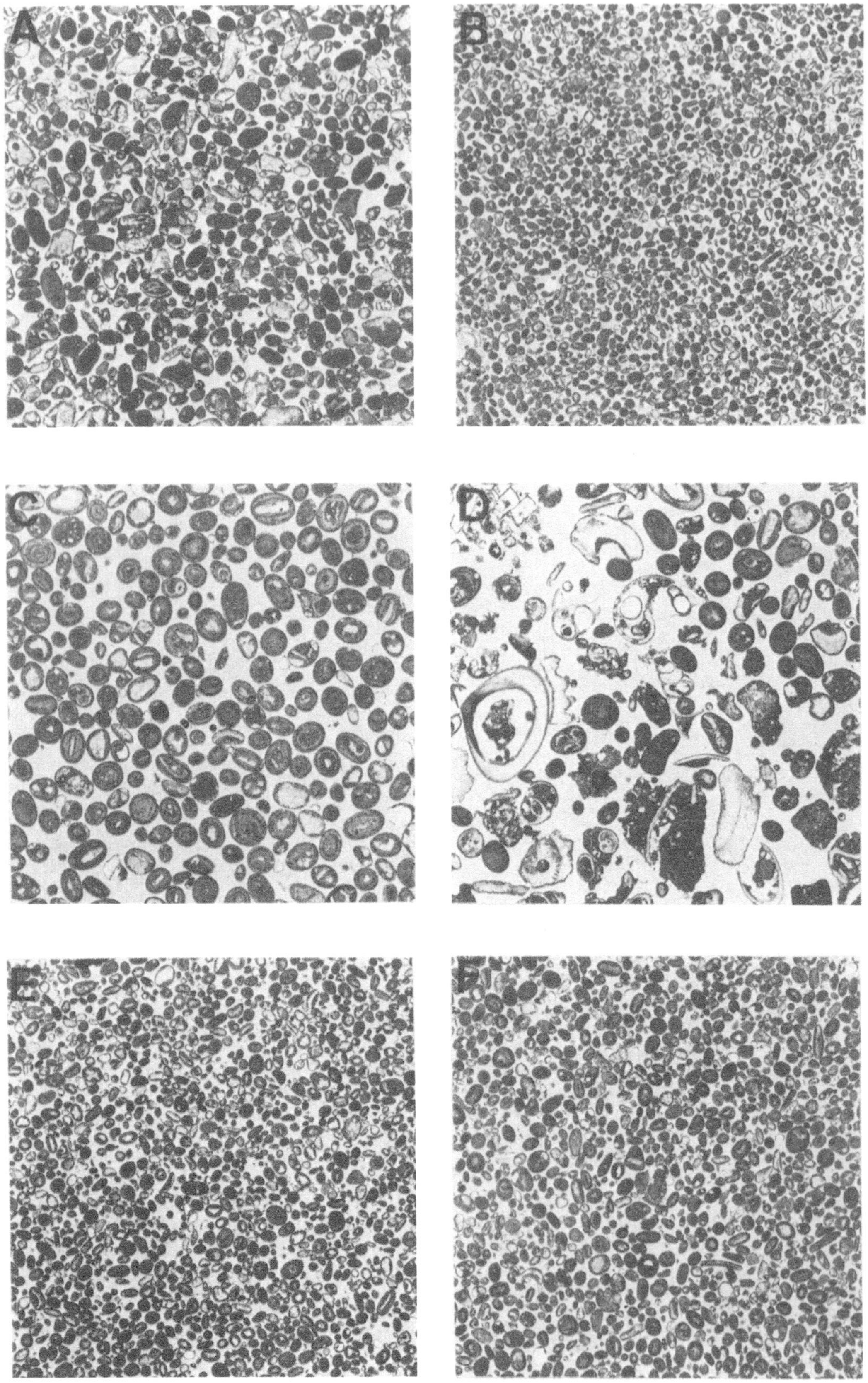

Legend to Fig. 5 A-F on the following page

Ooid distribution and current directions indicated by the morphology of the delta system suggests that the ooids tend to be retained near the levée by the opposed forces of seawards flowing tidal currents and onshore waves (Fig. 3). Grains that fall into the channel are mixed with skeletal debris and swept seawards, prograding the tidal delta. Other grains which eventually reach the periphery of the delta are subsequently swept back onto the adjacent beaches where they are mixed with skeletal grains. These heterogeneous beach sediments are transported along the beach and contribute to the lateral accretion of the barrier island,while some may ultimately be reincorporated into an adjacent tidal delta and begin a new cycle. Many ooids are swept from the beach to form aeolian dunes which attain heights of ca. 20 m on these barrier islands (see Evans et al, in this volume).

It is suggested that this complex system of sand movement may favour ooid formation. Although individual grains are not necessarily transported great distances and may frequently be buried temporarily within the beach or delta, they tend to remain for long periods within the delta system. The processes of ooid formation are discussed in a subsequent section.

2. Tidal bars in coastal channels

Elongate bars of oolitic sand occur in wide channels between the island of Yas and the adjacent shoreline of W. Abu Dhabi,and also between the island of Bahrain and the coast of Saudi Arabia (Figs. 6,7).

Yas Island :

The channel between Yas Island and the Arabian shoreline at Jebel Dhanna attains maximum depths of ca. 5 m and is partially blocked by a tail of sediment accreting from the lee side of Yas Island. This "broad sedimentary ridge is modified by a series of oolitic bars arranged "en échelon" parallel to the axis of the channel (Fig. 6), their orientation and morphology suggesting that they are strongly influenced by tidal currents. Individual bars emerge at low tide and are further modified by a complex system of ripples. Salinity in this area averages ca. 40‰,being somewhat lower than is associated with the oolitic deltas.

Bahrain Island :

The wide (20 km) channel between Bahrain and the shores of Saudi Arabia near Dahran also attains maximum depths of ca. 5 m and is partially blocked by a series of oolitic tidal bars. Individual bars measure up to 10 km in length and have

---

FIG. 5. Photomicrographs of typical oolitic tidal delta and related coastal barrier sediments :

A. Pelletoidal and skeletal sand seawards of delta (6 m depth).

B. Outer delta; fine oolitic sand (2 m depth).

C. Inner delta; coarse oolitic sand from levée bordering axial channel (0 m depth).

D. Axial channel; mixed oolitic and coarse skeletal grains (6 m depth).

E. Barrier beach adjacent to delta; mixed ooids with variable cortex thickness.

F. Aeolian dune 100 m from the beach; ooids with variable cortex thickness

ALL PHOTOS X 5

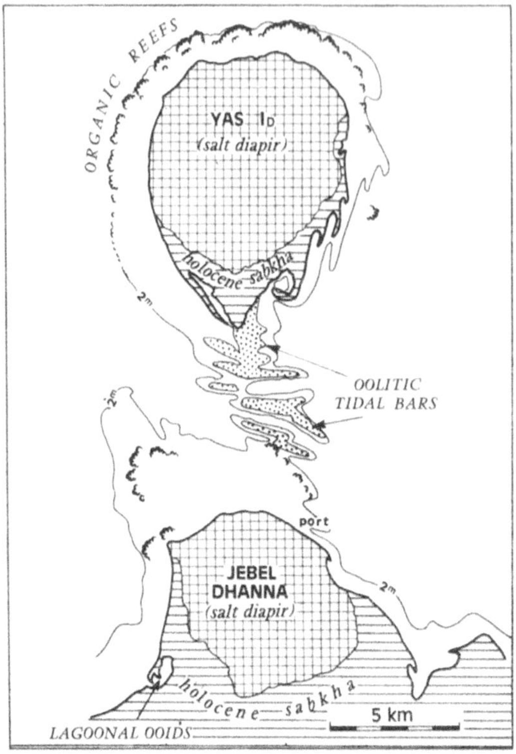

FIG. 6. Map showing distribution of oolitic tidal bars in channel between Yas Island and Jebel Dhanna, W. Abu Dhabi

amplitudes of 5 m. They are arranged en échelon, their axes being oriented parallel to the axis of the channel (Fig. 7). Several emerge as small islands while others constitute shallow banks which are awash as low tide. These bars are extensively lithified; a thin crust of hard limestone consisting of ooids cemented with fibrous aragonite overlies loose oolitic sands whose individual grains are highly polished and identical in appearance to the uncemented sediment elsewhere on the bar. Carbon dating (by Shell Research) of these latter ooids has given an age of 430 years (± 170). Salinities in the area are ca. 50‰. The local variations in ooid size and abundance have not been studied in detail in these areas. Sediments composing the axial parts of each bar consist essentially of ooids with a well developed cortex, but in both areas there is a marked increase in skeletal grains in the intervening channels. The progressive increase in ooids towards the axis of each bar suggests that these bars are the loci of ooid formation. The orientation and morphology of these bars and their localization within broad coastal channels indicates that the distribution of these oolitic sediments is influenced by bi-directional tidal currents.

3. Open coastal embayment and beach

The W end of the Trucial Coast at Sabkha Matti faces directly into the "shamal" winds and is frequently swept by waves and longshore currents. Near the frontier post at Sila (Fig.9) oolitic sands extend across a 500 m wide intertidal sand flat down to depths of ca. 2 m, where they grade rapidly seawards into skeletal sands. These intertidal and shallow infratidal oolitic sands are piled into a series of elongate ripples whose axes are oriented both parallel and obliquely to the shoreline (Fig. 8) and attain amplitudes of 1 m. Ooids on the wide intertidal sand flats are relatively small and occur in two modes (100-200 μ and 300-400 μ) within any given sample and are virtually uncontaminated by non-oolitic elements. These small ooids have unpolished surfaces. However, towards the adjacent beach berm the ooids become somewhat bigger (500-800 μ) and, together with the skeletal and quartz grains which dilute them, have a polished appearance. The mixed ooid-bioclastic beach, the principal site of ooid accumulation, is flanked landwards by a coastal plain also composed mainly of ooids, many of which are probably blown from the adjacent beach.

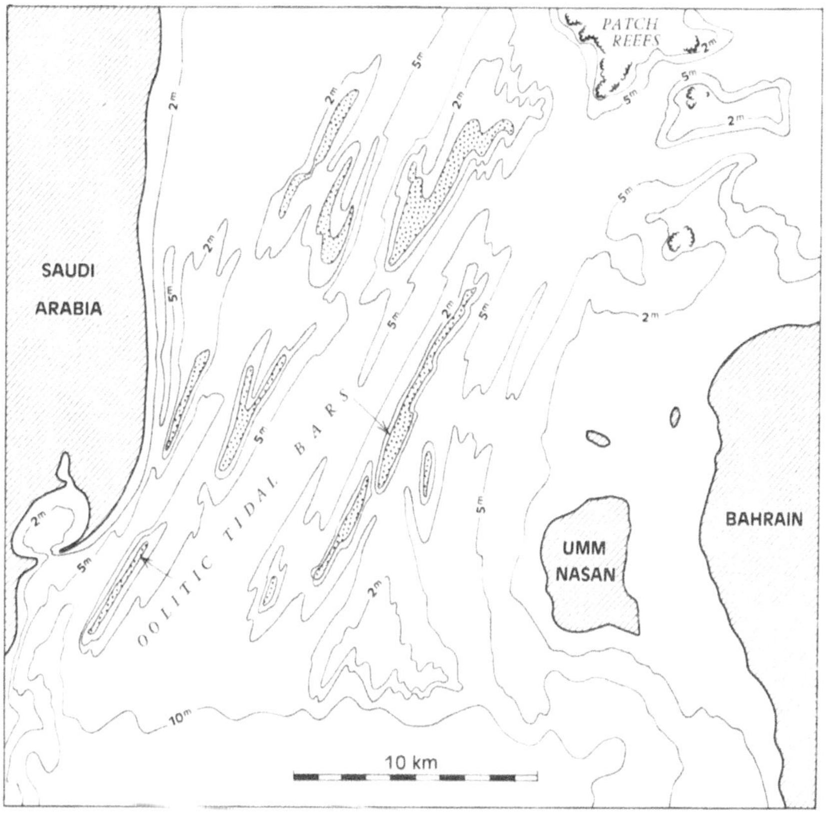

Fig. 7. Map showing distribution of oolitic tidal bars between Bahrain Island and coastal Saudi Arabia

4. Lagoons and leeward coasts

Although most ooids are forming in relatively agitated environments, significant amounts are forming also in highly protected settings both within coastal lagoons at Khor Odaid and near Jebel Dhanna (W. Abu Dhabi and on the leeward coast near Umm Said (SE Quatar, Fig. 1).

a) The oolitic sediments of Khor Odaid (Fig. 10) :

Khor Odaid is a large coastal embayment with average depths of 2-5 m. It is almost completely isolated from the open waters of the Persian Gulf by the lateral accretion of tidal flats along the SE coast of Qatar Peninsula, a narrow channel some 15 km in length and 1 km in width maintaining the connection with the adjacent Trucial Coast Embayment (Fig. 10). The presence of primary gypsum in the shallow infratidal areas in the extreme NW parts of this lagoon (pers. comm. M.W. Hughes Clarke and E.A. Shinn) and scattered living corals and echinoids near the S parts, indicates a considerable range of salinities.

FIG. 8. Aerial photograph of the W. end of Sabkha Matti showing mega-ripples and beach of oolitic and skeletal sand

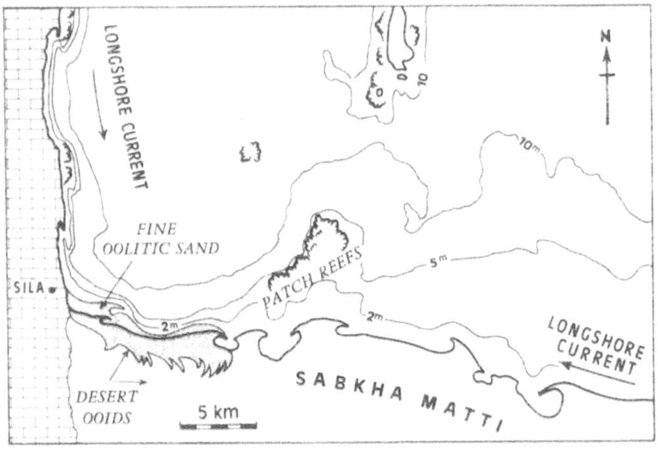

FIG. 9 Map of W. Sabkha Matti showing distribution of oolitic sand

Although the dominant sediment type within Khor Odaid is fine quartz sand derived from the surrounding aeolian dunes, the S shores consist mainly of fine oolitic sands. Individual ooids are relatively small (300 μ) their thin cortex enveloping a rounded quartz grain, each ooid being regular in shape. These white oolitic sands constitute the seawards fringes of the sabkha and the adjacent 100-800 m wide intertidal and shallow infratidal sand flats whose surfaces are modified by very vaguely defined ripples oriented obliquely to the shore. These sand flats, submerged

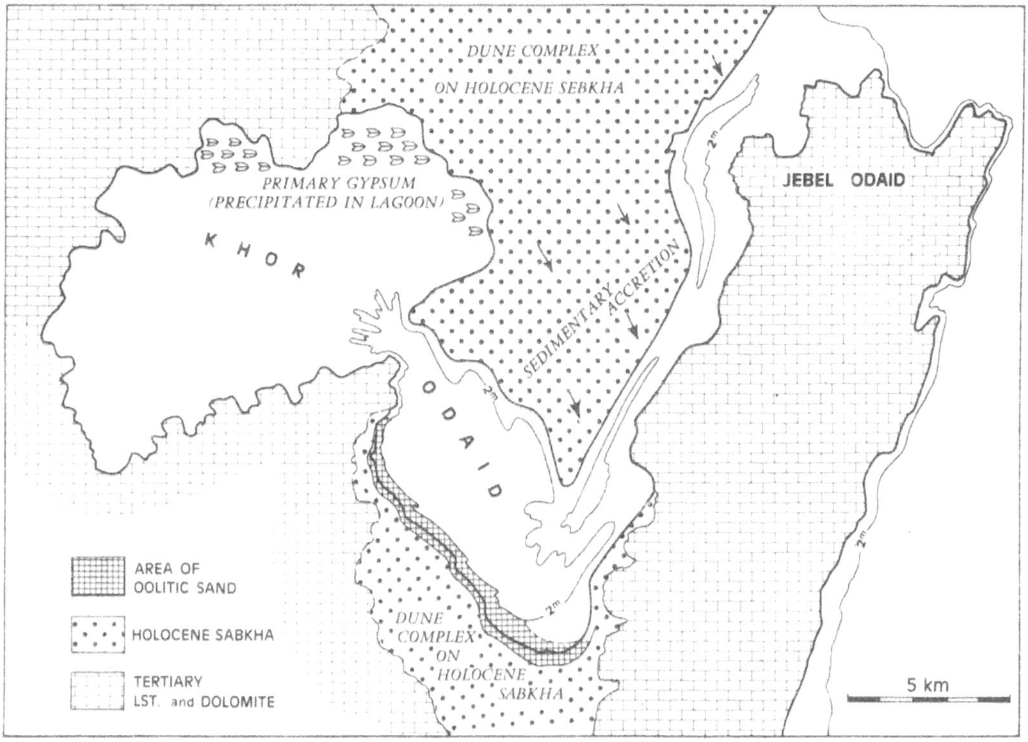

FIG. 10. Map showing distribution of oolitic and associated sediments within Khor Odaid lagoon, W. Abu Dhabi

to depths of about 1 m at full tide, terminate in a clearly defined accretion scarp whose base, 2-5 m below sea level, marks the rapid transition into muddy (carbonate) quartz sand rich in miliolid and peneroplid Foraminifera.

The localization of ooids within the intertidal and shallow infratidal zones probably indicates that most are forming within these areas. These oolitic sands continue laterally along the S shores of Khor Odaid for some 10 km (Fig. 10), being replaced locally by hard pellet sand. Accumulation of oolitic sand along the seawards edge of the sabkha has stimulated its progradation. However, although this sabkha measures several km in width, the oolitic sands constitute only its seawards margin. This clearly indicates that tneir formation is a relatively recent event post-dating the closing of the lagoon by lateral accretion of the quartz sabkha along its N shores (Fig.10). These ooids therefore are the product of a truely lagoonal environment. The degree of water agitation within the area formation is moderate, these S, windward, shores of Khor Odaid being affected frequently by small waves.

b) The coated grains W of Jebel Dhanna (Figs. 1,11) :

Coastal spits accreting westwards from the W side of Jebel Dhanna have almost completely enclosed a small coastal embayment most of whose floor is exposed at low tide. The main sediment type within this shallow lagoon is hard pellet sand which is covered by the habitual algal carpet near the upper limits of the intertidal zone. In the most protected part of the embayment, along the leeward side of the adjacent spit system, these pellet sands are replaced by beach rock and by a variety of coated grains. Virtually all coated grains occur in a narrow zone (some 5-10 m in width) which marks the upper limits of the intertidal zone and extends into the lower parts of the supratidal zone. These sediments are flooded by lagoonal waters and perhaps by waters fluxing through the coastal spit, especially during periods of strong onshore wind.

The coated grains vary in size from less than 1 mm to 30 cm and exhibit various morphologies. Relatively small grains (1-3 mm) with rounded nuclei are subspherical and in this respect comparable to the classical ooid morphology. Most grains, however, are irregular. This irregular external morphology most frequently reflects an irregular nucleus, individual laminae within the cortex maintaining a constant thickness on all sides of the nucleus (Fig. 16A); in this respect these grains differ from the classical ooids of the Abu Dhabi tidal deltas, which frequently exhibit a thinning of the cortex over local protuberances on the nucleus. Other grains have a highly irregular morphology which result mainly from local thickenings of the cortex giving the grain a "cauliflower" appearance. Although most are unattatched, others are firmly welded to the underlying beach rock pavement. The resemblance of these irregular forms to algal oncoids is very striking and it is evident that they cannot be described as "ooids". Is it stressed, however, that within a few square meters there is every gradation between ovoid (but unpolished) grains 1-2 mm in diameter, to highly irregular "cauliflower" forms measuring up to 30 cm. The latter invariably encrust fragments of beach rock. These aragonite incrustations are discussed in greater detail by Purser and Loreau elsewhere in this volume.

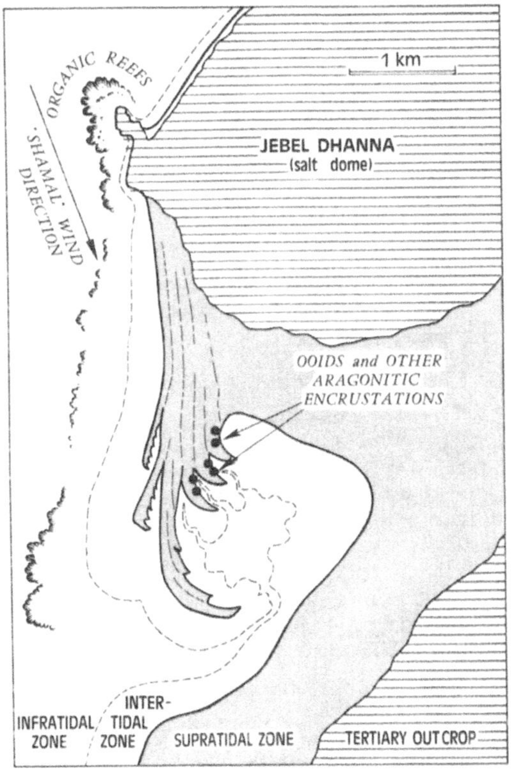

FIG. 11. Map showing the distribution of oolitic sediments within Jebel Dhanna lagoon, W. Abu Dhabi

c) The coated grains S of Umm Said (Fig.12) :

The SE Coast of Qatar Peninsula is a lee shore which has prograded laterally creating a 5-10 km wide Holocene Sabkha composed essentially of quartz sand. This sabkha, together with the pisolitic grains which are locally abundant on its surface, are described elsewhere in this volume by Shinn.

Coated grains,observed by Shinn,occur near the limit between the inter and supratidal zones close to the seawards edge of the evaporitic,sand sabkha. The environments in which they occur in many respects are comparable to those at Jebel Dhanna (discussed in the preceding section). It is not surprising therefore that the coated grains in these two areas are very similar in shape, internal structure and mineralogical composition. The coated grains at Umm Said, termed "pisoliths" by Shinn, range in size from 1-5 mm. They have unpolished surfaces and an external morphology reflecting the shape of the nucleus (Fig.16A). When the nucleus consists of a rounded quartz grain the morphology of the grain is typically ooid in character, but when the nucleus consists of an irregularly shaped, skeletal grain (which is most frequently the case) the shape tends to resemble that of an algal oncoid. The characteristic fibro-radial structure of these coated grains at Umm Said and Jebel Dhanna is discussed below.

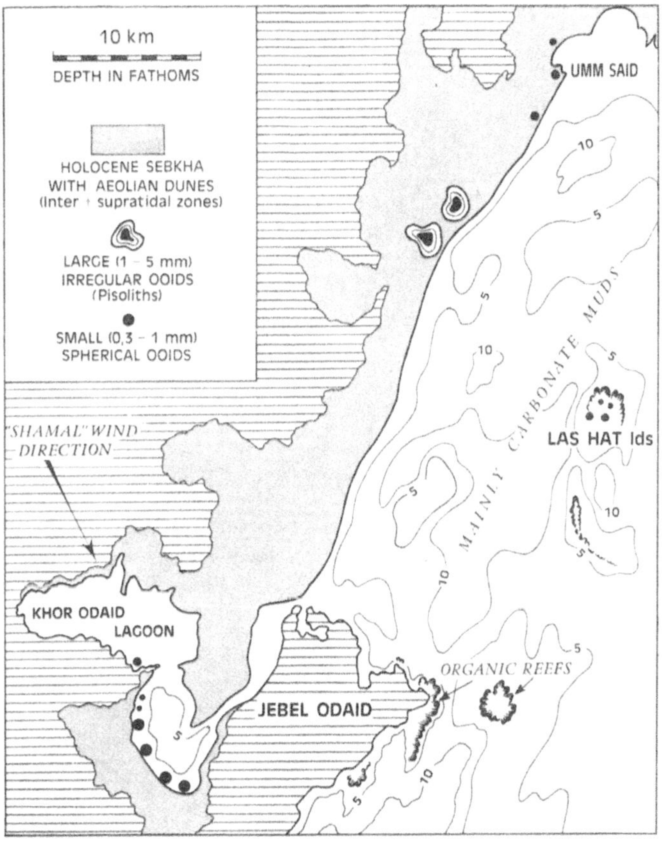

FIG. 12. Map showing regional setting of Umm Said and Khor Odaid lee coast and lagoonal ooids

DESERT OOIDS

Deserts bordering the shores of the Trucial Coast consist of low mesas of Tertiary limestone and ridges of oolitic limestone termed "miliolite". This latter formation is probably Pleistocene and appears to be a lithified aeolian deposit (see Evans et al, in this volume) comparable to the Pleistocene limestones which constitute many of the Bahamian islands. Much of Abu Dhabi and the adjacent Trucial Coast States also consist of unconsolidated aeolian dunes composed of both quartz and carbonate particles.

Dunes situated on the barrier islands of E Abu Dhabi consist, almost entirely, of ooids (as noted in a previous section) while the coastal plain near Sila (W. Sabkha Matti) in the extreme W corner of the Trucial Coast consists mainly of ooids for distances up to 4 km from the shore. The spectacular "sief" dune complex which covers much of the Rub al Khali was examined along the highway between Abu Dhabi and Buraimi (Fig. 1). Dune sediments in this area contain up to 50% rounded carbonate grains, including many ooids, for distances up to 40 km from the shores of the Persian Gulf. The widespread distribution of ooids within the deserts bordering the Trucial Coast embayment almost certainly is the consequence of on-shore shamal winds; ooids forming in the shallow coastal areas are transported to the adjacent beaches from which they are blown into the desert.

III. THE STRUCTURE AND ULTRASTRUCTURE OF PERSIAN GULF OOIDS

OOID MICROSTRUCTURE

The standard light microscope has been used mainly to facilitate the understanding of SEM studies. Is has aided in the separation of ooids from other grain types, the description of the ooid nucleus and the relative thickness of the cortex. Two principal types of ooids are readily defined : a widely distributed "classical type" whose cortex is composed of concentric laminae of tangentially oriented crystals alternating with laminae of cryptocrystalline, unoriented aragonite (Fig.13A,B). The second, somewhat unusual group, localized within the protected lagoon W of Jebel Dhanna and on the lee coast near Umm Said (Qatar) consist of alternating laminae of radially oriented aragonite and unoriented cryptocrystalline aragonite (Fig.16A).

1. Ooids having a "statistically tangential" structure :

In general, the form of the nucleus has little influence on the external morphology of the ooid, which tends to be ovoid or subspherical.

Within the tidal deltas the thickness of the cortex varies from several microns in small ooids (100-200 μ) which occur around the fringes of the delta where 95-99% of grains are coated, to 2/3 of the ooid radius within the larger ooids (0.5-2mm) on the levées bordering the axial channel where 100% of grains are coated. Ooids forming within the open coastal embayment near Sila(Sabkha Matti) have cortexes which measure 1/2 - 4/5 of the ooid radius. Oolitic sediments within this area include ca. 5% uncoated quartz or skeletal grains on the intertidal sand flats, 10% on the beach and 10-20% within the aeolian sediments some 4 km from the shore. Many ooids in this area exhibit empty or filled micro-perforations and surface depressions, those within the desert attaining 20-100 μ in diameter and are apparently circular in cross-section. They may communicate with each other and with the exterior of the grain.

The oolitic sediments of Khor Odaid lagoon contain ca. 80% coated grains and are characterized by a thin cortex which measures 1/5 of the ooid radius within 75% of the ooids and 3/5 within ca. 25% of ooids. The cortex of these small ooids is often perforated with empty micro-cavities 10-100 μ in diameter similar to those observed in ooids collected from desert environments near Sila.

The optical properties of the lighter-coloured laminae (seen in thin-slide) are comparable to those of Bahamian ooids. Thus, it was observed that Persian Gulf ooids exhibit a slight pleochroism, a black cross under polarized light and birefringence varying from one lamina to another within a given ooid between second order blue and yellow-orange. When measured in thin-slides (25-30 μ thick) this birefringence has a value of 0.030 ± 0.005, comparable to that measured by Illing (1954) for Bahamian ooids. The cause of this abnormally weak birefringence (standard aragonite being 0.155) of modern ooids is explained in "Appendix B" of this contribution.

The somewhat sombre layers which separate the light coloured laminae discussed above, together with the infillings of micro-perforations, have a cryptocrystalline structure which has no reaction under cross-polarized light. Treatment with 1% acetic acid containing methylene blue indicates that the clear laminae of oriented aragonite consist of very fine, alternating laminae of carbonate and organic material; within the cryptocrystalline layers the organic phase is diffused but more abundant.

2. Ooids having a "statistically radial" structure :

The shape of the nucleus determines the external morphology of the ooid (Fig.16A). The cortex is generally well developed, constituting 2/3 - 9/10 of the radius within ooids measuring 1-5 mm in diameter. It consists of grey or light brown laminae alternating with clear, unpigmented layers, individual laminae often being relatively thick (10-100 μ). In thin slides having a thickness of only 15 μ the more transparent laminae are seen to consist of micro-crystals 5-50 μ in length whose orientations are statistically radial,while the more sombre laminae consist of cryptocrystalline carbonate. In cross-polarized light the black cross, very rarely present, is observable only in the external laminae.

OOID ULTRASTRUCTURE : Morphology, mineralogy, chemical composition and cristallography of the particles of the cortex. (For definition see "Appendix A")

1. Gross mineralogical analysis :

Determination of the gross mineralogy is necessary in order to establish the aragonitic nature of the cortex of ooids having radial structure,as well as the composition of certain nuclei (quartz, dolomite, anorthose). It has demonstrated, almost invariably, the presence of both calcite and magnesian calcite. The calcite/aragonite ratios (Table I) were established as follows :

Standard samples of pure calcite and aragonite (Collection, Muséum National d'Histoire Naturelle) were powdered and mixed mechanically in varying proportions (1/99, 2/98, 3/97, 4/96, 5/95, 7.5/92.5, 10/90, 15/85 and 20/80). The curve demonstrating the concentration of calcite within the mixture was constructed as a function of

$$\frac{\text{Principal calcite peak}}{\text{"Background noise"}} \Big/ \frac{\text{Principal aragonite peak}}{\text{"Background noise"}}$$

; the error for calcite is ± 2% (or ± 3% when there is a peak of magnesian calcite); the maximum relative error for aragonite content being 3.5%.

2. Results based on electron microscope, X-ray diffraction and micro-analysis by X-ray energy emission spectrograph :

a) The cortex and surface of ooids having tangential structure : The SEM reveals two fundamental ensembles :

<u>Firstly</u>, a moderate-compact felt (Fig.13D,14A,B) of elongate, cylandric, anhedral grains with blunt ends (Fig.13D,14A) or subhedral needles with pointed ends (Fig.14B). These crystals measure 0.5-2 μ in length and 0.15-0.3 μ in diameter. While their orientation is <u>statistically</u> tangential with respect to the laminations, individual crystals may deviate from the tangential (observed on the surface of the grain,or on rare fractures provoked by brief ultra-sound treatment in distilled water). An average of 80-90% of individual crystals are oriented + 30° or - 30° with respect to the tangent while 10-20% have a more acute orientation and may attain a radial orientation, especially within the external lamina (Fig.15E). On certain external or fractured internal surfaces the orientation of these crystals seems to be haphazard, although one can not be certain that this fabric has not been induced during sample preparation.

FIG. 13

A. Ooids from levée adjacent to axial channel within tidal delta, showing brownish or grey cryptocrystalline zones alternating with clear, chamois-coloured laminae with preferred orientation.

(thin slide, X 29)

B. As for A; cross-polarized light, (X 29).

C. Ooid from internal part of tidal delta (SEM, X 54).

D. Surface of ooid (Fig.13C), showing slightly blunted battons of aragonite whose orientation is statistically tangential to the surface of the ooid. (SEM, X 20,700).

E. Ooid collected from desert near Sila, W. Abu Dhabi, (SEM, X 54).

F. Surface of ooid (Fig.13E) showing variable morphology of constituent aragonite particles, including blunt battons and subspheric nanograins. (SEM, X 14,400).

G. Surface of ooid collected from beach on Sadyat Island (E. Abu Dhabi) showing mucous (somewhat dessicated) similar to that observed on numerous ooids from divers localities. (SEM, X 3,600)

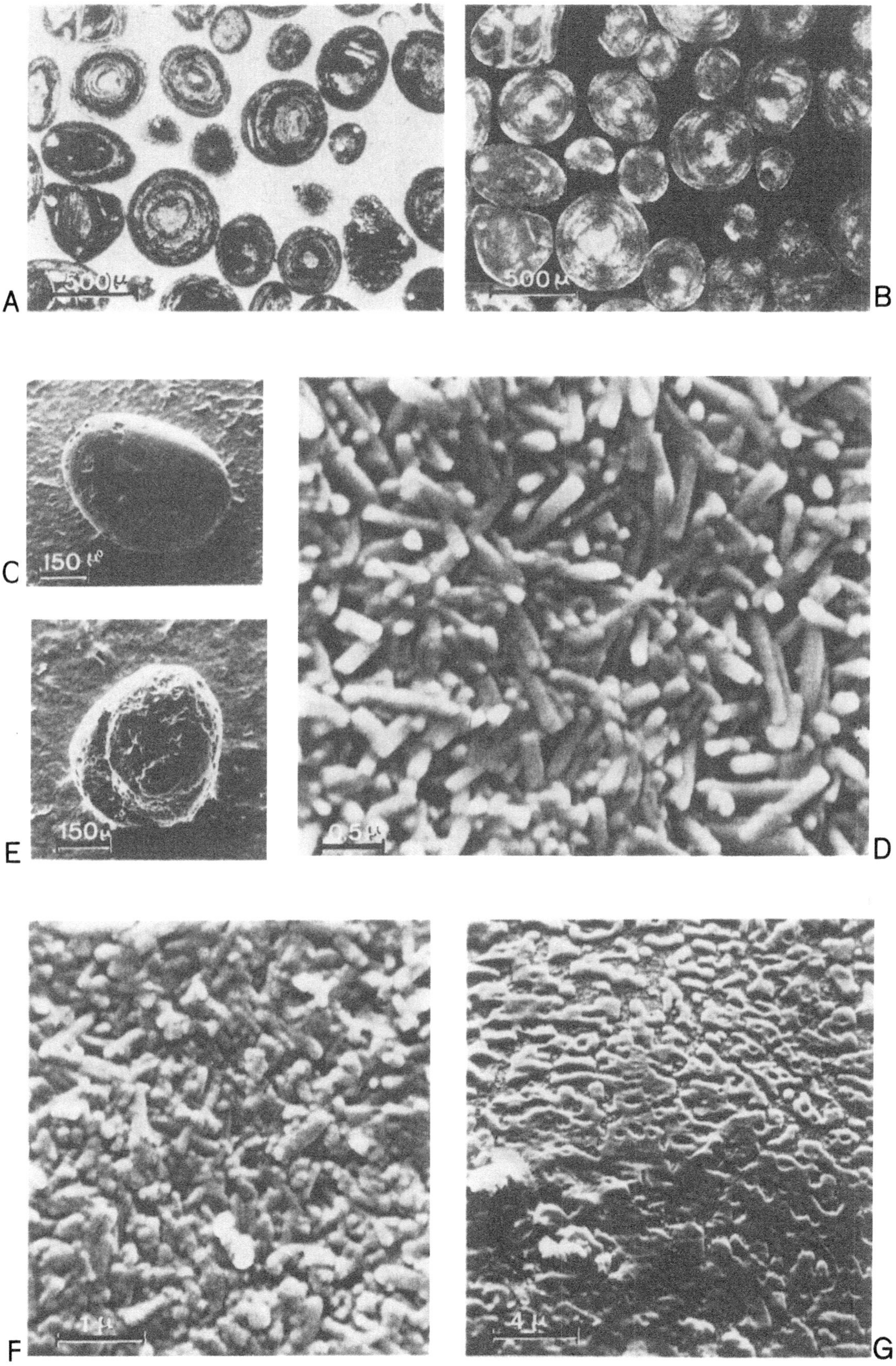

Legend to Fig. 13 A-G on the preceding page

FIG. 14

A. Surface of ooid from external part of tidal delta showing loosely packed felt of anhedral or subhedral battons with a statistically tangential orientation. (SEM, X 14,400).

B. Broken surface (directly below external lamina) of ooid from inner part of E. Abu Dhabi tidal delta showing porous felt of subhedral or anhedral, slightly oriented needles of aragonite. (SEM, X 18,000).

C. Section through ooid collected in desert near Sila (fresh surface cleaned by 5 seconds of ultra-sound treatment) showing statistically tangential orientation of the elongate, slightly worn battons of aragonite. (SEM, X 18,000).

D. Surface of ooid collected from beach berm at Sila (Sabkha Matti) showing assemblage of subspheric nanograins. (SEM, X 14,400).

E. Fractured surface of ooid from inner part of E. Abu Dhabi tidal delta showing an alternation of nanograins and elongate anhedral crystals with statistically tangential orientation. (SEM, X 15,750).

F. Surface of ooid from inner part of E. Abu Dhabi tidal delta showing a degree of coalescence of the constituent particles whose original morphology is uncertain. (SEM, X 14,400)

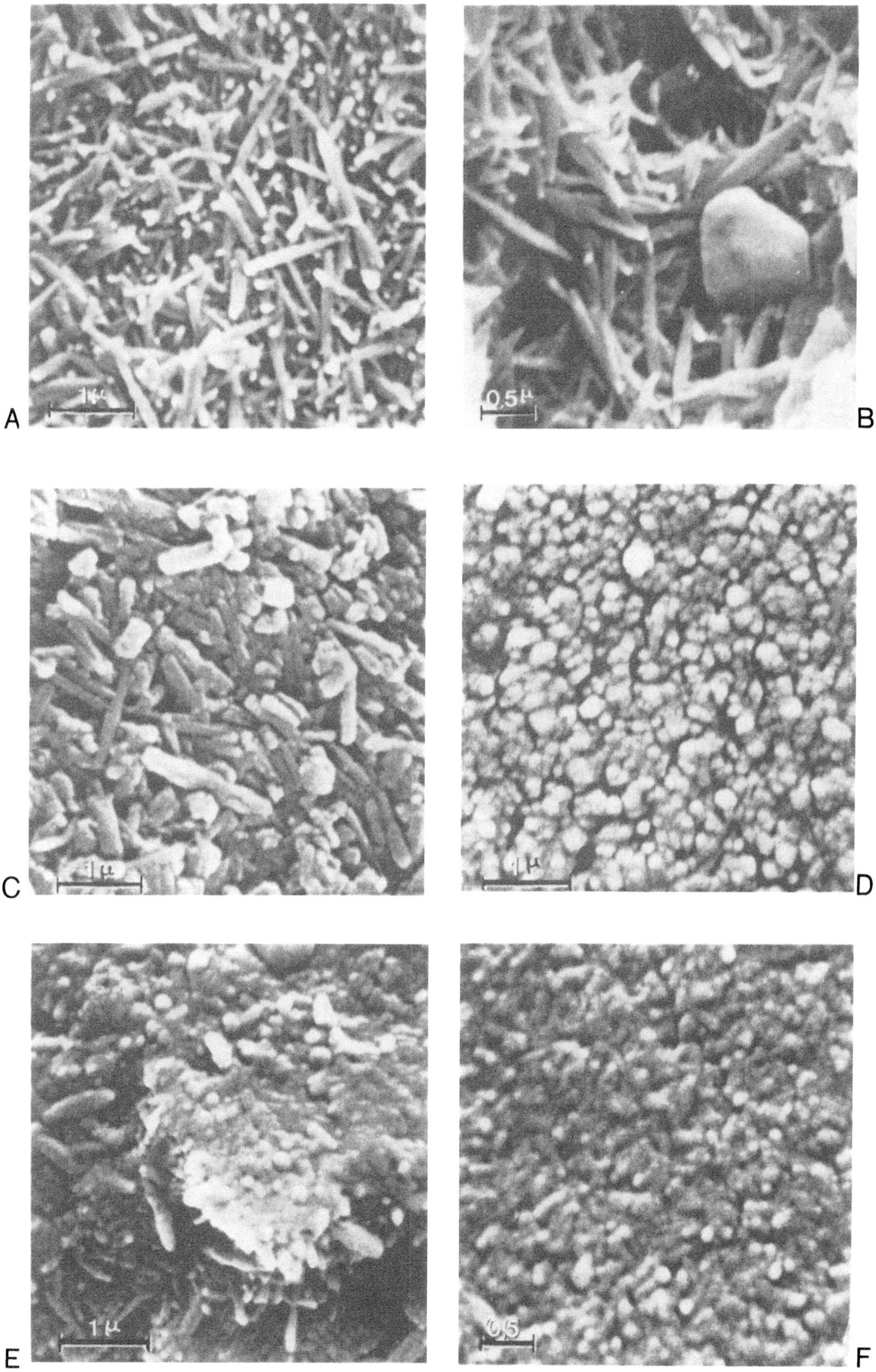

Legend to Fig. 14 A-F on the preceding page

FIG. 15

A. Section through cortex of ooid from desert at Sila(Sabkha Matti) showing relatively large, (10-50 μ) mutually interconnected vacuoles communicating with the surface of the ooid. (SEM, X 360).

B. As above, showing coalescence of constituent particles around edge of vacuole. (SEM, X 7,200).

C. Surface of ooid from inner part of tidal delta showing micro-perforations developed tangentially below external layer of the cortex (SEM, X 450).

D. Filling of micro-cavity or perforation by anhedral crystals within an ooid from beach, Sadyat Id. (SEM, X 1,440).

E. Surface of ooid collected from protected depression behind emergent coastal bar near inner part of tidal delta, Sadyat Id, showing radial crystal growth direction. (SEM, X 14,400).

F. Crystal "whiskers" on surface of ooid collected from internal part of E. Abu Dhabi tidal delta. (SEM, X 18,000)

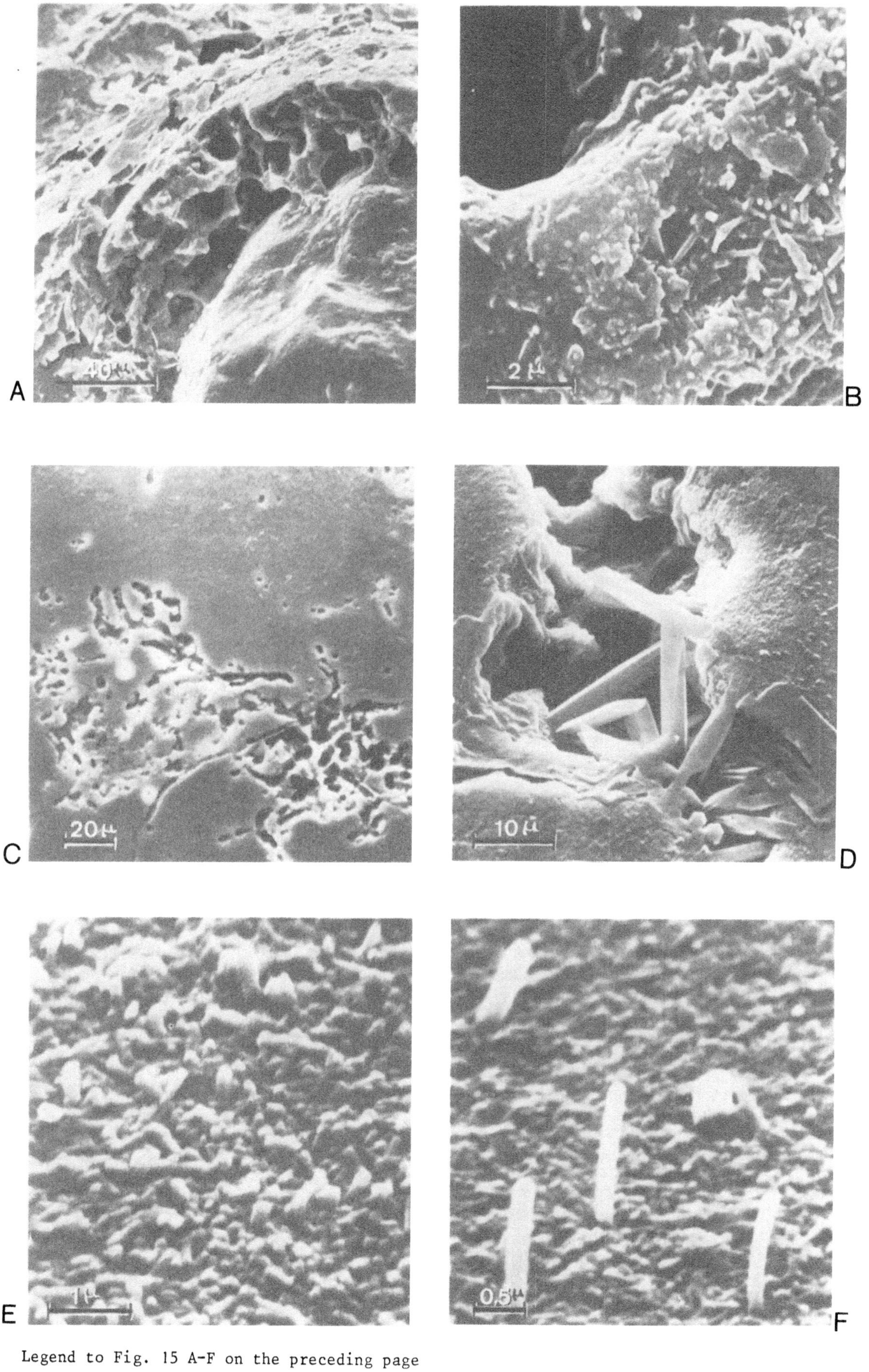

Legend to Fig. 15 A-F on the preceding page

FIG. 16

A. Large ooids collected from tidal flats near Umm Said (SE Qatar) showing alternation of greyish, non-pigmented, cryptocrystalline zones and thinner laminae of clear, microcrystalline aragonite with radial orientation. (Thin-slide,X 9).

B. Section through ooid (from Umm Said) showing radial orientation of component crystals. (SEM, X 90).

C. Section (Fig.16B) showing alternation of nanograined and fibro-radial fabrics. (SEM, X 1,440).

D. Section as in Fig.16C showing anhedral or tightly packed subhedral microcrystals with radial orientation. (SEM, X 720).

E. Section as in Fig.16D showing isometric, anhedral nanograins or intermediate-sized battons overlain by layer of radially disposed microcrystals with blunted extremities. (SEM, X 7,200).

F. Same section showing more or less compact assemblage of isometric nanograins within cryptocrystalline lamina. (SEM, X 14,400)

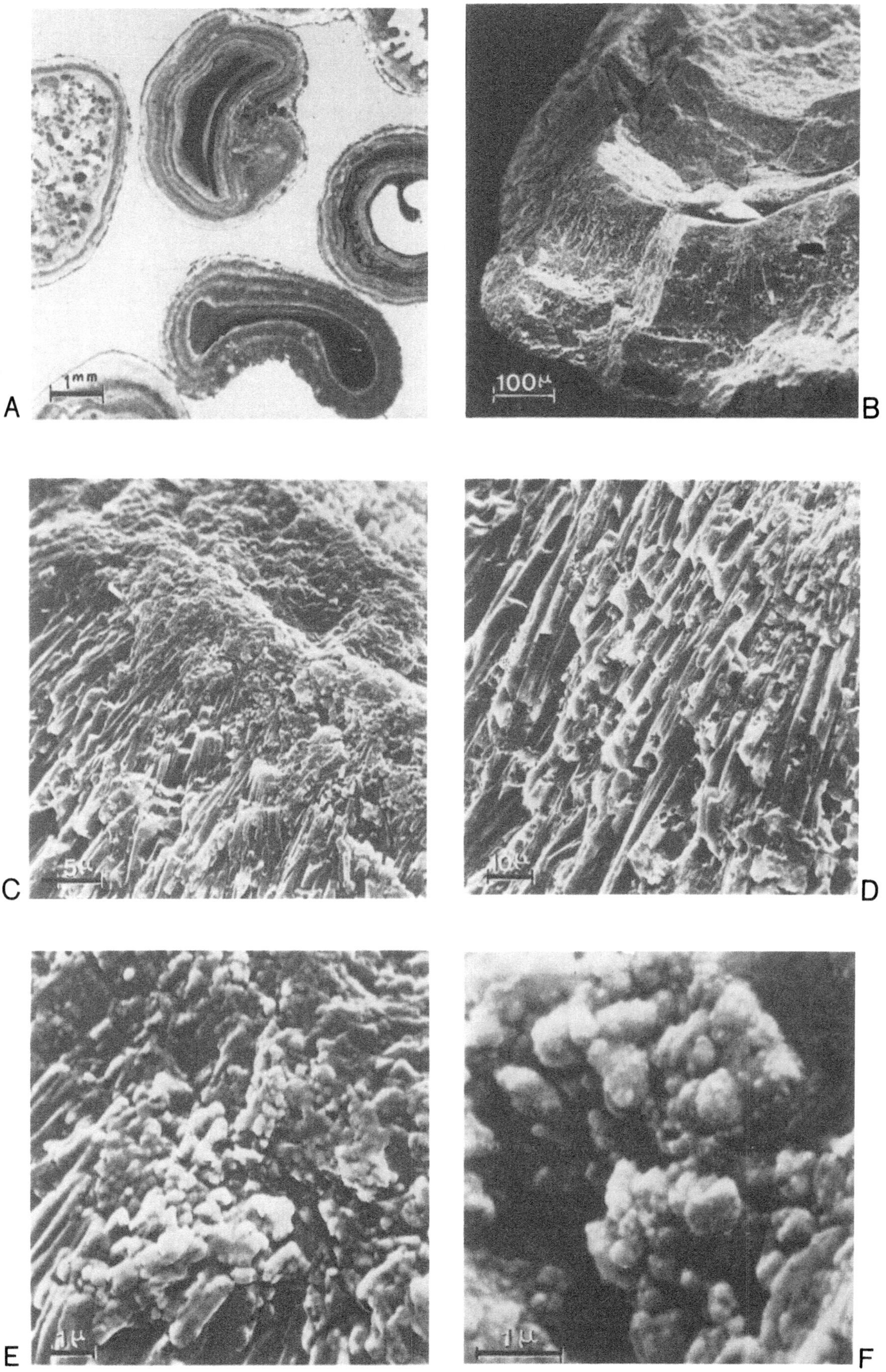

Legend to Fig. 16 A-F on the preceding page

FIG. 17

A & B. Anhedral and subhedral microcrystals within ooid from Umm Said exhibiting statistically radial orientation. (SEM, X 900 and 1,800).

C. Section through ooid from lagoon at Jebel Dhanna showing compact assemblage of isometric nanograins within cryptocrystalline lamina. (SEM, X 22,500).

D. Surface of an ooid from Umm Said tidal flat; the growth of the constituent anhedral and subhedral crystals normal to the surface is comparable to that of crystal growth in many beach rocks. (SEM, X 4,050).

E. Crystal growth within a pore between laminae of ooid from Umm Said showing crystal orientation statistically normal to the substrate i.e. statistically radial with respect to the cortex. (SEM, X 3,600).

F. Surface of ooid from lagoon at Jebel Dhanna showing statistically radial crystal orientation; the extremities of individual crystals are worn (?) or truncated, and tend to coalesce. (SEM, X 7,200)

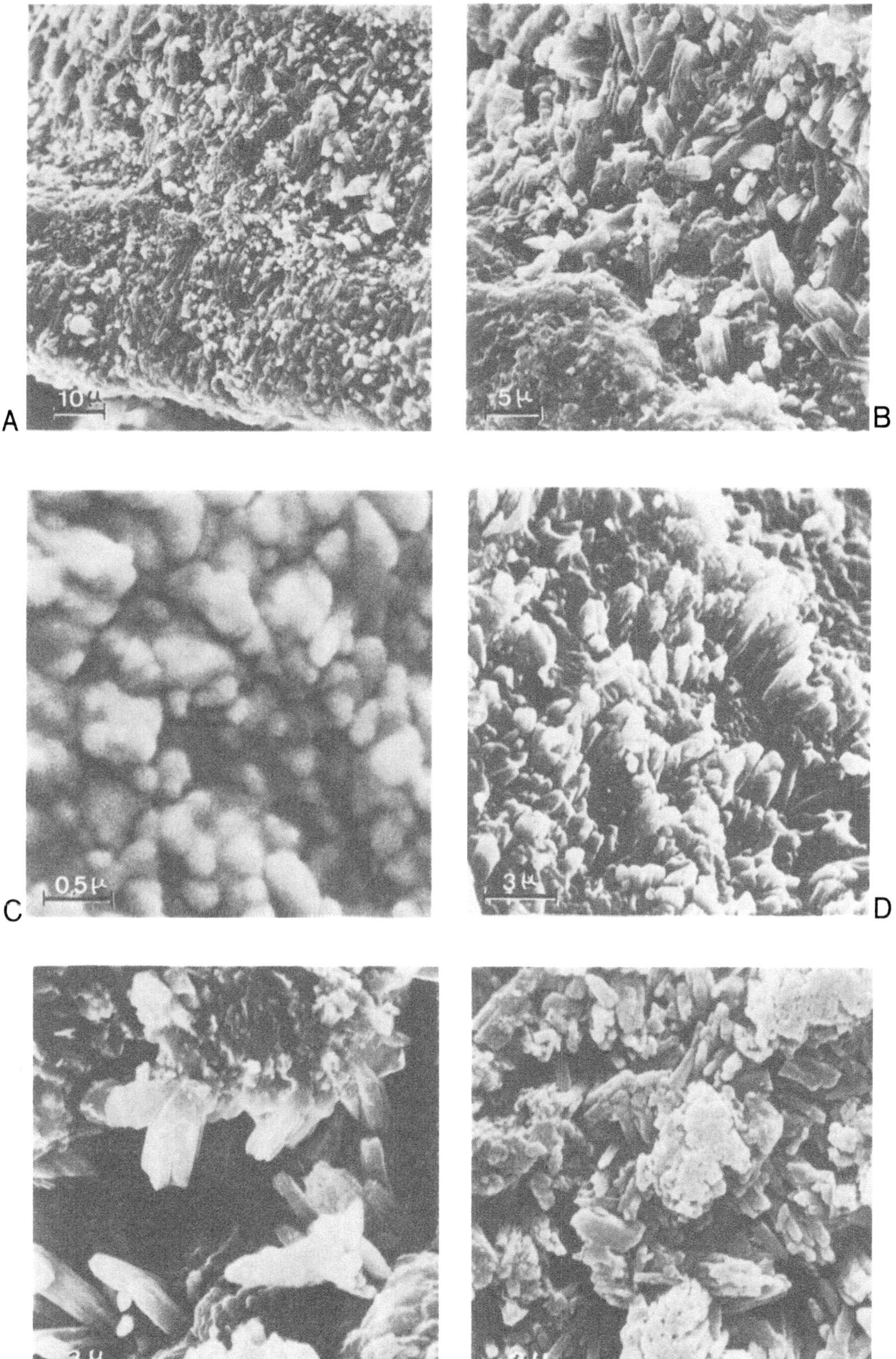

Legend to Fig. 17 A-F on the preceding page

TABLE I . MINERAL AND CHEMICAL ANALYSES OF OOIDS

		Bulk Mineral Analyses						Carbonate content soluble in cold acid	Mg^{++} and Sr^{++} concentrations in this carbonate		Strontium concentration in the aragonite
		Calcite/Aragonite		presence							
				Mg Calcite	Dol.	Quartz	Feld.	in %	Mg^{++} in ppm	Sr^{++} in ppm	in ppm
Ooids from Abu Dhabi tidal delta	1	1-3	99-97	+				95,4	3200	8860	9040
	2	10-12	90-88	+		+		83,5	7300	8350	9400
	3	5-7,5	95-92,5	+		+		91,7	3700	9170	9750
	4	3	97	+				95,1	2900	9080	9370
	5	5	95	+				94,5	6200	7880	8290
	6	8	92	+		+		94,0	4200	8130	8850
	7	4	96	+				94,5	3500	8630	9000
	8	2	98	+				94,7	3300	8920	9100
	9	5	95	+				90,8	4300	8640	9100
average											9100 ± 800
Ooids from open embayment at Sila (Sabkha Matti)	10	8	92	-	+	+	+	70,2		8150	8850
	11	15	85	+	+	+	+	54,9		7050	8300
	12	10	90	-	+	+	+	75,5		8090	8300
	13	5	95	-	+	+	+	62,8		9350	9840
	14	7	93	+		+	+	71,7		8650	9300
	15	3	97	-		+		76,5		8260	8500
	16	4	96	-		+		58,1		8910	9300
Adjacent desert (Sila)	17	5	95	-	+	+	+	49,4		8900	9380
average											9060 ± 800
average (after Kinsman 1969)											9590 ± 500
Ooids having radial structure	18	5	95	+	+	+		72,9		8380	8820
Bahamian ooids	19	2	98	+				95,1	2600	9710	9900
	20	1	99	+				95,4	2100	9510	9620
average											9760 ± 150
average (after Kinsman 1969)											9800

Persian Gulf lagoonal muds	21	10	90	+	71,3	3200	8470	9420
	22	5	95	+	86,8	10200	9560	10060
Hard Pellet sand	23	1	99	+	90,2	5700	9830	9930
Soft pellets	24	6	94	+	81,8	12400	9050	9620
average								9750 ± 330
average (after Kinsman 1969)								9360 ± 500

Note : High magnesian calcite is based on the presence of an additional peak between the main calcite ray and the main dolomite ray

Examination of these crystals by transmission E-microscope and by electron diffraction has confirmed their aragonitic composition, monocrystalline structure and the orientation of the C-axis parallel to the long axis of the crystal. Thus, the C-axis is statistically tangential - or statistically normal to the ooid radius - but its orientation may vary through 360° around this radius. For these reasons, X-ray diffractions on isolated ooids fail to demonstrate preferential crystal orientation. These orientations also contribute to the abnormally weak birefringence of the aragonitic cortex (see Appendix B).

Secondly, a compact assemblage of nano-grains (Fig.14D), each measuring 0.1-0.6 μ in diameter (average 0.25 μ), whose individual shapes are subspherical and therefore have no preferential orientation. Transmission E-microscope and electronic diffraction indicate that 80-90% of these elements are polycrystalline, one crystal within each grain generally being considerably bigger than the other non-developed nuclei. 10-20% of the nanograins are monocrystalline, measuring 0.1-0.3 μ while forms intermediate between nano and elongate also have been noted.

The two fundamental types of grains alternate (Fig.14E) within the cortex, as in Bahamian ooids. On the surface of certain ooids the packing of these particles is so tight that the outer layer (approx. 1 μ in thickness) tends to be coalescent and the initial morphology of particles cannot be distinguished (Fig.14F). An additional modification also occurs within the cortex of worn desert ooids (collected from near Sila, Sabkha Matti) whose external surfaces often exhibit tightly packed micro-crystals with blunt ends, certain being intermediate between elongate needles and sub-rounded nano-grains (Fig.13F). The aragonitic particles around the periphery of micro-perforations (measuring several tens of microns in diameter (Fig.15A)) tend to coalesce (Fig.15B). A very thin layer of mucous similar to that reported by numerous authors, has also been noted on the surface of ooids collected from the shallow infratidal areas (Fig.13G).

Perforations traversing laminae or localized along planes between laminae, are not uncommon (Fig.15C). Measuring several microns in diameter, they are more or less sinuous and bifurcating , and resemble the filamentous structures described by Dangeard (1936) and Nesteroff (1956). In the Persian Gulf ooids, however, these organisms appear to be perforant and not constructive.

The external surface of many ooids support isolated, euhedral or anhedral crystals (Fig.14B) exceeding 1 micron in size in addition to arcuate, whisker-shaped crystals (1-8 μ x 0.2-0.4 μ) oriented normal to the surface (Fig. 15F). They are invariably present, irrespective of whether or not the ooids are washed in distilled water, or the surfaces coated with gold, platinum, carbon, copper, or are uncoated, thus excluding the possibility that they are metallic growths originating during the preparation of the sample. Analysis by X-ray energy emission spectrometer coupled with the SEM has also eliminated the possibility that these curved crystals are NaCl. In spite of the fact that the Na peak could not be measured because of feeble concentration, tests show that the Cl peak may be measured even when present in very minor amounts; it was not present in the analyses of these crystals. On the contrary, these analyses demonstrated a Ca peak similar to that of the remainder of the cortex and the whiskers must therefore be regarded as calcite or aragonite.

b) The cortex and surface of ooids having radial structure :

Their cortex consists of alternating layers having the following characteristics :

i) - closely packed euhedral-subhedral crystals measuring 10-50 μ in length and 2-5 μ in diameter, which tend to taper towards pointed or rounded extremities. Their orientation with respect to the nucleus is normal (Figs.16B-D).

ii) - a mass of densely packed, relatively short (3-15 μ x 1-3 μ) crystals with flat terminations oriented radially or somewhat oblique (30-40°) to the radius (Figs.

17A,B). Figure 17D illustrates this fabric within the external layer of an ooid from Umm Said Sabkha, which exhibits a striking similarity to certain inter-granular marine cements, while Fig. 17E demonstrates it in a pore between adjacent layers within the cortex. The packing of crystals within the outer layer of the cortex is sometimes loose, the component crystals being statistically radial and tapered towards their extremities which are often worn or truncated (Fig. 16E).

iii) - a compact assemblage of sub-spherical nano-grains 0.1-0.6 μ in diameter (Figs. 16F, 17C), resembling those within ooids having tangential structure. The transition zone between layers of fibrous and sub-spherical grains often includes blunted, elongate crystals (Fig. 16E).

c) Chemical analysis of the carbonate fraction and strontium concentrations within the aragonite :

Ooids selected by hand were washed repeatedly in distilled water, dried at 105° (to establish dry analysed weight) and etched with cold HCl (I.N). The chloride solution was filtered to retain organic matter and the filter carefully washed. It is hoped that the Ca^{++}, Sr^{++} and Mg^{++} within the solution thus obtained are derived from the carbonate and that the fraction contributed by the interstitial water and organic matter within the grain is negligible.

The carbonates are measured by CO2 titration (error 1%). The previously established calcite/aragonite ratios (error3.5%) permit the calculation of the % carbonate contributed by the aragonite and the Sr. amount measured by Atomic Absorption (error 1%) is attributed to the aragonite (see Table I). The maximum error of this method is 5-6%, but the presence of Sr. within associated magnesian calcite may lead to Sr. calculations in excess for the aragonite. It is noted, however, that the Sr^{++} concentrations measured are slightly less than those recorded by Kinsman (1969) in Persian Gulf ooids and it is suspected that this difference is due essentially to dilution by ooid nuclei. Nevertheless, analyses of ooids from Sila (Sabkha Matti),whose nuclei are essentially non-carbonate, have yielded Sr. concentrations comparable to those of ooids from the Abu Dhabi tidal deltas whose nuclei are mainly aragonitic pellets (with Sr. concentrations of 9750 $\pm$ 300 ppm), coral fragments (7700 ppm according to Kinsman) and occasional gastropod fragments (2000 ppm). Thin slides indicate that the varying proportions of these different types of nucleus are such that their various Sr. concentrations would tend to compensate one another.

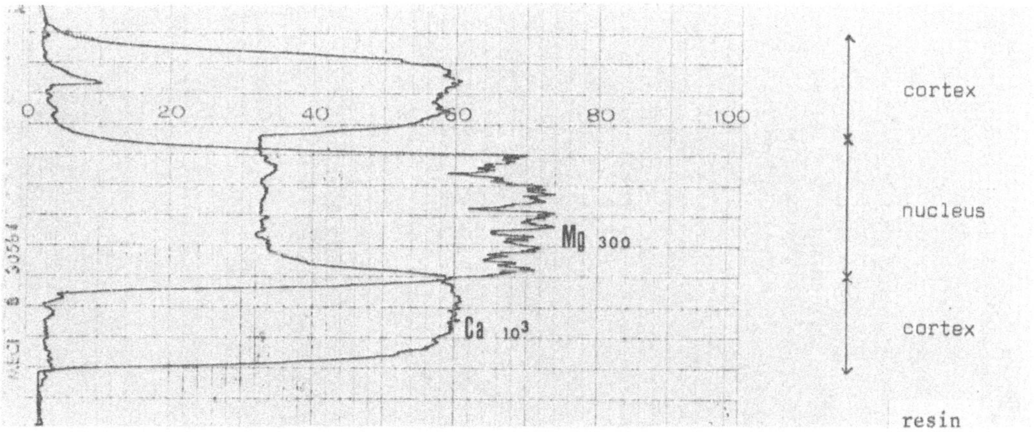

FIG. 18. Distribution of magnesium within one ooid demonstrated by microprobe, showing its substitution for calcium within the magnesian calcite nucleus (a foraminiferid)

Table 2 . Table of birefringences calculated for modifications due to α maximum angular displacement with respect to the tangent (25 and 30°), and for aragonitic thickness T_{Ar} of 89 to 95% of total thickness T of the slide

α	Bi_{max}	Decrease in birefringence due to angular displacement of crystal axes with respect to the tangent	Thickness of the aragonite	Calculated birefringence	Decrease in birefringence due to the haphazard orientation of the optic axes within the plane statistically tangential	Decrease in birefringence due to porosity
30°	0.116	0.039	100% (theoretic)	0.032	0.084	
		(= 25%)	T_{Ar} = 89% T = 26,7 µm	0.028	(54%)	0.004
			93% T = 28 µm	0.030		0.002
			95% T = 28,5 µm	0.031		0.001
25°	0.127	0.028				
		(= 18%)	100% (theoretic)	0.037	0.090	
			89% T = 26,7 µm	0.033	(58%)	0.004
			93% T = 28 µm	0.034		0.003
			95% T = 28,5 µm	0.035		0.002

The average Sr. concentration within the aragonitic cortex of ooids from the tidal deltas of Abu Dhabi is 9100 $\pm$ 800 ppm, that of ooids from the open embayment and beach at Sila (W end of Sabkha Matti) 9060 $\pm$ 800 ppm, ooids from the adjacent desert some 4 km from the coast having similar concentrations. Ooids with radial structure from the sabkha at Umm Said (SE Qatar) contain approximately 8800 ppm.

Carbonate muds from the Abu Dhabi lagoons and ooids from the Bahamas were analyzed (Table I) for comparison. They indicate that the cortex of Persian Gulf ooids have Sr. concentrations slightly inferior to those of the adjacent muds and pellets and significantly inferior to those of Bahamian ooids (9760 $\pm$ 150 ppm). The concentration of Mg is much more variable than that of Sr and seems to be closely related to the presence of magnesian calcite.

Micro-probe measurements were made in order to evaluate whether the two inter-laminated aragonite facies within the cortex (elongate crystals and nano-grains) contained different amounts of Sr, Mg and Ca. Although results are somewhat limited it is evident, nevertheless, that the presence of Mg coincides with a reduction in Ca (Fig.18), especially in ooids having a nucleus consisting of a foram or echinoderm fragment (composed of magnesian calcite). The Mg concentration within the aragonite cortex was too feeble to be detected and the lack of concentric zonation of Sr suggests that the two aragonite facies have similar concentrations.

3. Relation between ooid microstructure and ultrastructure.

a) Irrespective of the type of ooid, zones consisting of preferentially oriented, elongate crystals correspond, almost certainly, with the clear laminae (exhibiting optic orientation) seen in thin slide, while the laminae composed of nano-grains or non-oriented rods, correspond to the darker, cryptocrystalline laminae visible under light microscope.

b) Illing (1954) has suggested that the abnormally low birefringence in Bahamian ooids is due to the combined effects of imperfect crystal orientation and numerous pore spaces within the cortex, but did not demonstrate these phenomena.

In the light of the observations outlined in the preceding pages the abnormally low birefringence may be explained mathematically (for details see Loreau (1972), and "Appendix B" of this contribution) by taking into account the height of the pore space traversed by the light rays, the imperfect orientation of the constituent aragonite crystals with respect to the tangent of the ooid, and, especially, the haphazard orientation of these crystals in the statistical tangent plane - thus a haphazard orientation with respect to the direction of the light rays. The different calculated values of birefringence are listed on Table 2 : all are very similar to the birefringence determined directly by microscope, thus confirming these hypotheses.

c) The polished external surfaces of many ooids generally reflect a degree of coalescence of the component crystals whose initial morphologies are difficult to distinguish.

IV. RELATIONSHIP BETWEEN OOID STRUCTURE AND THEIR ENVIRONMENT OF FORMATION

THE PRINCIPAL CHARACTERISTICS OF THE NUCLEUS WITHIN A GIVEN ENVIRONMENT

There are two ooid provinces on the Trucial Coast, each having distinct nuclei. In the vicinity of the coastal barrier in E Abu Dhabi (tidal delta), beaches and associated aeolian dunes), isolated by lagoons from the adjacent continent, the ooid nucleus is generally carbonate, often in the form of a pellet. Lateral facies transitions to pellet sands indicates that these nuclei are essentially autochthonous. The rareness of quartz, feldspar and detrital dolomite nuclei demonstrate the ineffective nature of aeolian transport towards the sea in this area.

Table 3 . Relationship between the structure of the ooid cortex, especially the outermost lamina, and the nature of the environment

	Ooids characterized by radial structure	Ooids characterized by tangential structure			
Locality	Umm Said (SE Qatar) Jebel Dhanna (W. Abu Dhabi)	Sila, Sabkha Matti and tidal delta between Sadiyat Id. and Ras Gharab (NE Abu Dhabi)			Tidal delta between Sadiyat Id and Ras Gharab (NE Abu Dhabi)
Tidal zone	Limit between inter and supratidal	lower intertidal	intertidal and shallow infratidal		
Degree of water agitation	low, due mainly to protection by adjacent spits	moderate	moderate-high		high - very high (water locally milky)
Morphology of the milieu	essentially flat sabkha surface or minor depressions behind beach spit	shallow depression protected between major ripples	interbar depression  occasional grass	bar or levée; tidal channel	delta front (seewards)
Structure of the outermost layer of cortex	radial (95%) occasionally with slight abrasion of external crystal points	loose to more or less dense felt of elongate rods: occasional subhedral nanocrystals with radial orientation; or nanograined	more or less tangential orientation; or nanograined	generally dense felt of blunt rods with tangential orientation; crystals often with a compacted or coalescent appearance; or nanograined	loose felt of blunt rods

Table 4 . The relationship between cortex structure, especially the external layer, and divers beach and dune environments

Nature of the environment	Beach berm nearly always emergent	aeolian dune on barrier island (Sadiyat, NE Abu Dhabi)	aeolian dune in desert
Structure of the cortex	alternating laminae of crystals having statistically tangential orientation		damaged cortex with pitted surface, numerous (dissolution?) pores with coalescent lining, different from those produced by endolithic organisms.
Structure of the outer layer of cortex	usually nanograined, occasionally coalescent.	nanograined, coalescent rare shock-pits, slight abrasion of constituent elongate crystals. mixed.	abrasion of constituent surface crystals which have morphologies varying between elongate rods and nano-grains.

In W. Abu Dhabi, at Sila (Sabkha Matti) and within the lagoon of Khor Odaid, (Figs.9,10) nuclei are composed mainly of quartz and other terrigenous grains. These allochthonous nuclei probably result from the close proximity of the ooid-forming environment to the continental shore, there being no coastal barrier system.

THE DOMINANT STRUCTURE OF THE CORTEX (PARTICULARLY THE OUTER LAYER) WITHIN ANY GIVEN ENVIRONMENT :

1. Relationship between cortex thickness and environment :

Within the tidal deltas of E Abu Dhabi ooids on the seawards fringe are small (80-250 μ) and of two types. The dominant type contains a pellet nucleus and has a cortex only several microns in thickness; the second group has variable nuclei and a cortex which may attain 2/3 of the ooid radius. The ooids within the very shallow levées in the axial parts of the delta are considerably bigger (2 mm) and have a cortex measuring 3/4 - 4/5 of the radius. Between these two extremes, ooids have an average diameter of 500 μ and a cortex which is 1/5 - 3/4 of the ooid radius. Ooids from the open embayment at Sila exhibit a progressive increase in cortex thickness as one traverses the 500 meter wide intertidal sand flats towards the beach.

2. Relationship between the structure of the cortex and the milieu :

a) Intertidal and shallow infratidal zones : It is significant that the radial structure of certain ooids is present only in very protected settings. Ooids having several layers of radially oriented aragonite crystals, including the outer layer, are localized near the limit between the inter- and supratidal zones, an environment which is more humid than wet. On the W side of Jebel Dhanna (Fig.11) where they are protected from the open sea by a coastal spit system, these ooids are associated with aragonitic tufa and various types of irregularly coated grains resembling algal oncoids (discussed by Purser and Loreau in this volume). Ooids forming in protected depressions between mega-ripples within the tidal deltas, although having a statistically tangential structure, may also exhibit radially oriented, elongate, subhedral crystals on their surface (Fig.15E and Table 3).

The relationships between the tangential structure and the milieu is more difficult to establish. Is is possible, however, that the dominant structure of the outer layer is related to the environment in which the ooid was collected, irrespective of its genesis or degree of abrasion (Table 3). Thus, at Sila (W. Sabkha Matti) and within the tidal deltas where the sediments are arranged in series of mega-ripples and depressions (often with grass), the ooids from the depressions very frequently exhibit an external layer of loosely packed needles. On the adjacent ripple crests, however, the felt of needles is more densely packed suggesting a degree of mechanical compaction. The somewhat blunted and worn appearance of these external crystals may be due to abrasion or to local dissolution of their edges (which are less stable than their faces). Within the axial channel and adjacent levée the external layer of the cortex tends to have a coalescent structure. Finally, along the front of the delta the felt of crystals is relatively loose but it is not certain whether this is the result of construction or destruction of the structure of the cortex; the presence of crystal morphologies intermediate between needles and nanograins suggest abrasion. In this case the slightly muddy appearance of the water localized along the fringes of the delta might be due to abrasion of ooids rather than to precipitation.

b) Desert ooids (Table 4) : all ooids which have suffered aeolian transport possess alternating laminae similar to those of ooids having tangential structure. Those collected from the highest parts of the beach (on Sadiyat Id; and near Sila) have an outer layer which is essentially nanograined or sometimes tending to coalesce. Ooids from the adjacent aeolian dunes are similar but their constituent particles show traces of abrasion which is considerably advanced in ooids collected from the desert, 4 km from the shore at Sila; abrasion of the outer layer of nanograined or coalescent elements has exposed the underlying lamina, and the cortex has frequently been damaged by shocks sustained during transport. Vacuoles with coalescing peripheries are probably the consequence of dissolution by dew.

3. Features common to ooids having tangential structure irrespective of their environment :

a) the presence of whisker crystals on their surface whose precise moment of formation and origin is unknown; they could form within the marine milieu, during subsequent transport of the sample or during its drying, perhaps even within the vacuum during SEM manipulation.

b) the presence of mucous has been recorded by numerous authors including Kinsman (1964), who observed it to contain many small crystals. Although no direct proof of a common origin has been established, nano-grains seem to be related frequently to the presence of an organic substrate.

4. Aspects common to all ooids irrespective of milieu :

a) Cryptocrystalline laminae : these have been observed to consist either of nano-grains or as haphazardly oriented needles or rods.

b) nano-grains : these seem to be the common product of several different processes. They are frequently primary elements formed on organic substrates but may also develop within evaporitic inter and supratidal environments. Others seem to be the product of abrasion or slight dissolution of other types of nano-crystals.

c) micritization : this also results from several different processes and may be primary in origin (apparent micritization) due to the genesis of nano-grains or unoriented needles, or secondary due to local destruction of the cortex by micro-perforations (Bathurst, 1966) which are subsequently filled with autochthonous nano-grains or needles. Finally, micritic fabrics also result from mechanical or chemical destruction of the primary structure of the cortex.

V. MILIEUX AND MECHANISM OF OOLITIZATION

GENESIS OF ARAGONITIC PARTICLES :

The aragonitic particles constituting the classic ooid cortex are very similar to the particles of muds or other aragonitic sediments accumulating within the Abu Dhabi lagoons or tidal flats; two basic forms of strontium-rich aragonite occur: elongate crystals and subspheric nanograins, indicating that certain factors leading to ooid formation are basically those which govern the formation of most carbonate sediments. The feeble magnesium concentrations in these sediments suggest that the particles are not the product of skeletal reef abrasion; the strontium concentrations of the aragonite (9000-9700 ppm) indicate that they do not result from the breakdown of abundant cerithid gastropods (1500-2000 ppm) while the absence in the Persian Gulf of the Codiacean algae Penicillus, Halimeda etc. - significant aragonite contributors in the Caribbean - tend to preclude a direct algal origin. These results are moreover in keeping with the conclusions of Kinsman (1969). Although the problem concerning the organic or inorganic precipitation is not definitely solved, the relationship between the localization of organic matter (mucous) and the presence of nano-grained structure suggest, for these latter, mechanisms including (Loreau 1970, 1971) :

- heterogeneous nucleation due to abundant impurities and very localized increase in the Ca^{++} reserve, both caused by organic matter (this last process being described by Trichet (1967, 1971).
- crystal growth limited by the great number of evenly distributed nuclei which mutually interfere, and also by the exhaustion of the ionic reserve within a limited volume.

MILIEUX OF OOLITIZATION :

In spite of the fact that conditions favouring the precipitation of aragonite seem to be fairly widespread along the Trucial Coast, especially within the lagoons (Kinsman, 1964) ooids are forming only very locally and constitute less than 5% of coastal sediments. It is evident that a nucleus is a prerequisite for ooid formation, but potential nuclei are ubiquitous and other factors must be sought to explain the local formation of ooids. Furthermore, their distribution is determined neither by the water agitation alone nor by any specific salinity; ooids are forming on the tidal deltas where salinity averages 40 g/litre while those within the lagoons of W Abu Dhabi are associated with salinity exceeding 60 g/litre.

Three features appear to be common to most ooid-forming environments :

1. a ridge (ripple or levée) and depression morphology resulting in closely associated low and high energy environments,

2. a relationship between the structure of the external layer of the cortex and the milieu (Tables 3 and 4). Although there exists a mixture of structures within a given sample, presumably the result of transport from depressions to ridges, etc., the dominant fabric of the outer layer (40-95% of grains) seems to be related to the local morphology and the degree of agitation of the environment in which the sample was collected :

Water agitation :	← very feeble-feeble →		← moderate-strong →		← strong →
Morphology of the milieu :	← depressions	→	← ridges	→	front of delta
Dominant structure of external lamina :	radial orientation	porous felt of unoriented needles	felt of tightly packed tangentially oriented needles	coalescent structure	loose felt (in process of formation or destruction ?)

The distribution of the subspheric, nanogranular elements does not seem to be closely related to the milieu but rather to the presence of an organic substrate which frequently coats these grains and, more remotely, to the degree of water evaporation.

3. the presence of bi-directional currents. In common with Bahamian ooids, virtually all oolitic sediments forming in relatively agitated coastal environments of the SE Persian Gulf seem to be associated with a reversible current system which is generally tidal. The series of tidal deltas in E. Abu Dhabi are the most striking examples. These latter are located at the seawards ends of channels which are the sole connection between the coastal lagoons and the open sea and are therefore characterized by reversible tidal flows, as discussed in a previous section (Fig. 3). Further W, in Central Abu Dhabi, channels between barrier islands link the extensive Khor al Bazm lagoon with the open sea,but are not the only connection; current patterns are more complex, being dominated by a lagoonwards movement which transports sediment across the Great Pearl Bank barrier into the Khor al Bazm lagoon (discussed by Purser and Evans in this volume). These channels do not have oolitic tidal deltas.

The morphology of the ooid bars between Yas Island and W. Abu Dhabi and between Bahrain Island and coastal Saudi Arabia, also suggest the strong influence of bi-directional tidal currents, while oolitic sands in the open embayment of Sabkha Matti occur in an area of converging long-shore currents.

MECHANISMS OF OOLITIZATION :

Concerning the aragonite particles constituting the cortex, one should first consider whether they have grown "in situ" on the ooid (as suggested by most workers) or whether they have been formed elsewhere and added subsequently to the ooid in a manner comparable to snowball formation (Sorby, 1879). The radial disposition of certain aragonite crystals within the cortex and on the external surface of the ooid - not only within the lagoonal ooids but also on the surface of "classical" ooids forming within the tidal deltas (Fig.15E) - is the habitual orientation of aragonite crystals growing with their C-axis normal to the support; it is an indication that oolitization is initially the result of "in situ" crystal growth. In addition, certain subhedral aragonite crystals are seen to protrude through an underlying felt of aragonite rods whose individual crystal habits are distinctly different (Fig.15E). This relationship strongly suggests that the subhedral crystals have not been deposited on the surface of the ooid as such but are the product of "in situ" growth. Lastly, as already discussed, the primary nanograined structure is also formed "in situ".

It is probable that crystals may grow in all orientations - including tangential. However, the relationship between the milieu of more and more agitated water and the progressively tighter packing, plus increasingly perfect tangential orientation suggests that the radial or haphazard orientation is primary, being formed mainly while the ooid lies within the protected inter-bar depressions; individual crystals are subsequently broken or flattened to a tangential orientation while the ooid is reworked in a more agitated environment on an adjacent bar or beach. In contrast, in ooids within well-protected lagoons(Jebel Dhanna) or tidal flats (Umm Said) - although the crystals are distinctly bigger - the radial structure is retained. These quiet water ooids may attain unusually large sizes (Fig.16A) mainly because they do not suffer the abrasion which polishes and rounds ooids in agitated environments.

Bearing in mind the preceding observations, the process of ooid formation within an Abu Dhabi tidal delta (Figs.2-3) may be envisaged as follows. Between the fringe and the interior of the delta there is a clear increase in both ooid size and cortex thickness (Figs.4,5). "A priori", this gradient could be due to progressive growth or to mechanical sorting related to transport. The morphology of mega-ripples and channels clearly indicates a transport from the delta flanks towards the interior (Fig.3), probably by onshore waves, this transport being affected via a series of bars and inter-bar depressions. As already noted, SEM studies of the external layer of the ooid cortex suggest that the aragonite rods or nanograins grow on the ooid essentially while the grain is within the depression, these constituents being subsequently compacted and somewhat reoriented (tangentially) when the ooid is on the more agitated crests of the adjacent ripple or bar. The processes may have something in common with the observations of Weyl (1967) whose laboratory experiments using Bahamian ooids showed that the precipitation of carbonate on the ooid surface was not constant and tended to cease rapidly when the ooid remained in an agitated environment; a period of ooid stability was required before precipitation recommenced.

Transport from the delta flanks to the delta interior via ripples and depressions is probably relatively slow due to opposed bi-directional current systems which characterize the inner parts of the delta; near the axial channel seawards-flowing,ebb-tidal currents tend to counteract the shorewards wind-driven waves and surface currents. Ooids are therefore retained within the favourable oolitizing system. Most are ultimately ejected from this system as relatively large ooids with thick cortex, by falling into the axial channel down which they are transported to the delta front. They are deposited and partially abraded along the delta front and possibly within the channel, their deposition leading to progradation of the delta. Other ooids are lost from the system by being swept onto the barrier beach, especially along the NE side of the delta. These latter ooids may be blown from the beach to adjacent aeolian dunes, or be transported laterally along the beach to the NE by active longshore currents, contributing to the lateral progradation of the barrier island. Some ooids may, perhaps, be reincorporated into the adjacent delta system.

Finally, it should be stressed again that significant amounts of oolitic sand are forming in lagoonal settings. The ooids of Khor Odaid lagoon, where there is also a vague bar-and-channel morphology, have a classical but thin cortex while others at Umm Said and Jebel Dhanna are formed on the sediment surface with virtually no lateral movement and are characterized by well developed fibro-radial structure, large size and irregular shapes. These latter grains occur on evaporitic tidal flats, the processes leading to their formation possibly having something in common with those associated with the formation of "cave pearls".

VI. COMPARISON BETWEEN THE OOIDS OF THE PERSIAN GULF AND THOSE FROM OTHER MODERN CARBONATE PROVINCES

STRUCTURE AND SHAPE :

The statistically tangential structure of most Persian Gulf ooids is similar to that observed in Bahamian ooids by Illing (1954),Newell et al (1960), Bathurst (1967) and Loreau (1970), and also to ooids described by Rusnak (1960), Freeman (1962) and Behrens (in Frishman, 1969) along the coast of Texas. Similar structures have also been noted in ooids from the Gulf of Suez by Fabricius and Klingele (1970). The radial structure of ooids forming mainly in protected environments of the Persian Gulf, in contrast, differs from that of most ooids. Those from the Great Salt Lake (Utah) described by Earley (1938) and examined in detail by Fabricius and Klingele (1970), although partly calcitic, exhibit clearly defined radial structure which locally traverses individual laminae of the cortex suggesting that this fabric is partly a recrystallization effect; the fibro-radial aragonitic structure of certain Persian Gulf ooids is essentially primary. Loreau (1970) has also noted radially disposed crystals developed locally within the cortex of Bahamian ooids, while both Rusnak (1960) and Behrens (in Frishman, 1969) have established that these radial structures within ooids of the Texas coast were magnesian calcite. The irregular form of certain ooids in Lagune Madre (Rusnak, 1960) is also a feature common to Persian Gulf ooids forming in protected environments.

The large, fibro-radial ooids of the Persian Gulf most closely resemble "cave pearls", in size, shape and internal structure. Their concentric laminations of constant thickness enveloping irregularly shaped nuclei result in irregular external morphologies similar to the "cave pearls" illustrated by Hahne et al. (1968) from

Rhur mines, by Donahue (1969) and Baker and Frostick (1947) from caves. The radial disposition of aragonite crystals within the cortex of certain Persian Gulf ooids also has much in common with the micro-structure of "cave pearls" illustrated by these workers, but differs in detail; crystallites in the marine ooids of the Persian Gulf are often fibrous while those of many "cave pearls" appear to be more equant (see Fig. 13 in Hahne et al, 1968) probably reflecting, a different mineralogy. The unpolished surfaces of both Persian Gulf ooids having fibro-radial structure is also a feature common to many "cave pearls" (Donahue, 1969). It is also interesting to note that the large fibro-radial ooids of the Persian Gulf occur locally in association with marine tufas (described by Purser and Loreau in this volume) and in this respect may be compared with the association of "cave pearls" and cave tufa.

The morphology, size and structure, of the large, irregularly shaped, fibro-radial ooids are also comparable to the "vadose pisoliths" described by Dunham (1969) from the deserts of New Mexico although the characteristic vertical polarity of the concretionary cements (Dunham, Fig.18) has not been observed in the Persian Gulf ooids; it is common, however, in the associated aragonitic tufas encrusting adjacent beach rock and lithoclasts.

DISTRIBUTION :

Although ooids occur in surface sediments of the Persian Gulf in contrasting environments ranging from depths exceeding 100 m to aeolian dunes, their widespread distribution is readily explained in terms of post-glacial transgression and lateral transport. Modern ooid formation is almost certainly confined to the present shoreline at depths less than ca. 5 m. Depth, degree of water agitation, hydrography and, to a lesser degree, salinity, of these ooid-forming environments are very similar to the classical Bahamian oolitic systems and it is presumed that the basic factors leading to ooid formation have much in common - with one significant exception; the concentration of carbonate sediments (and ooids) along the margins of the Bahamian platforms has been explained in terms of increasing water temperatures and decreasing CO^2 of oceanic waters flowing on to the shallow banks (Black, 1933, Illing, 1954, Newell et al, 1960, etc.), but this hypothesis could hardly explain the distribution of Persian Gulf carbonate sediments,including,ooids which are remote from oceanic waters.

The major difference between the ooids of the Persian Gulf and those of the Bahamian platforms is not related to structure, size,or environment of formation, but to their distribution within the respective provinces. Because the architectural styles of the Persian Gulf and Bahamian platforms are diametrically opposed, the distribution of the oolitic sediments with respect to other sediment types and to the shoreline is significantly different. Persian Gulf ooids are forming along the shore where considerable amounts (possibly most) accumulate in intertidal or desert environments; their relation to deeper marine sediments is remote. Bahamian ooids are forming on platform margins which are often remote from the shore and only a minor fraction accumulates above sea level. On the contrary, the Bahamian ooids, because of their proximity to a clearly defined platform margin, are closely associated with oceanic environments into which some may be transported.

The distribution and environment of formation of Persian Gulf ooids has much in common with the oolitic sediments of Shark Bay, W Australia (Logan et al, 1970). This similarity probably is due to very similar morphologies of the two provinces, both being shallow depressions lacking major shelf edges.

The oolitic sand forming within the lagoonal and tidal flat environments such as Khor Odaid, Jebel Dhanna and Umm Said, may be compared with those of Bimini lagoon in the Bahamas (Bathurst, 1967) and to Lagune Madre on the Texan coast (Rusnak, 1960 and Freeman, 1962). The structure of certain lagoonal ooids, both in the Persian Gulf and in Lagune Madre is characterized by the fibro-radial orientation of parts of the constituent aragonite, and by irregular external morphologies. In the Persian Gulf, their size greatly exceeds that of those in other areas, and the highly irregular shapes of associated grains into which these ooids grade seems to be peculiar to the Persian Gulf. Their localization within very restricted, evaporitic tidal flats in W. Abu Dhabi suggests that their distribution may be determined by hot, arid, climatic conditions.

VII. CONCLUSIONS

1) Although most ooids in the Persian Gulf are forming in shallow, somewhat agitated infratidal environments, significant amounts are forming in lagoonal settings. These latter ooids may be small and have a thin cortex. Others are unusually big, 5 mm being common near Jebel Dhanna and on the sabkha at UmmSaid, and differ from the classical ooids in both form and structure.

2) SEM and divers micro-analyses have demonstrated the ultrastructure of Persian Gulf ooids :

- light coloured laminae (seen in thin section) within the cortex consist of a more or less dense felt of monocrystalline,aragonitic rods having a statistically tangential orientation;

- they alternate with sombre cryptocrystalline laminae, rich in organic matter, having an aragonitic,nanograined structure, or with non-pigmented micritic laminae consisting of non-oriented rods. The nano-grains (0,25 μ in average) are especially polycristalline (one crystal being bigger than the other non-developed nuclei). The strontium concentration in the aragonite is the same in the different types of structures.

- radial orientation may be seen on the surface of ooids in protected environments.

- the polished external surfaces generally consist of a structure which tends to be coalescent.

- the cortex of ooids forming in lagoonal environments or in tidal flats (Jebel Dhanna, Umm Said) has an essentially primary, aragonitic, radial structure, the crystals being generally bigger than in classical ooids.

3) The abnormally weak birefringence (0.030 instead of 0.155) is explained mathematically by taking into account :

- the height of the pore space traversed by the light rays,

- the statistical tangential orientation of C-axis of the crystals leading to the consideration of a new indicatrix (resembling an apple),

- and, especially, the haphazard orientation of these C-axis in the statistical tangent plane - thus a haphazard orientation with respect to the direction of the light rays.

4) A relationship exists between the nanostructure of the external layer of the cortex and certain characteristics of the milieu :

- in relatively agitated situations on the crests of ripples, levées and on certain beaches, the constituent aragonite crystals are oriented statistically tangentially with respect to the surface and are often tightly packed or coalescent;

- in less agitated environments between the crests of ripples, or on protected tidal flats, the constituent aragonite crystals are especially loosely packed and may exhibit a radial orientation.

5) The nanostructure realized by the aragonitic rods at the surface of ooids and the observed relationships between this nanostructure and the nature of the milieu lead to the hypothesis that the elongate aragonite crystals constituting the cortex of most ooids grow "in situ", initially in a radial or disoriented manner within the more protected micro-environments between ripples. Subsequent reworking of these grains on crests of adjacent ripples tends to modify this radial or haphazard orientation to a secondary tangential disposition, followed by mechanical compaction of the crystals, polishing and abrasion of the grains. Ooids which remain in protected areas, including lagoons and tidal flats, retain their radial structure (generally with bigger crystals).

6) Water agitation alone does not explain the distribution of ooids along the Trucial Coast for, although coastal sands are in motion along much of this shoreline, ooids occur only locally. Their localization seems to coincide with the presence of bi-directional tidal currents which pile the oolitic sand into a system of ripples and levées. Because the current regime is bi-directional, individual grains tend to remain within the oolitizing environment where they pass from an inter-ripple depression to the adjacent ripple crest, and so forth, this alternation favouring the creation of a laminated cortex by the mechanism outlined above.

7) The authors have confirmed that the ooids of the Persian Gulf are forming locally in very shallow coastal environments where they constitute pure oolitic sand. However, much of this sand is subsequently transported from the point of formation to accumulate either on adjacent beaches or as aeolian dunes which may contain ooids some 40 km from the shoreline. Relict ooids occur down to depths of 100 m in the axial parts of the basin where they were formed during a lower sea level. Thus, although at any given moment ooid formation is very local and is closely related to specific environments, within a very brief period of geological time (less than 20.000 years) oolitic sediments have been dispersed over wide areas, most of which are not environments of ooid formation. These facts should be kept in mind when interpreting ancient oolitic formations whose regional extent may not necessarily reflect the magnitude nor the distribution of ooid forming environments.

APPENDIX A

(J-P. Loreau)

"ULTRASTRUCTURE" : DEFINITION AND METHOD OF STUDY

The petrology of carbonate sediments, more and more frequently, requires an understanding of the structure on a micro-scale, for it is evident that it is at this scale - the size of the crystal - that certain phenomena of sediment formation are expressed.

The study of ooids or other sediments which consist of minute crystals, often only a few microns (or less) in size, is generally incomplete and lacks precision unless the electron microscope is used. The SEM in particular has helped to define the morphologies of the constituent particles whose mineralogical composition may be established globally by X-ray diffraction. However, these observations, although they permit the definition of unknown micron-scale fabrics and the morphology of the crystals (including their size), are not sufficient to permit an accurate interpretation, it is also useful - sometimes indispensable - to know with precision the nature - i.e. the mineralogical, chemical and crystallographic properties - of the elementary particles. Thus, mineralogy is established in addition by electronic diffraction on individual particles; chemical analyses tend to specify the nature, concentration and localization of the various ions which substitute for calcium utilising atomic absorption spectrometer, microprobe and X-ray energy emission spectrometer coupled with the SEM; the crystallographic study, using electronic transmission microscope and electronic diffraction is made in order to determine crystal facies, whether individual elements are mono- or polycrystalline, the orientation of their crystal axes with respect to the morphology, the twinnings and the superstructures.

Finally "ultrastructure" (which is not a scalar term) is defined as the group of morphological, mineralogical, chemical, crystallographic characters of the elementary particles of the deposit. The demonstrated ultrastructures imply mechanisms such as nucleation, appearance of a given polymorphic variety and crystal growth which may be specified (Loreau 1971, Loreau and Rimsky in press).

Furthermore, for the description of structure it is sometimes useful to note that the observation has been localized by light microscope (ex. microstructure); similarly, it may be useful to summarize the morphological characteristics defined by SEM under the name of "nanostructure" or "nanofabric".

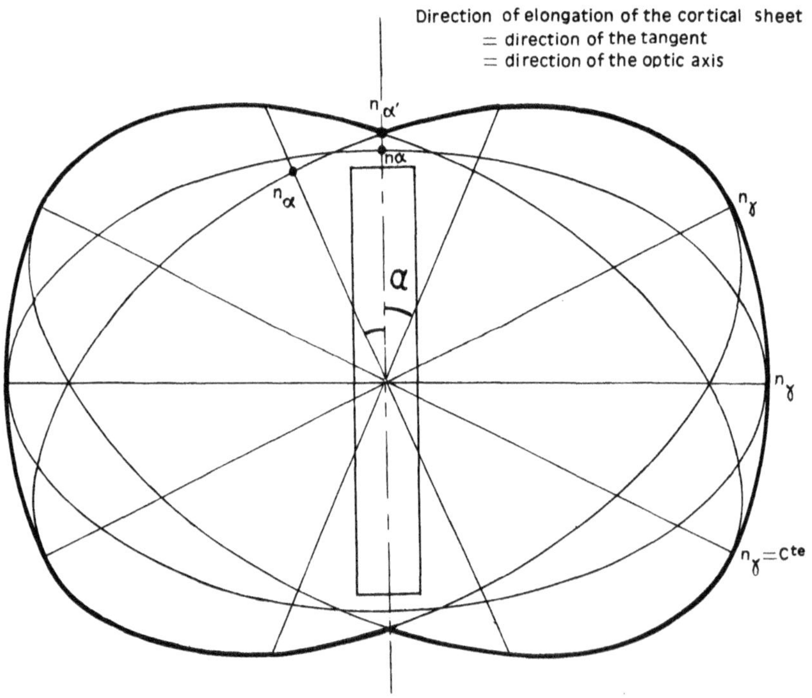

FIG. 19. Scheme showing the position of the ellipsoids for nanocrystals in the plane of the thin-slide (optic axis ⊥ to incident light ray), and for the angular displacement of ± 25° with respect to the tangent

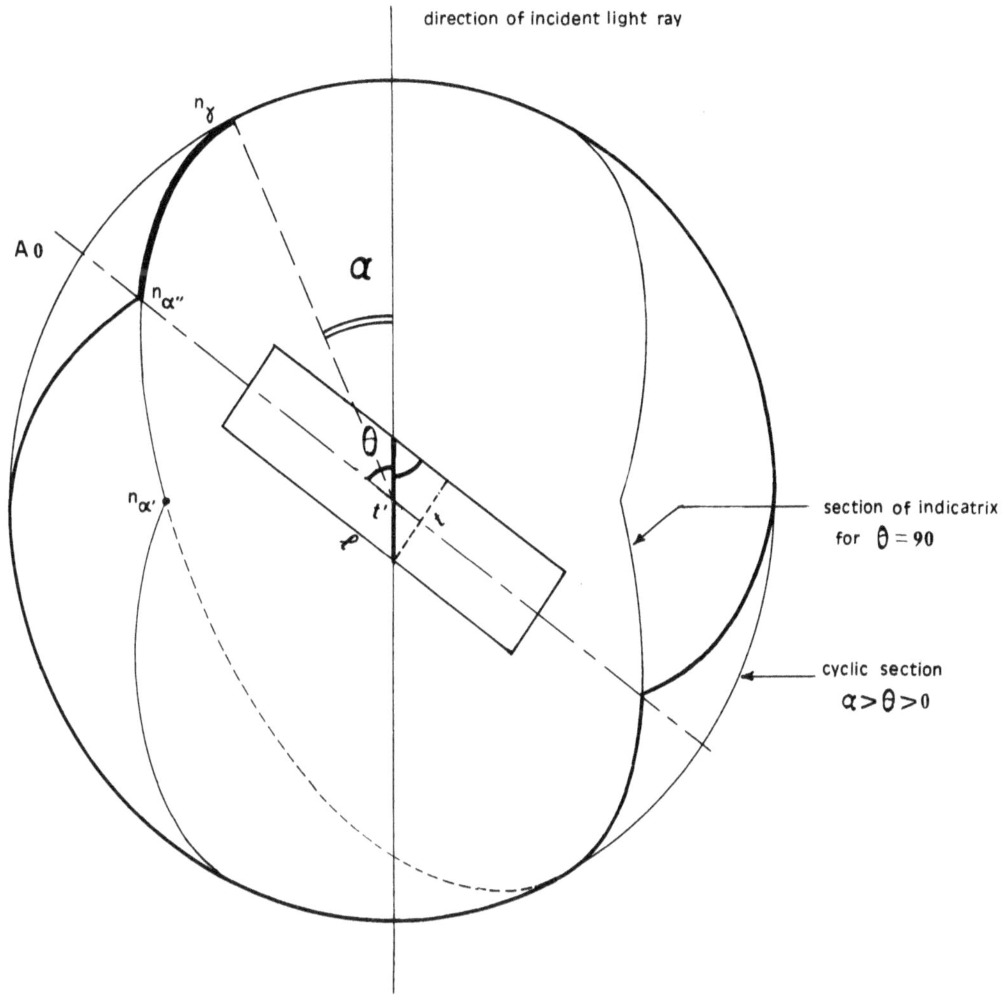

FIG. 20. Scheme showing the position of a nanocrystal in a plane perpendicular to the thin-slide passing through the optic axis and intersection of this plane with the indicatrix

APPENDIX B

(J-P. Loreau)

DETAILS OF THE CALCULATION OF THE BIREFRINGENCE IN THE CORTEX OF OOIDS.

The explanation of the abnormally low birefringence (0.030 as against 0.155 for standard aragonite) is discussed only briefly in the extensive literature on recent ooids.

Based on the studies outlined in the preceding text, this anomaly is explained essentially in mathematical terms. The calculation takes into account

1) the magnitude of the inter-crystalline porosity traversed by the light rays,
2) the imperfect orientation of the C-axis of crystals with respect to the tangent of the cortex sheets, and
3) the haphazard orientation of these C-axes within the plane statistically tangential to the cortex sheet and perpendicular to the thin slide.

1) Calculation of the thickness of inter-crystalline empty space traversed by the light rays within the thickness of the microscope slide.

SEM photographs (Fig.14A, B) clearly illustrate the loosely-packed fabric of aragonite crystals within the cortex of many ooids. The density (gcm^{-3}) of the ooids examined, in effect, is 2.60-2.64 (average 2.62 $\pm$ 0.02) being very comparable to the values given by Lacroix (in Illing, 1954) i.e. 2.58-2.70. Within ooids composed of > 90% aragonite, where the cortex represents 2/3 - 4/5 of the radius (approx. 50 times the volume of the nucleus), the density of the nucleus (quartz 2.64, dolomite 2.86, aragonite 2.94, and calcite 2.71) has only a minor influence on the measured density of the ooid as a whole. This density indicates that the volume of the aragonite is apparently equal to $\frac{2.62}{2.94}$ of the volume of the ooid cortex i.e. 89.1%. There exists, therefore, a empty volume equivalent to 10.9%; if it is presumed at first to be homogeneous, 10.9% total thickness T of the microscope slide represents the thickness T_v traversed by the light rays in either resin or air (isotropic media).

If $T = T_{Ar} + T_v = 30\ \mu \Longrightarrow T_v = 3.3\ \mu$

and $T_{Ar} = 26.7\ \mu$

(T_{Ar} = Thickness of aragonite)

2) Calculation of the maximum apparent birefringence of the nanocrystals within the plane of the thin slide (optic axis normal to the incident light ray).

SEM observations, together with electron diffraction measurements, show that the tangential orientation of the C-axis is only statistical and that this orientation may vary up to 30° with respect to the tangent (if one takes into account

80-90% of the crystals). n_γ and n_β being very close, the aragonite is considered as being uniaxial, in order to simplify the calculation. If one considers the new indicatrix (Fig.19) which is the statistical result of the position of each individual indicatrix, it is evident that n_γ remains constant while n_α increases to a value n_α' - the intersection of the two extreme ellipsoids -. Let α be the maximum angle between the optic axis and the tangent and consider the intersection of an extreme ellipsoid with a plane passing along the optic axis. From the equation of the ellipse, one may deduce the following approximate expression :

$$(n_\gamma - n_\alpha') = (n_\gamma - n_\alpha) \sin^2 \left(\frac{\pi}{2} - \alpha \right)$$

which is equivalent to the expression for the maximum birefringence in the new indicatrix.

With $\alpha = 30°$

$$(n_\gamma - n_\alpha') = 0.155 \times \sin^2 60$$

$$Bi_{max} = 0.116$$

The angular displacement of the nanocrystals with respect to the tangent leads to the consideration of a statistical indicatrix (resembling an apple) and a maximum birefringence Bi_{max} (Table 2) which is lower than that expected for aragonite.

3) Path difference calculated for nanocrystals within any plane of the thin section.

Because each nanocrystal of aragonite may occupy any position normal to the ooid radius, each optic axis forms an angle θ which varies with respect to the incident ray of polarized light (Fig. 20). A plane containing the optic axis A_o and the incident ray cuts the indicatrix along an intersection, the thicker part of which (on figure 20) belongs to the initial ellipse (thus allowing calculations).

If t is the thickness of a crystal, t' the thickness of the aragonite traversed - with $t' = \frac{t}{\sin \theta}$ -, the path difference for each crystal is expressed by :

$$\delta = t' \times Bi_{max} \times \sin^2\theta$$

For each crystal t' varies between the minimum thickness t and the longer diagonal (practically its length) when θ varies from $\frac{\pi}{2}$ to arcsin $(\frac{t}{\ell})$, and the birefringence varies from Bi_{max} to 0 when θ varies $\frac{\pi}{2}$ to α , and stay zero when $\alpha \geqslant \theta \geqslant 0$ because the cyclic section is obtained (Fig.20).

One may define three sectors of crystal orientation (I, II, III, see Table 5) within a plane normal to the thin section with respectively three average values of the traversed thickness ($\overline{t'}_I$, $\overline{t'}_{II}$, $\overline{t'}_{III}$) , three average values of the birefringence ($\overline{Bi}_I$, $\overline{Bi}_{II}$, $\overline{Bi}_{III}$) , three numbers of corresponding crystals (n_I , n_{II}, n_{III}) and three partial path differences $\Sigma\delta_I$, $\Sigma\delta_{II}$, $\Sigma\delta_{III}$. The calculations are briefly exposed in Table 5 :

Table 5 . Summary of the variations of Bi_θ and t' with respect to θ , and average values of the crystal thickness traversed, of the birefringence, of the approximate crystals number, and of the path difference for the three respective sectors of crystal orientation

θ	$\frac{\pi}{2}$	α	$\arcsin(\frac{t}{\ell})$	0
t'	t' → (increasing)		$\simeq \ell$	ℓ
Bi_θ	Bi_{max} → (decreasing)	0	0	0
	I	II	III	
$\bar{t}'_\theta$	$\bar{t}'_I = t \dfrac{\int_{\frac{\pi}{2}}^{\alpha} \frac{1}{\sin\theta} d\theta}{\alpha - \frac{\pi}{2}}$	$\bar{t}'_{II} = t \dfrac{\int_{\alpha}^{\arcsin(\frac{t}{\ell})} \frac{1}{\sin\theta} d\theta}{\arcsin(\frac{t}{\ell}) - \alpha}$	$\bar{t}'_{III} \simeq \ell$	
$\overline{Bi}_\theta$	$\overline{Bi}_I = Bi_{max} \dfrac{\int_{\frac{\pi}{2}}^{\alpha} \sin^2\theta \, d\theta}{\alpha - \frac{\pi}{2}}$	$\overline{Bi}_{II} = 0$	$\overline{Bi}_{III} = 0$	
n_θ	$n_I = \dfrac{\frac{\pi}{2} - \alpha}{\frac{\pi}{2}} n$ cf.N.B.	$n_{II} = \dfrac{\alpha - \arcsin(\frac{t}{\ell})}{\frac{\pi}{2}} n$	$n_{III} = \dfrac{\arcsin(\frac{t}{\ell})}{\frac{\pi}{2}} n$	
$\Sigma\delta_\theta$	$\Sigma\delta_I = n_I \times \bar{t}'_I \times \overline{Bi}_I$	$\Sigma\delta_{II} = n_{II} \times \bar{t}'_{II} \times 0 = 0$	$\Sigma\delta_{III} = n_{III} \times \bar{t}'_{III} \times 0 = 0$	

N.B. : If n is the approximate number of crystals within the thickness of aragonite T_{Ar} (in the thin section), there exists :

$$T_{Ar} = \frac{\frac{\pi}{2} - \alpha}{\frac{\pi}{2}} n \times \bar{t}'_I + \frac{\alpha - \arcsin(\frac{t}{\ell})}{\frac{\pi}{2}} n \times \bar{t}'_{II} + \frac{\arcsin(\frac{t}{\ell})}{\frac{\pi}{2}} n \times \bar{t}'_{III} \qquad (1)$$

$$n = \frac{\pi \times T_{Ar}}{2[(\frac{\pi}{2} - \alpha)\bar{t}'_I + (\alpha - \arcsin(\frac{t}{\ell}) > \bar{t}'_{II} + \arcsin(\frac{t}{\ell})\bar{t}'_{III}]} \qquad (2)$$

The path difference $\Sigma\delta_I$ (for the crystals whose C-axes make an angle with the light ray from $\frac{\pi}{2}$ to α) is thus :

$$\Sigma\delta_I = \frac{\frac{\pi}{2} - \alpha}{\frac{\pi}{2}} \; n \times t \; \frac{\int_{\frac{\pi}{2}}^{\alpha} \frac{1}{\sin\theta} d\theta}{\alpha - \frac{\pi}{2}} \times Bi_{max} \frac{\int_{\frac{n}{2}}^{\alpha} \sin^2\theta d\theta}{\alpha - \frac{\pi}{2}} \qquad (3)$$

The path difference $\Sigma\delta$ for the total thickness of aragonite :

$$\Sigma\delta = \Sigma\delta_I + \Sigma\delta_{II} + \Sigma\delta_{III} = \Sigma\delta_I$$

And the path difference Δ for the total thickness of the thin section :

$$\Delta = \Sigma\delta_I + (T_v \times Bi_v)$$

with T_v = Thickness of "vacuum"

and Bi_v = Birefringence of the air or the resin filling the vacuum space, i.e. birefringence which is zero.

$$\Delta = \Sigma\delta_I$$

The final calculated birefringence Bi_c is thus :

$$Bi_c = \frac{\Delta}{T} = \frac{\Sigma\delta_I}{T} \qquad (4)$$

with

$T = 30\ \mu m \longrightarrow T_{Ar} = 26.7\ \mu m$ and $T_v = 3.3\ \mu m$

$\alpha = 30° = \frac{\pi}{6}$

$\left.\begin{array}{l} t = 0.2\ \mu m \\ \ell = 1\ \mu m \end{array}\right\} \longrightarrow \text{Sin}\ (\frac{t}{\ell}) = 0.2$ and $\arcsin\ (\frac{t}{\ell}) = \frac{\pi}{15}$

The calculations load to :

$\overline{t'_I} = 0.25\ \mu m$, $\overline{t'_{II}} = 0.58\ \mu m$, $\overline{t'_{III}} \simeq \ell = 1\ \mu m$

$\overline{Bi_I} = Bi_{max} \times 0.70 = 0.116 \times 0.70$

$n = \frac{2403}{37.5} \simeq 64$ nanocrystals

$\Sigma\delta_I = 0.856\ \mu m$

$Bi_c = 0.028$

If the magnitude of the inter-crystalline porosity is less important as a result of a tighter packing of the nanocrystals within certain laminae, this packing being associated with a more "tangential" crystal orientation, (± 25°) the calculation of maximum birefringence within the denser layer may be modified according to equation (1), i.e. 0.127 . The thickness of the aragonite is estimated as 95% of the thickness of the slide and the final birefringence is calculated in this case from equations (2), (3), (4). These various values, listed in Table 2, are very comparable to those determined directly by microscope.

Thus, the assemblage of very minute particles (individual thickness ca. 0.2 μm to 1 μm), which number approximately 70 in the thickness of the slide and have similar statistical orientation, reacts as a monocrystal on a very local scale. However, the optic properties differ from those of standard mono-crystalline aragonite, particularly with respect to the birefringence which is only 16-23% that of standard aragonite. Indeed it has been shown (above) that one must take into account :

1) the slight variation (25-30°) in the orientation of the nanocrystals with respect to the direction of the tangent to the cortex. This variation leads to a new indicatrix (in the form of an apple) and involves a maximum birefringence which is less than that of standard aragonite; the decrease is of the order of 23%.

2) the statistical orientation of the nanocrystals within the thickness of the microscope slide, each optic axis being oriented at any angle with respect to the incident ray of polarized light. Furthermore, the form of the new indicatrix gives rise to cyclic sections before the optic axis of each crystal within the slide is parallel to incident light ray. For these reasons the birefringence is further diminished by by approximately 55%.

3) the existence of inter-crystalline spaces (in fact, filled with resin or air, lacking birefringence) which lower birefringence by only 0.7-2.6%.

The variation in the degree of compaction of the constituent nanocrystals observed during SEM studies is associated both with a variation in the magnitude of the microporosity and with varying degrees of preferred orientation of these crystals. These modifications result in the variation of birefringence from one lamina to another within a given ooid.

The Precipitation of Aragonite and Its Alteration to Calcite on the Trucial Coast of the Persian Gulf

B. D. Evamy[1]

ABSTRACT

Along the Trucial Coast of the Persian Gulf aragonite is believed to be precipitating actively in certain of the sandy intertidal zones in response to the direct evaporation of marine water. Depending upon sub-environment, the mineral occurs as an intergranular, pore-destroying cement in rock, or as a glossy crust of tangentially-oriented crystals around both rock, when it appears like a layer of lacquer paint, and grains (ooliths). Since, in the latter case, the precipitation of aragonite is not a pore-destroying process, intertidal trends may be found in the fossil record from porous oolite to tightly cemented, non-oolitic skeletal sand.

Where aragonite has precipitated as cement, the resulting intertidal rock may be recognized in the fossil record from a number of consequent diagenetic features, such as cemented algal mat structures, polygonal cracks, "teepee" structures and arcuate ridges resembling cave dams. A further characteristic is the presence of abundant lithoclasts in the intertidal rock which have depositional components similar to those of the enclosing medium.

The change from an environment where aragonite can be precipitated to one where calcite is stable can cause early aragonite cement to alter to calcite with a cryptocrystalline texture or one pseudomorphing the earlier aragonite. Whether early aragonite cement is leached or replaced by calcite must depend primarily on the climate at the time of deposition and soon after.

I. ENVIRONMENT OF ARAGONITE PRECIPITATION

Along the Trucial Coast of the Persian Gulf aragonite is believed to be precipitating actively in certain of the sandy intertidal* areas and supratidal* "surf-spray" zones. The mineral occurs as an intergranular, radially-oriented cement in rock (Fig. 1A), and as crusts of tangentially oriented crystals around both rock (Fig. 1B) and grains (Fig. 1C), the latter therefore being termed ooliths.

---

[1] Shell Research B.V., Rijswijk, The Netherlands.

* The intertidal zone is here, though not universally, considered to represent the area exposed between normal or lunar tides, including spring tides. The supratidal zone is understood to comprise the area flooded by marine water exclusively during abnormal or storm tides.

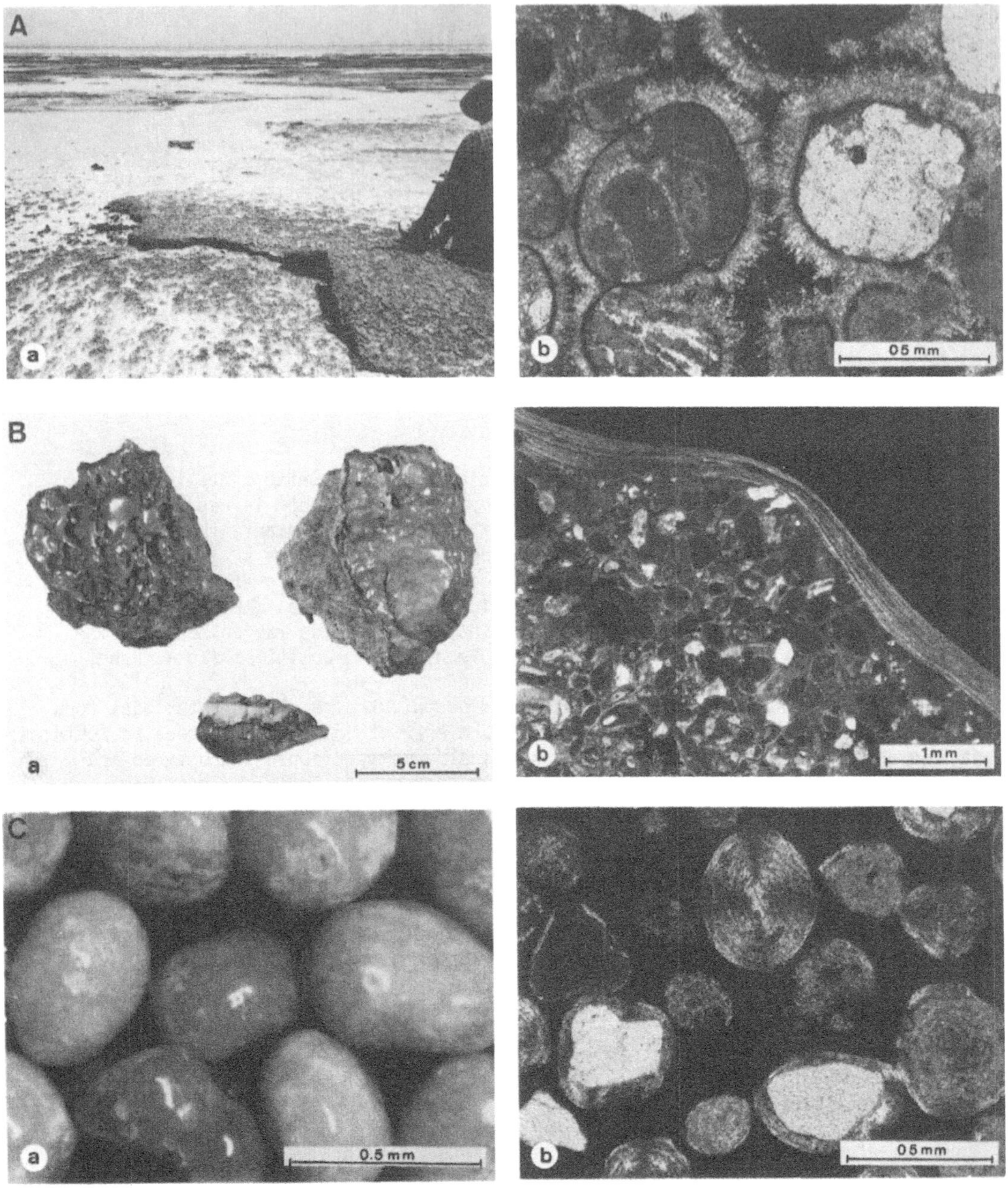

Legend to Fig. 1 A-C on the following page

The aragonite to be discussed is believed to have precipitated in response to the direct evaporation of marine water, which, except at considerable oceanic depth, is everywhere saturated or even oversaturated with respect to calcium carbonate. Even at saturation, however, the concentration of dissolved calcium carbonate in marine water is so low that filling an intergranular pore with aragonite requires the evaporation of at least 30,000 equivalent volumes of sea water. Environments in which marine water may be evaporated and then replenished are the intertidal zones and supratidal surf-spray zones, especially in warm climates. The highest rate of evaporation in these zones takes place at the surface, where insolation and wind desiccation reach a maximum. Precipitation of aragonite in the inter- and supratidal zones may, therefore, be expected where marine water is prevented from flowing out of near-surface pores. This requirement is fulfilled a) in the surf-spray zone (the most common location of beachrock) owing to the effects of surface tension in loose sand, and b) on partially-cemented rock, where the intergranular cement prevents complete downward percolation of the spray. The water is consequently retained in isolated pools and evaporated between successive high tides.

Marine water can also be retained in surface pores of very gently sloping and, as a result, extremely wide intertidal flats where the water table, even at lowest tide, is only just below the surface. Some of the most gently-sloping intertidal flats in the Sabkha Matti area (W. Abu Dhabi), are up to 2 km wide (Fig. 1A). As in the surf-spray zone, the marine water left in isolated ponds and filling surface pores is liable to be evaporated between tides.

II. FACTORS CONTROLLING ARAGONITE MORPHOLOGY AND THEIR SIGNIFICANCE

The morphology of the aragonite crystals, interpreted as having been precipitated in the tidal zones of the Trucial Coast, seems to depend upon whether the sandy sediment is a) organically bound and anchored by algae, b) inorganically bound by pre-existing cement, or c) free to move under the influence of waves.

A. Organically-bound sediment

Algal mats showing irregularly-wrinkled or polygonal structures are widespread along the Trucial Coast. Kendall and Skipwith (1968, 1969) described examples from Khor al Bazam Lagoon, Central Abu Dhabi and noted their association with beachrock. Taylor and Illing (1969) supposed that cementation was initiated through the stabilization of loose sediment by algal colonies. However, they found that this process was uncommon in the examples studied by them around Qatar and correctly ascribed its absence to the predominance of mud rather than sand.

Fig. 1. MORPHO-TYPES OF PRECIPITATED ARAGONITE IN THE TIDAL ZONES OF THE TRUCIAL COAST

A: Intertidal rock of the Sabkha Matti area, Trucial Coast. (a) field photograph, (b) thin section showing acicular aragonite crystals lining interparticle pores (crossed polarized light); M 7 (0-6 cm).

B: Crust of tangentially oriented aragonite needles on rock in the surf-spray zone, Umm al Qawain, Trucial Coast. (a) hand specimen showing crust and distinctive surface gloss, (b) thin section showing analogy of crust to oolitic coatings (crossed polarized light); Evm 410.

C: Oolitic sand from west coast of Sabkha Matti, Trucial Coast. (a) grains under reflected light showing characteristic surface gloss, (b) thin section of ooliths showing outer coats of tangentially oriented aragonite needles (crossed polarized light); Evm 569

Along the coast of the Sabkha Matti, algally bound sand is widespread. Here complete transitions have been found from carbonate sand bound by algal mats with irregularly-wrinkled (Fig. 2A-a) or polygonal (Fig. 2B-a) structures to cemented rock with these structures preserved (respectively, Figs. 2A-b and B-b), but the algal binding decayed. Similarly, uncemented, but algally-bound sand exhibiting ripple marks has been traced laterally into cemented rock with the pattern of ripples preserved (Fig. 3A). In each case the cementing material is acicular (needle-like) aragonite (Fig. 1A-b) filling the primary intergranular pores to a greater or lesser degree, according to position in the transition.
Penecontemporaneous reworking of the intertidal rock has given rise to an abundance of lithoclasts, which are strewn widely over the rock surface. In the fossil record, the depositional components of these lithoclasts would presumably be essentially similar to those of the host rock.

It is a characteristic of the intertidal beachrocks of the Trucial Coast that they are extensively bored by mollusks and other organisms. In the fossil record, therefore, they would strongly resemble the bored hard-grounds commonly found at the tops of regressive cycles. Supratidal sediment deposited above such intertidal rock may be eroded towards the end of the same cycle or at the beginning of the next, whilst the intertidal rock is preserved owing to its penecontemporaneous lithification.

Whereas algally bound muds are frequently recognizable in the fossil record as stromatolites, it is difficult to imagine a laminated structure being preserved in similarly-bound sands. Consequently, other evidence for sands having been bound by algal mats is of considerable importance in environmental interpretation. The preservation of the irregularly-wrinkled and polygonal structures of algal mats in grainstone (Figs. 2A-b and B-b) by early cementation and lithification could well provide such evidence in ancient rocks. Additional features of purely diagenetic origin permitting the recognition of this type of intertidal zone in the fossil record could be the presence of numerous flat lithoclasts that closely resemble the host rock, and the structures described in section III (large-scale polygonal cracks and "teepee" structures Fig. 3, and arcuate ridges resembling cave dams Fig. 4A).

B. Inorganically-bound sediment (cemented rock)

On coasts where relatively impermeable rock occurs in the surf-spray zone, the exposed surface of the rock tends to acquire a crust of tangentially oriented aragonite needles which, in thin section, resembles (Fig. 1B-a) the aragonite coatings of ooliths (compare with Fig. 1C-b). The crusts of tangentially-oriented aragonite over rocks are seen in hand specimens (Fig. 1B-a) to have a distinctive gloss, resembling that of ooliths (Fig. 1C-a). The crusts vary in colour from battleship-grey, seen near Ras al Khaimah and Umm al Qawain (NE Trucial Coast) to buff, observed on the west side of Sabkha Matti (W. Trucial Coast). In the field, the widespread distribution of such coated rocks in the surf-spray zone makes the entire surface appear as though it had been smeared with a thick layer of lacquer paint. According to Fairbridge (1957) such crusts, first noted by Darwin, encrusting basalt boulders on Ascension Island, are quite common in the warmer latitudes. Those coating certain of the beachrocks of Bikini have been illustrated by Emery et al. (1954).

Rocks exhibiting crusts of tangentially-oriented aragonite tend to have outer, case-hardened zones immediately underlying the crusts. In these zones the intergranular pores are almost completely filled by microcrystalline aragonite; the fine crystallinity is believed to represent rapid precipitation. In contrast, the interiors of the same rocks are characterized by open pores, or pores only slightly filled with acicular cement, the coarser crystallinity presumably indicating slower precipitation. This situation suggests that crusts of tangentially-oriented aragonite tend only to become deposited in response to evaporation of marine water at the very surface of rather impermeable rock. Poor permeability, as a factor favouring the development of crusts of tangentially oriented aragonite, is indicated also by certain beachrocks collected from the salt-dome island of Halul, off the E coast of Qatar.

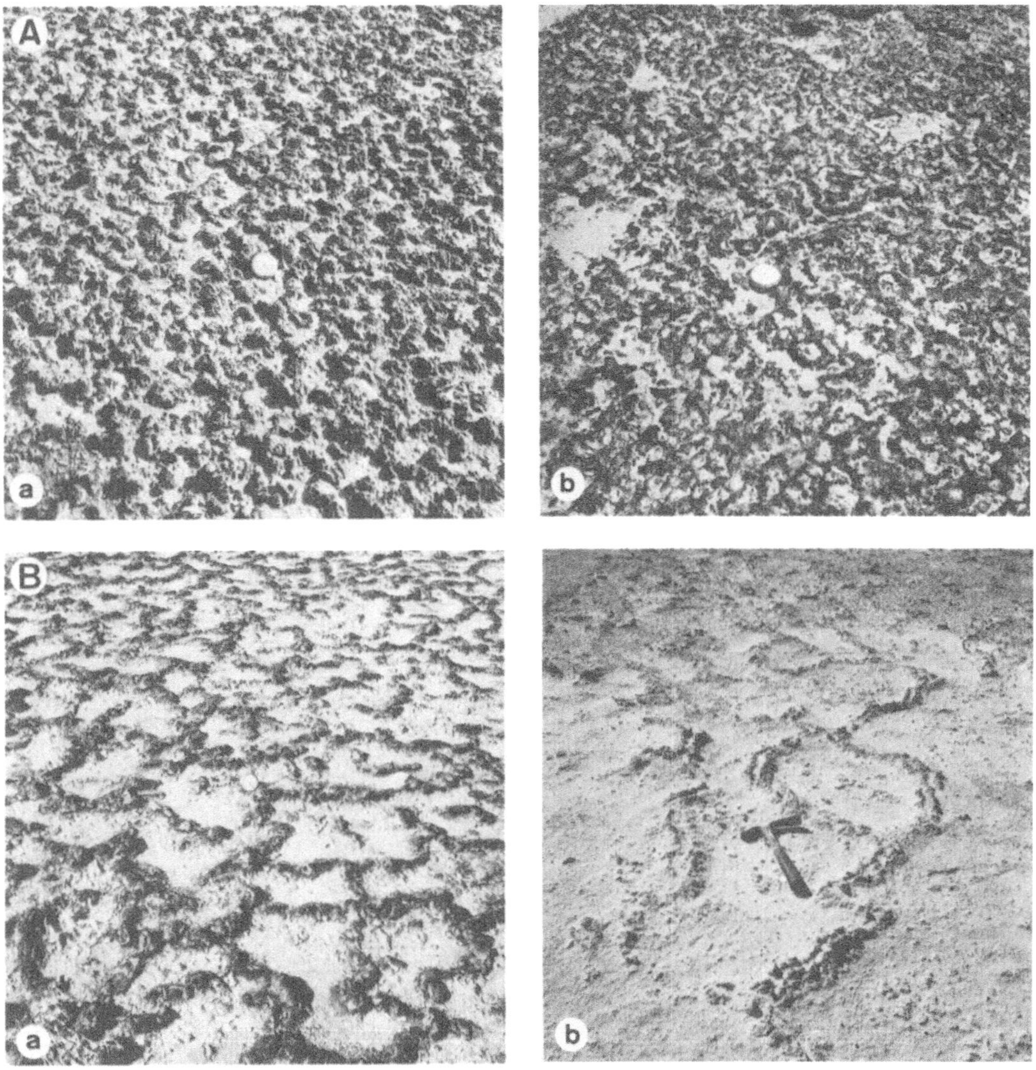

Fig. 2. ALGAL-MAT STRUCTURES AND THEIR CEMENTED EQUIVALENTS, SABKHA MATTI AREA, TRUCIAL COAST

A: Irregularly-wrinkled algal mat (a) living, (b) decayed but with sediment cemented so that algal-mat structure is preserved.

B: Algal mat, exhibiting polygonal structure, (a) living, (b) decayed but with sediment cemented so that algal-mat structure is preserved.

These beachrocks are aragonite-cemented, mixed bioclastic and lithoclastic grainstones. The bioclasts are sub-Recent and have been made porous at their margins by boring algae. In contrast, the lithoclasts are derived from a Cambrian series of relatively impermeable, dolomitized stromatolites. The lithoclasts are selectively coated by glossy grey envelopes of tangentially oriented aragonite whilst the bioclasts are cemented together by radially-disposed aragonite druse.

C. Unbound sediment

The wide intertidal zones of Sabkha Matti are not cemented into beachrock in all regions. On the W side of the sabkha, near Sila, there exist intertidal zones up to 1 km wide made up of loose sand. The sand, however, is nowhere bound and anchored by an algal mat and consists for the most part entirely of oolite (Fig. 1C) discussed by Loreau and Purser, in this volume. The grain size of the oolite, which is not found more than 1 km immediately offshore, increases landwards from the shallow subtidal zone, being greatest in the steeply inclined beach berm. This suggests that aragonite has indeed precipitated from the water, probably both in the intertidal and shallow subtidal environments in the form of oolitic coatings on grains (Fig. 1C-b)

Fig. 3. TEEPEE AND ASSOCIATED STRUCTURES, SABKHA MATTI AREA, TRUCIAL COAST

Intertidal rock cracked into large-scale polygons and exhibiting ripple marks (A), embryc teepee structure (B) and mature teepee structure (C)

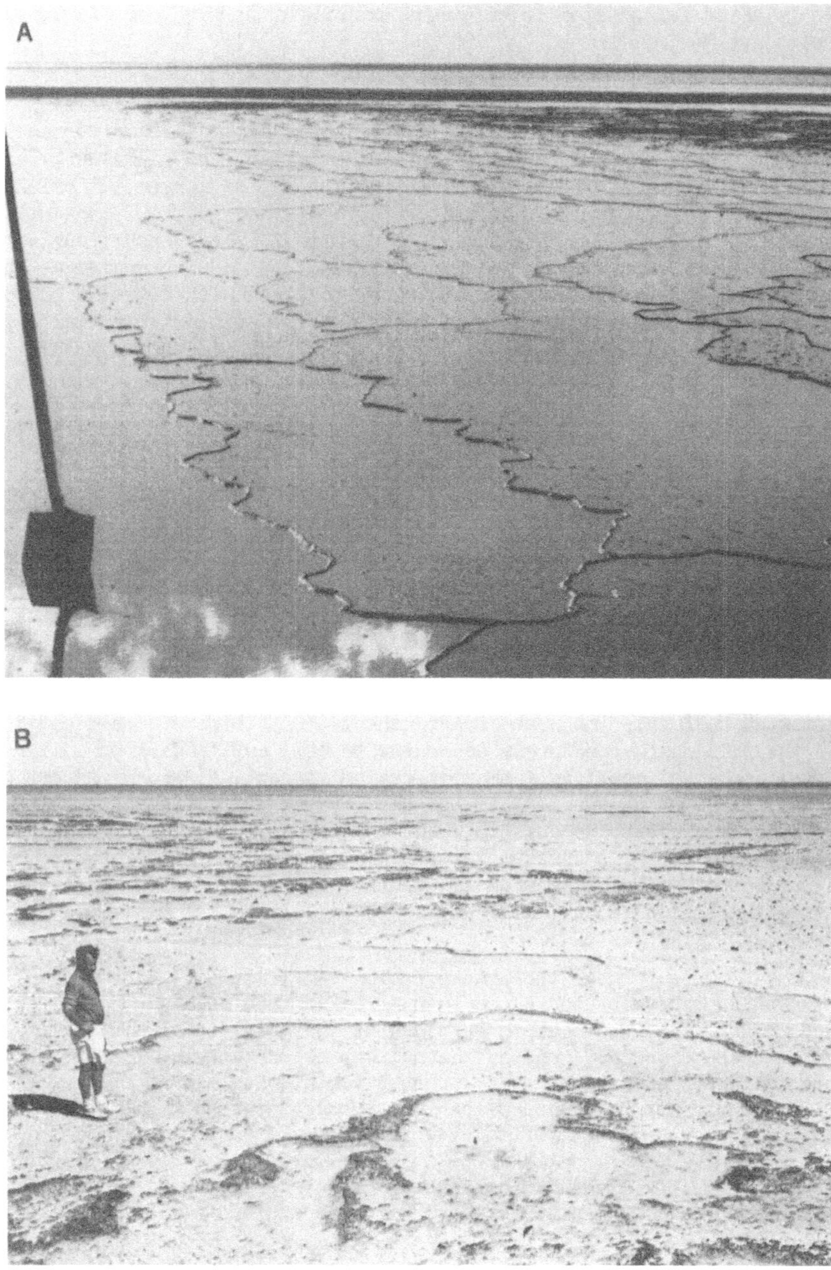

Fig. 4. ALGAL DAMS IN THE INTERTIDAL ZONE; UNLITHIFIED AND LITHIFIED

A. Algal mat locally buckled into elongate, irregularly aligned ridges. Sabkha Faishakh area, Qatar.

B. Arcuate ridges, which are consistently convex seawards, and which resemble cave dams, commonly overlie the penecontemporary rock of the Sabkha Matti area, Trucial Coast

and not as cement. Unlike the cement, which is believed to be precipitated at low tide when the grains are stationary, the aragonite of the oolitic coatings is probably precipitated below water level during periods of rising and falling tide when the grains are moving.

Shearman et al. (1970) recognized Recent ooliths as being composed of alternating concentric layers of organic mucilage and aragonite needles, thus explaining the tangential orientation of the latter. The high-energy environment was considered possibly to add to the definition of the tangential orientation of the aragonite. It is significant, therefore, that the crusts of tangentially-oriented aragonite on rock (Fig. 1B) appear to form in the high-energy surf-spray zones of the Trucial Coast. These rocks are probably also coated with mucilage, and the added bombardment by grains thrown up in surf, together with desert wind polish, causes precipitated aragonite needles to adopt a well-defined tangential orientation. This problem is discussed further by Loreau and Purser (in this symposium).

III. SOME DIAGENETIC STRUCTURES ASSOCIATED WITH ARAGONITE PRECIPITATION

A. "Teepee" structure

Where almost fully cemented, the intertidal rock in the Sabkha Matti area has locally been heaved into spectacular "teepee" structures (Fig. 3C). This type of structure was so named by Adams and Frenzel (1950) because of its resemblance to the American-Indian tent. It is suggested that the Trucial Coast "teepee" structures could be initiated by the cracking of the intertidal rock into large-scale polygons, illustrated by figure 3A, as predicted by Folk (oral communication, reported in Kendall and Skipwith, (1969)). These polygonal slabs of rock generally overlie uncemented sediment. The work of Artjushkov (1963) offers an explanation for the origin of the polygonal structures. According to him, any loading on a light medium (underlying loose sediment) by a heavier medium (intertidal rock) induces gravitational instability, which can lead to the development of polygonal structures in the upper medium. In the case in question, the upper medium is rigid, so the polygonal structures would appear as cracks. The cracking might also be caused, or at least aided, by repeated thermal expansion and contraction, which must take place as the generally dark coloured intertidal rock is alternately heated by insolation and cooled by water.

Once formed, the cracks would tend to become filled by loose sediment, either from above by normal sedimentation or from below due to upward movement of the underlying loose sediment in response to the gravitional instability already discussed. The effect of renewed thermal expansion, coupled possibly with the force of crystallizing aragonite during further cementation of the cracked rock, is likely to first buckle the polygonal slabs into embryo "teepees" (Fig. 3B) and later thrust them one over another, giving rise to more mature structures of this type (Fig. 3C).

Shinn (1969) has described and interpreted similarly impressive structures associated with submarine cemented Holocene sediments within the Persian Gulf.

B. Arcuate ridges

A further feature of the well-cemented intertidal zones of Sabkha Matti is the occurrence of arcuate ridges (Fig. 4B) which tend to be convex seawards. They probably originate through the cementation of sand, previously bound by an algal mat, the latter having been buckled locally into a series of ridges, such as those shown in figure 4A. The convexity of the cemented ridges towards the sea gives them the appearance of cave dams. Modification of buckled algal mats to intertidal cave-dam structures would be yet another indication of the rapid aragonite precipitation which seems to characterize this environment.

IV. THE REPLACEMENT OF ARAGONITE CEMENT BY CALCITE

The change from an environment where aragonite can be precipitated to one where calcite is stable, without intervening leaching, seems to cause early aragonite cement to alter marginally to cryptocrystalline calcite. This situation is exemplified by the grainstone shown in figure 5A. Except at their points of contact with other grains, the aragonite particles, for example the algally-bored mollusk fragment (a), are surrounded by an early aragonite cement (b). Owing to the selectivity of carbonate cementation, the clean feldspar grain (c) and quartz grains (d) have not acted as nuclei for cement. The aragonite cement is altered to cryptocrystalline calcite (e) next to the remnant pore space (f).

The pore-filling aragonite is interpreted as having formed in a marine environment. The rock in which it occurs now crops out in the supratidal part of the Holocene coastal complex. It is considered that meteoric water periodically occupying the remnant pores permitted the observed marginal alteration of the aragonite needles to cryptocrystalline calcite*.

A more complete alteration of former aragonite druse to finely crystalline calcite is believed to be represented in figure 6. The absence of this calcite (a) at the contacts between depositional grains (b) and its uniform development (c) where original pore space remains suggest that it is a replacement of former drusy crystals, believed to have been aragonite. It is interesting to note that only the former druse has been altered to cryptocrystalline calcite (a and c, Fig. 6); aragonite grains, such as ooliths (d) and mollusk fragments (e), entirely withstood the change. In other samples, what are believed to have been former aragonite druses have been altered to microcrystalline dolomite, whilst depositional components have apparently been unaffected. Thus, crystals of aragonite cement seem to be more susceptible to alteration than depositional grains of aragonite. This is probably because firstly the cement is in more direct contact with the pore fluids that cause the diagenesis, and secondly, it is not as likely to be protected from alteration by algal mucilage such as that surrounding and pervading ooliths and other carbonate grains.

It is difficult to distinguish depositional grainstones (Dunham, 1962) cemented completely by microcrystalline carbonate from depositional packstones. This is the case for many of the abundant, lithified, grain-supported rocks with fine-grained matrices which are found along the Trucial Coast, and in cores from shallow wells in the Persian Gulf. The problem of identifying the diagenetic origin of certain microcrystalline carbonate matrices has also been raised by Purser (1969) from his studies of the Middle Jurassic carbonates of the Paris Basin. The replacement of penecontemporaneous aragonite cement by cryptocrystalline calcite could explain the apparently paradoxical association, common in intraformational carbonate lithoclasts, of grain types representing high-energy environments (e.g. ooliths) with mud matrices normally indicative of quieter water conditions.

The alteration of aragonite cement to calcite in a calcite-cementing environment need not necessarily yield cryptocrystalline calcite, as it does in the case illustrated by figure 6. In figure 5B, for example, a gastropod (a) supports a rim of acicular crystals (b). The most recent cement, however, nearest to the remnant pore space, is composed of clear equant crystals (c). Immediately underlying the contact between the acicular and equant crystals is a zone of cryptocrystalline carbonate (d). Whereas all these cement types are now calcite, it is believed that formerly the acicular and cryptocrystalline carbonates (b and d) formed a druse of aragonite. When the diagenetic environment changed from one of aragonite to one of calcite stability, the aragonite druse altered to cryptocrystalline calcite marginal to the pore space, to yield a fabric such as that shown in figure 5A. The fresh-water diagenetic environ-

*It is not certain whether this calcite is non-magnesian; if magnesian, it could have formed as a beach-rock cement. Ed.

ment then caused the remnant porosity to become progressively filled with clear, sparry crystals of calcite cement (Fig. 5B-c), and the remaining unaltered aragonite to be pseudomorphed by acicular calcite (Fig. 5B-b).

V. CONCLUSIONS

The porosity distribution in the carbonate sands of the Sabkha Matti area W Abu Dhabi, seems to depend upon the morphology of precipitated aragonite. In the higher-energy environments, where the grains are kept in motion, this aragonite tends to precipitate as oolitic coatings, and the interparticle porosity is preserved. In the lower energy zones, algal mats stabilize the grains which than become susceptible to cementation by radially-oriented aragonite, thereby destroying the original porosity.

Whether the unstable aragonite cement of beachrock is leached or replaced by calcite must depend primarily on climate. The Persian Gulf area, where replacement seems to be common, is much more arid than the Bahamas, where leaching may predominate.

The change from an environment where aragonite can be precipitated to one where calcite is stable, without an intervening significant phase of leaching, seems to cause early aragonite cement to alter to calcite, which commonly has a mud texture, but in some instances retains the drusy texture of the previous aragonite.

ACKNOWLEDGEMENTS

The author is indebted to Shell Research B.V. for permission to publish this paper. He would like to thank his colleagues at the Koninklijke/Shell Exploratie en Produktie Laboratorium for assistance and guidance during its preparation, particularly Messrs. B.H. Purser, M.W. Hughes Clarke and R.R.J. Davilar.

Fig. 5. ALTERATION OF ARAGONITE CEMENT TO CRYPTOCRYSTALLINE CALCITE

A. Grainstone containing a prominent, algally-bored mollusk fragment (a) which has nucleated drusy aragonite cement (b). Neither the clean feldspar grain (c) nor the clean quartz grains (d) have seeded cement. The aragonite cement has been marginally altered to cryptocrystalline calcite (e) next to remnant pore space (f).
Thin section, plane polarized light, x 50, Evm 610, Quaternary, Sabkha Matti, Trucial Coast.

B. Grainstone containing a prominent gastropod (a), which is rimmed by acicular crystals of cement (b). The most recent cement, however, next to the remnant pore space, is composed of clear equant crystals (c). Between the acicular and equant crystals is a zone of cryptocrystalline carbonate (d). The present mineralogy of all the textures of cement shown is calcite.
Thin section, crossed polarized light, x 50, Evm 370, Quaternary, Ras Hanjurah, Trucial Coast

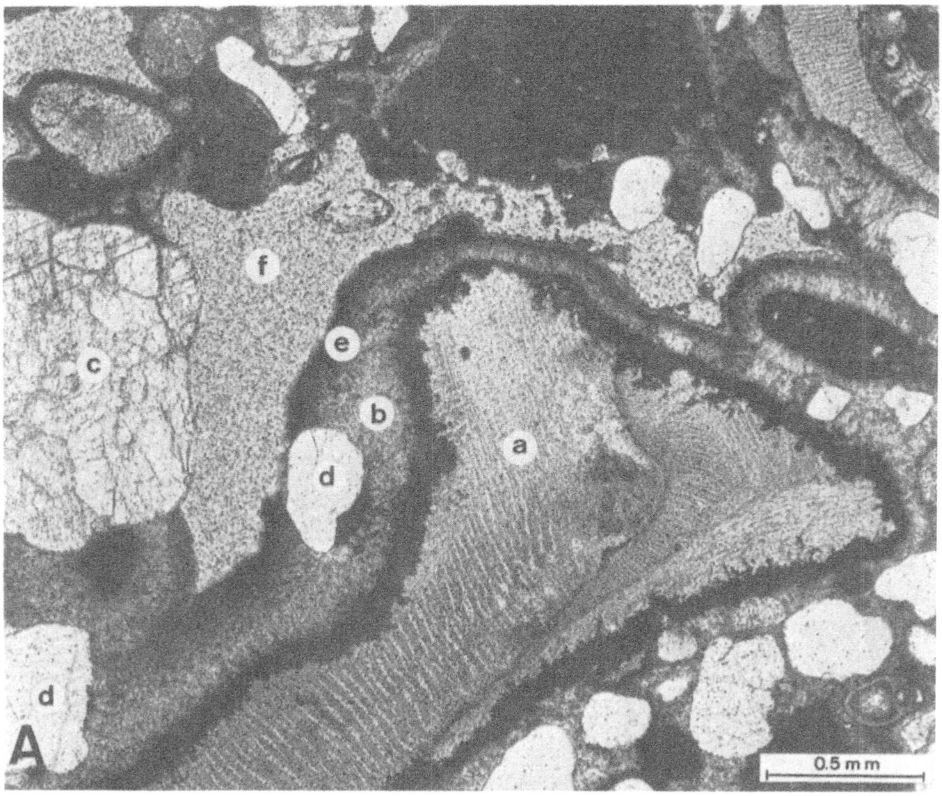

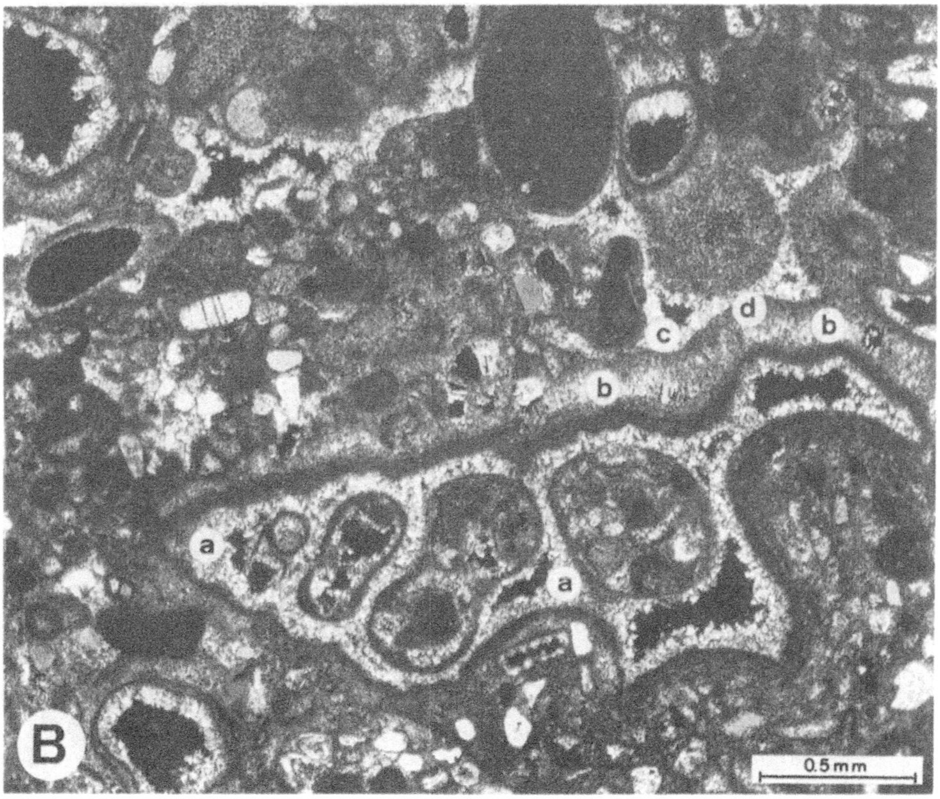

Legend to Fig. 5 A and B on the preceding page

Fig. 6. CRYPTOCRYSTALLINE CALCITE CEMENTS

Grainstone cemented by cryptocrystalline calcite (a). The distribution of this calcite, absent at the contacts between depositional grains (b) and uniformly developed where original pore space remains (c), suggests it to be a replacement of former drusy aragonite. Nevertheless, the aragonite mineralogy of the depositional components, for example the ooliths (d) and the mollusk fragment (e) is entirely unaltered. Where cryptocrystalline calcite entirely fills pores, as shown in the small rectangular field of view, a depositional grainstone may acquire a pseudo-packstone texture. Thin section, above: plane polarized light; below: crossed polarized light, x 100, M 2 (4.30 m), Quaternary, Sabkha Matti, Trucial Coast

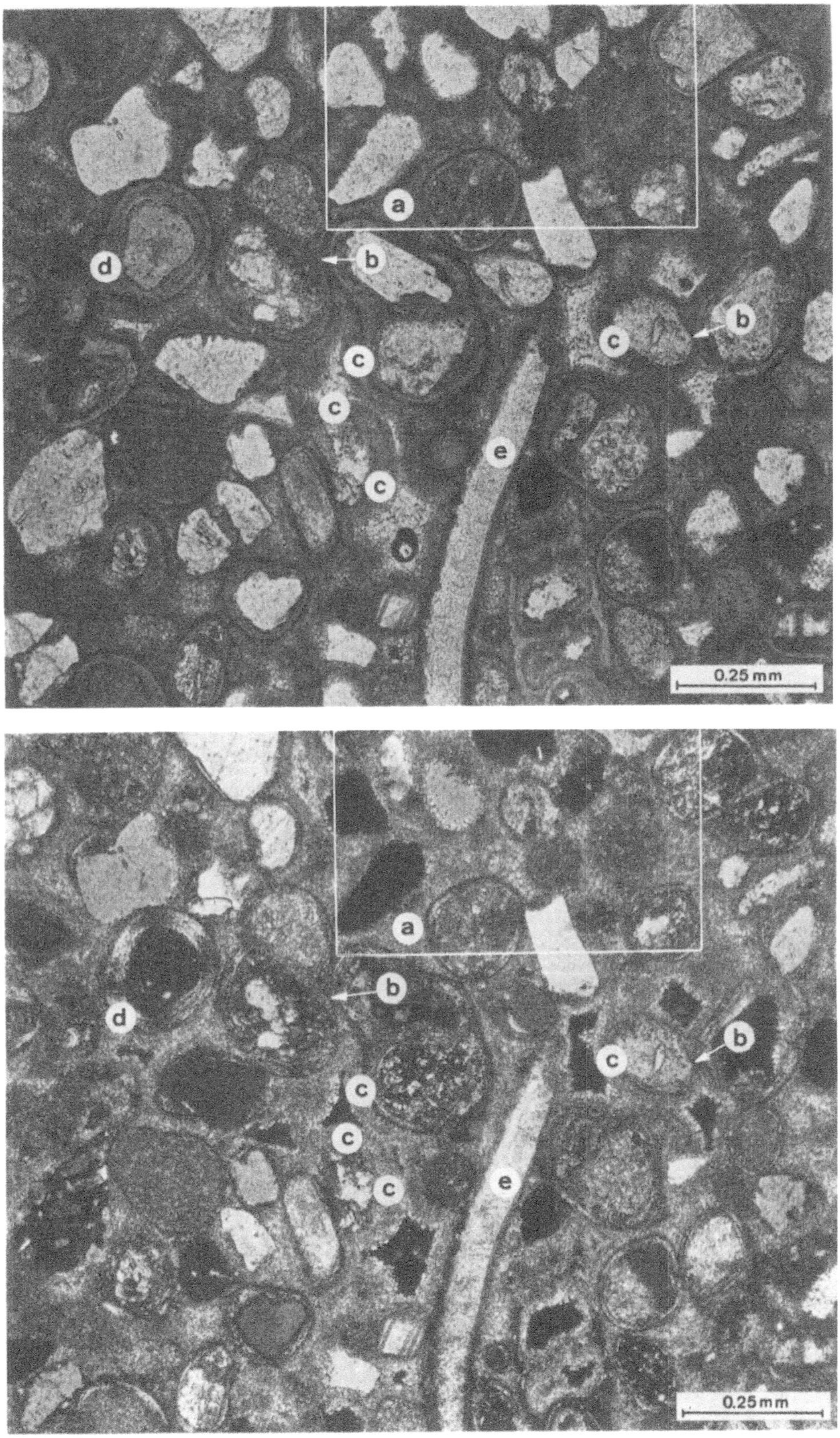

Legend to Fig. 6 on the preceding page

Aragonitic, Supratidal Encrustations on the Trucial Coast, Persian Gulf

B. H. Purser[1] and J.-P. Loreau[2]

ABSTRACT

Strontium-rich aragonite crusts exhibiting various morphologies coat both discreet particles and beach rock within the supratidal zone of the Trucial Coast. They are best developed in protected lagoonal settings, although polished "pelagosite" crusts are common within the "splash zone" exposed to wave action.

The morphology, both of coated grains and associated aragonite sheets, changes from smooth to highly irregular as one crosses into the upper parts of the supratidal zone. The latter frequently have dripstone morphologies. Micro-dripstones are developed on the under surfaces of many beach rock layers and are composed partly of detrital sediment which is deposited on the roof of small caverns by a rising water table to produce "sedimentary dripstone".

Although many of these aragonitic encrustations resemble lithified algal stromatolites, their dripstone morphologies, nanostructures and invariably lithified character strongly suggests that they are formed by physico-chemical precipitation from sea water. Criteria are given which may permit the distinction between supratidal tufas (here termed "coniatolites"), lithified stromatolites, and cave tufas.

I. INTRODUCTION

That the inter and supratidal zones of the Persian Gulf are environments of considerable diagenetic activity is clearly indicated by the widespread occurrence of dolomite and associated evaporite minerals. Most of these diagenetic products are related to intense evaporation of interstitial sea water. Possibly the most widespread mineral, however, is aragonite precipitated in the form of mud (Kinsman, 1969), ooid cortexes, and beach rock cements. Aragonite occurs also as laminated crusts on hard substrates, its various morphologies, structures, and processes involved in its formation being the subject of this contribution.

Aragonitic crusts have been noted by Evamy (in this volume) at various places on the Trucial Coast. Those studied by the present authors occur in the W parts of Abu Dhabi (Fig. 1). All are confined to a precise environment - the supratidal zone and the uppermost limits of the intertidal zone. While most are developed within very protected lagoonal settings or on wide tidal flats, certain types encrust the "splash zone" on exposed cliffs.

---

1 Lab. de Géologie Historique, Université de Paris Sud (Orsay)

2 Lab. de Géologie, Muséum National d'Histoire Naturelle, Paris.

The morphology of these aragonitic encrustations is exceedingly variable. Those described by Evamy are highly polished, giving a superficial impression of "grey enamel paint", while others have porous surfaces which are highly irregular, closely resembling certain spring tufas or travertines. These crusts, invariably composed of aragonite, have two fundamental styles. They may coat hard substrates, especially beach rock, and thus extend as a narrow band along the lower limits of the supratidal zone. Aragonite may also coat rock fragments and mollusks, and thus occur as discreet grains whose sizes range from less than 1 mm to more than 30 cm. These coated grains almost invariably occur together with the sheet-crusts, the two types exhibiting identical surface morphologies. Both are clearly the product of a common process, their granular or sheet geometries being determined by the substrate available.

These supratidal aragonite crusts are of interest, not only because they seem to characterize a very precise marine environment, but also because they closely resemble both algal stromatolites and certain non-marine encrustations, including cave tufa. Their recognition may therefore be of some significance, especially in ancient analogues. Although they could easily be misinterpreted as algal stromatolites, this would not be disastrous in terms of environment. However, to confuse them with certain non-marine encrustations could lead to the creation of non-existant unconformities and excessive degrees of paleo-emergence.

The fieldwork upon which these results are based was carried out in May, 1971. Its organization was greatly facilitated by the Abu Dhabi Petroleum Company (ADPC), to whom the authors express their sincere thanks.

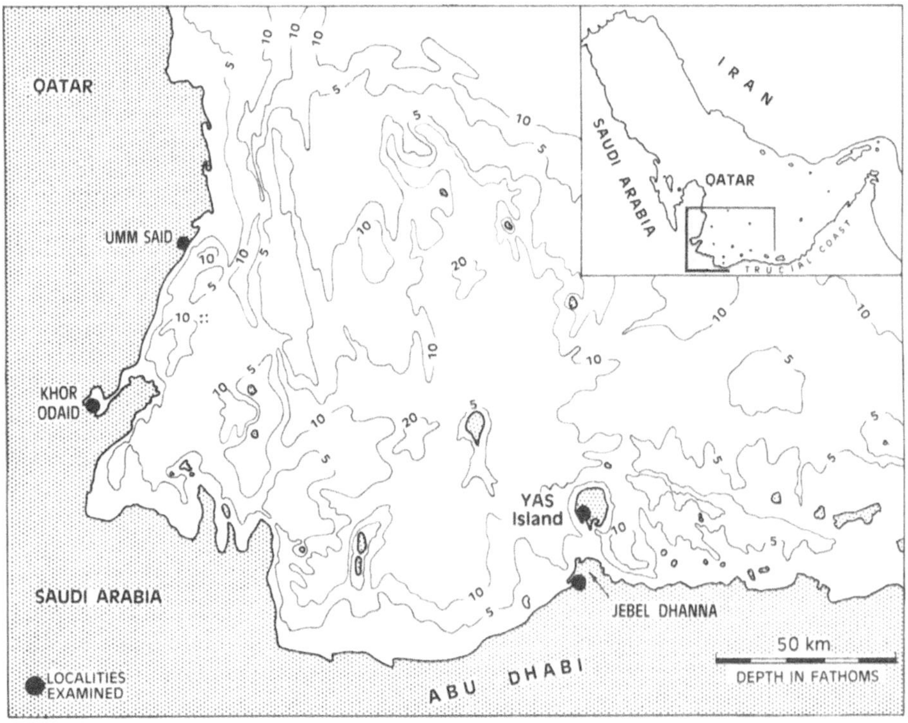

Fig. 1. Map showing the distribution of localities where aragonitic encrustations have been studied

II. NATURE AND DISTRIBUTION OF ARAGONITIC ENCRUSTATIONS

A. KHOR ODAID LAGOON

This 5-10 km wide coastal lagoon situated at the SE extremity of Qatar Peninsula (Figs. 1 and 2) is almost completely isolated from the open waters of the Persian Gulf to which it is connected by a narrow channel. It is not surprising therefore that its waters precipitate gypsum at the more secluded, leeward side, its SE, windward shores consisting essentially of the oolitic sand discussed by Loreau and Purser elsewhere in this volume. These southern shores are flanked by 100-500 m of intertidal sand flats, and by a low beach berm (30-100 cm high) frequently lithified to beach rock. This beach grades imperceptibly into wide (> 2 km) quartz sand sabkhas with high aeolian dunes. These sabkhas decrease in width towards the W as the mesas of Tertiary dolomite approach the shore; the extreme SW shores comprise a series of small, lunate bays whose intertidal sand flats and beach berm abut directly against the cliffs of Tertiary dolomite. The beach berm and adjacent cliffs flanking these isolated embayments are coated locally by various types of aragonitic crust (Fig. 2).

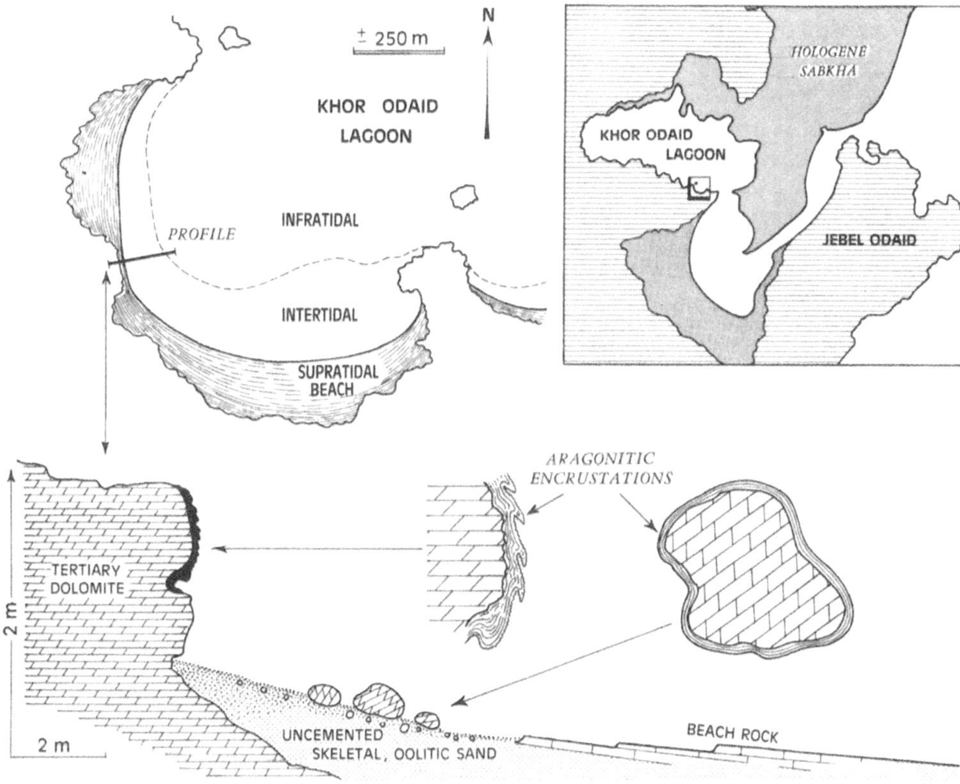

Fig. 2. Khor Odaid locality: Sketches showing the distribution of aragonitic crusts, and their lateral variation in morphology

<u>Coated grains:</u> Sand-sized quartz grains are coated with polished aragonite to form ooids, discussed by Loreau and Purser (in this volume). Locally along the SW shores of Khor Odaid, blocks of beach rock or Tertiary dolomite which have fallen from the adjacent cliffs, are also coated with a pellicle of polished aragonite (Fig. 3A).

These coated blocks may attain 50 cm in diameter and are completely enveloped by the blue-grey laminated cortex of constant thickness. These boulder-sized "pisoliths" are embedded within the oolitic and skeletal sands of the beach but are not welded to adjacent blocks or grains. The micro and ultrastructure of these aragonitic coatings is discussed below.

These sub-rounded cobbles and boulders coated with enamel-like aragonite similar to that discussed by Evamy (in this volume) are a variety of coated grain having much in common with classical ooids. They are, of course, abnormally big, their large size precluding any significant movement or transport with which the process of "oolitization" could be associated. Field relationships and shape clearly indicate, however, that these blocks of Tertiary dolomite and Holocene beach rock, which serve as "nuclei" for these coated grains, have rotated on the beach. This turning, either by waves, or by gradual movement down the meter-high berm, probably permitted their complete envelopment by 0.5-1 mm thick aragonitic cortex.

Polished aragonite crusts: The pellicle which locally coats loose blocks of dolomite occurs more frequently as a sheet encrusting beach rock. These grey or brownish "pelagosite" crusts (see Evamy, in this volume) constitute a band 1-5 m wide which extends laterally along the limits of the inter and supratidal zones somewhat exposed to onshore waves (Fig. 3B). Although the crusts follow the micro-relief of the beach rock surface they tend to be somewhat thicker within local depressions where they may attain 2 cm in thickness. The surface of these crusts is invariably polished, although this polish is somewhat dulled along the landwards limits of the crust due possibly to local dissolution by dew or to numerous microperforations (Fig. 6F).

The authors' observations agree with those of Evamy concerning the relationship between the presence of "pelagosite" and the nature of the rock substrate: this crust is developed only on non-porous rock surfaces. At Khor Odaid this selective coating is strikingly expressed where clasts of beach rock have been re-incorporated into a subsequent generation of beach rock. The clasts of first generation beach rock are dense and their exposed surfaces coated with a beige pellicle of polished aragonite while the somewhat porous, second generation beach rock, lacks the external pellicle of "pelagosite".

Light microscope study indicates that the "pelagosite" crusts are composed of laminae whose individual thickness rarely exceeds 50μ (Fig. 4A). Optical study with quartz wedge showed that in 80 % of cases these laminations have a negative optic elongation (comparable to the cortex of classical ooids), and therefore must be composed of crystallites oriented tangentially to the substrate. SEM examination reveals that the polished surface has a coalescent fabric (Fig. 6E) while the laminae in vertical section are seen to consist of at least five nanofabrics:

- a compact mass of nanograins which coalesce to form flat, polycrystalline lamellae oriented tangentially to the substrate, is the most frequent component (Figs. 4F,G);
- a compact mass of non-coalescent nanograins is also very frequent (Figs. 6B,C);
- compact assemblages of subhedral crystals each measuring 0.1μ in diameter and less than 1μ in length are statistically tangential to the substrate; they are associated with a suite of lamellar crystals arranged parallel to each other, and radially with respect to the substrate (Figs. 5C,D);
- compact series of subhedral or lamellar crystals which often exhibit a very pronounced radial orientation (Fig. 5B);
- a loose, non-oriented felt of euhedral crystals each measuring 0.2-0.4 μ in diameter and 1-3 μ in length, generally located between compact lamellae (Figs. 4D,E).

DETAILS CONCERNING FOLLOWING FIGURES

Fig. 3. Polished, aragonitic, "pelagosite" encrustations:

A. coating boulders of Eocene dolomite, intertidal zone, Khor Odaid Lagoon.

B. coating Holocene beach rock, Khor Odaid Lagoon

Fig. 4. Micro and nanostructures of "pelagosite" crusts:

A. Section through smooth aragonite (pelagosite) encrusting Tertiary dolomite, lower supratidal zone, Khor Odaid Lagoon, showing fine lamination composed essentially of tangentially oriented crystallites (thin-slide, crossed nicols, X 45)

B. Fragment of pelagosite crust from Khor Odaid showing characteristic laminated structure (SEM, X 15)

C, D. Pelagosite crust showing dense laminations separated by layer of loosely packed micro-crystals. (SEM, X 495 and X 4,500)

E. Details of felt of unoriented, euhedral crystals between dense laminae (SEM, X 8,100)

F, G. Nanostructure of compact laminae showing polycrystalline pseudo-lamellae (each composed of coalescent nanograins) oriented tangentially to substrate. (SEM, X 3,780 and X 7,560)

Fig. 5. Nanostructures of "pelagosite" crusts:

A. Section through pelagosite crust showing more or less dense lamination of varying fabric. NB: B and C on photo show location of photos B and C. (broken surface, SEM, X 360)

B. Individual lamina consisting of subhedral crystals oriented radially to the substrate. (SEM, X 4,500)

C. Individual lamination consisting of elongate anhedral or subhedral crystals each 0.1μ in width and less than 1μ in length, whose orientation is statistically tangential with respect to the substrate. (SEM, X 18,900)

D, E. Stereo-pair (detail of photo C) demonstrating more precisely the orientation of the nanocrystals which are both statistically tangential to the support and somewhat parallel with respect to each other. (SEM, X 12,150)

Fig. 6. Nanostructures of "pelagosite" crusts:

A. Section through pelagosite crust from lower supratidal zone, Khor Odaid Lagoon. (Fractured surface, SEM, X 72)

B. As for photo A permitting localization of photo C. (SEM, X 4,050)

C. As for photo B showing the dense mass of non-coalescent nanograins each measuring 0.1 - 1μ (SEM, X 14,400)

D. External surface of polished pelagosite crust from Khor Odaid. (SEM, X 14)

E. External surface of pelagosite showing somewhat coalescent nanostructure. (SEM, X 16,200)

F. Pelagosite from landwards limits of pelagosite zone at Khor Odaid whose somewhat unpolished surface is partially due to numerous micro-perforations (SEM, X 54)

Fig. 3. "Pelagosite" crusts, intertidal zone, Khor Odaid

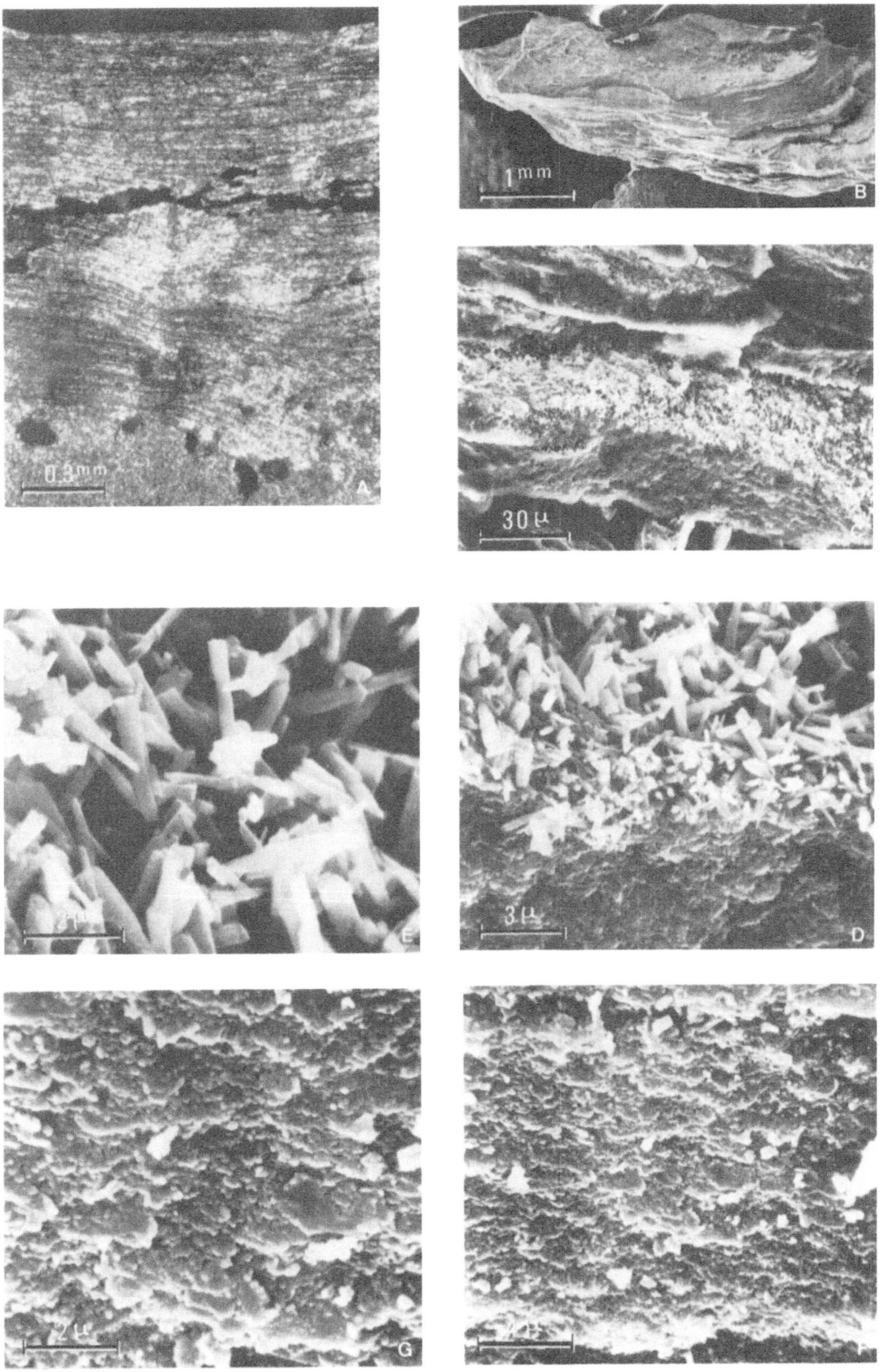

Fig. 4. Micro and nanostructures of "pelagosite" crusts

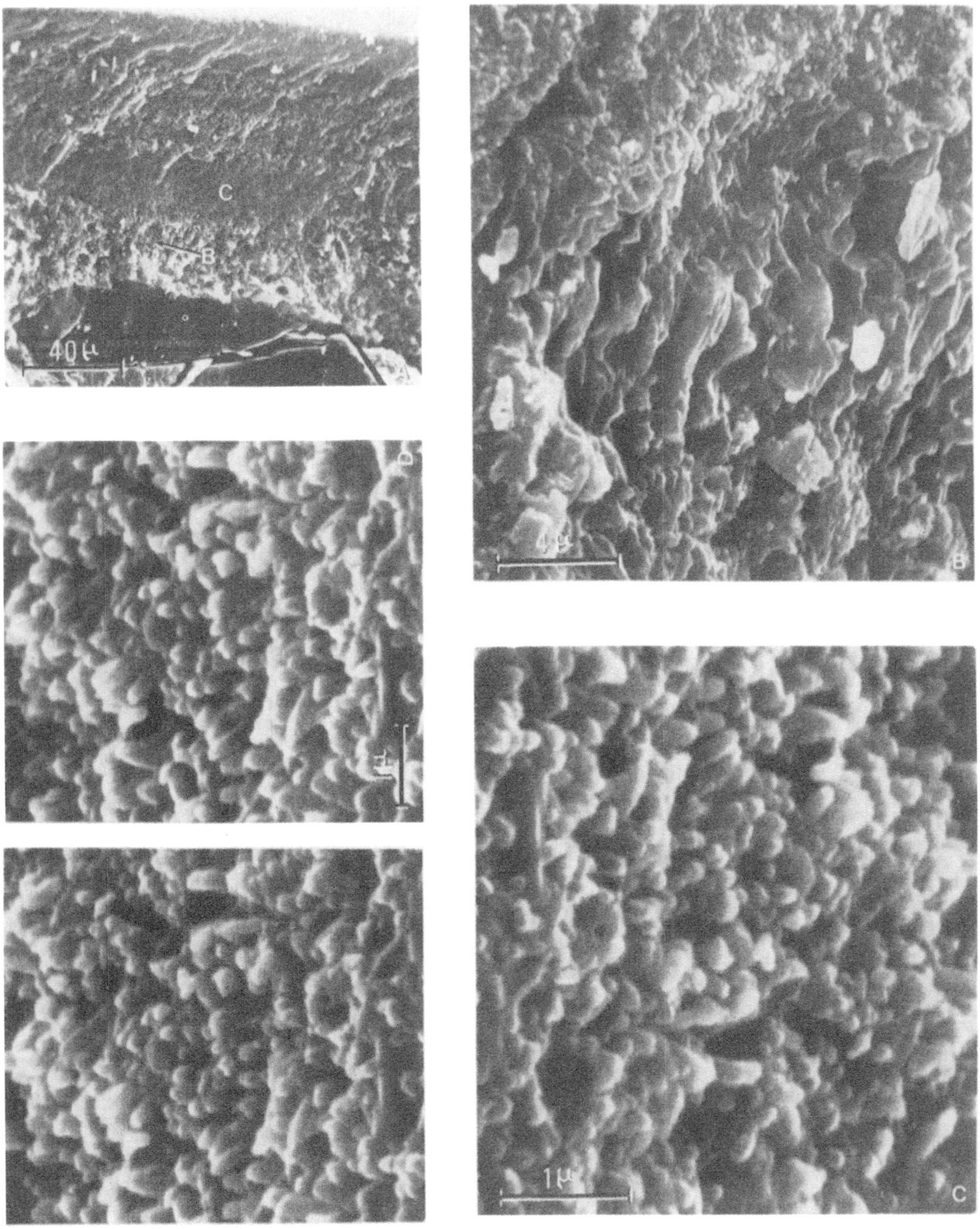

Fig. 5. Nanostructures of "pelagosite" crusts

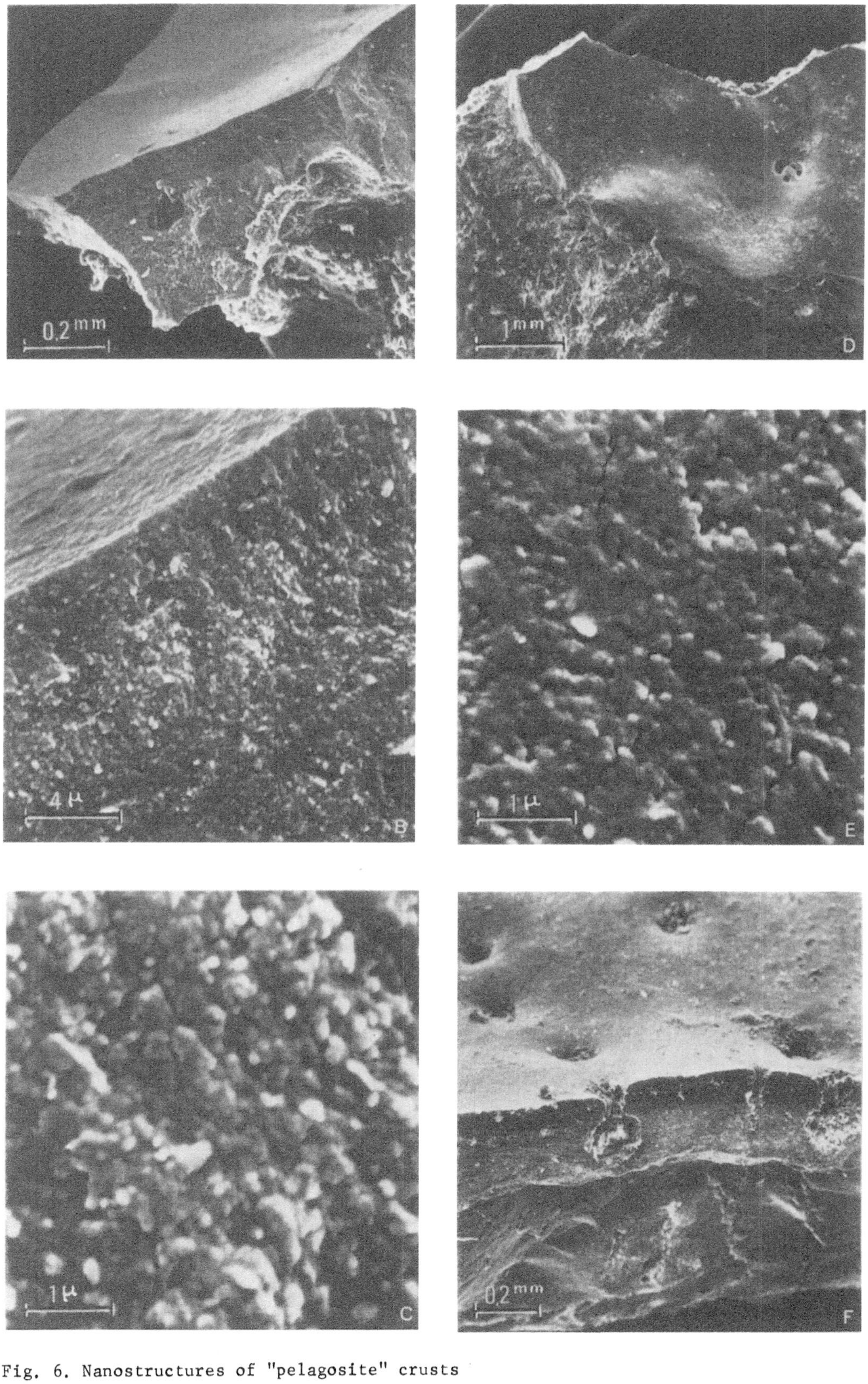

Fig. 6. Nanostructures of "pelagosite" crusts

DETAILS CONCERNING FOLLOWING FIGURES

Fig. 7. Field photos of irregular aragonite crusts at Khor Odaid:

A. Low cliff of Tertiary dolomite coated with a highly irregular crust of dark brown, unpolished aragonite.

B. The external morphology of the crust recalls that of certain cave or spring tufas.

Fig. 8. Details of morphologies of crust on cliff at Khor Odaid:

A. A reticulate system of ridges related to the inclination of the substrate, and lobate stalactites.

B. The aragonitic crust with a lobate, drip-stone morphology where the substrate is overhanging.

C. Ledge decorated with small stalagmites of aragonite.

Fig. 9. Micro and nanostructures of crusts on cliff at Khor Odaid:

A. Section through irregular aragonite crust on Tertiary dolomite in upper part of supratidal zone in Khor Odaid Lagoon (see Fig. 7), showing the micritic fabric and trapped ooid and quartz grains. (thin slide, polarised light, X 45)

B. As for photo A, with crossed nicols.

C. Section through supratidal aragonitic crust at Khor Odaid showing typical, fine laminations. Optical measurements indicate that component crystallites are oriented normally to the substrate. (thin slide, X 45)

D. Photo showing relatively large crystals (possibly calcite) developed between laminae or fissures within the crust.
NB: X-rays analyses have shown traces of calcite in these crusts. (thin slide, crossed nicols, X 180)

E. Alternating light and dark laminations within upper supratidal zone at Khor Odaid. Optic measurements indicate that component crystallites are oriented normally to the lamination. (thin slide, crossed nicols, X 450)

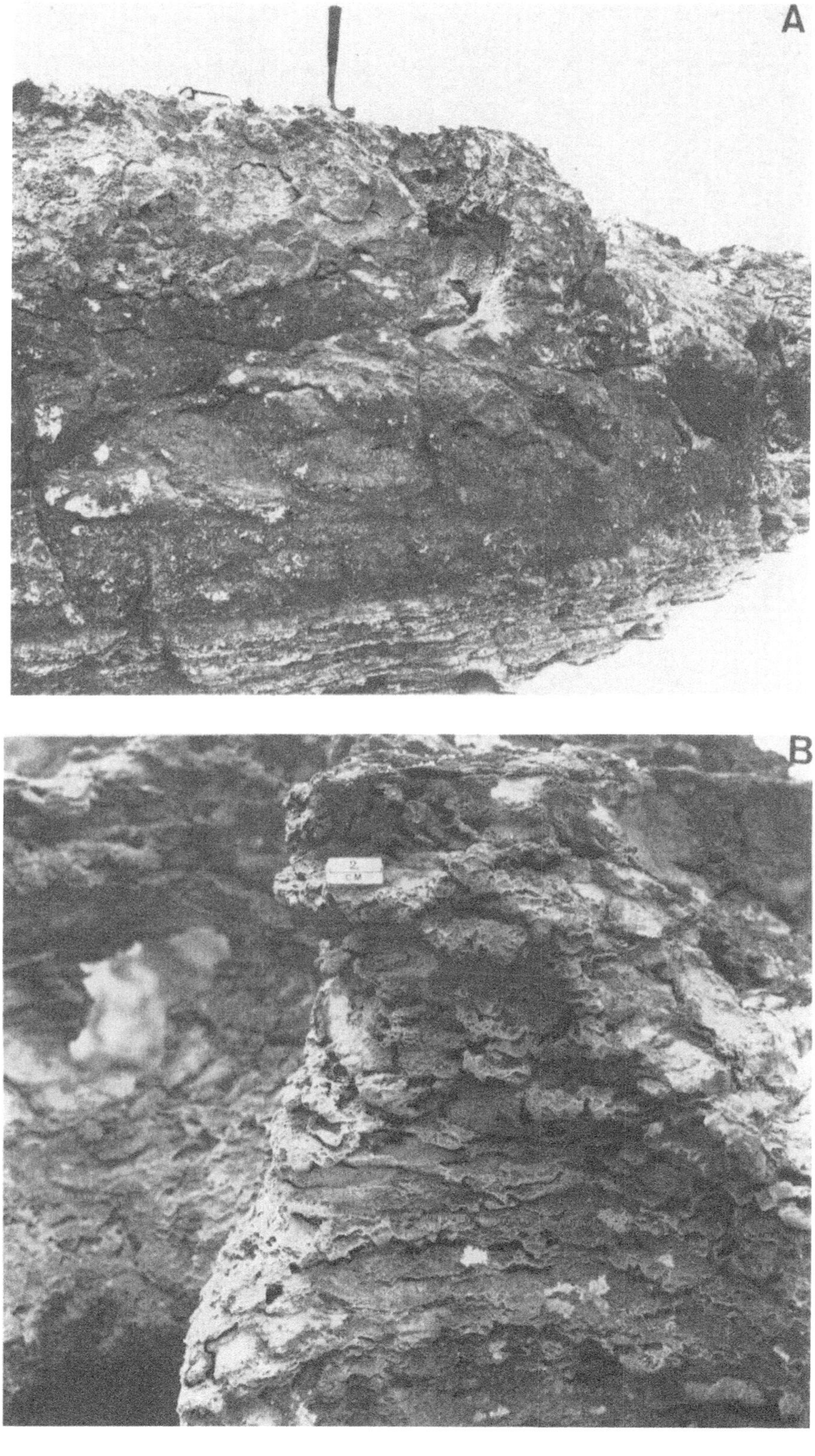

Fig. 7. Field photos of irregular crusts, Khor Odaid

Fig. 8. Detailed morphology of irregular crusts coating cliff at Khor Odaid

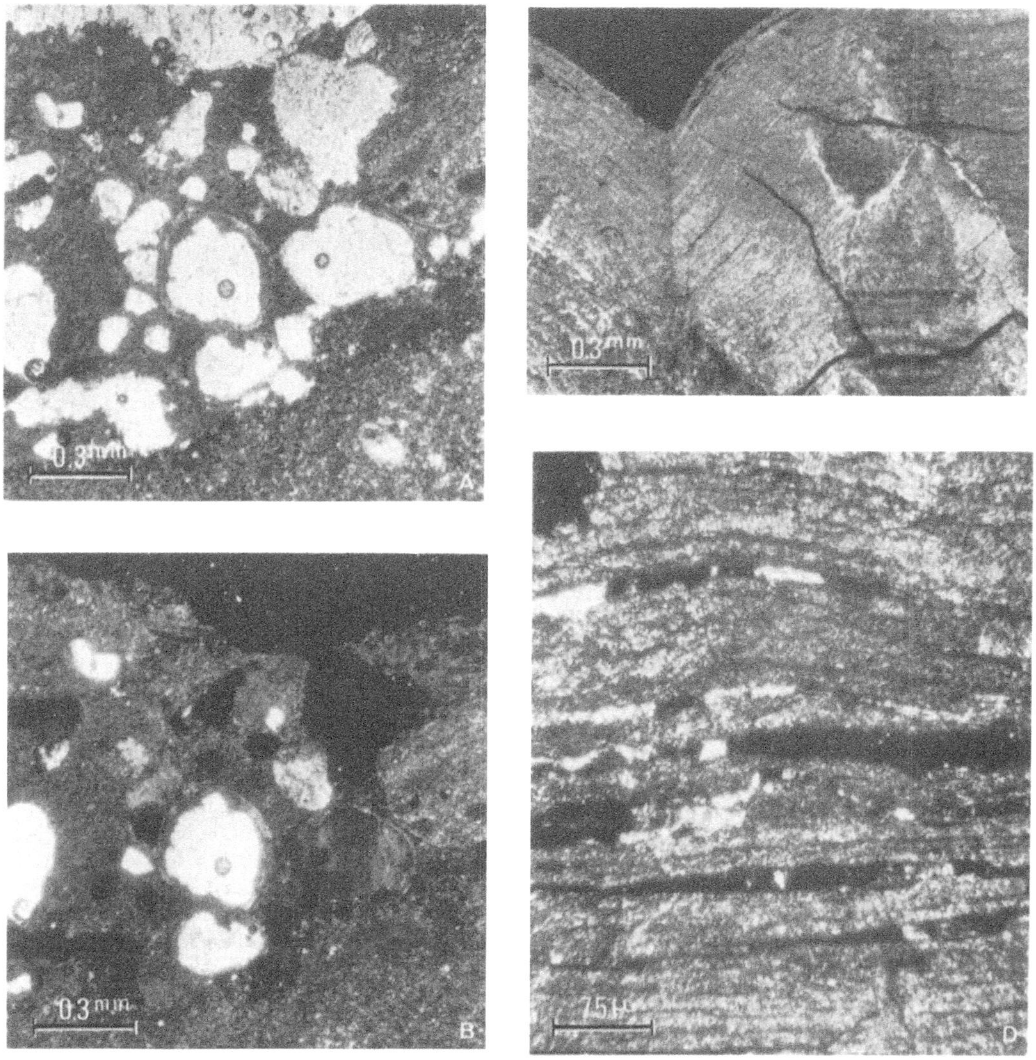

Fig. 9. Micro and nanostructures of crusts on cliff at Khor Odaid

This nanofabric differs somewhat from that of the classic ooid cortex in the more compact assemblages of crystallites and in the abundance of nanograins and lamellar shaped crystals. The tangential orientation of these elements may be due to the streaming action of running water, these crusts occurring, almost invariably, in exposed coastal settings in the lower part of the "splash zone" where sea water tends to run in a fairly constant direction down the inclined slope. The more brownish crust along the landward limit of the pelagosite zone is generally more heterogeneous with frequent micritic structure and trapped clasts, including quartz and ooids (Figs. 9A, B).

Irregular aragonite crusts: The polished crusts and coated grains are limited to the lowest part of the supratidal zone and the "splash zone" on exposed cliffs. In certain localities within Khor Odaid however, cliffs of Tertiary dolomite abutting directly onto the beach are coated by a highly irregular crust of dark brown, unpolished aragonite (Fig. 7A). These latter crusts seem to be limited to the highest parts of the "splash zone" approximately 1-1.5 m above the top of the beach berm. Their external morphologies recall those of certain cave or spring tufas (Fig. 7B). Where the substrate is sub-vertical this 1-20 mm thick crust has a characteristic crenulated morphology (Fig. 8A), the reticulate system of ridges being related to the inclination of the substrate; most extend laterally around the inclined surface and are distinctly asymmetric in cross-section, their lower sides being much the steeper. Where the substrate is overhanging the aragonitic crust assumes a lobate, dripstone morphology (Fig. 8B), individual stalactites attaining 5 cm in length. Adjacent ledges may be decorated with small stalagmites of aragonite (Fig. 8C), these morphologies confirming the influence of dripping water.

The microstructure of these unpolished, irregular crusts coating the low cliffs of dense Tertiary dolomite is complicated, and quite different from that of the smooth "pelagosite" crusts. In thin slides the very irregular crust near the top of the cliff (Fig. 8A) consists of porous, vaguely branching masses of carbonate somewhat resembling the structure of travertine. Lower down the cliff the smoother crust is composed of very thin sombre laminae alternating with clear laminae of fibrous crystals (Fig. 9C). Minute, sheet-like cracks are filled with bigger crystals (Fig. 9D). SEM examination of these lower crusts shows that they are composed mainly of elongate lamellar crystals which tend to be oriented radially with respect to the substrate . Their aragonitic composition and close proximity to the shore indicates that the parental waters from which these crusts have precipitated are almost certainly marine. It is therefore of interest to note that the classical dripstone effects, normally associated with continental environments, may also occur on a small scale within an environment which is essentially marine.

B. JEBEL DHANNA LAGOON

The salt diapir which has formed the volcano-like promontory of Jebel Dhanna has been modified along its seawards flanks by a series of spits of bioclastic sand which have accreted laterally enclosing a shallow lagoon (Fig. 10) 1 - 2 km in width. The lagoon is virtually dry at low tide and its surrounding flats of pellet sand are coated locally with a narrow strip of typical algal mat marking the upper limits of the intertidal zone. The algal mat is a flat sheet several mm in thickness developed on soft sediment. It is not lithified.

The lagoonal flanks of the protecting spits are extensively lithified, differing degrees of beach rock cementation of the inclined bedding producing a series of arcuate ridges which slope towards the adjacent intertidal sand flats. These gently inclined layers of beach rock and associated gravels are heavily encrusted with unpolished aragonite. As in Khor Odaid lagoon, the morphology of these crusts is highly variable, exhibiting a progressive change as one passes up the inclined layers of beach rock (Fig. 11).

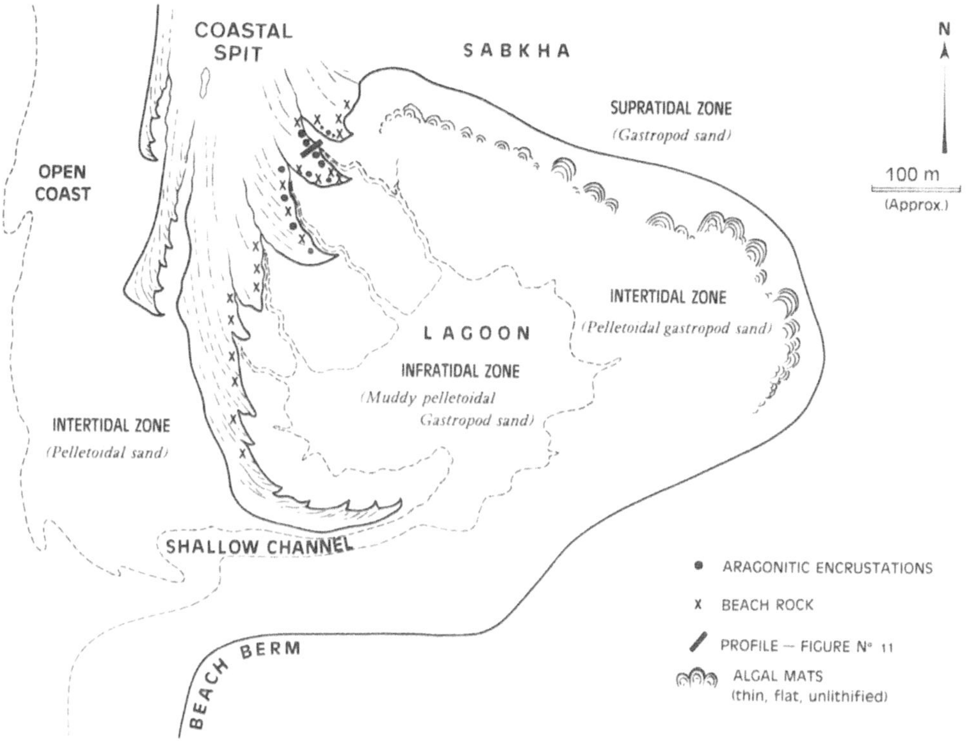

Fig. 10. Map of Jebel Dhanna locality showing the distribution of aragonitic crusts

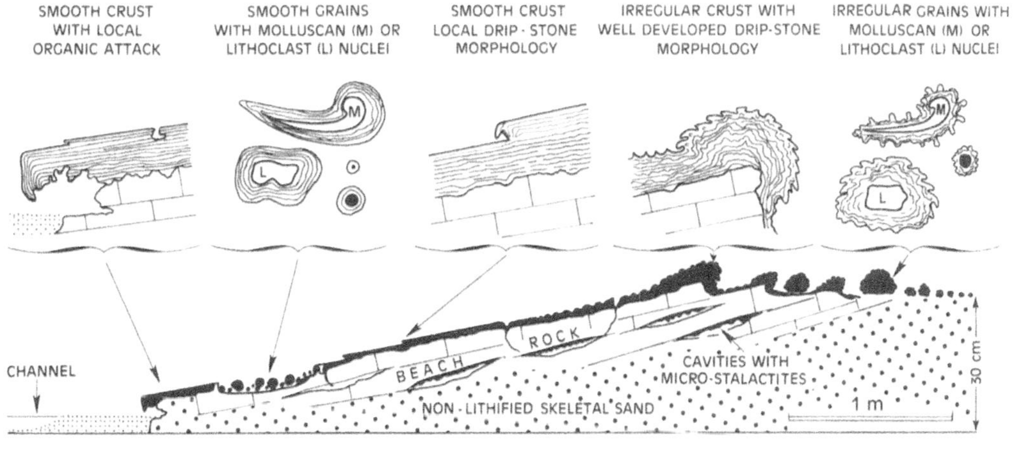

Fig. 11. Profiles across the supratidal zone at Jebel Dhanna locality showing progressive changes in morphology of aragonitic crusts

Coated grains: Detrital grains, including rounded quartz sand, sub-rounded gravel-sized clasts of beach rock, and irregularly shaped mollusks, are completely enveloped in a layer of earthy looking, laminated aragonite. The size and shape of these grains is determined by two main factors : the character of the nucleus, and the micro-environment in which the grain occurs

Quartz grains are small and rounded and the resultant coated grain has a similar external morphology. These grains have been discussed by Loreau and Purser in this volume, and may be regarded as ooids or pisooids, although they lack the characteristic polish and fine laminations of classical marine ooids. Larger nuclei, consisting of fragments of beach rock, or various mollusks, are less regular in shape (Fig. 13B, D). They attain diameters of 30 cm. Those within the lowermost part of the supratidal zone have smooth, slightly chalky surfaces. Others, including numerous cerithid gastropods concentrated within shallow depressions in the higher parts of the supratidal zone (Fig. 14) are highly irregular in shape; size and general morphology, although determined by the nature of the nucleus, is modified on a smaller scale by a complicated system of closely spaced ridges and short lumps. These decorations give these grains a distinct "algal" appearance (Fig. 15).

The cortex of the coated grains at Jebel Dhanna varies both in thickness and structure depending on the size of the nucleus and position within the supratidal zone. Sand-sized nuclei have a regularly laminated cortex up to 1 mm in thickness composed of radially disposed, acicular crystals (Loreau and Purser, Fig. 17F, in this volume), or 0.3 μ sized nanograins of aragonite. Their structure, identical to that of pisooids occurring on the tidal flats near Umm Said in SE Qatar (see Shinn, in this volume), is discussed by Loreau and Purser elsewhere in this symposium. Gravel or cobble-sized nuclei, consisting of large mollusks or rock fragments, have aragonitic coatings ranging in thickness from ca. 1 mm to more than 2 cm (Fig. 15B). These coatings exhibit irregular concentric laminations closely resembling those of algal oncoids, but differ somewhat from these latter in that many laminae may be traced completely around the nucleus. These laminations tend to be more regular in grains occurring in the lower parts of the supratidal zone. Grains in the upper parts of the supratidal zone, whose characteristic surface decorations have been noted above, tend to have very irregular laminations. Close inspection suggests that these are earlier surface morphologies buried in subsequent cortical layers.

Polished sections through these irregular cortical layers show that certain of the larger depressions truncate cortical laminations indicating that the depressions are deepened by dissolution or abrasion (Fig. 15B). This local destruction almost certainly contributes to the complexity of the micro-relief characteristic of these grains.

Relatively small (1-5 mm) ooids and pisooids are coated with non-oriented, nanograined aragonite and lighter-coloured layers of euhedral, radially oriented aragonite crystals (see Loreau and Purser, in this volume). The size of these crystals is more comparable to that of acicular beach rock cement than to the much smaller rods of aragonite composing the cortex of classical ooids. Irregularly shaped grains, including the numerous cerithid gastropods in the upper parts of the supratidal zone (Fig. 14), are coated with aragonite whose main nanostructure consists of subhedral, elongate or flattened crystals (Fig. 16A, B) which seem to form by the coalescence of smaller isometric nanograins. Although there is a lower degree of preferred orientation relative to the ovoid grains in the lower parts of the supratidal zone, individual laminae may consist mainly of radially oriented (Fig. 16C), and more rarely, of tangentially oriented crystals. Individual protuberances on the surface of these grains generally consist of non-oriented masses of nanograins and felts of somewhat rounded rods 1μ in length.

Perforations averaging 100μ in diameter and 1 mm in depth are frequent and are sometimes partially filled with organic material (Fig. 16D, E). Other smaller perforations are partially filled with euhedral crystals which have grown within this micro-environment (Fig. 16F). Algal filaments are relatively rare. Insoluble residues, however, include varying amounts of organic material whose precise nature has not yet been determined; much of this material is finely granular and is probably derived from the fungi-like specks which are particularly abundant within the shaded depressions on the surface of these grains.

<u>Aragonitic crusts:</u> The aragonitic layers which coat discreet grains also encrust the adjacent surfaces of gently inclined layers of beach rock, in this respect re-

sembling the aragonitic encrustations at Khor Odaid. However, although there is a comparable lateral zonation from smooth crusts in the lower parts of the supratidal zone (Fig. 11) to irregular, tufa-like encrustations in the higher parts of this zone, in Jebel Dhanna Lagoon there are no polished "pelagosite" sheets. These latter develop mainly on rock substrates exposed to a certain degree of wave agitation; the beach rock developed along the lagoonal side of the coastal spit at Jebel Dhanna is apparently too protected to favour the development of these polished aragonite crusts.

The aragonitic layers coating the beach rock within the protected setting of Jebel Dhanna Lagoon nevertheless exhibit a progressive lateral variation in morphology identical to that of the associated grains (Figs. 11, 12, 13). Crusts near the lower limits of the supratidal zone are relatively smooth (but not polished). As one proceeds up the inclined beach rock surface the smooth crusts are replaced progressively by a more crystalline aragonite with characteristic micro-ridge morphology and cuspate lobes which are oriented down-slope (Fig. 13A) suggesting the influence of running water. This morphology is reminiscent of crusts at Khor Odaid, and certain non-marine tufas. The highest part of the beach rock, some 30 cm above the surface of the adjacent intertidal sand flats, has a complex morphology identical to that of the associated coated grains. This morphology is particularly well developed along the up-dip edges of beach rock layers (Fig. 12A). SEM study indicates that local surface depressions are deepened by perforations (Fig. 17B, C) contributing to the micro-relief of these crusts.

The crusts coating beach rock have nanofabrics essentially similar to those of the associated coated grains; composed mainly of nanograins and various, somewhat coalescent crystals which constitute oriented lamellae, certain layers nevertheless consist of radially oriented fibrous aragonite resembling beach rock cement. Local depressions contain non-lithified mud, probably of allochthonous origin (Fig. 17D, E, F, G).

TABLE 1 Analyses of aragonitic encrustations and comparison of their strontium concentrations with those of marine ooids and lagoonal muds

	% carbonate soluble in cold HCL	Aragonite calcite ratio	Mg in soluble carbonate (in % $MgCO_3$)	Strontium in soluble carbonate (in ppm)	Strontium in aragonite (in ppm)
Radially structured pisooids from Umm Said (SE Qatar) tidal flat	72.9	95/5		8380	8820
"Pelagosite" crusts from Khor Odaid	74	95/5	10	8590	9040
Irregularly shaped coated gravels from upper supratidal zone Jebel Dhanna	86	90/10	13	8420	9350
Average for all types of supratidal encrustation					9070 ± 300
Tidal delta ooids E Abu Dhabi					9100 ± 800
Aragonite muds Abu Dhabi lagoons					9700 ± 300

DETAILS CONCERNING FOLLOWING FIGURES

Fig. 12. Field photos showing inclined layers of beach rock encrusted with aragonite; supratidal zone, Jebel Dhanna:

A. Upper supratidal encrustations on up-dip edges of beach rock layers and around clasts of beach rock; note irregular morphologies.

B. Lower supratidal encrustations on inclined layer of beach rock showing relatively smooth surface with lobate irregularities oriented down slope suggesting dripstone origin

Fig. 13. Contrasting morphologies of crusts in upper and lower parts of supratidal zone, Jebel Dhanna locality:

A. Upper supratidal zone; irregular morphology of crust on beach rock layer.

B. Upper supratidal zone; irregular coating on mollusks.

C. Lower supratidal zone; smooth coating on beach rock.

D. Lower supratidal zone; smooth coating on mollusks and rock clasts

Fig. 14. Field photo of natural, uncemented accumulation of mollusks (mainly cerithid gastropods) each encrusted by an irregular pellicle of beige coloured aragonite

Fig. 15. Large, irregular shaped coated grains from upper supratidal zone, Jebel Dhanna locality.

A. Vertical view of external morphology.

B. Section showing fragment of beach rock (nucleus) coated with laminations which are cut locally by surface depressions (arrows) indicating that surface morphology is due partly to dissolution

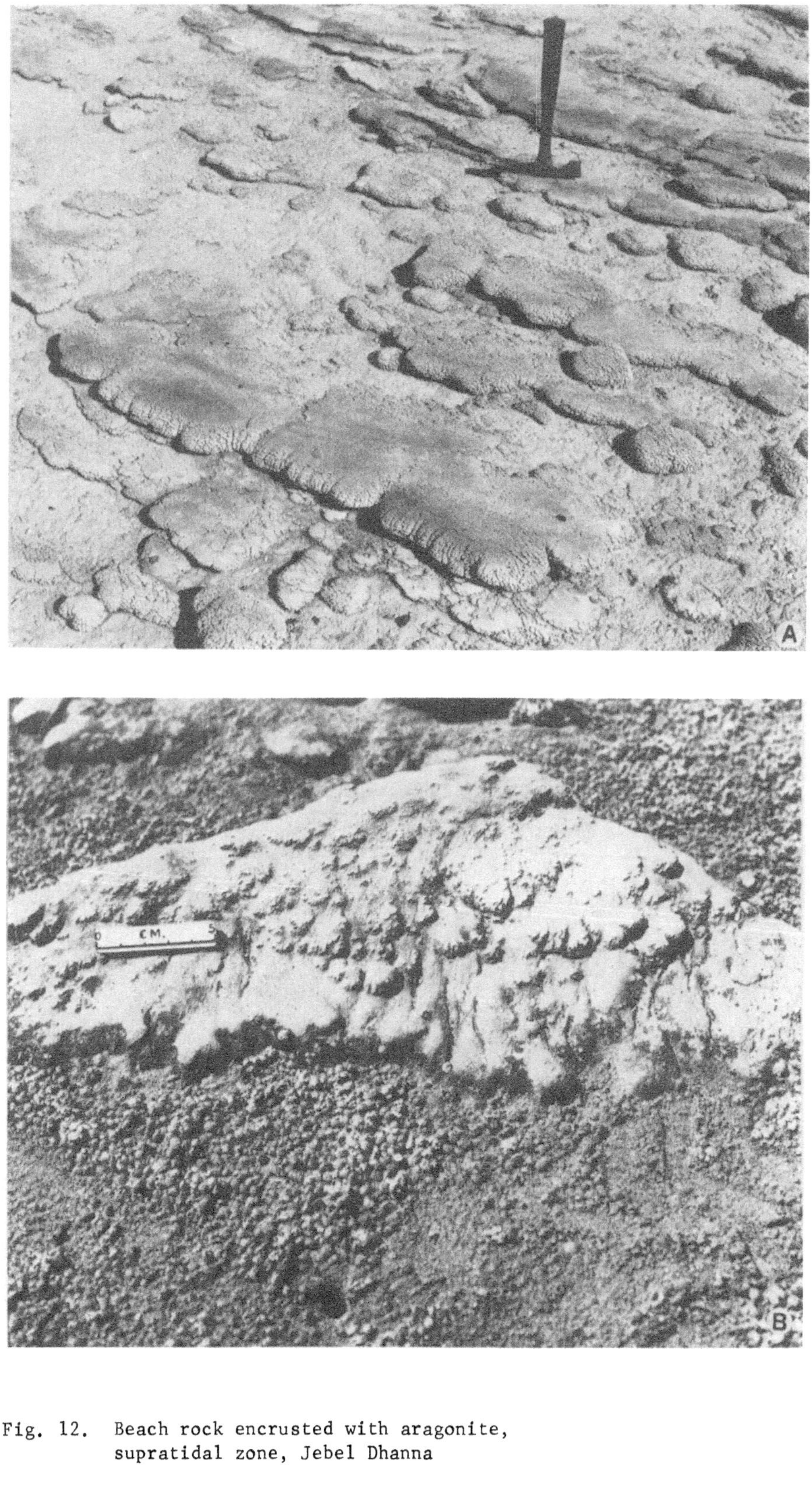

Fig. 12. Beach rock encrusted with aragonite, supratidal zone, Jebel Dhanna

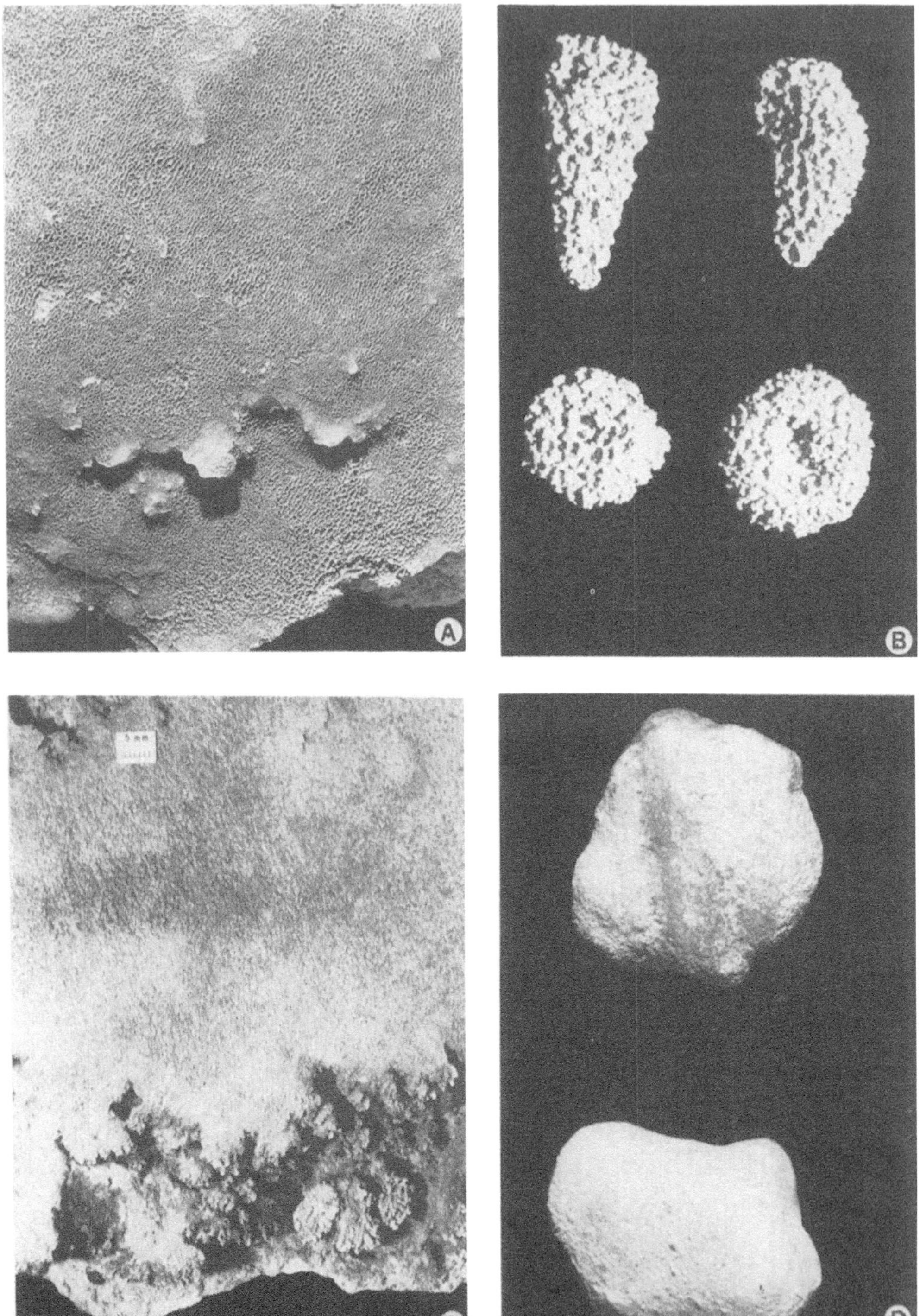

Fig. 13. Contrasting morphologies of crusts in upper (A, B) and lower (C, D) supratidal zones

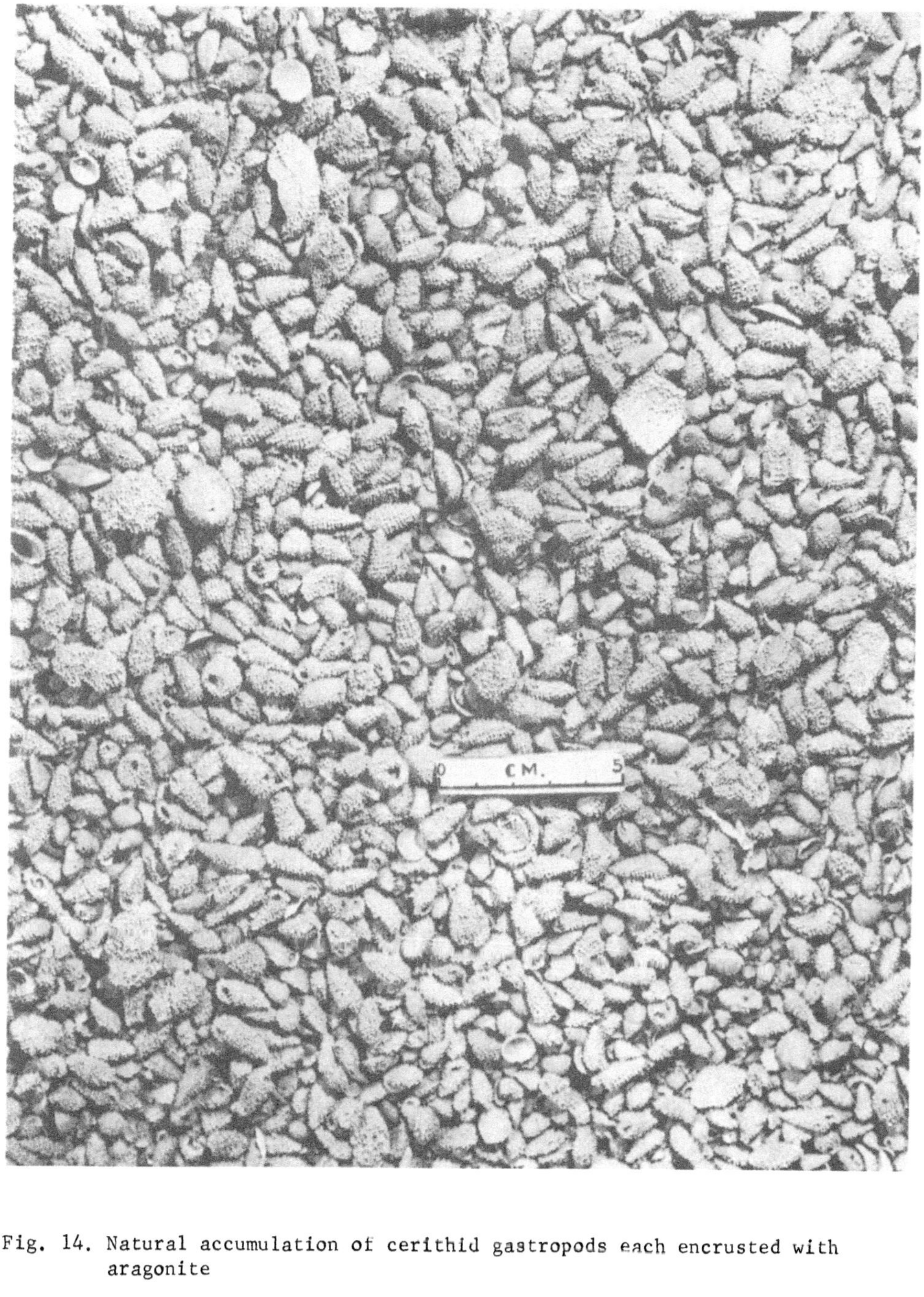

Fig. 14. Natural accumulation of cerithid gastropods each encrusted with aragonite

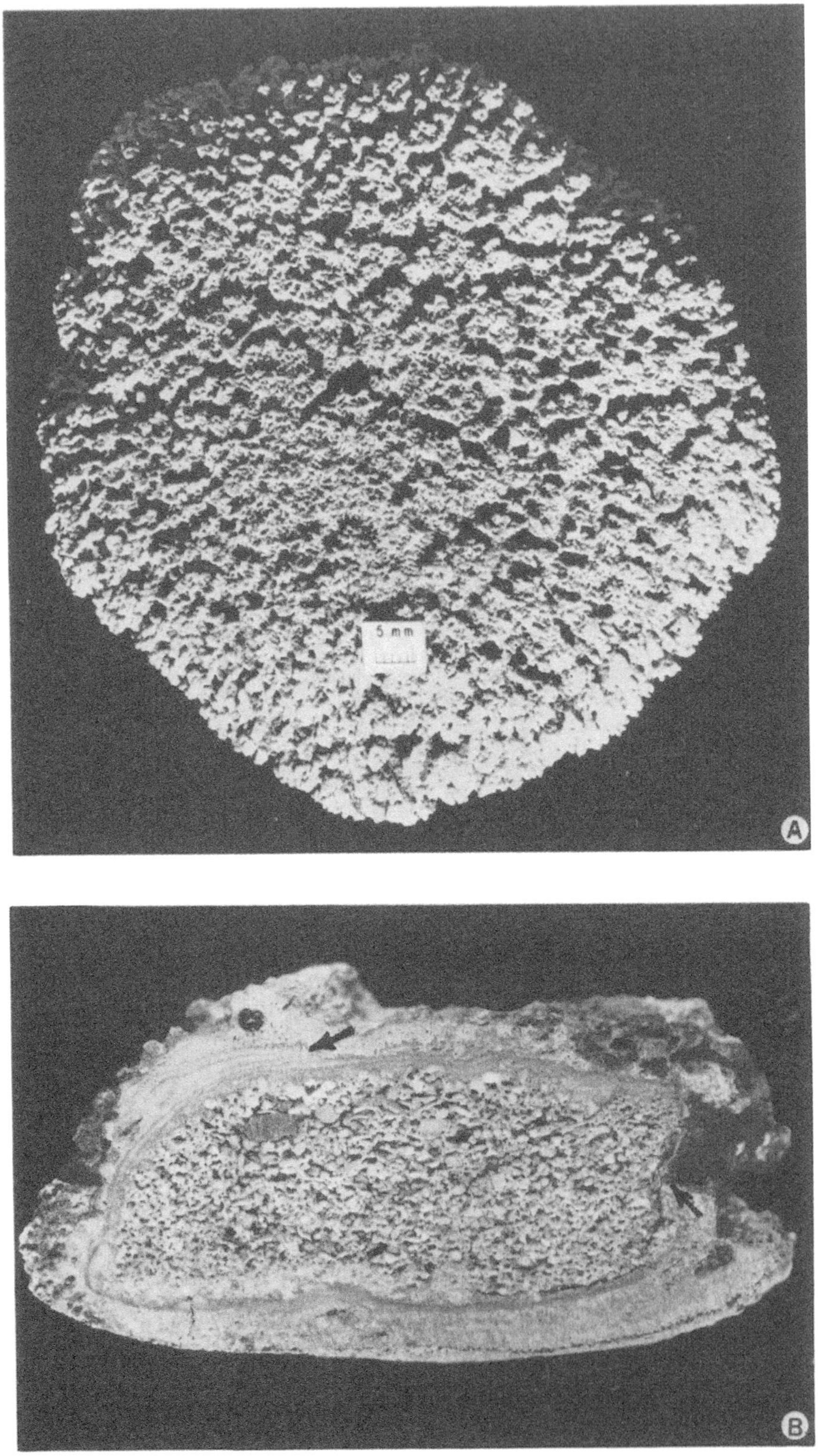

Fig. 15. Irregular, coated grains, upper supratidal zone, Jebel Dhanna

DETAILS CONCERNING FOLLOWING FIGURES

Fig. 16. Nanostructures and morphologies of aragonite crusts in supratidal zone, Jebel Dhanna locality:

A. Section through aragonite encrusting cerithid gastropod, upper supratidal zone. (broken surface, SEM, x 810)

B. Detail of external edge of photo A, showing flattened, subhedral crystals which appear to be composed of smaller, coalescent nanograins. (SEM, x 3,600)

C. Inner part of crust illustrated in photo B, showing radially oriented, euhedral crystals. (SEM, x 1,800)

D. External surface of irregularly coated grain from upper part of supratidal zone, showing numerous perforations averaging 0.1 mm in width and 0.5 mm in depth. (SEM, x 45)

E. Detail of photo D, showing perforation occupied by alga or fungi. (SEM, x 180)

F. Perforation several microns in diameter, partially filled with euhedral crystals which have grown within this micro-environment. (SEM, x 3,600)

Fig. 17. Details of the morphologies of aragonitic crusts within the supratidal zone, Jebel Dhanna locality:

A. Surface morphology of the aragonite encrusting beach rock in the upper supratidal zone, showing typical micro-ridges and depressions. NB. B = localization of photos B and C; D = localization of photos D, E, F, and G. (SEM, x 16)

B,C. Details of surface morphology; depressions have been deepened by perforations. (SEM, x 90; x 495)

D,E. Oblique view of outer surface and section through a micro-depression lined with unlithified mud which resembles that within the adjacent lagoon. (SEM, x 40; x 80)

F,G. Detail of mud within micro-depression showing fine skeletal debris and blunt rods of aragonite each measuring ca. 1μ in length. (SEM, x 3,825; x 8,100)

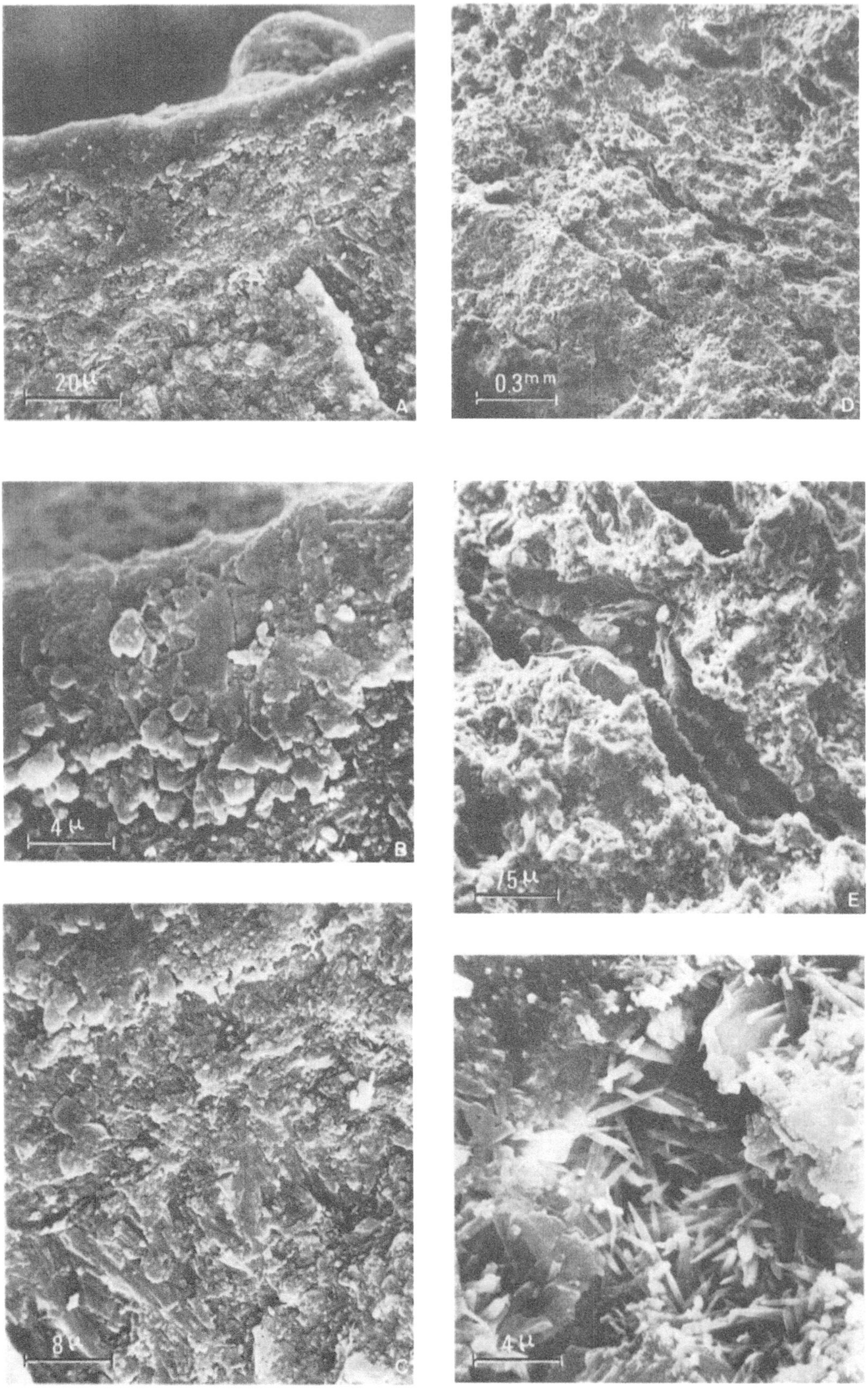

Fig. 16. Nanostructure and morphology, supratidal crusts

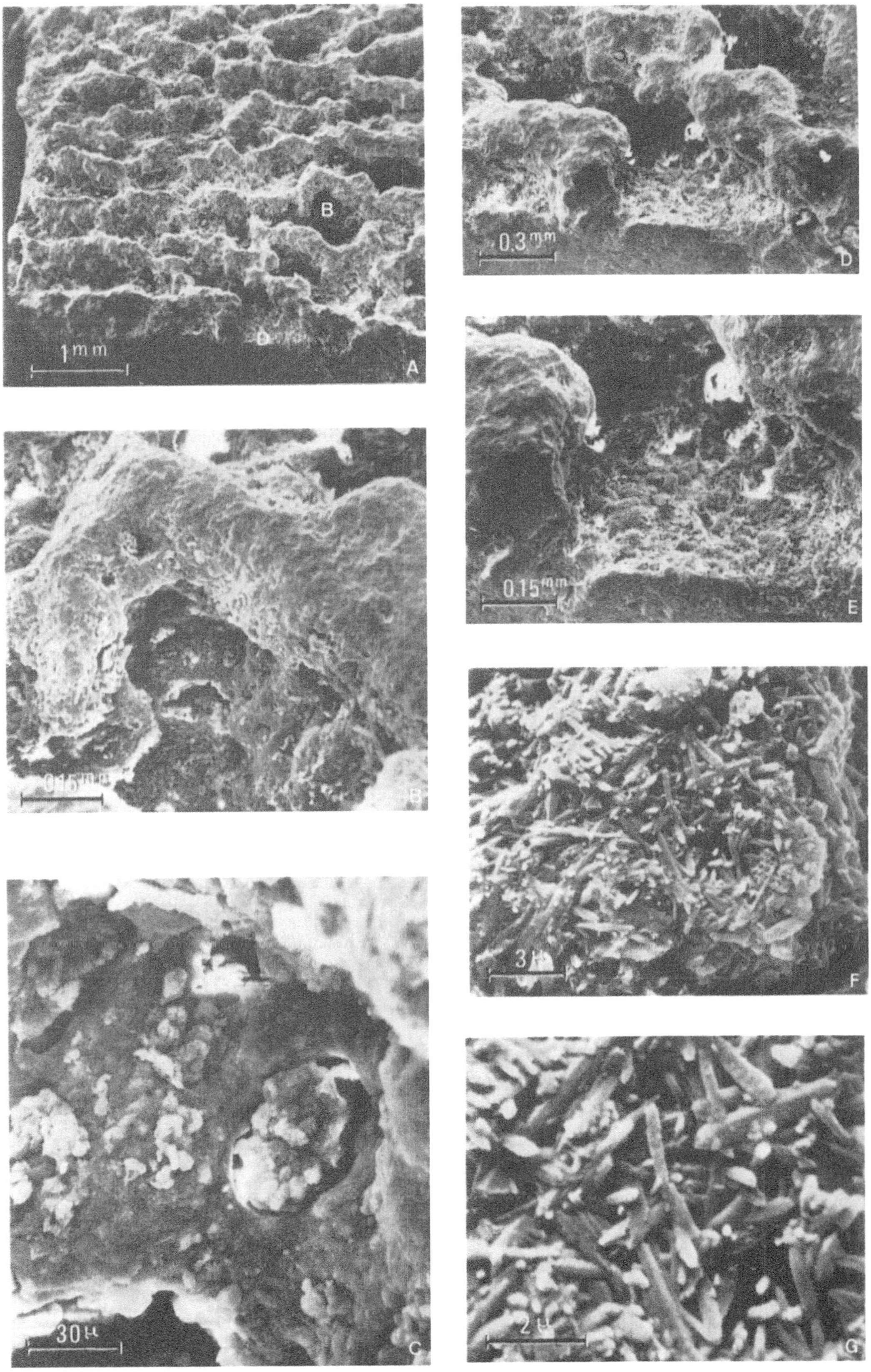

Fig. 17. Morphologies of crusts, supratidal zone, Jebel Dhanna

Processes involved in the formation of aragonitic crusts:

Table 1 indicates that the strontium concentrations within the aragonite composing these crusts are very similar to those within the Abu Dhabi ooids and lagoonal muds. This indicates that the crusts are virtually in equilibrium with sea water from which the aragonite has precipitated. However, it is not possible to establish from these data whether this precipitation is totally physico-chemical or partially bio-chemical.

The principal problem, however, is to determine whether these encrustations were initially algal or whether, on the contrary, they have been formed essentially by precipitation and lithification of aragonite which has been inhabited subsequently by divers micro-floras. The following observations suggest that the latter phenomenon is the more plausible:

1. These crusts are most fully developed within the upper parts of the supratidal zone; algal mats, a characteristic feature of the Trucial Coast, attain maximum development in the upper parts of the intertidal zone, those around Jebel Dhanna Lagoon being 20 - 30 cm topographically lower than the lithified crusts.

2. The aragonite invariably encrusts a hard substrate - either a discreet grain or a layer of beach rock; algae, in contrast, do not require a hard substrate.

3. These crusts are invariably lithified, there being no soft, spongiastrome equivalents.

4. SEM examination indicates that algal or fungal filaments tend to perforate these crusts (Fig. 17B, C) and are not comparable to the felt associated with algal stromatolites.

5. Individual laminae may be traced around the nucleus, especially on grains in the lower parts of the supratidal zone; algal oncoids generally show a marked asymmetry with discontinuous laminations which are thickest on the upper part of the grain (Logan et al., 1964).

6. The dense, often coalescent nanostructures, and laminae composed of radially oriented crystals with no apparent organic matter, suggest a physico-chemical precipitation independant of the observed algae or fungi.

7. Although superficially resembling algal growths, these crusts locally have a preferred downward orientation suggesting a dripstone effect (Fig. 12). The sub-parallel series of micro-ridges developed in the upper parts of the supratidal zone often have steeper down-slope flanks resembling the tufas of Khor Odaid Lagoon, and certain non-marine spring tufas.

These observations suggest that these crusts are not lithified algal constructions but are the consequence of aragonite precipitation onto a hard substrate. That chemical precipitation is not the sole process involved in their construction is indicated by the presence of abundant mud, silt, and sand-sized detrital grains, often concentrated within small lenses. The manner in which these sedimentary particles are incorporated within the crust is apparent when one examines their irregular surfaces; depressions between the complicated system of ridges and protuberances are filled partially with loose detritus including foraminiferal tests, and it is clear that these depressions are small, but effective, sediment traps.

The sequence of formational processes:

The processes involved in the formation of supratidal aragonite crusts at Jebel Dhanna are expressed on Fig. 18. They include:

- the formation of a hard nucleus, often a fragment of beach rock or mollusk.
- precipitation of aragonite in concentric layers around this nucleus. Although

bio-chemical processes cannot be excluded, it is more probable (in view of the observations outlined in the preceding section) that the aragonite is precipitated directly from sea water; these grains and beach rock crusts are immersed in sea water either during exceptionally high tides within the lagoon, or by seepage through the coastal spit.

- local destruction of the cortex or crust, especially within the small surface depressions between the tufa-like ridges. This leads to a deepening of these depressions and an increase in surface relief. The abundance of micro-flora within these depressions suggests that this destruction is organic.
- local trapping of sediment within the surface depressions, the subsequent precipitation of aragonite resulting in the incorporation of sediment into the cortex.

The internal structure, composition, and processes involved in the formation of these aragonitic sheets are almost certainly identical to those associated with the coated grains; the two distinct styles - grains and sheets - would seem to be determined by the substrate available.

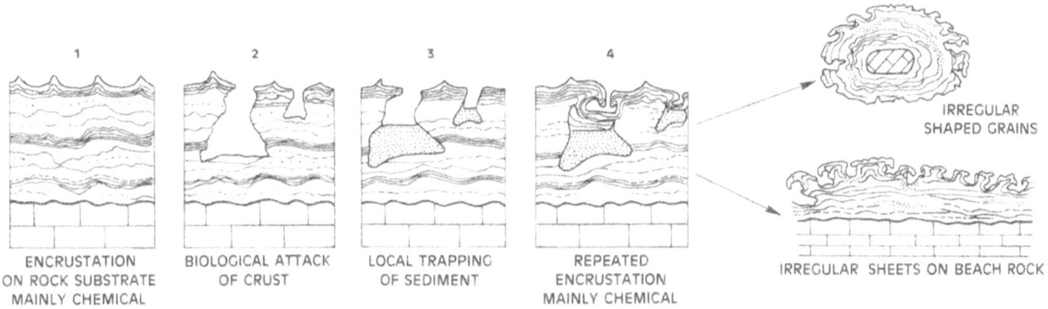

Fig. 18. Sketch summarizing the processes involved in the evolution of the irregular, aragonitic encrustations

Stalactitic aragonite crusts under beach rock:

The layers of beach rock around Jebel Dhanna Lagoon are approximately 30 cm thick and underlaid by loose, bioclastic sand and gravel which are removed locally by burrowing crabs to create small, sheet-like cavities below the beach rock (Fig. 11). The roofs of virtually all these cavities are decorated with micro-dripstone of two basic types.

"Crystalline micro-dripstones" composed of superimposed laminae of honey-coloured fibrous aragonite (Fig. 19A) are common below beach rock in the lower parts of the supratidal zone. Their structure is clearly radial, consisting of relatively large, fibrous crystals or aragonite closely resembling the intergranular cements within the beach rock substrate (Fig. 20). This type of encrustation tends to be lobate, individual lobes attaining 2 - 5 mm in length, and closely resembles the micro-dripstone developed within fissures along bedding planes of limestones outcropping in non-marine environments. Scattered detrital grains, especially faecal pellets and Foraminifera, very frequently occur within these irregular laminae of fibrous aragonite.

The origin of this marine dripstone seems to be essentially physico-chemical, precipitation from sea water percolating downwards through the beach rock following marine flooding seeming to be the obvious explanation. However, the presence of detrital grains (Fig. 20A), often of considerable size (1-2 mm), incorporated within these laminations suggests an alternative explanation; a rising water table associated with tidal fluctuation would have a similar effect and would also tend to float certain porous grains from the floor of the micro-cavity and plaster them against the adjacent roof. The subsequent fall in water level would leave certain of these grains attached to the lower surface of the beach rock (i.e. the roof of the cavity), either due to surface tension of the wet surface, or to the precipitation of aragonite which would weld these grains to the surface. The falling water-table would leave droplets of sea water on the roof from which precipitated aragonite would create micro-stalactites (Figs. 19, 20).

"Sedimentary micro-dripstone" occurs beneath beach rock layers in the upper parts of the supratidal zone. It differs from "crystalline micro-dripstone" in having a brown or grey, earthy appearance due to the detrital composition of these encrustations. Downwards oriented lobes may attain 5 mm in length and are composed of mud, silt, or sand-sized grains of which faecal pellets are the most common (Fig. 19B). These curious sedimentary accumulations are completely lithified by inter-granular, fibrous or micritic cement, almost certainly aragonitic in composition. Their origin must be comparable to that discussed above, being related to a fluctuating water-table which - strange at it may seem - results in the "upwards" deposition of sediment.

The micro-morphology of these sedimentary encrustations is quite variable. Stalactitic lobes seem to be composed mainly of sand-sized detrital particles but micro-dripstone composed of detrital carbonate mud exhibits a more complex morphology. This latter consists of an intersecting system of narrow, crenulate ridges and closely-branched masses of carbonate resembling travertine structure. These crusts attain 5 mm in thickness. It is possible that this fine, earthy textured carbonate is derived from muddy waters either during periods of water level fluctuation, or during periodic flooding. Precipitation within the micro-cavity, cannot, however, be excluded.

C. VARIOUS OTHER SITES

Aragonitic encrustations have been studied in some detail around the supratidal fringes of Khor Odaid and Jebel Dhanna lagoons, where they are best developed. However, similar phenomena have been noted at various localities along the Trucial Coast and adjacent islands indicating that the processes involved, although active only locally, nevertheless are not "accidental"

Umm Said, SE Qatar (Fig. 1):

Shinn, and Loreau and Purser have described (elsewhere in this symposium) the characteristic pisooids which are locally abundant within the upper intertidal and supratidal zones of this wide, lee coast sabkha. Their perfectly laminated cortex, consisting mainly of radially oriented acicular aragonite, is comparable to that of the rounded grains within the supratidal zone at Jebel Dhanna.

Yas Island, W. Abu Dhabi (Fig. 1):

This spectacular salt dome island has a fringing reef on its windward side and a series of spits which are accreting down-wind from its flanks and leeward shores. These narrow ridges of bioclastic sand have enclosed two lagoons in a manner almost identical to those at Jebel Dhanna. The lagoon protected by the spit on the SW side of the island includes virtually all types of aragonitic encrustation found in the identical environment at Jebel Dhanna; beach rocks along the leeward side of the spit are encrusted with both smooth and highly irregular, cauliflower-like masses of aragonite, while local depressions between inclined layers of beach rock contain various types of coated grain. It is very probable that many of the coastal lagoons which characterize much of the Arabian shoreline have similar encrustations.

Sabkha Matti, W. Abu Dhabi (Fig. 1):

The wide tidal flats which characterize the central parts of the open embayment at Sabkha Matti are extensively lithified. These beach rocks, unlike those

DETAILS CONCERNING FOLLOWING FIGURES

Fig. 19. Micro-dripstone within cavities below layers of beach rock:

A. "Crystalline micro-dripstone" under beach rock, showing smooth, lobate, micro-stalactites and scattered skeletal grains (arrow) incorporated into the dripstone. Lower supratidal zone, Sabkha Matti.

B. "Sedimentary micro-dripstone"; each earthy-textured lobe consists mainly of faecal pellets cemented with micritic carbonate; upper supratidal zone, Jebel Dhanna locality.

C. Detail of individual "sedimentary micro-stalactite" in photo B, showing faecal pellets (x 9).

Fig. 20. Structures and morphology of crystalline micro-dripstone:

A. Typical micro-morphology of crystalline micro-dripstone; rounded lobes are due to aragonite coating detrital grains which are thus incorporated into the dripstone (SEM, x 18)

B. Detail of photo A, showing acicular aragonite crystals oriented statistically normal to the substrate (the roof of the micro-cavern), and coalescent nanostructure near the ends of micro-stalactites. (SEM, x 180)

C. Details of crystalline micro-dripstone, showing typical euhedral crystals whose extremities are truncated locally; individual crystals measure 1 - 3μ in diameter and several microns in length, and tend to resemble inter-granular cement within the overlying beach rock. (SEM, x 9,000)

D. Section through crystalline micro-dripstone below beach rock from lower supratidal zone, Sabkha Matti. (fractured surface, SEM, x 18)

E,F,G. Aragonitic micro-dripstone illustrated in photo D showing details of radially oriented crystal fabric. (SEM, x 180, 450 and 900).

Fig. 19.

Micro-dripstone morphologies developed in cavities below beach rock

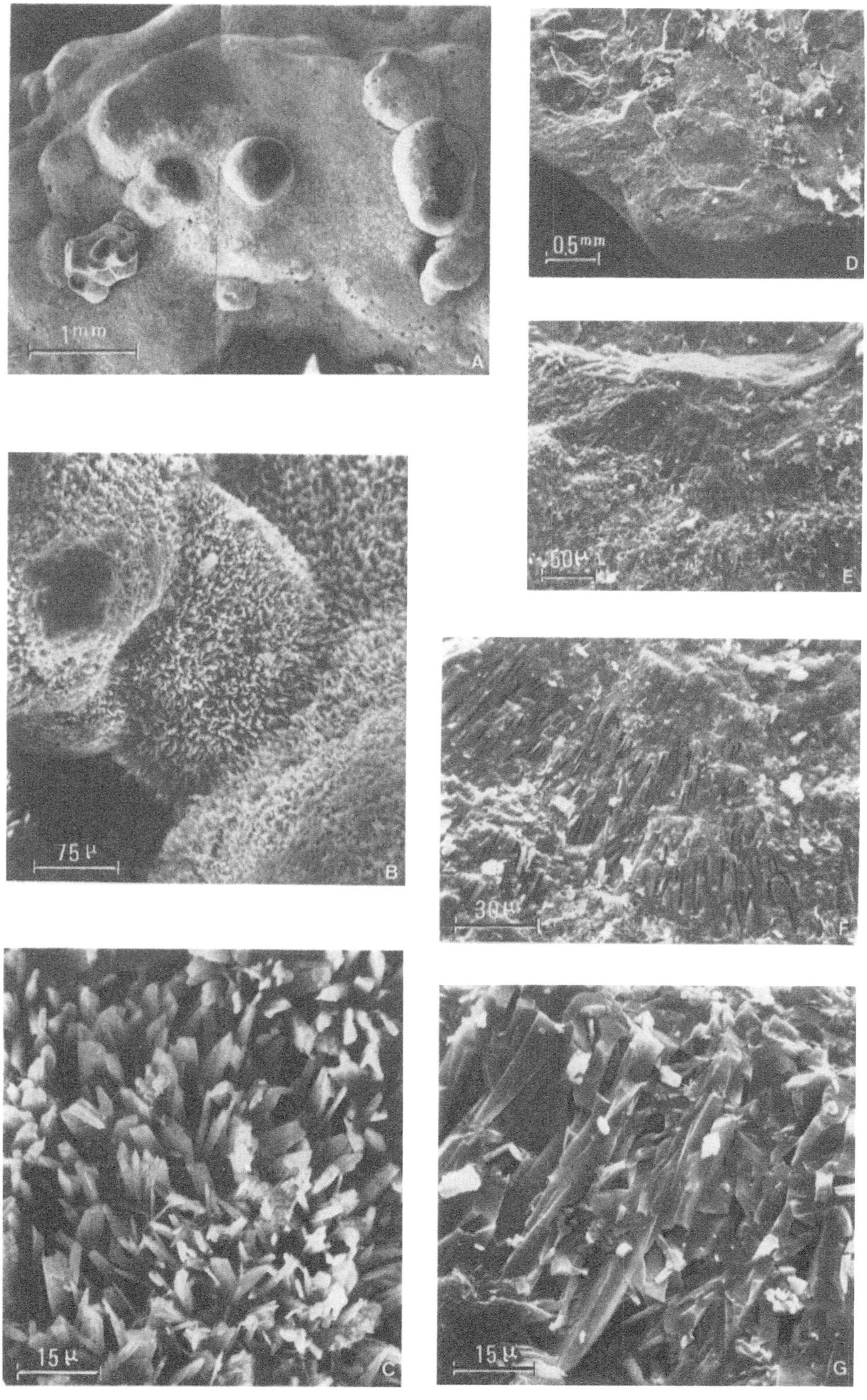

Fig. 20. Structure and morphology, crystalline micro-dripstone

within the lagoons on Yas Island and at Jebel Dhanna, are exposed to the open sea and are swept frequently by waves. At several localities their surfaces are encrusted with polished "pelagosite" crusts of light brown aragonite. Sheets of beach rock are locally buckled and thrust into the teepee structures discussed by Evamy (in this volume), this superficial deformation creating numerous cavities between the rigid limestone and the loose sediment below. Many of these cavities along the lower limits of the supratidal zone are lined with "crystalline micro-dripstone". In at least one locality the tightly cemented beach rock is encrusted with polished "pelagosite" on its upper surface and by lobate "crystalline micro-dripstone" on its lower surface.

III. DISCUSSION AND CONCLUSIONS

Certain of the phenomena described in the preceding section have been reported by various workers. In particular, the polished aragonite crusts occurring in less protected environments seems to have been noted by Darwin (Fairbridge, 1957) on Ascension Island, and later by Emery et al. (1954) on Bikini; it would seem to be a fairly common feature in sub-tropical and tropical shoreline environments. Various types of supratidal aragonite crust have been reported by Davies (1970) from Shark Bay, W. Australia, and by Newell and Rigby (1957) from the coast of Andros Island. Although descriptions of these crusts do not permit an effective comparison with those of the Trucial Coast, it is nevertheless clear that they occur in a similar supratidal environment and generally encrust rock substrate. They should not be confused with somewhat similar crusts associated with soil development in non-marine environments.

The highly irregular surface morphologies including dripstone effects, together with the associated grains of varying dimensions and shapes seem to be reported for the first time in this contribution. Their morphologies, frequently resembling algal stromatolites and certain non-marine encrustations, merit further discussion.

A. FEATURES COMMON TO ALL ENCRUSTATIONS:

Certain features are common to virtually all encrustations examined along the shores of the Trucial Coast;

- they are composed of nanograined, lamellar or acicular, strontium-rich aragonite.
- they are confined to the upper parts of the intertidal, and supratidal zones.
- in any given locality there is generally a progressive change in morphology from smooth to irregular crusts as one passes towards the higher parts of the supratidal zone.
- all encrust a hard rock substrate, generally beach rock.
- at any given locality aragonite may encrust both discreet grains and rock outcrops, the characteristic lateral change in morphology across the supratidal zone affecting both styles of encrustation.

B. COMPARISON BETWEEN ENCRUSTATIONS AND THE CLASSIC OOID CORTEX:

The significance of coated grains:

At certain localities, notably at Jebel Dhanna, grains range from sand to cobble sized particles whose shapes vary between sub-rounded to highly irregular. Round, sand-sized nuclei encrusted with laminated aragonite, frequently with radial structure, may be classified as ooids. They may, however, be distinguished from classical marine ooids in that their cortical laminae maintain a constant thickness on all sides of the nucleus whose form is reflected in the external morphology of the grain; classical ooids from agitated marine environments have a cortex which varies in thickness around the nucleus tending to produce an ovoid morphology even when the nucleus is somewhat angular in shape.

Sand and cobble-sized nuclei appear to be coated with aragonite in an equally efficient manner. However, the cortex on large nuclei is composed essentially of unoriented, nanograined aragonite exhibiting irregular laminations, many of which can be traced completely around the nucleus. Their formation would not seem to require any significant degree of movement. These larger grains could be classified as pisooids, especially when their shape is fairly regular.

When the available nucleus is irregular in shape - such as a mollusk or angular rock clast - the subsequent coating preserves this morphology. This irregular shape may be further amplified within the upper parts of the supratidal zone by the development of a complex system of micro-ridges, knobs and depressions. The external morphology, internal structure, and size of these irregular grains has much in common with that of oncolites (oncoids). However, it should be recalled that the term "oncolithi", as defined originally by Pia (1926) denotes a grain of algal origin. Because the irregularly-shaped coated grains discussed in this contribution are probably not algal in origin a term is obviously needed to describe "oncoid-like" objects whose origin is not necessarily algal; the authors propose the term "coniatoid" (Greek Koniatos = lime encrusted) for these grains, and the term "coniatolite" for all the hard sheet-like, carbonate encrustations described in this contribution.

The association of coated grains and aragonite sheets:

It is of considerable interest to note that the aragonite may not only encrust sand-sized particles to produce classical ooids, but may also encrust outcrops as sheets. However, although the resultant pellicle of laminated "pelagosite" has a microstructure very similar to that of the ooid cortex, there are significant differences in the nanostructure; pelagosite crusts have a very compact and variable nanostructure (Figs. 4,5,6) which contrast with those of the ooid cortex, the latter having only two types of more or less loosely packed nanostructure (Loreau and Purser, Figs. 13 and 14, elsewhere in this volume). It is probable that these differences in nanostructure will be obliterated when the aragonite converts to calcite.

C. SIMILARITIES AND DIFFERENCES BETWEEN CONIATOLITES AND STROMATOLITES

It should be evident from the preceding descriptions that the aragonitic laminations have much in common with lithified algal stromatolites. Similarities include:

- external morphology, with numerous surfaces inclined at angles exceeding 40°.
- irregular internal laminations which reflect the external morphology.
- detrital sediment trapped within these laminations.
- a common inter-supratidal environment of formation.
- variation in morphology as one crosses the supratidal zone.

Characteristics which may permit the distinction between aragonitic crusts essentially of physico-chemical origin, and algal stromatolites are the following:

- because aragonitic crusts (coniatolites) develop on hard substrates the contact may often be discordant (a "hard ground" or unconformity); grains directly below the crust may be truncated or bored. Stromatolites most frequently encrust a soft substrate, although certain oncoids may encrust a hard skeletal nucleus.
- coniatolites frequently have a dripstone micro-morphology with downward growth around edges, and may be suspended from lower surfaces.
- coniatolitic crusts, although they may fracture, do not exhibit the dessication features characteristic of many stromatolites.

- because coniatolites are always lithified they do not include birds eye (fenestrate) structures typical of many stromatolites.
- although coniatolite crusts effectively trap detrital sediment, these grains are generally isolated or occur in small pockets where they have been trapped in local surface depressions. Algal stromatolites are frequently composed mainly of detrital material which may constitute distinct layers.

The distinction between marine tufa, or coniatolites, and algal stromatolites may not always be easy, or even possible - but this should not prevent the attempt.

D. SIMILARITIES AND DIFFERENCES BETWEEN MARINE AND NON-MARINE TUFAS:

The marine encrustations examined are very similar to those found around springs or within caves, their formation, in spite of the contrasting environments, probably having much in common. The association of sheet-like crusts, often with irregular dripstone morphologies, and discreet coated grains, is reminiscent of the association of tufa and cave pearls. These latter, in common with the coated grains found in the supratidal environment, have irregular shapes which reflect the form of the nucleus. The marine, coniatolitic encrustations differ from cave and spring tufa in the following manner:

- they are composed of relatively pure, nanograined or fibrous aragonite, cave tufa most frequently consisting of somewhat larger, bladed crystals of calcite (Donahue 1969, Hahne et al., 1968, etc.).
- the marine tufas have numerous micro-perforations probably made by endolithic algae or fungi; although micro-perforations occur in cave tufa they seem to be much rarer.
- aragonitic tufa generally, but not invariably, encrusts a marine substrate such as skeletal gravel or beach rock, and its setting is marine.

Both marine encrustations and those within caves and other terrestrial environments, are forming in a vadose environment above the water table, and consequently both types exhibit dripstone effects. The discovery of marine vadose encrustations indicates that one should not automatically equate vadose fabrics with a non-marine environment. The confusion may lead to the creation of non-existant unconformities and to excessive degrees of paleo-emergence.

E. OCCURRENCE OF CONIATOLITIC ENCRUSTATIONS IN ANCIENT LIMESTONES:

That these marine encrustations are not the bizarre products of an exceptiona modern environment is demonstrated by the occurrence of almost identical fabrics within certain Bathonian limestones of Bourgogne (Purser, in press). Their presence within a conformable sequence of marine carbonate confirms episodic supratidal emergence indicated by the numerous dripstone cements within these marine gravels. Stromatolitic-like encrustations in Urgonian carbonates near Marseille described by Masse (1969) also seem to have much in common with the marine tufas of the Abu Dhabi coast. These supratidal encrustation probably exist in many ancient carbonate sequences, certain probably having been recorded as stromatolites.

Geochemistry of Tidal Flat Brines at Umm Said, SE Qatar, Persian Gulf

K. de Groot[1]

ABSTRACT

The landward part of the 7 km wide sabkha at Umm Said, SE Qatar, is filled with a stagnant brine virtually saturated with halite. Recent dolomite occurs in the sabkha sediments, the quantity being fully accounted for by the amount of Mg^{++} ions lost from the interstitial brine.

The existence of a reflux system in the seaward parts of the sabkha was established. It was not, however, possible to give any unequivocal demonstration of the effect of this potential system for dolomitization. Although both a reflux mechanism and Recent dolomite formation occur in this tidal flat, the first process has apparently not influenced the second sufficiently to permit the demonstration of reflux dolomitization.

INTRODUCTION

Refluxing of hypersaline waters through carbonate sediments has been postulated by Adams and Rhodes (1960) as a mechanism for the formation of dolomite. This process involves flowing of sea water into,or over, coastal sediments in hot, arid climates. Evaporation results in the precipitation of gypsum and the Mg^{++}/Ca^{++} ratio in the water is thus increased. The dense, Mg^{++} - rich water transforming calcite or aragonite to dolomite, flows downwards through the sediment and returns ultimately to the sea.

At some localities where Recent dolomite has been found, including the Pekelmeer on the isle of Bonaire,(Deffeyes et al.;1965), and Sabkha Faishakh on the W coast of Qatar (Illing et al., 1965), a study of the water chemistry has been made to evaluate whether the reflux model was operating. Studies by van der Poel (unpublished Shell Research report) indicated that refluxing of hypersaline water did not seem to be occurring in Sabkha Faishakh and subsequent studies in the Pekelmeer did not support earlier calculations. On the contrary, all water analyses made, both in Faishakh and in the Pekelmeer sediment, showed that hypersaline water in the sediment in which the dolomite was probably forming overly less dense, and thus less saline, water. Van der Poel concluded from his data obtained on Sabkha Faishakh that the poor vertical permeability of the sediment apparently prevented vertical fluid flow and thus reflux.

---

1 Shell Research B.V., Rijswijk, The Netherlands.

Subsequent work by Murray (1969) on the hydrology of South Bonaire furnished evidence for influx of seawater into the Pekelmeer through permeability conduits in the underlying rock. However, during early summer a major reflux event is suggested: heavy brine from the Pekelmeer then flows back to the sea through the same conduits that supply the sea water during most of the year. Thus reflux dolomitization of the underlying rock is quite possible, although Murray did not report mineralogical data showing effective dolomitization of these rocks.

Studies by E.A. Shinn on the siliciclastic Umm Said Sabkha (described elsewhere in this volume) offered the opportunity of studying the water chemistry of a sabkha which, because of its homogeneous, sandy nature, was likely to have a greater vertical permeability than the areas previously examined (discussed above). It was considered, furthermore, that the distribution of Recent dolomite might give an indication of the effect of any existing reflux system.

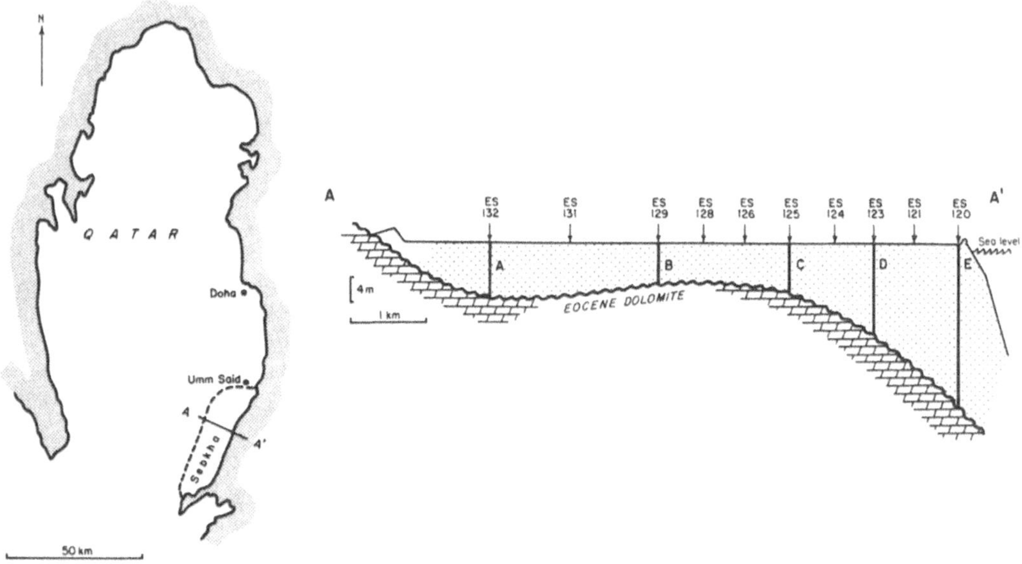

Fig. 1. Profile across 7.5 km wide sabkha south of Umm Said, Qatar. Section based on wells A, B, C, D, E. Water sample pits ES 120-132 are shown by arrows

ANALYTICAL AND SAMPLING METHODS

Sampling

A series of five, continually-cored wells was drilled across the Umm Said Sabkha, their locations being illustrated by Shinn elsewhere in this volume. Between these wells shallow pits were dug for water sampling, (indicated by arrows, Fig. 1), sabkha brines flowing into these pits from the sides and from below. Water level was about 50 cm below the sabkha surface.

Water analysis

While some water samples were taken from the pits, most were obtained from cores. After opening the core tubes in the laboratory, the bottom 10 cm was immediately transferred to a filter centrifuge for water extraction. This water was analyzed for chloride, calcium and magnesium and, in a few cases, also for sulphate. Analysis

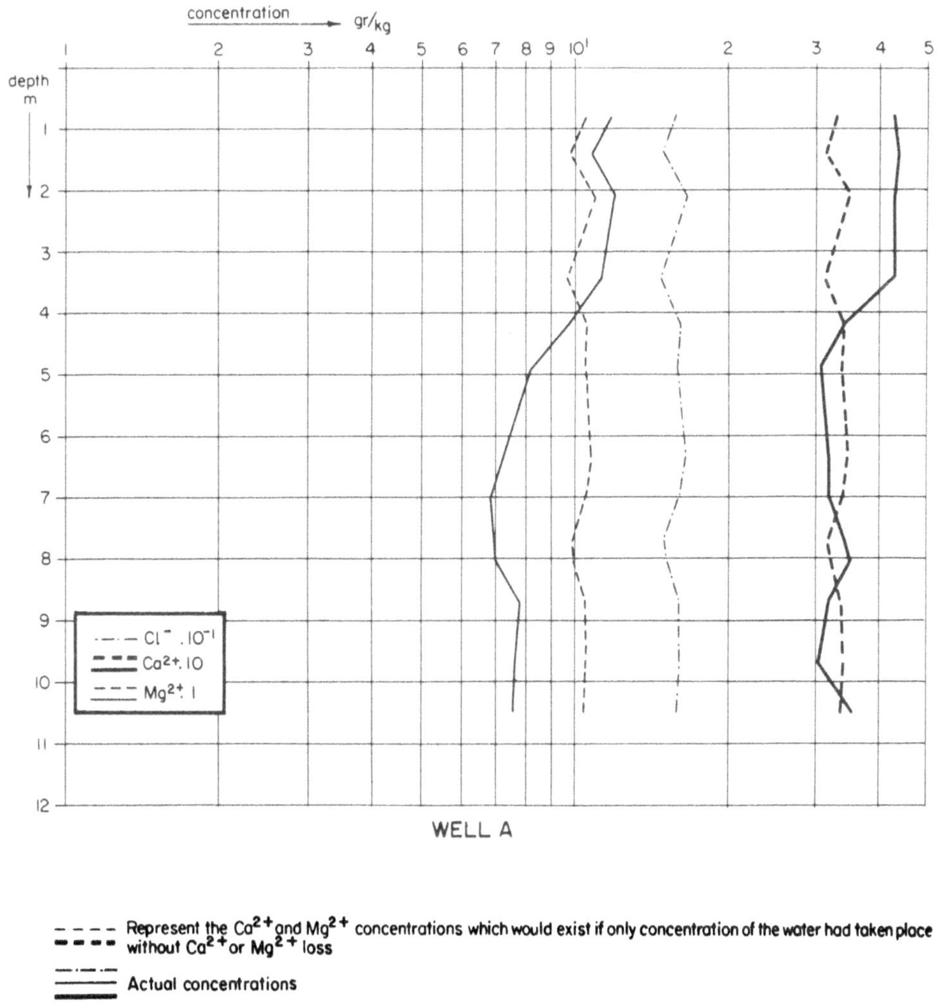

Fig. 2A. Comparison of concentrations of Ca^{2+} and Mg^{2+} with the Cl-concentration in the interstitial water as a function of depth

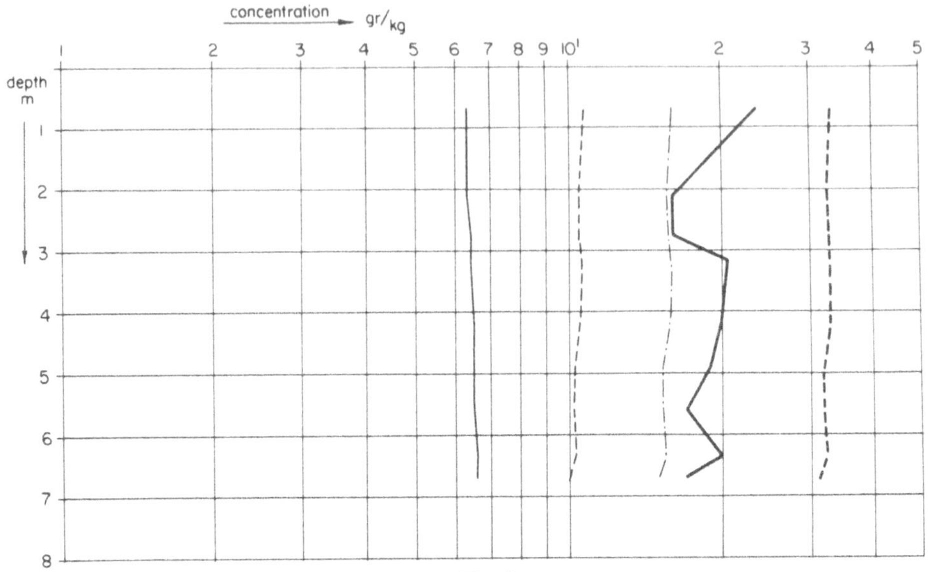

Fig. 2 B WELL B

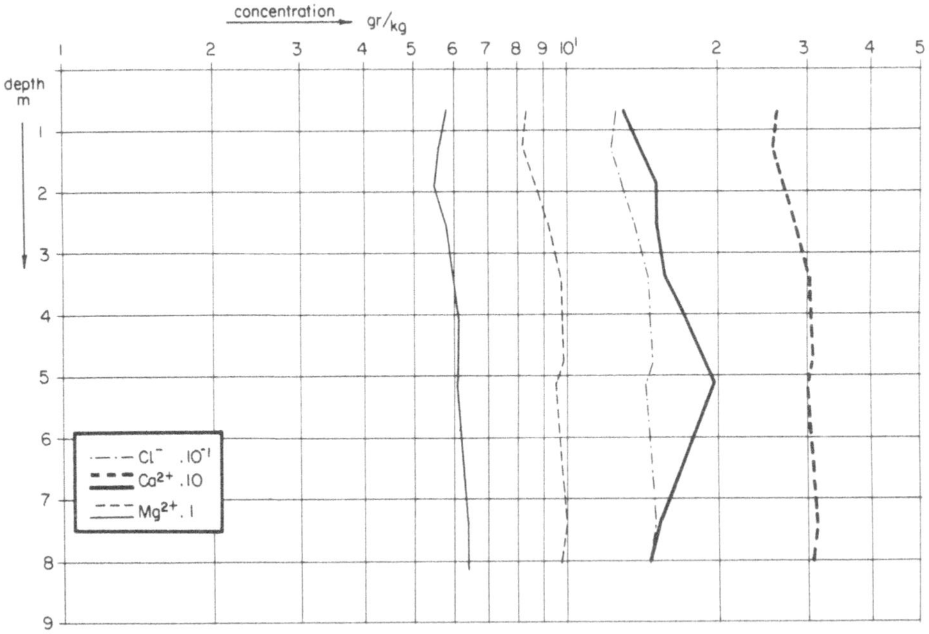

Fig. 2C WELL C

Represent the Ca^{2+} and Mg^{2+} concentrations which would exist if only concentration of the water had taken place without Ca^{2+} or Mg^{2+} loss

Actual concentrations

Fig. 2B and C. Comparison of concentrations of Ca^{2+} and Mg^{2+} with the Cl^--concentration in the interstitial water as a function of depth

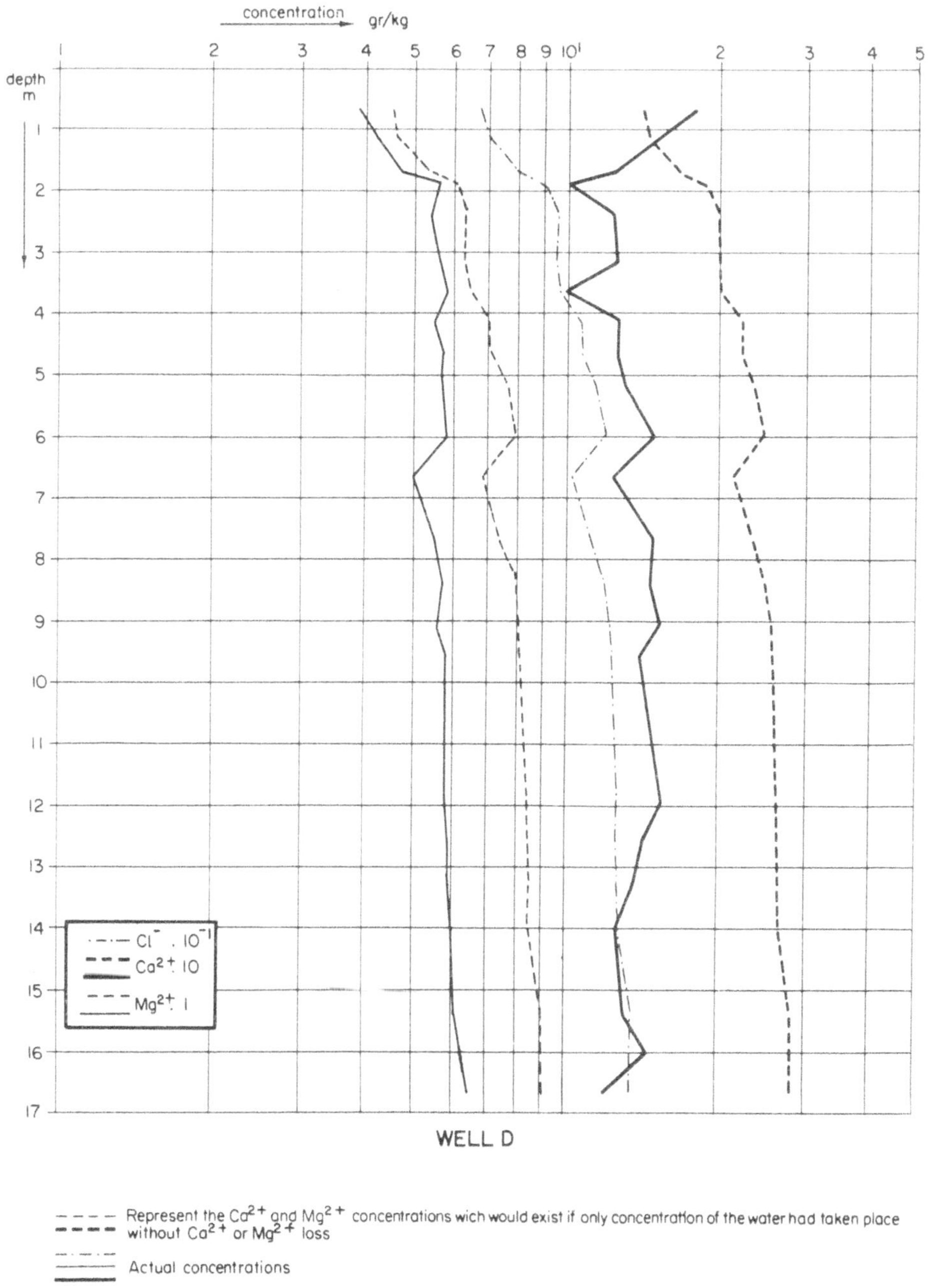

Fig. 2D. Comparison of concentrations of Ca^{2+} and Mg^{2+} with the Cl^{-}-concentration in the interstitial water as a function of depth

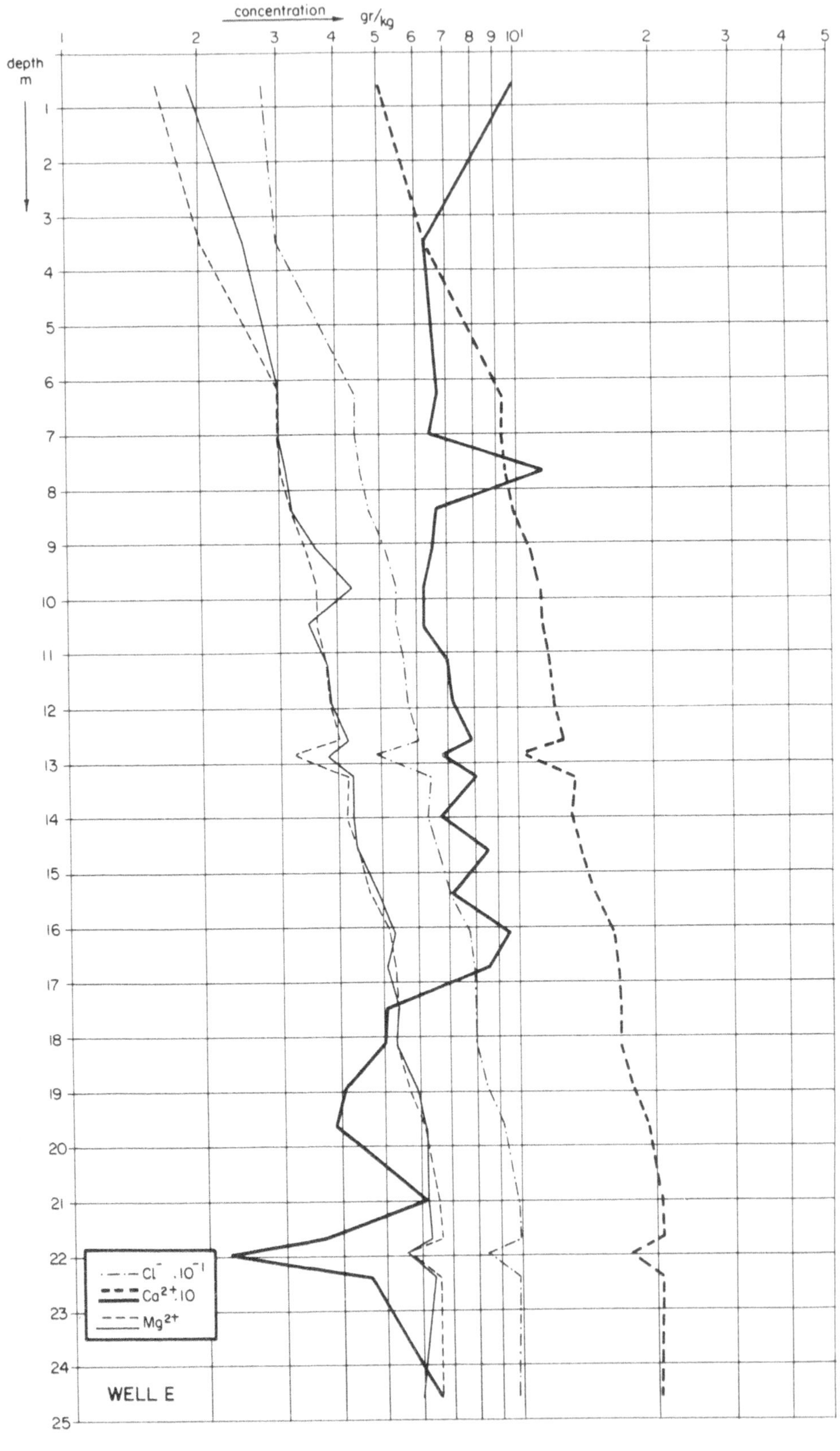

Legend to Fig. 2E on the following page

of carbonate or bicarbonate was not attempted as the carbonate/bicarbonate ratio and the absolute amounts of these ions in the cores could have changed due to biological action during transport.

Chloride was determined by automatic potentiometric titration with $AgNO_3$. The sum of calcium and magnesium was determined by EDTA titration. Calcium was determined by flame-photometry, as the complexometric titration of calcium in the presence of large amounts of magnesium gives rise to large systematic errors. Sulphate was determined with lead nitrate in an acetone-water mixture after chloride removal and an ion exchange procedure to replace all cations by H^+.

Analysis of the sediment

Some data were obtained by X-ray diffraction analysis. The figures given for the amounts of the various minerals present have a total analytical error of $\pm$ 15% (relative). X-ray diffraction was also used to determine the amount of $CaCO_3$ in dolomite which showed that all analyzed Eocene dolomite in Qatar has the composition ($Ca_{0.50 \pm 0.005} \cdot Mg_{0.50 \pm 0.005} CO_3$), whereas Recent dolomite contains more Ca, with up to 57 mole % $CaCO_3$ (Illing et al., 1965). The amount of Ca present in the dolomite has thus been used to distinguish Eocene from Recent dolomite.

RESULTS AND DISCUSSION

Chlorinity and density distribution of the interstitial water as evidence of reflux

Water analyses are given in Table I and Figures 2a to e. Chlorinities of the interstitial waters are high, especially in wells A, B and C where the water is saturated with halite. A chlorinity profile constructed from the sample points is presented in figure 3. It shows that the distribution of chlorinity is essentially different from that constructed from interstitial water measurements made by van der Poel on Sabkha Faishakh, W Qatar, a locality where dolomite is forming. It shows dense water overlying lighter water.

In the permeable Umm Said Sabkha, between well B and the sea, heavy water is overlaid by less dense water, a situation which can be explained best by the reflux theory: sea water flowing in during occasional marine flooding and by lateral seepage through the upper layer of sediment, evaporates from the sabkha surface and is concentrated. It tends to sink into the sediment and flow back towards the sea through the deeper parts of the sabkha.

In a tidal-flat filled with saline water which is being concentrated by surface evaporation, the water lost by evaporation will generally be compensated by inflow of adjacent sea water. As a result water density increases landwards and the free water level becomes somewhat lower. One can envisage, in the absence of flow, a horizontal level within the sabkha above which the weights of the water columns are equal everywhere, or, in other words, a level at which a line of constant hydrostatic pressure is horizontal. At this level no horizontal flow can take place. Above this level, horizontal flow, once begun, will be landwards; below this level the flow will be seawards if no permeability barriers are present. Figure 4 demonstrates the conditions under which reflux is theoretically possible.

Fig. 2E. COMPARISON OF CONCENTRATIONS OF Ca^{2+} AND Mg^{2+} WITH THE CL^--CONCENTRATION IN THE INTERSTITIAL WATER AS A FUNCTION OF DEPTH

- - - - - - Represent the Ca^{2+} and Mg^{2+} concentrations which would exist if only concentration of the water had taken place without Ca^{2+} or Mg^{2+} loss

— · — · — Actual concentrations

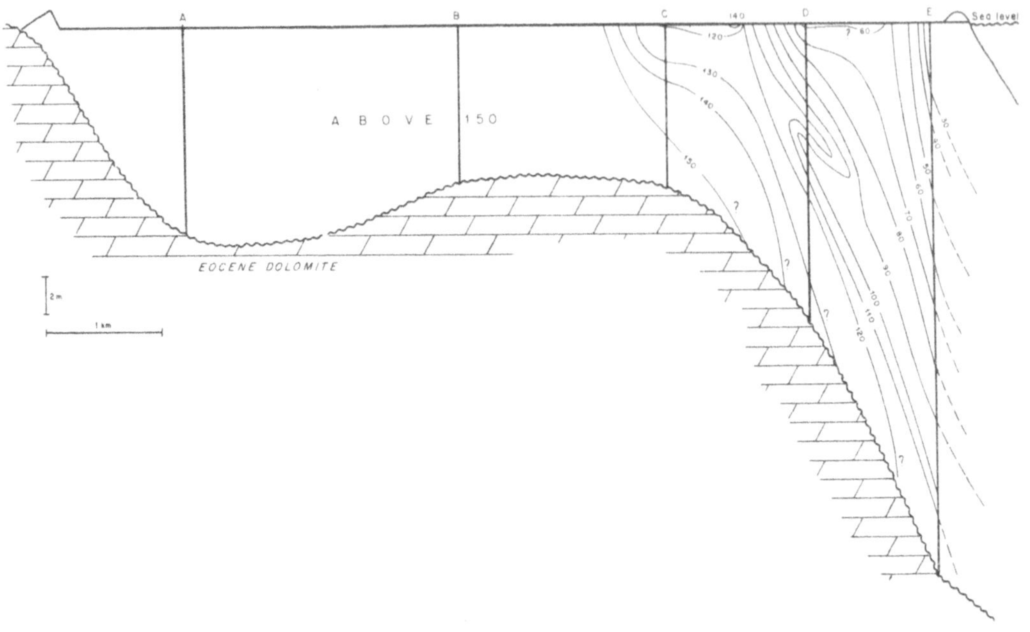

Fig. 3. Isochlorinity lines (in parts per thousand) of the interstitial water in the sabkha

In the Umm Said sabkha the free water level was not measured exactly in the field, but was found to vary very little: it was about 50 cm below the horizontal sabkha surface everywhere. Assuming a linear drop of 50 cm in the water level from well E to well B (which is probably excessive) and assuming vertical flow resistance to be negligible, one can readily calculate the hydrostatic pressures at various depths. In Table 2 the hydrostatic pressures at 8 m and (where possible) at 16 m depth are listed. It is clear that under the assumed conditions the horizontal line of constant pressure is present above 8 m in the Umm Said sabkha. One can calculate the rate of flow in the sediment under influence of the hydrostatic pressure differences, using Darcy's law. For instance, the rate of flow between well B and C at 8 m depth becomes:

$$\frac{\Delta p \cdot k}{\eta \cdot \ell} = \frac{0.154 \times 20}{2 \times 1.77 \times 10^5} = 8.7 \times 10^{-6} \text{ cm/sec} = 274 \text{ m/year},$$

using the hydrostatic pressure difference Δp at 8 m = 0.154 kg/cm^2 .

the viscosity η of the brine = 2 centipoise

a (uniform) permeability k of the sediment = 20 darcies

the distance ℓ between well C and B = 1770 m.

In Table 3 a number of possible flow rates between the wells, under the influence of hydrostatic pressures only, are listed for uniform permeabilities of 1 and 20 darcies respectively. In the coarse, upper parts of the sabkha permeabilities of 10 darcies are likely, so that rates of seaward flow of a few tens of metres per year are possible. If the drop in water level going from E to B is less than 50 cm, the calculated flow rates will be higher.
These calculations suggest therefore that a seaward flow of hypersaline water exists in the deeper part of the sabkha between wells C and E. However, no significant circulation seems to take place in the most inland part of the tidal-flat between well A and C, as only minor chlorinity or density gradients were recorded.

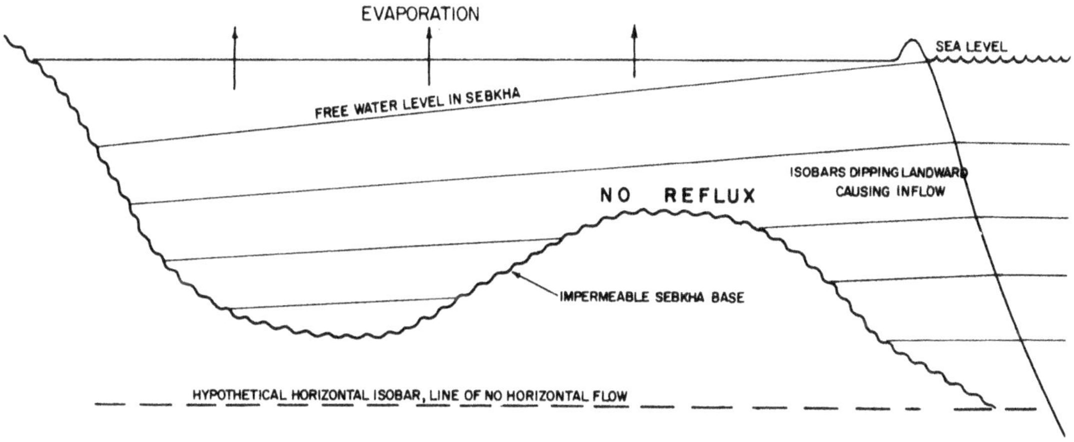

Fig. A NO REFLUX IN SEBKHA, INFLUX BALANCES EVAPORATION

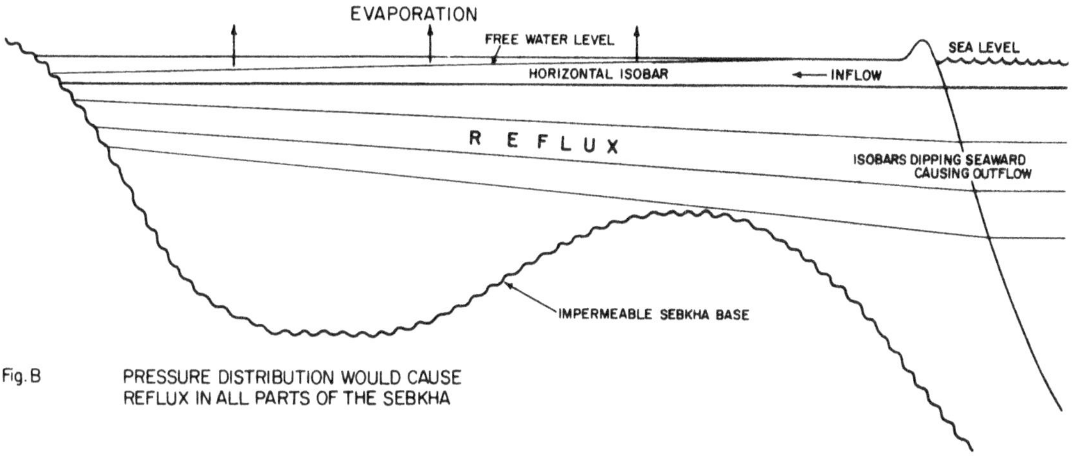

Fig. B PRESSURE DISTRIBUTION WOULD CAUSE REFLUX IN ALL PARTS OF THE SEBKHA

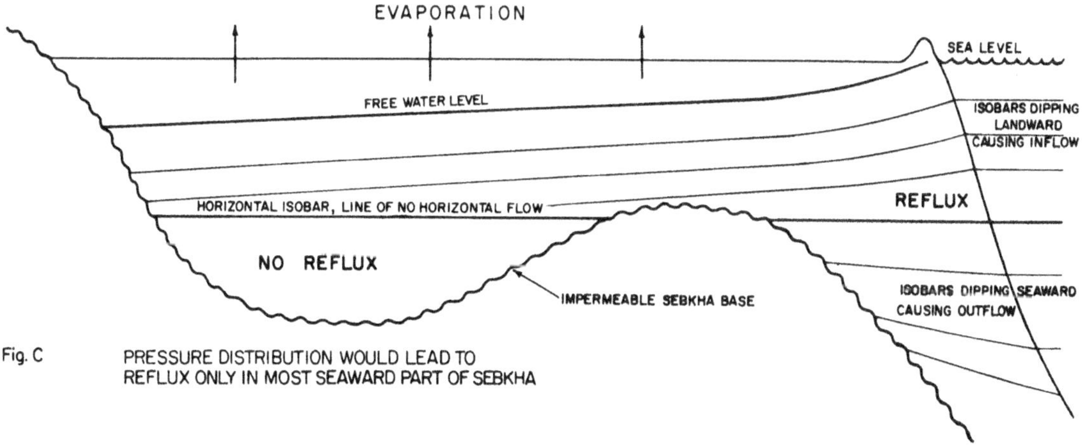

Fig. C PRESSURE DISTRIBUTION WOULD LEAD TO REFLUX ONLY IN MOST SEAWARD PART OF SEBKHA

Horizontal flow will take place in the direction of the lower pressure

Fig. 4. Possible hydrostatic pressure distributions in stationary model of Umm Said Sabkha

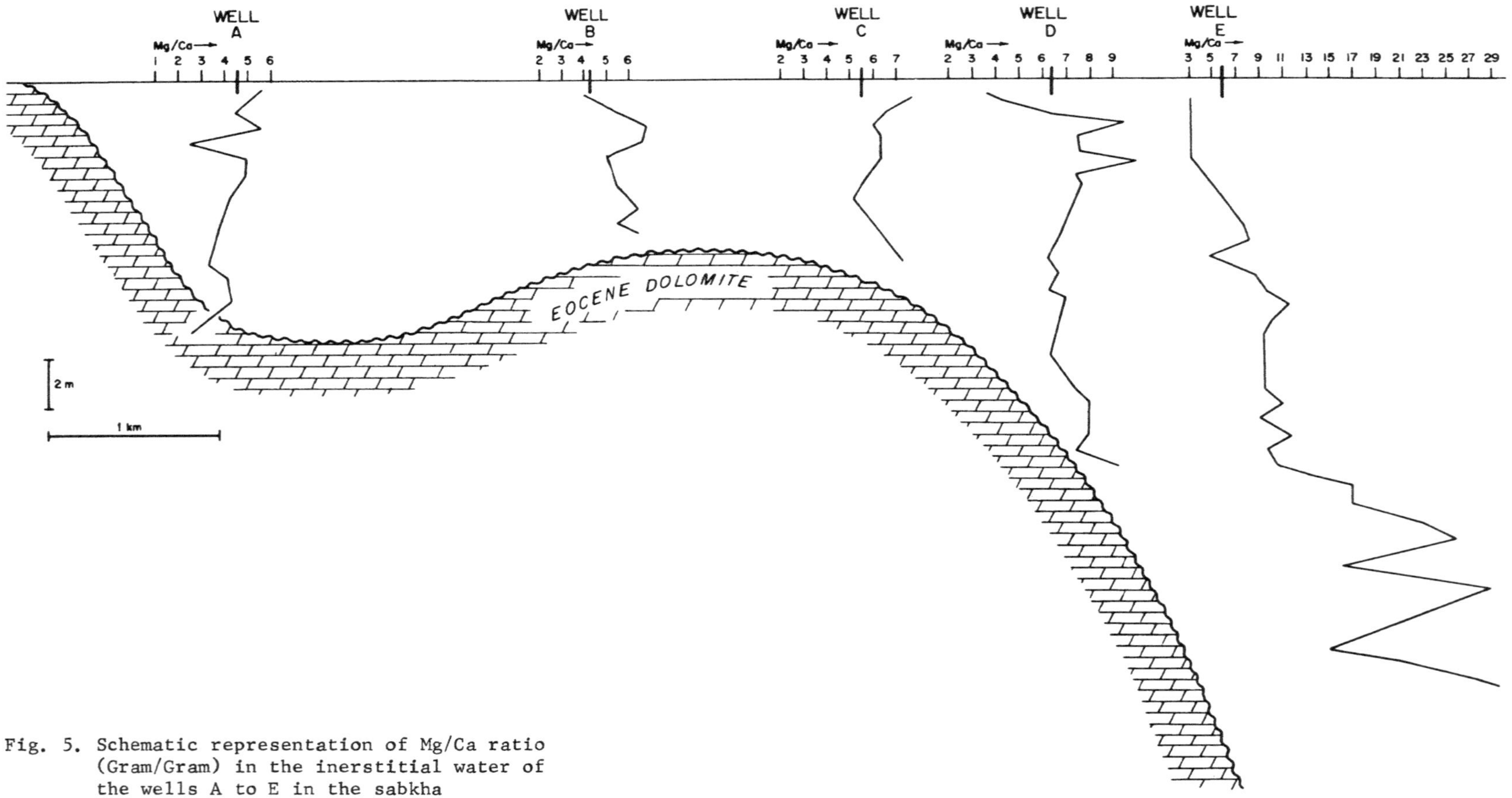

Fig. 5. Schematic representation of Mg/Ca ratio (Gram/Gram) in the inerstitial water of the wells A to E in the sabkha

Concentrations of Ca^{++} and Mg^{++}

A plot of the Mg/Ca ratio (Fig. 5) shows that a rather large variation exists along the traverse. The lowest values are encountered in wells A and B where the highest salinities were recorded, while the highest Mg/Ca ratios are present in the water in the lower part of well E, nearest the present coast.

High chlorinities in wells A and B probably indicate that this water has already lost gypsum (or anhydrite); in wells A, B and C gypsum has been observed (by Shinn). Although the precipitation of $CaSO_4$ should have resulted in high Mg/Ca ratios, this ratio is found to be low. This must have been caused by dolomite formation because no other Mg-salts have been found by X-ray diffraction analysis. Because the process of dolomite formation is very slow, the low Mg/Ca ratios measured indicate that water supply to this inland part of the sabkha is extremely limited.

In figures 2A - 2E the dotted lines give the amount of Ca and Mg which would be present if the sea water had been concentrated without any Ca and Mg loss by precipitation of $CaCO_3$, gypsum or dolomite. It is clear that in all wells the waters must have lost a considerable amount of Ca^{++} as gypsum, calcium carbonate or dolomite. The trend is reversed in the top part of well A. In view of the very high chlorinities measured there it is clear that the observed halite precipitation from the water is responsible for the increase in the Mg/Cl and Ca/Cl ratios. In the bottom part of well A the Ca/Cl ratio is high in spite of the Ca loss as gypsum, indicating that even this water must have lost chloride. In all wells except E and the top part of well D, Mg has therefore been lost from solution, as dolomite. The fact that the Mg loss is noticeable again indicates that the residence time of the water in the sabkha is about the same, or longer, than the time necessary to form the dolomite present, as was already observed in wells A and B.

The presence and age of dolomite

When the 50 cm deep sample pit was dug at location B, Shinn (verbal communication) noted that, after a certain amount of water had flowed into the pit, this water turned milky. While water continued to enter the pit a rim around the milky water seemed to remain clear. This observation gave the impression that a precipitate was being formed in the pit itself. Analysis of a large sample of the suspension collected in the pit showed that it consisted mainly of well-ordered, and Ca-rich (55 mole % Ca) dolomite. Dating of the sample by E.L. Martin of Shell Research, Houston showed this dolomite to be 5000 years old. Fine dolomite crystals were also found in wells A, B and C. (More chemical work on this problem is reported in Appendix A.) It is concluded that the milky water observed in the pit contained fine grained sediment derived from surrounding interstitial waters. In a thin layer the suspension appears clear; this probably explains why no turbidity is seen when there is little water in the pit. It may also explain why the water appears clear around the edges of the pit.

X-ray diffraction analyses of sediment samples have been carried out to obtain a more complete picture of the dolomite distribution in the sabkha. As shown in Table 4, the fraction $< 16\mu$ of a number of samples from the quartz-rich sediments contains amounts of dolomite varying between 5 and 40%. This dolomite is present partly, or exclusively, as Ca-rich dolomite (53-55 mole % Ca), except in the bottom part of well E, and is in this respect similar to the Recent dolomite reported by Illing et al. (1965), from Sabkha Faishakh.

X-ray analyses have also been made of some samples from the carbonate "Unit 4" present in the transgressive part of the sabkha (see Shinn, Fig. 5 in this volume), the results of which are presented in Table 5. All dolomite in the fraction $< 16\mu$ of these samples is Ca-rich, but the amounts found are small. The coarser fraction has larger amounts of dolomite, but this is probably derived from adjacent Eocene dolomites as the composition is $Ca_{0.50}\,Mg_{0.50}\,CO_3$. Furthermore, these crystals are too large to have formed in 5000 years, the approximate age of the sabkha.

It is concluded that Recent dolomite occurs in rather small amounts all over the sabkha. In the quartz-rich sediment the highest concentrations of dolomite are found in well B where the dolomite has a ^{14}C age of 5000 years. The concentration of dolomite in the carbonate layers varies erratically.

Rôle of reflux in the distribution of dolomite

The amount of Mg^{++} ions necessary to form the dolomite present in wells A, B and C has been estimated. The amount of the fraction $< 16\mu$ in this sediment is approximately 1% (average of 15 determinations). The percentage of Recent dolomite in this fraction averages 20%, so that 0.2% of the sediment is Recent dolomite. Assuming a porosity of 30% there exists per 1000 cm^3 of sediment 700 cm^3 or about 2 kg of solid including 4 g of dolomite. To form this dolomite from 300 cm^3 of interstitial water requires $4 \times 0.12 \times \frac{1000}{300} = 1.6$ g Mg^{++} per litre of water. In the wells A, B, C and even in the bottom part of well D, the amount of Mg^{++} actually lost from the interstitial waters is of the same order (see Fig. 2), so that extra water supply is not necessary to account for the dolomite found in this part of the sabkha.

As the reflux process was shown to be inactive in the inland part of the sabkha (the part west of well C), the dolomite in that area seemingly has not been formed via a reflux process.

The area where refluxing seems to be active has relatively less dolomite than the inland part of the sabkha and, at least in well C and the bottom part of well D, it has been demonstrated that a reflux system is not necessary to explain the limited amount of dolomite present. In the quartz sediment in the lowest part of well E no Recent dolomite has formed, possibly because this sediment is too young to allow time for dolomitization.

Dolomite present in the carbonate layers in the sabkha has to be disregarded as proof of the effectiveness of reflux for dolomitization; stromatolites and mudcracks have been found in this layer indicating that it could have formed as a supratidal crust.

In summary, the refluxing of hypersaline waters, although a potential dolomitizing process, is not necessary to explain the distribution and limited amounts of dolomite in the Umm Said sabkha.

CONCLUSIONS

A study of the dominantly quartz sand sabkha at Umm Said, SE Qatar has revealed that the landward part of the sabkha is filled with brine saturated, or almost saturated, with halite. This brine is probably almost stagnant, and has lost Mg^{++}. This loss of Mg^{++} can fully account for the small amount of Recent dolomite found in the sediment, which also contains sufficient fine grained calcium carbonate to act as a carbonate source for the dolomite. Dolomitization is probably very slow in this part of the sabkha and is seemingly not the result of a refluxing system.

The demonstration of Recent dolomite at Umm Said is admittedly of secondary interest in view of earlier discoveries of this mineral around the Persian Gulf, and elsewhere (Florida, Bahamas, Australia); the fact, however, that research has demonstrated a reflux mechanism potentially capable of effecting wisespread dolomitization, is regarded as significant.

Table I

Analyses of interstitial water from cores

Well A

No.	Depth, m	Cl g/kg	Ca g/kg	Mg g/kg
1	0.62 - 0.72	157.94	4.346	13.603
2	1.36 - 1.46	149.30	4.375	11.287
3	2.03 - 2.13	165.39	4.288	13.980
4	2.76 - 2.86	-	-	-
5	3.40 - 3.50	147.81	4.278	12.364
6	4.15 - 4.25	161.68	3.440	9.720
7	4.82 - 4.92	159.27	3.064	8.121
8	5.51 - 5.61	-	-	-
9	6.23 - 6.33	164.53	3.219	7.281
10	6.94 - 7.04	160.18	3.205	6.859
11	7.63 - 7.73	149.69	3.366	6.888
12	8.00 - 8.10	151.65	3.459	6.954
13	8.66 - 8.76	160.55	3.213	7.808
14	9.61 - 9.71	161.09	3.048	7.707
15	10.43 -10.53	158.48	3.546	7.707
16	11.14 -11.24	-	-	-

Well B

No.	Depth, m	Cl g/kg	Ca g/kg	Mg g/kg
1	0.63 - 0.73	159.94	2.672	6.338
2	2.04 - 2.14	156.45	1.565	6.262
3	2.70 - 2.80	157.00	1.633	6.364
4	3.36 - 3.46	157.56	2.111	6.377
5	4.17 - 4.27	156.95	2.032	6.465
6	4.87 - 4.97	153.42	1.925	6.525
7	5.59 - 5.69	152.77	1.728	6.460
8	6.28 - 6.38	155.19	1.977	6.612
9	6.65 - 6.78	150.60	1.744	6.611

Well C

No.	Depth, m	Cl g/kg	Ca g/kg	Mg g/kg
1	0.63 - 0.73	124.98	1.268	5.810
2	1.22 - 1.33	122.79	1.379	5.569
3	1.84 - 1.94	129.10	1.488	5.457
4	2.51 - 2.61	135.79	1.495	5.797
5	3.35 - 3.45	145.04	1.566	6.046
6	4.04 - 4.14	145.84	1.726	6.070
7	4.76 - 4.86	147.40	4.281	4.528
8	5.10 - 5.20	143.91	1.972	6.147
9	7.35 - 7.45	150.16	1.532	6.367
10	7.98 - 8.08	146.41	1.475	6.419

Well D

No.	Depth, m	Cl g/kg	Ca g/gk	Mg g/kg
1	0.66 - 0.76	66.943	1.780	3.948
2	1.11 - 1.19	69.312	1.504	4.176
3	1.61 - 1.71	77.745	1.241	4.739
4	1.86 - 1.96	90.002	0.993	5.634
5	2.37 - 2.47	95.014	1.215	5.424
6	3.08 - 3.18	94.323	1.230	5.603
7	3.59 - 3.69	95.795	0.970	5.770
8	4.08 - 4.18	104.96	1.245	5.518
9	4.61 - 4.71	106.55	1.236	5.655
10	5.15 - 5.25	113.03	1.295	5.745
11	5.92 - 6.02	119.03	1.476	5.803
12	6.59 - 6.69	102.00	1.218	4.986
13	7.67 - 7.77	111.15	1.475	5.492
14	8.34 - 8.44	118.75	1.455	5.742
15	9.06 - 9.16	121.87	1.515	5.569
16	9.49 - 9.59	122.53	1.402	5.756
17	11.85 -11.95	127.29	1.538	5.777
18	12.54 -23.64	124.83	1.422	5.821
19	13.23 -13.33	126.06	1.362	5.857
20	13.95 -14.05	126.64	1.259	5.985
21	15.35 -15.45	134.99	1.307	6.142
22	15.97 -16.07	133.90	1.437	6.290
23	16.62 -16.72	133.72	1.201	6.542

Well E

No.	Depth, m	Cl g/kg	Ca mg/kg	Mg mg/kg
1	0.52 - 0.62	28.107	982.3	1896
5	3.44 - 3.54	30.006	633.5	2544
9	6.22 - 6.32	44.087	671.0	3007
10	6.94 - 7.04	44.052	643.1	3024
11	7.66 - 7.76	44.546	1141.5	3108
12	8.34 - 8.44	46.609	660.4	3187
13	9.05 - 9.15	51.409	651.0	3560
14	9.75 - 9.85	54.011	618.0	4311
15	10.41 - 10.51	53.698	624.6	3515
16	11.09 - 11.19	56.070	700.9	3782
17	11.84 - 11.94	57.030	723.5	3899
18	12.53 - 12.63	60.185	779.2	4198
19	13.21 - 13.31	63.651	803.3	4346
20	13.93 - 14.03	63.415	665.9	4267
21	14.54 - 14.64	65.606	846.6	4372
22	15.34 - 15.44	70.367	712.8	4866
23	16.06 - 16.16	76.855	945.1	5258
24	16.69 - 16.79	78.245	853.0	5112
25	17.47 - 17.57	78.943	514.1	5356
26	18.11 - 18.21	80.251	505.6	5302
27	18.84 - 18.94	83.784	413.5	5889
28	19.54 - 19.64	92.725	387.9	6157
29	19.86 - 19.91	38.367	478.9	2362
30	20.95 - 21.05	97.827	618.6	6165
31	21.67 - 21.77	98.556	359.5	6349
32	22.37 - 22.47	97.636	464.9	6360

33	22.90 - 23.00	31.587	400.7	2226
34	24.49 - 24.59	96.887	656.6	5993
35	25.02 - 25.12	72.790	363.2	4577
36	25.86 - 25.90	105.68	366.2	6277
37	26.54 - 26.64	113.99	426.3	6393
38	27.25 - 27.35	113.363	277.7	6583
39	27.74 - 27.84	109.26	-	-
40	28.66 - 28.76	119.77	-	-
41	29.16 - 29.26	116.43	-	-

Table 2

Hydrostatic pressure in the wells along the traverse
(difference in water level between well E and B; 50 cm)

Well	Pressure at 8 m	Pressure at 16 m	Distance
B	1.10 kg/cm^2	-	BC = 1786 m
C	0.98 "	-	"
D	0.95 "	1.88 kg/cm^2	CD = 1203 m
E	0.89 3	1.76 "	DE = 1054 mm

Table 3

Possible water flow rates in metres per year at depths of 8 and 16 m for various permeabilities*)

	Permeability: 1 darcy	20 darcies
Flow between B and C at 8 m depth	+ 11	+ 218
" between C and D at 8 m depth	+ 4	+ 80
" between D and E at 8 m depth	+ 8.5	+ 170
at 16 mdepth	+17.2	+ 344

*) Seaward flow direction is taken as positive

Table 4

X-ray diffraction analysis of fraction < 16μ of samples from the upper quartz-rich part of the sabkha*)

		Aragonite	Calcite	Mg-Calcite	Dolomite	% $CaCO_3$ in dol.	Other minerals (Quartz, felds.)
A1	0.68- 0.76 m	30	15	5	20	54	30
A3	1.61- 1.71 m	40	15	20	15	51;54	10
A5	2.37- 2.47 m	40	10	20	15	50;54	15
B1	0.63- 0.73 m	35	10	5	40	55	10
B3	2.70- 2.80 m	40	10	10	30	54	10
C2	1.22- 1.32 m	40	10	15	20	55	15
C4	2.51- 2.61 m	25	35	-	15	55	25
C6	4.04- 4.14 m	30	20	15	15	55	20
D2	1.36- 1.46 m	30	25	15	15	53	30
D8	5.51- 5.61 m	40	20	10	10	50;55	20
D15	10.43-10.53 m	25	20	5	10	53	40
E1	0.52- 0.62 m	50	10	-	5	50;53	35
E10	6.94- 7.04 m	30	20	20	15	50;53	15
E17	11.84-11.94 m	50	10	-	5	50	35
E27	18.84-18.94 m	10	25	10	5	50	50

*) For the dolomite the amount of $CaCO_3$ in the lattice is also included. When two figures are listed two separate dolomite peaks could be distinguished on the X-ray diffraction pattern. The dolomite containing 50 mole % $CaCO_3$ is probably of Eocene age.

Table 5

Amount of dolomite as found by semi-quantitative X-ray diffraction of samples taken from the carbonate layers present in the various wells

Sample well	Depth	Fraction < 16μ		Fraction 16-53μ	
		% Dolomite	% $CaCO_3$ in dol.	% Dolomite	% $CaCO_3$ in dol.
A	3.69- 3.72 m	5	55	-	-
B	6.40- 6.43 m	5	55	10	50
C	6.00- 6.03 m	15	55	5	50
D	13.40-13.43 m	5	54	20	50
E	24.63-24.66 m	15	55	25	50

APPENDIX I

POSSIBILITY OF RAPID DOLOMITE FORMATION: SOME EXPERIMENTAL EVIDENCE

E. A. Shinn has observed that water flowing into a pit near well B suddenly turned milky. The milkiness appeared to be due to significant amounts of (ordered) dolomite in the water. Following this observation a number of experiments have been carried out to test the possibility of instant ordered dolomite formation. It was presumed that, if dolomite formed in a few seconds by some inorganic mechanism*), this process could be repeated in the laboratory.

Milky water sample ES 129 from a pit dug near well B was filtered and Ca^{++} and Mg^{++} were added in amounts equivalent to 2, 5 and 10 grams of dolomite formed per litre of water. The water was then flushed with CO_2*). Bicarbonate was added to the rather acid solutions, again in amounts equivalent to 2,5 and 10 grams of dolomite and the solutions were allowed to lose their CO_2 slowly to the air so that precipitation could occur. The precipitates which formed after some hours standing were filtered off and analyzed by X-ray diffraction (see table A-1). No immediate dolomite formation occurred and only a mixture of aragonite and Mg-calcite was found. The water from which the carbonate minerals precipitated was always different from the original water, to which Ca^{++} and Mg^{++} had been added, as only a small amount of the added Mg^{++} precipitated.

The results of these experiments are not surprising, as experimental evidence from Baron (1960) and measurement of growth rates of natural dolomite by Peterson et al. (1966) show that dolomite formation is an extremely slow process, in which the solid-state diffusion of Ca^{++} and Mg^{++} in the carbonate lattice are probably rate-determining steps. It is interesting to note that the lower the initial Mg/Ca ratio in the experimental solutions used, the higher the amount of Mg-calcite formed, and the lower the amount of Mg in the calcite lattice. At high supersaturations and in solutions with high Mg/Ca ratio, Mg effectively blocks the surface of calcite nuclei formed, so that aragonite, unhindered by Mg-absorption, can grow freely (de Groot and Duyvis, 1966). Only at low supersaturations, when the rate of carbonate formation is slow, can a Mg-containing $CaCO_3$ form from solutions with a high Mg/Ca ratio.

Table A - 1

Precipitates obtained from sample ES 129 water after addition of extra $CaCl_2$, $MgCl_2$ and $NaHCO_3$

Starting material	Composition of precipitate			
	Aragonite	Mg-Calcite	% $MgCO_3$ in Mg-calcite	
ES 129 + ions equiv. to 2 gr of dolomite/kg solution	95%	5%	12	Mg/Ca
ES 129 + 5 g "	75%	25%	10	ratio in
ES 129 + 10 gr "	65%	35%	6	solution decreasing ↓

* The only possible mechanism seemed to be release of CO_2 when the water came into contact with the air. pH measurement to evaluate the magnitude of a possible CO_2 escape was not successful in the field because electrode equilibration took the same time as the turbidity development in the water.

In another experiment the author used a saturated halite solution containing all major ions in the same ratio as they are present in sea water, except for Ca^{++}, Mg^{++} and SO_4^{--}; Ca^{++} and Mg^{++} were present in a 6 times higher concentration than in sea water (so that Mg^{++} concentration was equal to that in waters from pit ES 129). Sufficient SO_4^{--} was added to saturate the solution with gypsum. Flushing with CO_2 and subsequent flushing with air resulted in formation of aragonite. After standing (in the solution) for about 2 weeks, the precipitate changed to a mixture of aragonite (60%), Mg-calcite (20%) and a very Ca-rich, disordered dolomite (20%), containing about 65% $CaCO_3$. A disordered, very Ca-rich dolomite can be formed by recrystallization of a $CaCO_3$ phase in a few weeks.

These experiments, which accept the determined age of 5000 years for the dolomite, show that the ordered, Ca-rich dolomite present in pits dug in the sabkha could not have formed by instant precipitation.

Some Aspects of the Diagenetic History of the Sabkha in Abu Dhabi, Persian Gulf

P. Bush[1]

ABSTRACT

Sedimentary accretion around the landward margins of coastal lagoons in Abu Dhabi has produced a low accretion plain, the "sabkha" . Concentrated sea water from the adjacent lagoon seeps into the sabkha sediments both laterally via the intertidal zone, and vertically through the supratidal zone during exceptional marine flooding. Aragonite and gypsum are precipitated within the sediment from both brines, increasing the Mg:Ca ratio and entraining the dolomitization of aragonitic mud.

A buried algal mat prevents the vertical penetration of brine. Where all available aragonite above this permeability barrier has been exhausted (due to dolomitization) the elevated Mg/Ca ratios resulting from the continued precipitation of gypsum lead to the formation of magnesite. Where the buried algal mat dies out, towards the inner parts of the sabkha, brines percolate into and dolomitize underlying lagoonal sediments. Nodular anhydrite forms mainly above the algal mat partly by the dehydration of pre-existing gypsum.

1. INTRODUCTION

The nearshore, lagoonal and coastal plain sediments of the Trucial State of Abu Dhabi, have been the subject of considerable study since 1960. The results have been published in various papers and theses by Bush, Butler, Cuff, Evans, Kendall, Kinsman, Murray, Skipwith, Shearman and Twyman.

The Trucial Coast embayment is a shallow shelf area approximately 60,000 sq. km in extent. The deposits of the shelf are mainly skeletal sands with occasional areas of carbonate mud. Towards the landward margin of the embayment, in an area extending from Ras Ghanada in the E to Jabal Dhanna in the W, a line of islands and shoals protects extensive lagoons (fig. 1). Within these lagoons, deposits of pelletal carbonate muds and sands and skeletal and pelletal carbonate sands are accumulating on the banks, and in the channels respectively. (Evans, Kinsman & Shearman, 1964; Kinsman, 1964 a,b; Kendall and Skipwith, 1968 b; Evans & Bush, 1969). Algal mats have developed on the intertidal zone sediments in sheltered areas particularly along the mainland shore (Kendall & Skipwith, 1968, 1969 a), and these pass landwards into an extensive flat coastal plain - the sabkha. Much of this plain is normally supratidal but the seaward parts adjacent to the lagoon are occasionally covered by sea water from the lagoon during stormy weather. This coastal plain is composed dominantly of carbonate sands and muds together with varying amounts of evaporitic minerals.

---

1 Dept. of Geology, Imperial College of Science and Technology, London.

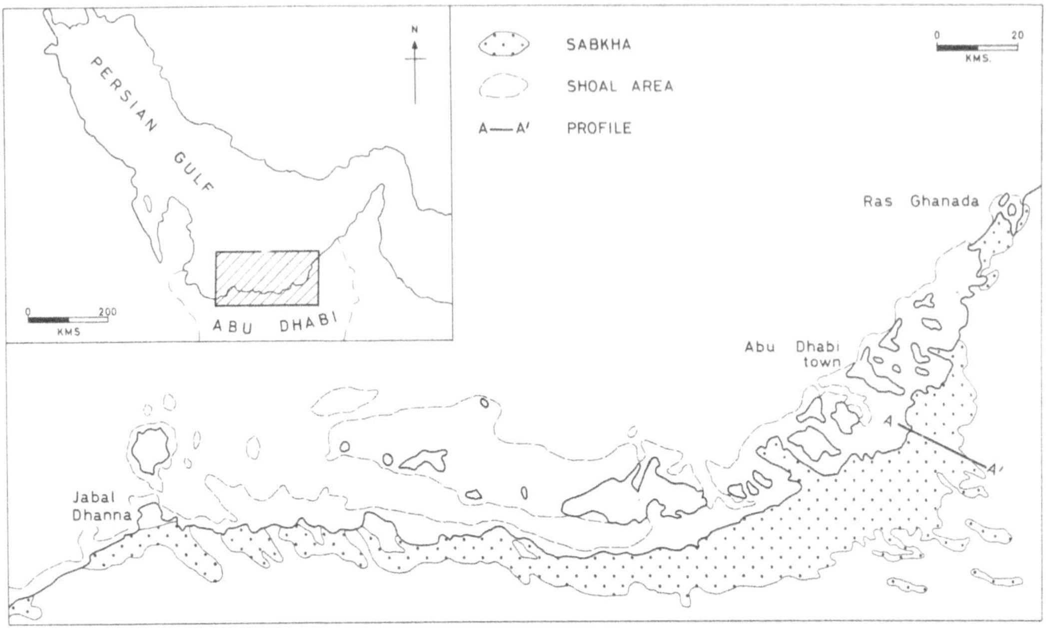

Fig. 1. Location of Abu Dhabi Area

2. THE DEVELOPMENT OF THE SABKHA PLAIN

The sequence of events that has given rise to the sabkha plain has been deduced from evidence provided by pits and boreholes sunk in a number of areas (Evans, Kendall & Skipwith 1964; Kinsman, 1964 b; Butler 1965, 1969, Evans et al., 1969). In addition, radio-carbon dating has made it possible to date these events.

Approximately 7,000 years B.P. the sea transgressed over the site of the present sabkha plain, which at that time was an area covered by sub-aerial dunes composed of quartzose carbonate sand. The extent of this transgression varies from place to place. In the area to the SW of Abu Dhabi Island, with which this paper is concerned (fig. 1 Line A-A'), the maximum transgression reached a point between 5 and 6 kilometers landward of the present low water mark by approximately 4,000 years B.P., and a beach ridge developed at the margin of the lagoon so formed (fig. 2). Landward of this ridge, the dunes were deflated to the level of the water table, as their source of sediment had been submerged beneath the waters of the lagoon. Seawards of the beach ridge the original aeolian, quartzose carbonate sand was re-worked during transgression and then gradually buried under newly formed skeletal carbonate sand which contained considerable amounts of re-worked aeolian sand in its lower parts. It was finally covered with grey muddy carbonate sand as the lagoonal environment became established. Deposition continued until the lagoon became very shallow and further accumulation of sediment was prevented by the action of waves and currents. A lithified crust, cemented with high magnesium calcite, formed on the surface of the sediments in the area of non-deposition (a similar crust is found on the lagoon floor at the present time). Sediment was transported by waves and currents to be deposited at the margins of the lagoon to form intertidal flats; this resulted in the lateral filling of the lagoon and progradation of the coastline.

An apparent fall of sea level by approximately one metre occurred between 4,000 B.P. and 3,750 years B.P. This resulted in some of the sediments of the inner lagoon becoming intertidal instead of subtidal. The new intertidal area was colonized by an algal mat. At first the algal mat grew out over the lithified crust, but later it continued to grow over the sediments pushed to the margins of the lagoon by the waves and currents. As this algal mat grew seawards it was slowly covered by sediment carried onto its surface by wind and the occasional storm. About 1,000 years B.P. this algal mat ceased to grow and was buried by intertidal sediments, but the plain continued to prograde seaward and finally a new algal mat developed and has continued to grow until the present day. Fig. 2 summarizes the sedimentary sequence produced by these events.

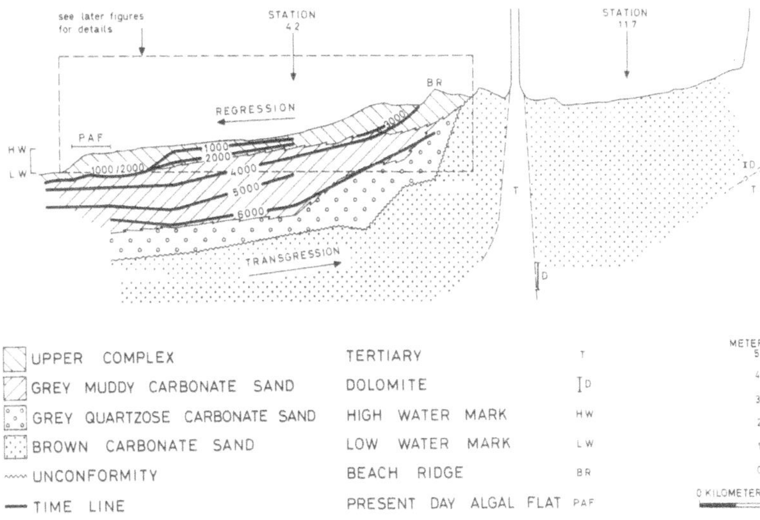

Fig. 2. Profile along line A-A' figure 1

3. THE DEVELOPMENT OF THE DIAGENETIC MINERALS

The climatic conditions (table I) in the area give rise to a high rate of evaporation from both the land and sea surfaces. This evaporation results in concentration of the sea water in the lagoon (Fig. 3) and lowering of the level of the ground water in the sabkha. The ground waters of the sabkha are replenished by lagoon water entering the sabkha sediments both laterally across the lagoon-sabkha boundary and vertically downwards through the surface of the sabkha during flooding at times of extra high tides when lagoon water submerges the outer part of the sabkha plain. Butler (1968) has called this latter process "flood recharge".

TABLE I

Climatic factors: The values quoted are for the normal range of temperature; more extreme conditions can occur

RANGE OF AIR TEMPERATURE	16° - 44° C
RANGE OF GROUND SURFACE TEMPERATURE	15° - 53° C
AVERAGE RAINFALL	3.7 cm/annum.
AVERAGE EVAPORATION	124 cm/annum.
HUMIDITY	30 % by day 90 % by night.

Table II
Chemical composition of waters entering the sabkha sediments, laterally from the lagoon margin, and vertically through the sabkha surface, compared with standard sea water

	Milliequivalents / litre				
	Ca^{++}	Mg^{++}	SO_4^{--}	Cl^-	Mg^{++}/Ca^{++}
STANDARD SEA WATER	21	109	58	559	5.2
LAGOON WATER (SABKHA MARGIN)	41	206	110	1099	5.1
FLOOD WATER (SABKHA SURFACE)	12	2819	395	5399	234.9

The composition of the water entering the sediments laterally from the lagoon shows little change in the ratio of the various ions. However, as a result of evaporation its concentration has doubled (table II). In contrast, the water entering the sediments vertically through the surface (fig. 3) has lost much of its calcium, and some of its sulphate. This difference in the composition of these two waters can best be explained by the fact that most of the floods that cover the surface of the outer sabkha consist only of very thin sheets of water which are driven by strong onshore winds. Such shallow sheets of water, by virtue of their thinness, and because of the high air temperature and low daytime humidity, are subject to rapid evaporation. This, together with the dissolution of the soluble salts of magnesium, sodium, and potassium from surface crusts results in an increase in the concentration of these ions in solution: calcium carbonate and calcium sulphate are precipitated during concentration, and thus the final waters are richer in magnesium, sodium and potassium and depleted in calcium.

(a) Diagenetic minerals produced by evaporation.

Aragonite.

Some aragonite is precipitated in the lagoon (Kinsman 1964b; Bush 1970.) and even more during the early stages of concentration of the brines both within the sabkha sediments, and also on its surface (as predicted by the work of Usiglio (1924)), (Fig. 4).

High Magnesium Calcite.

Crusts found on the floor of the inner lagoon are cemented by fine grained, high magnesium calcite.

Gypsum.

Gypsum is not precipitated from the open lagoonal waters in the Abu Dhabi area, and is therefore never found on the floor of the lagoon. The first evidence of precipitation of this mineral is found just beneath the surface algal mat, in the intertidal sediments (Fig. 5).

Celestite.

Some celestite is precipitated as a result of simple concentration of sea water, but much of it is formed from the strontium released during the dolomitization of aragonite. (Evans and Shearman, 1964.).

Anhydrite.

Much work has been carried out to determine the origin of Recent anhydrite in sediments. Kinsman (1966) summarized the physico-chemical data available on the stability relations between anhydrite and gypsum, as regards temperature and a_{H_2O}

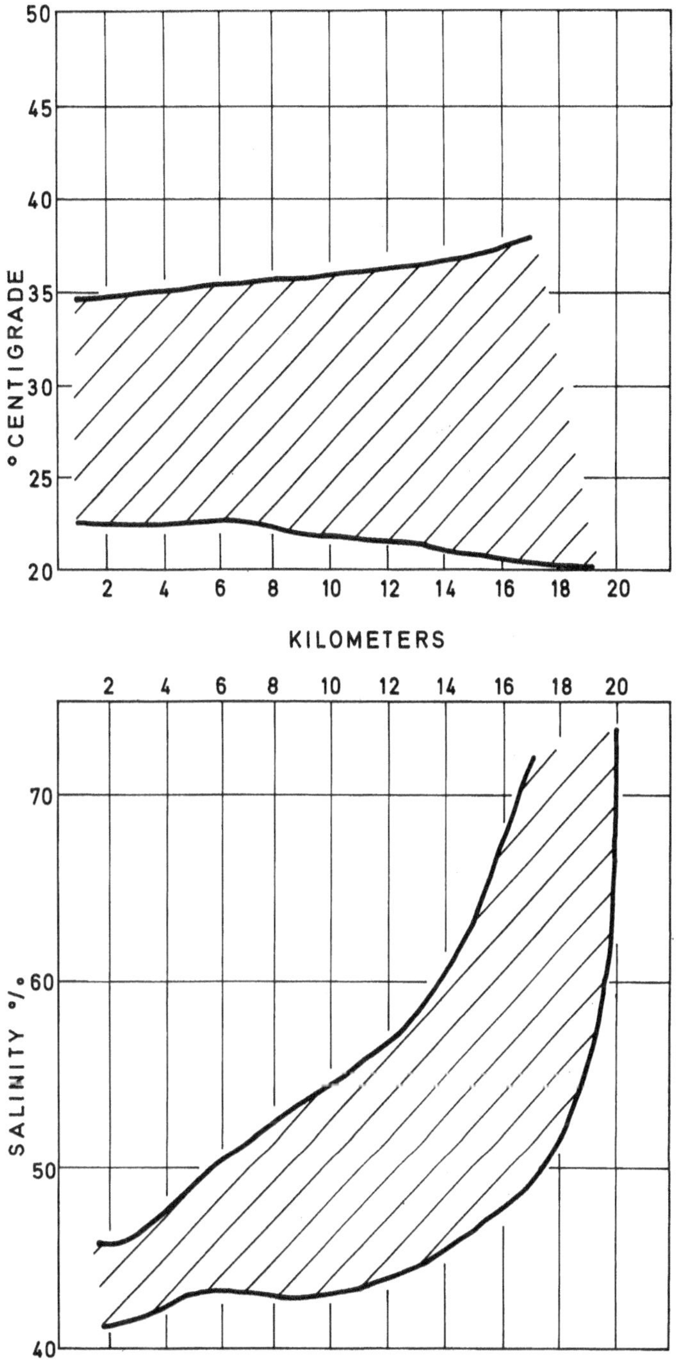

Fig. 3. Surface salinity and temperature of lagoonal and nearshore waters. Zero is approximately 4 km from the shore, the graph showing the variations in water properties from that point to the landward margin of the lagoon. Data from winter, spring and summer are combined

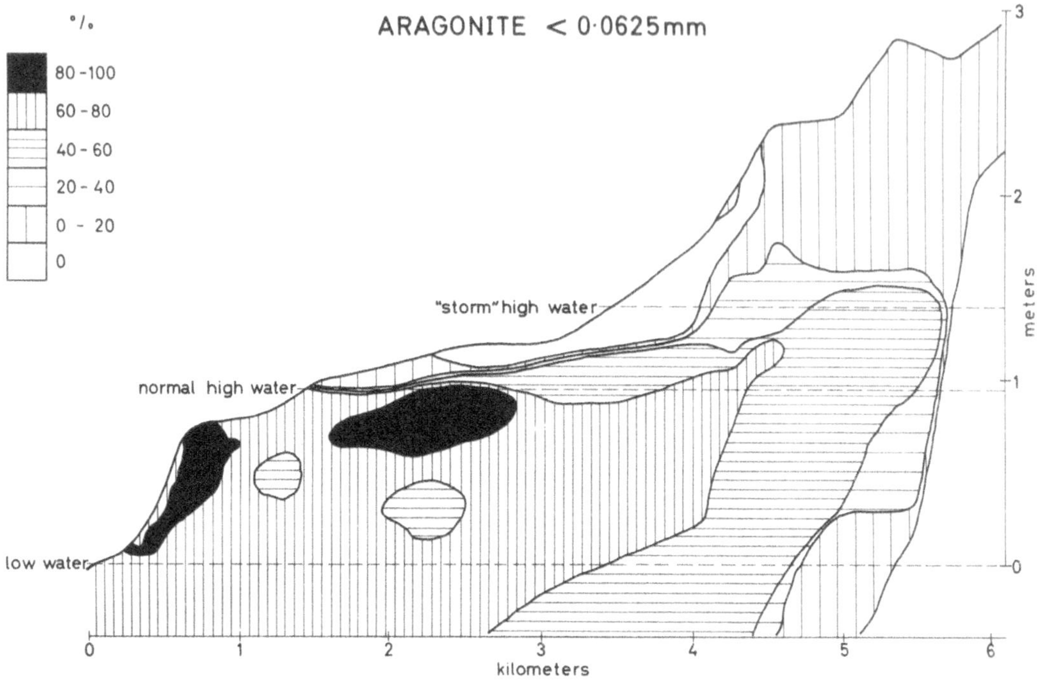

Fig. 4. Distribution of aragonite in the sediments finer than 0.0625 mm

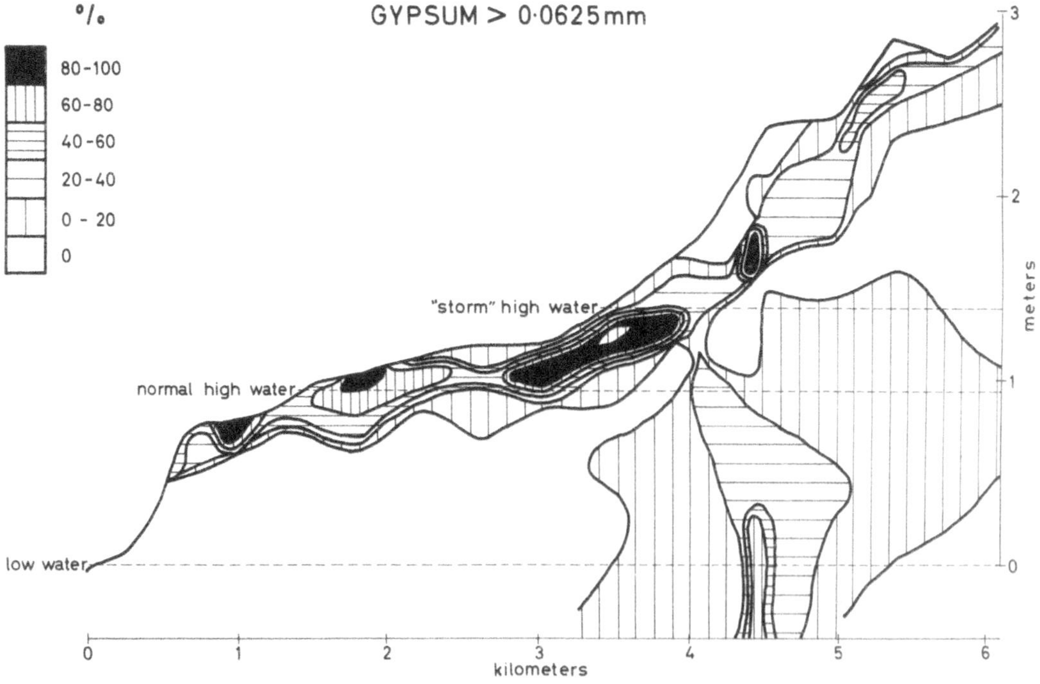

Fig. 5. Distribution of gypsum in the sediments coarser than 0.0625 mm

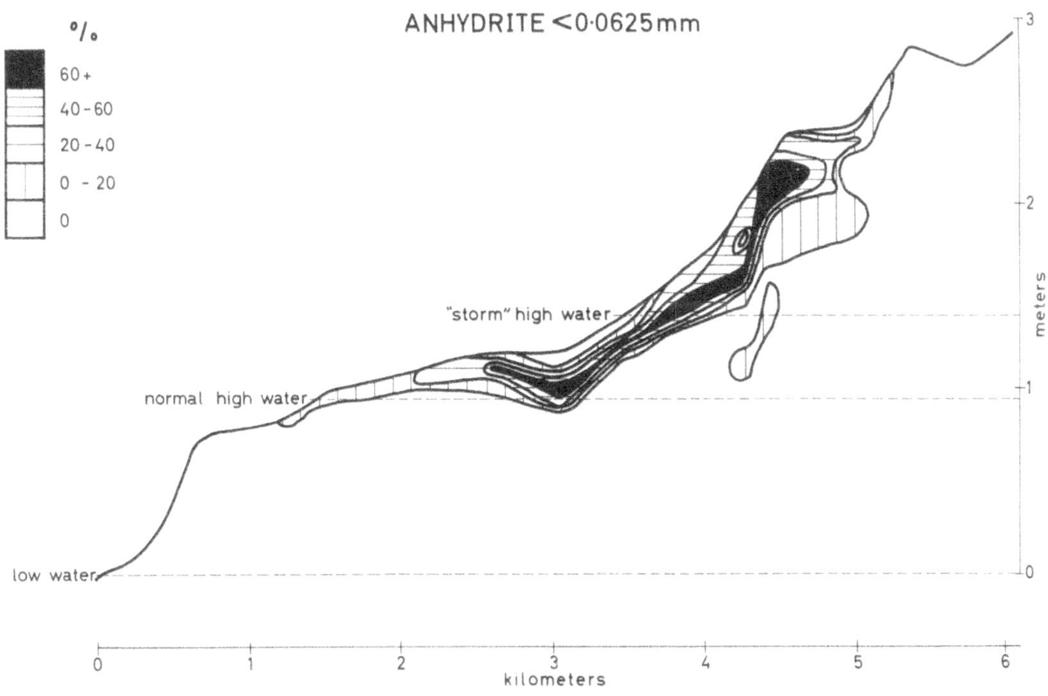

Fig. 6. Distribution of anhydrite in the sediments

and the concentration of brines consisting of concentrated sea water. He showed, using this data, that the gypsum and anhydrite found in the sabkha were in contact with brines with which they are theoretically in equilibrium. He considered, for physical, chemical and textural reasons, that the anhydrite found in the sabkha sediments formed as a primary precipitate and not by the dehydration of gypsum as suggested by Kerr and Thompson (1963) and Murray (1964). However, as a result of later field observations, Kinsman (1969) reported anhydrite pseudomorphing gypsum, and this has also been observed by the writer in the sediments of the sabkha of Abu Dhabi (Fig. 11). The anhydrite pseudomorphs after gypsum, found in the sediments of the sabkha of the latter area, were discovered in anhydrite deposits which lay close to the lagoon margin. In this location, the anhydrite was associated with unaltered gypsum. Cuff (1969) using detailed X-Ray diffraction techniques, has shown that the anhydrite found in the sabkha has many crystallographic defects. He has succeeded in dehydrating gypsum to anhydrite in the laboratory using brines of a similar composition to those found in the sabkha (personal communication). The anhydrite thus produced has similar crystallographic defects to those found in the anhydrite occurring naturally in the sabkha. This evidence supports Cuff's earlier (1969) suggestion that the sabkha anhydrites must have had a gypsum precursor.

In order to explain the presence of pseudomorphs of anhydrite after gypsum and the observed displacive habit of anhydrite nodules, it seems necessary to invoke dehydration of gypsum, in contact with brine, to anhydrite, followed by precipitation of further anhydrite on the nucleii formed. This could occur when the lateral progradation of the intertidal sediments into the lagoon led to an increase in the concentration of the brines in contact with the gypsum earlier formed (Fig. 6).

Halite.

Halite occurs in ephemeral crusts on the surface of the sabkha. This, and the absence of the more soluble salts of magnesium and potassium, have made it necessary to propose processes which prevent concentration of the brines to the level where considerable proportions of the more soluble salts would accumulate. Two main mechanisms have been suggested by several workers: "sub-surface reflux" (Deffeyes et al., 1965; Kinsman, 1966.) and surface flooding or "flood recharge" (Butler 1969). The removal of surface crusts of halite by flooding, as described by Butler, has been observed by the author after flooding produced by both rain and the occurrence of very high wind-reinforced tides. However, the distribution of dolomite and gypsum adjacent to the landward end of the algal mat (Figs. 5 & 8) strongly suggests that dense brines produced at, or near the surface of the sabkha, pass downward i.e., there is "subsurface reflux". The aeolian sand below the lagoonal sediments (Fig. 2) is thought to act as a good aquifer, particularly as it crops out on the margin and floors of the channels in the adjacent lagoon (Evans & Bush, 1969).

(b) Diagenetic minerals produced by reaction between brines and sediment.

Dolomite.

In 1962 Wells reported the presence of dolomite in sediments of Recent age on the coast of the Qatar Peninsula. This discovery was followed by the recognition of dolomite in the Recent sediments of the Trucial Coast (Curtis et al., 1963). Since that time much work has been carried out on the relationship between brine chemistry and dolomite formation (Shinn, 1964; Shinn et al., 1965; Illing et al., 1965; Deffeyes et al., 1965; Kinsman, 1964b, 1969, Butler, 1965, 1969; Bush, 1970). A common process emerges from these studies: the magnesium to calcium ratio of the brines is increased as the result of the precipitation of aragonite and gypsum, and when this ratio reaches approximately 10:1 dolomitization of fine grained aragonite occurs to produce fine grained dolomite (less than 10 μ in size).

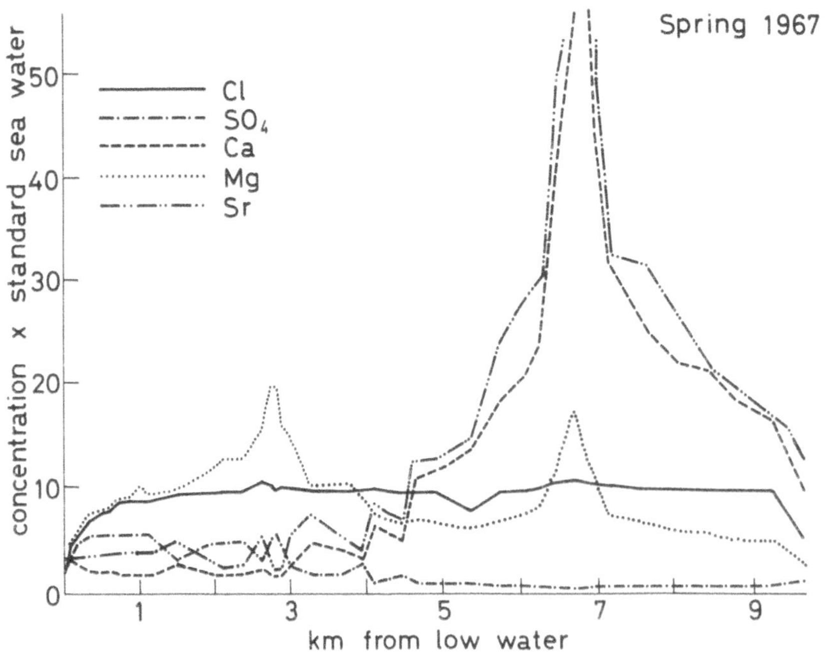

Fig. 7. Chemical composition of sabkha brines

The magnesium content of the sabkha brine rises sharply, and at the same time the calcium content falls in a landward direction away from the lagoon (Fig. 7). The magnesium-calcium ratio exceeds 12, half a kilometre from low water and diagenetic dolomite can be detected in the sediments. However, despite this precipitation of dolomite, the concentration of magnesium in the brines continues to rise until it reaches a maximum value at a point about 2.8 kilometres from low water. This continued increase results from two factors: firstly, much of the brine which reaches the sediment lying at a distance of up to 3 kilometres from the low water mark is introduced by "flood recharge"; it is therefore not in contact with aragonite until it percolates into the sediment: secondly, the concentrated brines that pass laterally through the sediments are held mainly in a zone above the algal mat, as the latter has a very low vertical permeability. Landwards from about 1.5 kilometres from the lagoon margin the sediments above the algal mat contain very little fine grained aragonite (i.e. smaller than 0.0625 mm), (Fig. 4); there is therefore little, or none, of this mineral available for dolomitization. At the point where the buried algal mat ceases to be a vertical permeability barrier the dense, magnesium-rich brines flow downward into the underlying aragonitic lagoonal sediments and dolomitize them. The calcium released by this dolomitization is combined with sulphate from the brines and is precipitated as gypsum (Fig. 4). Support for this hypothesis is provided by the distribution of dolomite adjacent to the landward end of the buried algal mat (Fig. 8).

Magnesite.

Magnesite, occurring in very small quantities, was first described from the sabkha sediments by Kinsman (1965b). The magnesite found by the writer in the area studied occurs exclusively in the sediments which occur above the algal mat, between 2 and 4 kilometres from the lagoon margin. This zone coincides with the area of high magnesium concentration in the brines. The presence of up to 50 % magnesite in the material finer than 0.0625 mm, together with the relatively low concentration of dolomite and the almost total absence of aragonite, suggests that the dolomitization process has been carried one stage further, and magnesite has been produced by the following reaction:

$$Mg^{++} + 2CaCO_3 = CaMg(CO_3)_2 + Ca^{++}$$

$$Mg^{++} + CaMg(CO_3)_2 = 2Mg(CO_3) + Ca^{++}$$

Huntite and Polyhalite.

These two minerals have been described from the sediments of the sabkha of the Trucial Coast but they have not been detected in the sediments of the area studied (Kinsman, 1965b; Butler, 1969).

CONCLUSIONS

Landward of a series of barrier islands and lagoons lies a flat, salt-encrusted plain or "sabkha" , formed by both earlier denudation of aeolian deposits and progradation of lagoonal sediments following a marine transgression which reached the area about 7,000 B.P. Low rainfall and a high rate of evaporation have resulted in the loss of water from both the lagoons and the sabkha groundwaters and has led to their concentration. Concentrated sea water from the lagoons enters the sabkha sediments laterally (i.e. horizontally) at the lagoon margin, and vertically when it has swept across the sabkha surface during exceptionally high tides.

The water percolating through the sediments horizontally from the lagoon margin is concentrated while it is in contact with and surrounded by sediment. On the other hand, water passing over the sediment surface is concentrated by evaporation in contact with the atmosphere. In both cases the waters precipitate aragonite and gypsum

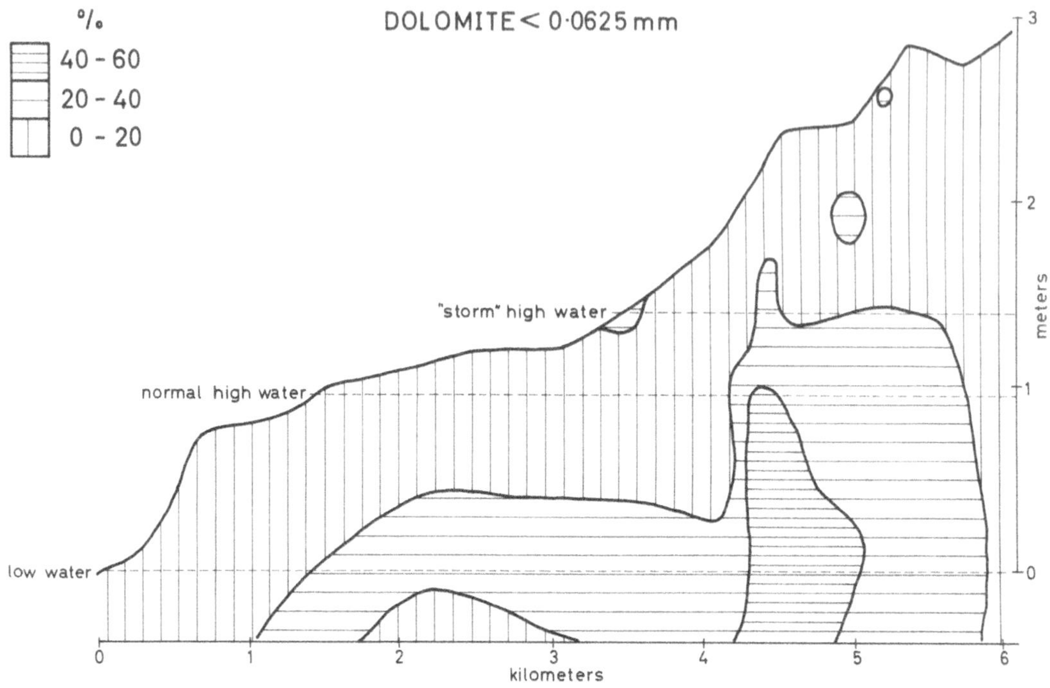

Fig. 8. Distribution of dolomite in the sediments finer than 0.0625 mm

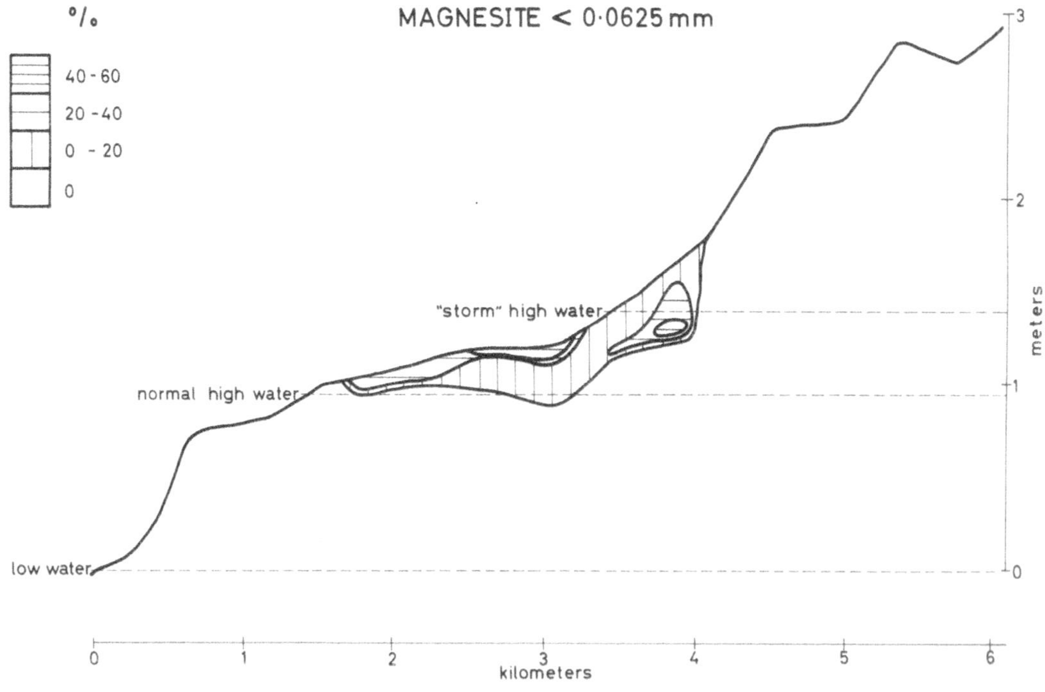

Fig. 9. Distribution of magnesite in the sediments finer than 0.0625 mm

from solution, but the surface brine also frequently precipitates some halite. Some of this concentrated surface brine drains back to the lagoon but some soaks into the sediment. After precipitation of aragonite and gypsum the brine has an enhanced magnesium-calcium ratio and is thus capable of dolomitizing fine grained aragonite. Once the available aragonite is exhausted the magnesium-calcium ratio of the brine rises still higher as calcium sulphate continues to be precipated, and either magnesite forms instead of dolomite, or the brines migrate until more aragonite is encountered.

A buried algal mat produces a horizontal impermeable layer within the sediments of the sabkha and prevents the dense, concentrated, surface brines from sinking downward into the underlying sediment. However, the algal mat dies out landward and thus at this point the brines are able to percolate downward into the underlying sediment. The fine grained aragonite in the sediments above the algal mat has nearly all been converted, first into dolomite then in part to magnesite. Where the brines have passed downwards into the aragonitic lagoonal sediments below and beyond the landward limit of the algal mat, further dolomitization has taken place. It seems probable that these downward-moving dense brines ultimately return to the lagoon through the underlying aeolian sands (see Fig. 10).

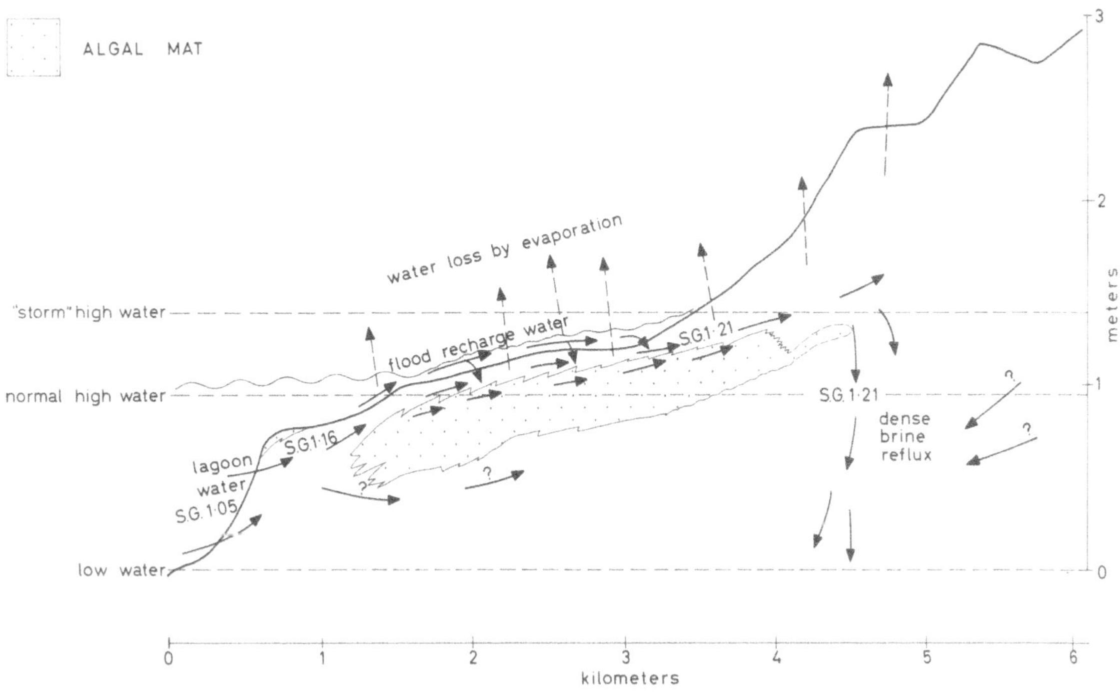

Fig. 10. Supposed paths of water movement in the outer parts of the Abu Dhabi sabkha

Nodular anhydrite occurs mainly in the sediments found above the algal mat. Some of this anhydrite is produced by dehydration of gypsum, as the concentration zones of the brines migrate seawards following the prograding sediments, and some is precipitated directly from the brines.

The lack of extensive deposits of the more soluble salts appears to be due to the process of surface flooding followed, on the one hand, by the draining of the waters back across the surface into the lagoon, and on the other hand, by subsurface reflux, which again probably leads to movements of the waters back into the lagoon.

The studies made to date indicate that large variations in physical and chemical conditions are possible in this type of coastal environment, as witnessed by the varied nature of the Recent and ancient sediments of this type.

ACKNOWLEDGEMENTS

This study was supported by the Natural Environment Research Council, Great Britain, Mobil Research and Development, Dallas, U.S.A. and Shell Research, Rijswijk, The Netherlands. The author would like to thank these organisations for their assistance.

The writer would particularly like to thank Graham Evans who made this research possible, and other colleagues from Imperial College who gave help and advice at all stages in its development.

Fig. 11. Pseudomorphs of anhydrite after gypsum.
a) Very fine grained anhydrite with no additional growth outside original crystal boundary.
b) Coarser grained anhydrite with some growth outside original crystal boundary.
c) Coarser grained anhydrite with additional growth beginning to obscure original crystal outline

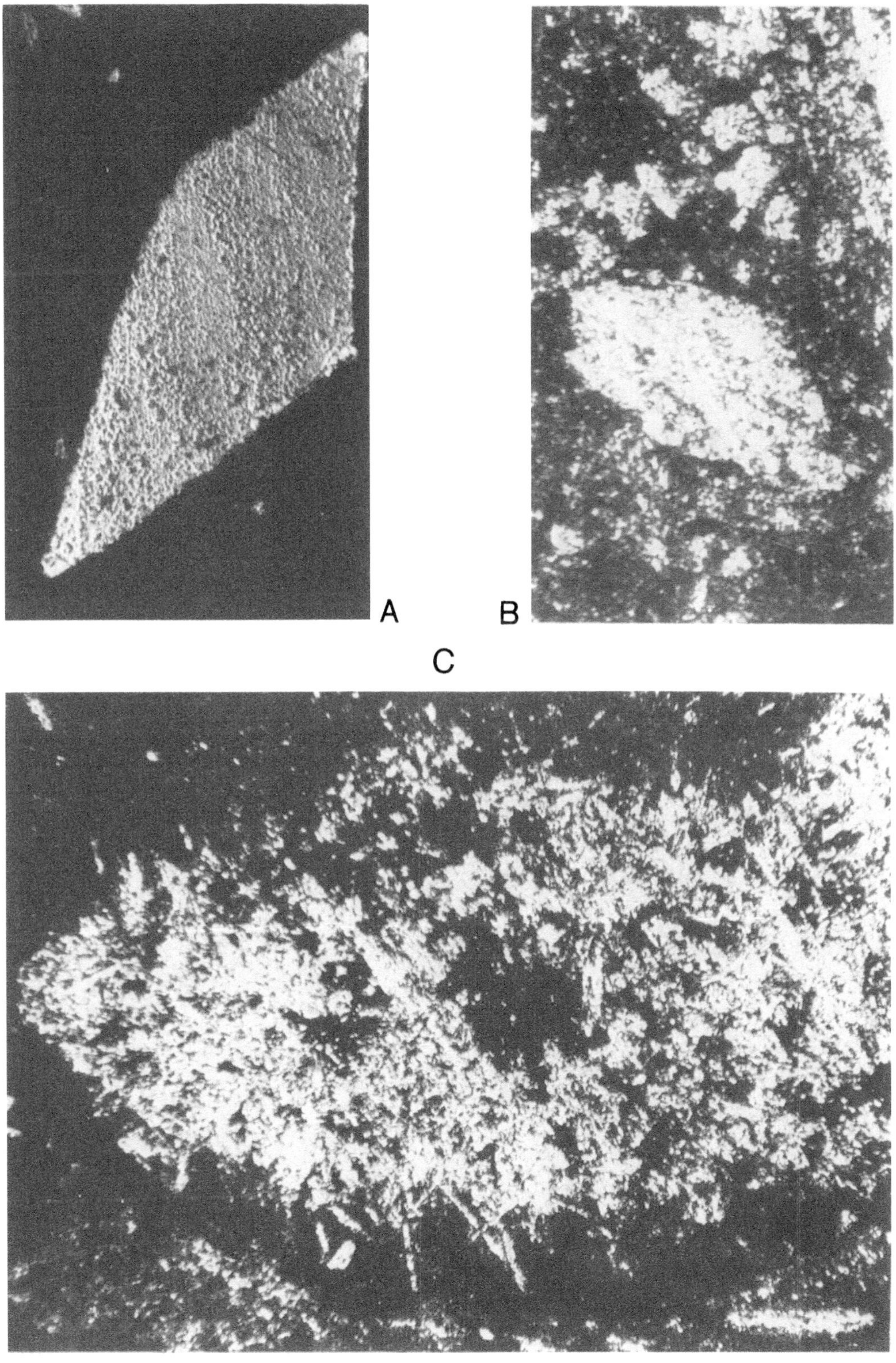

Legend to Fig. 11 A-C on the preceding page

Progress Report on Dolomitization - Hydrology of Abu Dhabi Sabkhas, Arabian Gulf

K. J. Hsü and J. Schneider[1]

ABSTRACT

The authors evaluate the dolomitizing potential of three hydrologic systems -- "seepage reflux" (Adams & Rhodes, 1960), capillary action (Friedman & Sanders, 1967), and "evaporative pumping" (Hsü & Siegenthaler, 1969). Laboratory experiments demonstrate the efficiency of the latter process as a potential mechanism of water flow necessary for dolomitization, and show that this flow is not due to capillary forces but to reduction in pore pressure due to evaporation of interstitial vapour.

In an attempt to confirm the validity of laboratory experiments a series of pits was excavated across the sabkha SE of Abu Dhabi Island where manometers were inserted into various horizons within the sabkha. Field experiments were complicated by the fluctuation of both sea level (due to tides) and water level within the sabkha. This did not prevent the discovery of local hydraulic gradients oriented vertically upwards, confirming the validity of the evaporative pumping model. Manometric measurement made on horizons below an impermeable limestone crust within the sabkha suggest that continental groundwater is the source for evaporative pumping, although seawater cannot be excluded as a source for the more coastal parts of the sabkha.

INTRODUCTION

Carbonate diagenesis involving material exchanges requires transport of ions in solution by moving waters. Dolomitization, in particular, depends on a continued supply of magnesium by "interstitial solutions". Moving interstitial waters below the saturation zone under land are, by definition, groundwaters. Hence the senior author was attracted to the hypothesis of "groundwater-dolomitization" when he was assigned by the Shell Research Company (Houston) to investigate the origin of dolomites. His enthusiasm was dampened somewhat by material-balance and flow-rate calculations based upon the groundwater data from humid regions; the groundwaters do not seem to contain enough magnesium in shallow horizons and apparently do not flow fast enough in deeper horizons for extensive dolomitization (Hsü, 1966, p. 137). Meanwhile, it became clear that Recent dolomites occur mainly in arid regions. Our knowledge of classical hydrology in humid regions led us astray.

Recognizing the common association of dolomites with evaporites, Adams and Rhodes (1960) postulated a mechanism of dolomitization by "seepage reflux". It was one of the several competing hypotheses derived from the study of ancient rocks and

1 Geological Institute, Swiss Federal Institute of Technology, Zurich, Switzerland.

received a considerable boost after Deffeyes, Weyl and Lucia (1965) carried out a brief study of coastal hydrology on an arid island. They found Recent dolomites on the edge of Pekelmeer, a supersaline lagoon, on Bonaire Island, Netherlands Antilles. Their hydrological estimates convinced them that the tidal inflow to the lagoon was not balanced by the outflow, and the evaporated, supersaline brine must have seeped through the bottom of Pekelmeer and dolomitized carbonate sediments on its reflux path. Their demonstrative experiment made the hypothesis even more attractive: brines constituting a density head could be shown to percolate easily downward through permeable sediments. After this demonstration of an apparent proof in an actualistic setting, the seepage reflux hypothesis was catapulted into the limelight. Various indications that Recent and ancient dolomites were formed in an arid climate were cited, or implied, as supporting evidence in favor of seepage reflux.

The reflux hypothesis assumed that the source of the descending saline water was an evaporating lagoon in an supratidal environment. Yet Recent dolomites were being reported from regions where no "pekelmeer" were present, including Andros Island in the Bahamas (Shinn, et al., 1965), playas of West Texas (Friedman, 1966), and sabkhas around the Persian Gulf (Illing, et al., 1965). The idea suffered a further setback when Lucia (1967), one of the original proponents, made a boring through the sediments under the Pekelmeer and failed to find any indication of seepage reflux.

An alternative to the reflux mechanism postulated the movement of dolomitizing solution by capillary action in the vadose zone (e.g., Shinn, et al., 1965); the genesis of supratidal dolomites has been compared to that of caliche, formed where "capillary upward movement of seawater .... becomes concentrated by evapotranspiration at the sediment-air interface" (Friedman and Sanders, 1967, p. 280).

A third mechanism was proposed by the senior author and his associate, and has been termed "evaporative pumping" (Hsü and Siegenthaler, 1969). Basing our work upon a theoretical analysis and simple experiments, we demonstrated the existence of an upward "Darcy-flow" of subsurface waters under evaporative conditions. The essence of our proposal is based on the fact that solar energy indirectly causes an otherwise hydrostatic system to become hydrodynamic when upward movement of groundwater is necessary to replace evaporative loss. Contrary to the hydrodynamics of humid regions, where the hydraulic gradient is commonly oriented in a horizontal direction, the hydraulic gradient under sabkhas should be directed vertically upward.

During the last few years, the tides seem to have turned against the seepage-reflux hypothesis. However, there is still considerable doubt as to whether the evaporative pumping movement is identical, or not, to the well-known phenomena of capillary draw, and whether the authors' model is any different from that previously postulated by Friedman and Sanders (1967). We have, therefore continued our experimentation. Furthermore, we obtained an opportunity to test our ideas in the field through a study of the groundwater hydrology of the Abu Dhabi sabkhas. In this article a more precise definition of evaporative pumping is attempted, and some preliminary field data in support of the hypothesis is presented.

Most of the experiments were performed by Christoph Siegenthaler and constituted a part of the Experimental Geology Program at the Swiss Federal Institute of Technology, Zurich. The field investigation was financed by the American Petroleum Institute as API Project 141. Perry Roehl of the Union Oil California and Bob Dunham of Shell Development Company have been most helpful in their support. Logistical assistance has been rendered by various branches of the Abu Dhabi Government and by the Abu Dhabi Petroleum Company. A special acknowledgements is due to Major B. Williams, Abu Dhabi Police Force, who helped in a thousand ways.

We are indebted to all who have worked on the Abu Dhabi sabkhas, to Doug Shearman and his Imperial College associates, who reconnoitered the region, and to David Kinsman, Geoffrey Butler, Graham Evans and their co-workers, who carried out detailed stratigraphical, mineralogical and geochemical studies. Kinsman and Evans kindly made available to us their unpublished maps, sections and diagrams. Their pace-setting permitted us to delve immediately into the dolomitization hydrology problems.

The authors wish particularly to express their thanks to Graham Evans of Imperial College, London, for his encouragement and whole-hearted support after he learned that we had been granted the privilege of going to the Abu Dhabi sabkha to conduct an investigation which he himself would have liked to undertake.

EVAPORATIVE PUMPING EXPERIMENTS

The authors have carried out a series of laboratory experiments on evaporative pumping hydrology. The experimental set-up is shown diagrammatically in Figure 1 a. Glass cylinders were filled with sediments ranging in size from very coarse sand to very fine silt. Each cylinder had an inlet near the base and was fed from a level-regulator (R). The height of water level within the various regulators is given in Table 1; minus values denote levels beneath the sediment surface and plus values denote those above. The water table within the sediment columns is also given. The difference between the two is Δh, namely the difference between the hydraulic head at the bottom and that at the top of water saturated, sediment columns. Table 1 also shows flow rates and evaporative conditions.

The water table within the sediment columns was the same as the level in the regulators before evaporation started, when the pore water was static. After the lamp L was turned on, the water table remained at about the same height in cylinders filled with coarse or medium sand, but fell appreciably in cylinders E and F, which were filled with silt. The level in E could not be exactly determined but the one in F was estimated on the basis of color tracers to be about - 15 cm.

Our conclusions are:

(1) The flow rate through sediments of very different permeability was approximately the same, the rate being fixed by the need to replace evaporative loss.

(2) The flow rate was apparently not related to the size of capillaries, nor is the existence of a vadose zone a pre-requisite for hydrodynamic flow. The medium-grained sand in cylinder G was submerged under water, the upward movement through such coarse sediment therefore could not be related to capillary action in a vadose or partially saturated zone.

(3) Depression of the groundwater table at the top of the water-saturated silt is consistent with Darcy's law, which postulates that the movement through the water saturated sediment requires the establishment of a hydraulic gradient (dh/dl). The Darcy equation states:

$$\frac{dh}{dl} = q \frac{\eta}{k \rho g}$$

where q is the flow rate, k is the sediment permeability, ρ is the density of water and η its viscosity, g is the gravitational acceleration. Take the cylinder F for example: for a flow rate of 1.94 cm^3/cm^2/day ($2.1 \cdot 10^{-3}$ cm/s), the computed gradient, (dh/dl), should be 1·3. The observed Δh for F was 5 cm, giving an observed (dh/dl) of 1.0, in sufficiently close agreement with the computed result. The Darcy Equation also explains the apparent lack of difference between the level of the regulator and water table in the sand filled columns. For example, the computed (dh/dl) for cylinder G is only $3 \cdot 10^{-4}$, where Δh should be $6 \cdot 10^{-3}$ cm. This minute difference in height is too small to be detected by our crude measurements.

Our former colleague, C. Siegenthaler, was still not convinced that capillary action did not play any rôle in evaporative pumping. He carried out another series of experiments in 1970, and the set-up is shown in Figure 1 b. In cylinders a and b he placed fine sand and silt ranging from 0.08 to 0.2 mm in diameter. In cylinders

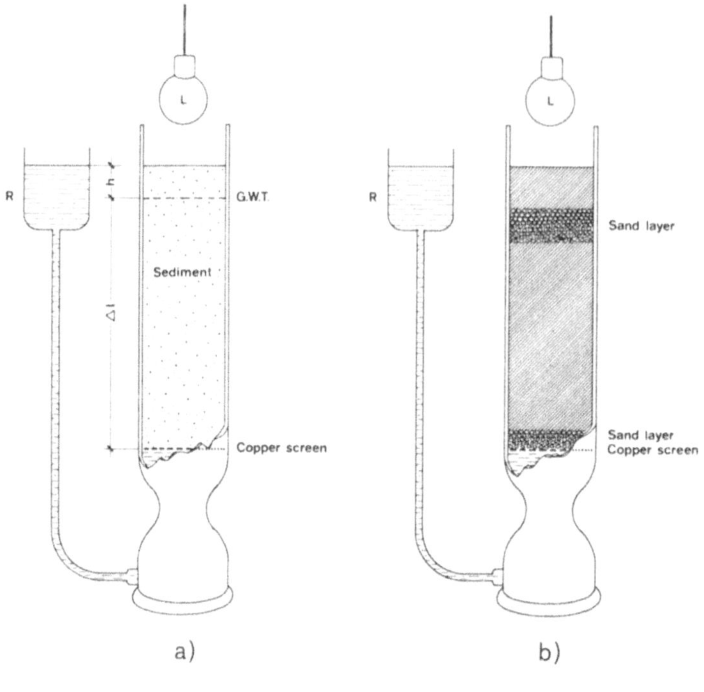

Figure 1. Evaporative Pumping Experiments. Schematic diagrams showing the experimental set-up. Water flows from a reservoir (R), with level regulated, upward through a column of sand to replace evaporative loss.

1a) Experimental set-up when each sediment column was homogeneous, but sediments of different permeability filled different columns and the reservoir level was adjusted to different heights (see text).

1b) Experimental set-up when very coarse sand was inter-layered between silty fine sand in some columns (see text)

c and d, he placed a coarse sand (1.0-1.5 mm in diameter) layer, 5 cm thick, sandwiched between fine sand layers, the top of the coarse sand layer was 2 cm deep (see Fig. 1 b). In cylinders e and f he placed a similar 5 cm-thick coarse layer, except that the top of the coarse sand was 6 cm deep. It was envisioned that capillary forces in fine sand might have been sufficient to draw water upwards, but the coars sand layers in cylinders c, d, e, and f should serve as barriers to capillary action. Thus, if the capillary forces do play a decisive rôle, the flow rate through the first two cylinders should be considerably faster than that through the other four cylinders.

The experiment was carried out over a period of two months. The level within the regulator was permitted to drop from 0 cm to - 22.5 cm to evaluate the possible effect of fluctuating water tables on evaporative loss. The height of sediment column was 23 cm in cylinders a and b, and 24 cm in cylinders c, d, e and f.

The results are shown in Table 2. Whatever the rôle of the capillary fringe might have been, the size of capillaries definitely did not affect the rate of evaporative pumping. The flow rate was practically the same for all the cylinders. Furthermore, the evaporative pumping rate remains constant even when the groundwater

TABLE 1

Results of Critical Experiments on Rate of Evaporative Pumping

	A	B	C	D	E	F	G	H	I	J	K
Lamp power (watts)	60	60	60	60	60	60	60	60	60	15	15
Lamp Height (cm)	2	2	2	2	2	2	2	2	1	3.5	3.5
Sediment type	v.cs. sand	v.cs. sand	med. sand	med. sand	v.f. silt	v.f. silt	med. sand	water only	med. sand	med. sand	med. sand
Sediment size (mm)	1.2	1.2	0.3	0.3	0.01	0.01	0.3	-	0.3	0.3	0.3
Sediment height (cm)	20	20	20	19	20	20	20	-	20	19	19
water level in Reservoir (cm)	-10	-2	-10	-2	-2	-10	+1	-	+1	-10	-2
water table in sediment (cm)	~ -10	-2	~ -10	-2	?	-15	submerged	-	submerged	~ -10	~ -2
Δ h (cm)	negl.	negl.	negl.	negl.	?	~ 5	negl.		negl.	negl.	negl.
Permeability (millidarcies)	8.1×10^5	8.1×10^5	7.3×10^4	6.9×10^4	15.4	15.4	7.3×10^4		7.3×10^4	7.3×10^4	7.3×10^4
Evaporation rate= (cc / cm^2 day)	1.21	2.48	2.54	2.24	1.94	1.94	2.03	3.48	4.11	0.236	0.236
Linear flow rate cm / day	3.0	6.2	6.3	5.6	4.8	4.8	5.1	3.48	10.3	0.59	0.59
Evaporating Condition (relative)	strong	strong	strong	strong	strong	strong	strong	strong	v. strong	weak	weak

table falls to 22.5 cm below the sediment surface. The presence of a partially saturated zone a few tens of centimeters thick apparently is not sufficient insulation to retard evaporative losses under these experimental conditions.

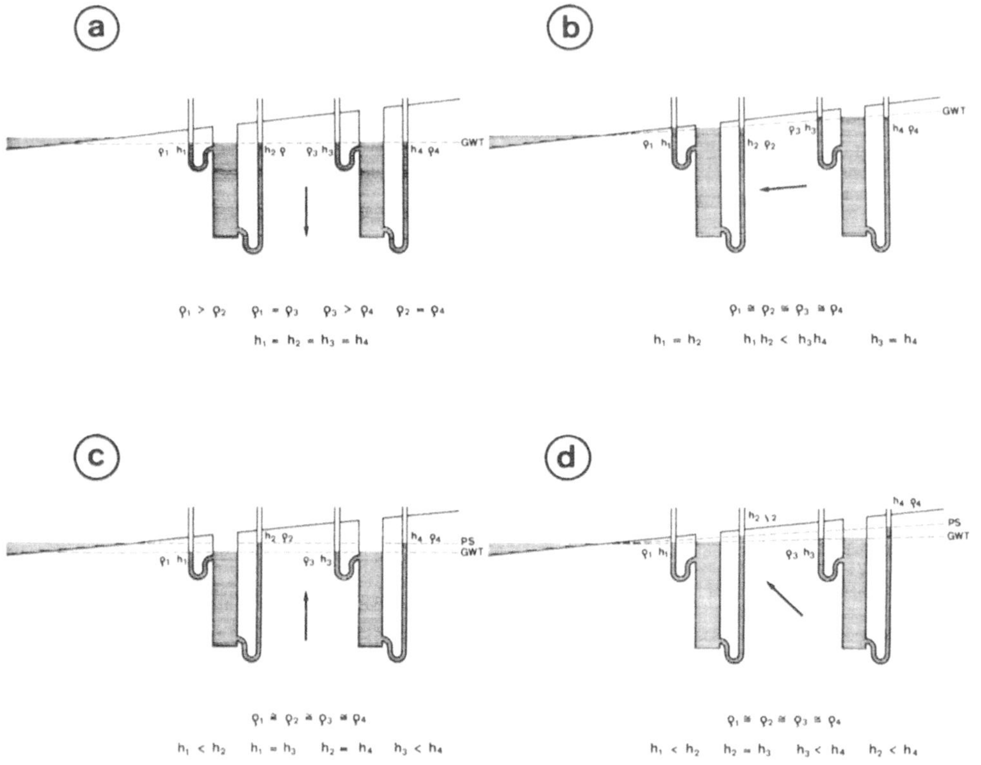

Fig. 2. Four Possible Models of Sabkha Hydrology
a) Seepage reflux model. Descending brines.
b) Capillary fringe model. Seaward-flowing groundwater.
c) Evaporative Pumping from a Seawater Reservoir.
d) Evaporative Pumping from a Continental Groundwater Reservoir.

PS = Piezometric surface; GWT = Groundwater table

TABLE 2

Average flow rate (cm^3/hour)

Date	Hour	Cylinder No. a	b	c	d	e	f	waterlevel
2. 19.	11oo							
21.	21oo	7.1	6.8	5.9	8.1	6.5	6.2	
	24oo	6.6	6.6	5.6	7.9	6.5	6.0	0 cm
23.	153o							
24.	10oo							
25.	9oo	6.7	7.8	5.2	7.8	6.3	6.1	
26.	915	7.1	8.3	6.0	7.5	6.5	6.5	
	16oo							1

Table 2 continued

28. 3. 2. 4. 5. 6. 9.	11oo 17oo 11oo 15oo 9oo 12oo 10oo	7.2 7.4 - 7.2 7.0 7.4	0.6* 0.6* 0.0* 8.1 7.4 7.6	5.6 5.6 6.2 5.6 4.8 -	7.3 8.0 - 8.3 7.8 7.9	0.6* 0.7* 0.0* 7.2 6.7 6.9	5.5 5.5 3.8 6.1 5.2 5.1	-2.5 cm
10. 11. 12. 13.	16oo 10oo 21oo 16oo 10oo	7.0 7.8 8.2 7.4 7.2	7.3 8.3 6.4 7.4 7.8	7.2 8.1 6.8 7.1 8.1	7.8 7.8 8.6 7.9 8.1	6.5 6.9 7.3 7.1 8.6	5.2 6.7 7.7 6.6 6.4	2 -5.0 cm
15. 16. 17. 18.	20oo 10oo 11oo 123o 16oo 19oo 2215 10oo 12oo 1645 2315 113o	7.2 7.9 6.9 7.8	7.5 7.1 4.2 AX X 7.8	7.5 7.5 7.5 7.6	8.2 8.6 7.9 8.2	8.2 7.5 7.9 7.8	7.0 7.1 6.9 8.2	3 -7.5 cm
  19. 20. 22.	1645 2215 83o 14oo 223o 113o 1245 193o	6.8 7.9 7.5	7.3 7.9 7.8	7.5 7.4 7.7	8.2 7.9 8.3	7.7 7.2 7.9	8.6 8.1 8.8	4 -10.0 cm
23. 24. 27.	12oo 15oo 20oo	6.7 7.4 7.5	7.9 7.6 7.9	7.6 7.8 7.5	8.5 8.7 8.6	7.6 7.8 7.7	7.3 7.8 7.7	5 -12.5 cm
31. 4. 2.	10oo 12oo	7.4	7.8	7.6	8.3	7.6	7.0 7.0	6 -15.0 cm
3. 4. 6.	10oo 16oo 23oo	7.3 7.5 7.8	7.5 7.5 7.5	6.8 7.0 7.2	0.22** 0.16** 0.2 **	7.6 7.5 7.5	6.8 6.3 6.3	7 -17.5 cm
13.	12oo	-	7.2	7.1	0.3 **	7.1	5.7	8 -20.0 cm
16. 19.	16oo 20oo	7.6 -	7.2 7.2	7.0 6.8	0.26** 0.26**	6.5 6.8	5.9 5.5	9 -22.5 cm
standard-deviation		0.4049	0.8106	0.9481	0.3599	0.6440	1.0974	
Mean		7.3467	7.3467	6.8093	8.1077	7.2200	6.5382	
Number of measurements		30	30	32	26	30	34	

* Lamp above the column went out; the very low flow-rate indicates water-supply rate under normal atmospheric conditions.

** The sediment was cracked and the cracks filled with air, which was apparently sufficient to stop the evaporative pumping mechanism.

The few very low evaporative rates given in Table 3 were, on the whole, due to the malfunction of a lamp; these low rates are representative of evaporation under almost normal atmospheric conditions. However, the lower rates for cylinder d after April 2 were a notable exception. In that case, the bulb was not burnt out, but the sediment was badly cracked and the cracks were filled with air of normal atmospheric pressures. This fact, together with the observation that evaporative rate was independent of capillary sizes, led Siegenthaler to conclude that the fluid movement above the water table was not governed by capillary forces, but was induced by a potential difference by a reduction in pore pressure when water vapour was being evacuated by evaporation (Siegenthaler, informal report, 1970). We are continuing our experimentation to further check this hypothesis.

Whatever the cause of fluid movement through the undersaturated zone, the experiments have clearly demonstrated that intensive evaporation led to the establishment of a hydraulic gradient in vertically upward direction. The crucial test lay in the field, and a systematic study of the sabkha hydrology became essential.

FOUR MODELS OF DOLOMITIZATION OF HYDROLOGY

Before our departure to Abu Dhabi, we made a theoretical analysis of what could be expected in the field. The four models of dolomitization postulated three radically different directions of water movement in the saturated zone: vertically downward, lateral, and vertically upward. The rate of water movement is probably too small to be measured. However, the direction of movement could be determined on the basis of observed hydraulic gradient. Field determinations of potentiometric surfaces should therefore provide the answer. Our analysis of the problem is shown diagramatically in Figure 2, and can be explained briefly as follows:

(1) Seepage reflux model (Figure 2a): The model postulates vertically descending supersaline brines (see Deffeyes et al., 1965). Higher than the normal density of evaporated brines (ρ_1, ρ_3) provides the hydrodynamic head, and the movement is governed by the relation (Hubbert, 1940; Deffeyes et al., 1965):

$$q = k \cdot \frac{gh}{\eta} \cdot \frac{\Delta\rho}{\Delta\ell}$$

where $\Delta\rho$ is the difference in density between the hypersaline and the normal seawater, h is the depth of the column underground where the supersaline water has penetrated, and $\Delta\ell$ is the length of the reflux path.

According to this model, if a hole is dug on the sabkha, the water table should be approximately at the main sea level or slightly below. If we continue to dig, the manometer height should remain the same, although the water at deeper levels (in the direction of hydrodynamic (flow) should have a density (ρ_2, ρ_4) somewhat less than that nearer the surface. Furthermore, the water table height in all the holes should be about the same.

(2) Capillary fringe model (Figure 2b): This model postulates vadose water ascending vertically by capillary action. The nature of groundwater movement has not been clearly defined. However, it seems to have been implied that groundwater flows seaward and the flow rate is :

$$q = k \frac{\rho\, g}{\eta} \left(\frac{dh}{dl}\right)$$

where (dh/dl) is the slope of the seaward dipping water table.

According to this model, if holes are dug on the sabkha, the groundwater table should be at various levels above high tide level (h_1, h_2, h_3, h_4). However, the height of the potentiometric surface should remain at the same height at any given spot, even if one inserts a manometer into a deeper horizon within the sabkha sediments. A density difference between the brines at various levels probably exists, but the role of $\Delta \rho$ is insignificant compared to that of Δh.

(3) Evaporative pumping of sea water (Figure 2c): This model postulates vertically ascending brines. The movement of the water above the water table could be due either to capillary action, or to pressure difference caused by the evacuation of water vapours from the pore space. The movement of water below the groundwater table is also upward, as the hydraulic gradient points vertically upward, and the flow rate, as we have discussed, is:

$$q = k \frac{\rho q}{\eta} \left(\frac{dh}{dl}\right) .$$

According to this model, if holes are dug on the sabkha, the water table should be found at about the same height (h_1, h_3), slightly below high tide level. However, if we dig deeper and place a manometer into a deeper horizon, the water should rise above the water table (h_2, h_4), but not higher than the highest tide level. A density difference between the brine probably exists, but the role of $\Delta\rho$ in hydrology is insignificant.

(4) Evaporative pumping of groundwater (Figure 2d): This model postulates steeply ascending water (the arrow in the figure would be nearly vertical if the vertical scale of the diagram were not disproportionately exaggerated). The movement of the water above the water table could be due either to capillary action, or to pressure difference caused by the evacuation of water vapours from the pore space. The movement of water below the groundwater table is also upward, and the flow-rate is :

$$q = k \frac{\rho g}{\eta} \left(\frac{dh}{dl}\right)$$

According to this model, if holes are dug on the sabkhas a seaward-dipping water table should be found at various heights (h_1, h_3) above high tide level. However, if we dig deeper and place a manometer into a deeper horizon, the water should rise to a height (h_2, h_4) higher than the water table.

Thus, when we set out for Abu Dhabi, we intended not only to make routine measurements of evaporation rate, groundwater table, etc., but also to evaluate whether or not the hydraulic gradient under the sabkha was oriented vertically upward

SABKHA HYDROLOGY IN ABU DHABI

Meteorological and hydrological measurements were carried out on an Abu Dhabi sabkha, mainly by one of us (JS), during November and December, 1971 and January, 1972. This sabkha is located ESE of Abu Dhabi town. Extensive mineralogical, geochemical and sedimentological investigations of the sediments in this area had been completed by Kinsman and associates (1969, 1971). They had also made a preliminary hydrological investigation, and found a seaward-dipping groundwater table. This convinced Kinsman (written communication) that an evaporative pumping mechanism was not operative. He postulated, instead, that dolomitizing "brines move downward and seaward through the sediment wedge" (Kinsman, 1969, p. 830).

Our current results have led us to a somewhat different conclusion. The main observations are:

(1) Air-evaporation was high, even in the winter months when temperatures rarely exceeded 40° C. Measurements with a Piche-type instrument indicated evaporation rates of 1 to 3 cm per day.

(2) Similar rates were found for interstitial waters in sandy sediments near the surface. These measured rates constitute a rough estimate of the maximal evaporative loss from the sabkha sediments during the winter months.

(3) Precise determination of evaporative loss from muddy sediments has not yet proved possible, but it appeared to be somewhat less than that from sand.

(4) The level of the groundwater table was variable. It was found at depths ranging from a few tens of centimeters to more than a meter below the sediment surface.

(5) Daily variation of groundwater table height was very small, ranging in millimeters. Diurnal tidal effect was not noticeable.

(6) A wind-driven storm tide in December (19th-21st) covered the coastal half of the sabkha, penetrating a few kilometers beyond the normal high tide strand line for three days. The water table in the flooded area was about the same before and after the flood, tidal recharge of groundwater proving insignificant, at least in this event.

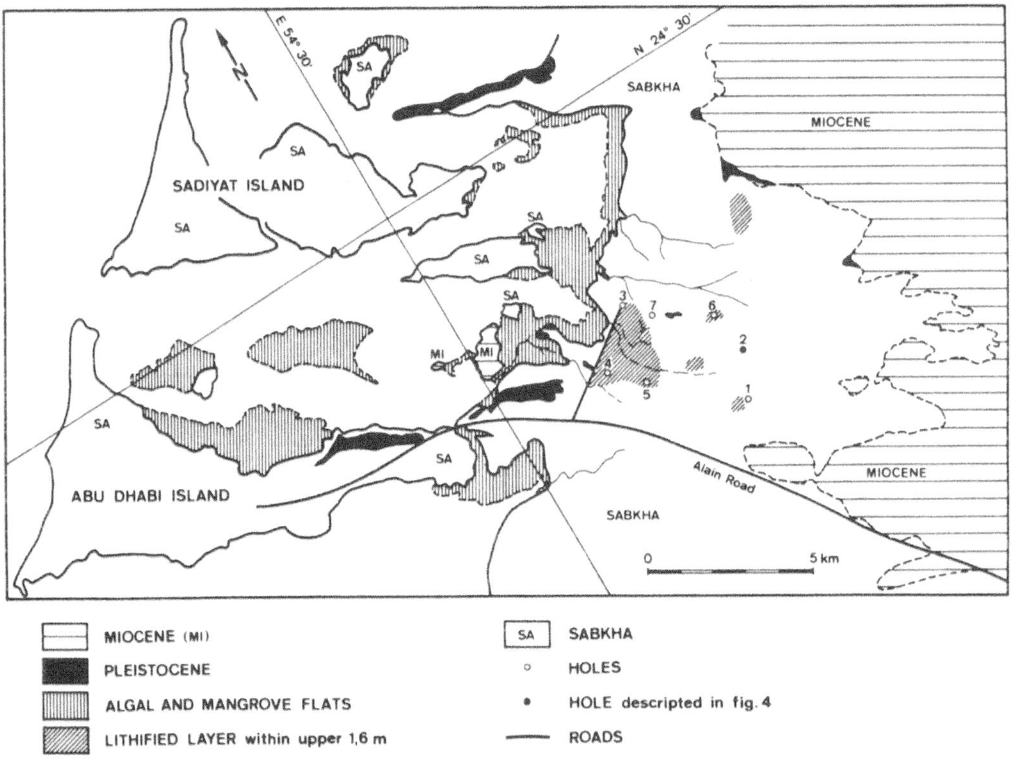

Fig. 3. Location Map of Abu Dhabi, showing the distribution of holes where a cemented layer has been encountered 1 or 2 meters below the surface. At location 2 the vertically-upward oriented hydraulic gradient was measured precisely

The crucial question regarding the height of the water table relative to sea level was difficult to resolve because both levels fluctuated:

Halite deposited in holes dug by previous workers afforded clear-cut evidence that the height of the water table had changed during the last few years; salt deposits up to 1 m thick have been found in these holes. The top of the salt serves as a fossilized record of the maximum stand of the groundwater. In all parts of the sabkha, the water table we recorded lay appreciably below that maximum stand, more than 1 m below in places. Consequently we were not surprised that our results were quite different from those communicated to us by Kinsman (personal communication); the water table has apparently fluctuated by some decimeters during the last few years.

The change of sea level is mainly tidal. The normal tidal range is more than 1 m. The December storm tide had a water mass 60 cm deep and was driven 2 m above the high spring tide level.

We have only been working on the sabkha for 3 months, and do not yet possess sufficient data to come to a definite conclusion on the height of the water table relative to the highest level of a tide that could effectively recharge the undersaturated zone. Our results so far show that the water table, on the whole,was higher than the lowest tidal level. However, the water table under parts of the coastal sabkha was slightly lower than high spring tide level and appreciably lower than the high storm tide level of December 1971. Our tentative conclusion is that both the sea and the saline groundwaters have contributed to replacement of the evaporative loss from sabkha sediments.

The water table has a very gentle seaward dip, as Kinsman noted. The most significant discovery,however, was proof that the hydraulic gradient was oriented almost vertically upward in several places, exactly as our evaporative pumping models had predicted.

It has not been easy to determine the change of manometer height with depth. When a hole was dug water level would rise, and a day or two would elapse before the water level in the hole stabilized. The gradual filling was mainly a transient response to seek equilibrium level, but the possibility that part of the rise was related to the fact that the hole was being progressively deepened was not excluded.

During our reconnaissance in December we noticed cemented but friable layers in the inner parts of the sabkha where near-surface sediments are mainly detrital sand. The crust was found at the surface,or at levels some decimeters below, constituting an obstacle to our digging. We found an even harder, deeper layer, but could not penetrate it. Finally, in January one of us (JS) succeeded in penetrating the crust at locality 2 (Figure 3), where surface elevation is 2.75 m above low spring tide level. After digging down to the hard layer a water table was established for the permeable layer above, near the top of saturation zone, at a level of 111,4 cm (Figure 4). The water was then quickly drained from the hole, and it was possible to dig through the second crust, which acted as a relatively impermeable cover for the underlying water-bearing sediments. As soon as the crust was broken, water rushed to levels higher than - 111,4 cm. However, the water level in the hole rose little because brines from the deeper horizon immediately seeped into the porous sediments above the zone of saturation. When a pipe was inserted through the crust as a manometer the height of the piezometric surface of the lower permeable horizon was found to be at -63.4 cm. The difference Δh in manometer height was 48.0 cm. The distance Δl, the top of the second aquifer to the water table is 90 cm. The (dh/dl) in the vertical direction is thus calculated to be about 0.5 (Figure 4). Using the Darcy formula, and assuming a flow rate of 1 cm per day to replace evaporative loss, we find the permeability:

$$k = q \frac{\eta}{\rho g \left(\frac{dh}{dl}\right)} = \frac{1}{8{,}6 \times 10^{3}} \times \frac{2 \times 10^{-2}}{1 \times 10^{3} \times 0.5} = 5 \times 10^{-7}\ \mathrm{cm}^{2} = 500\ \mathrm{m.d.}$$

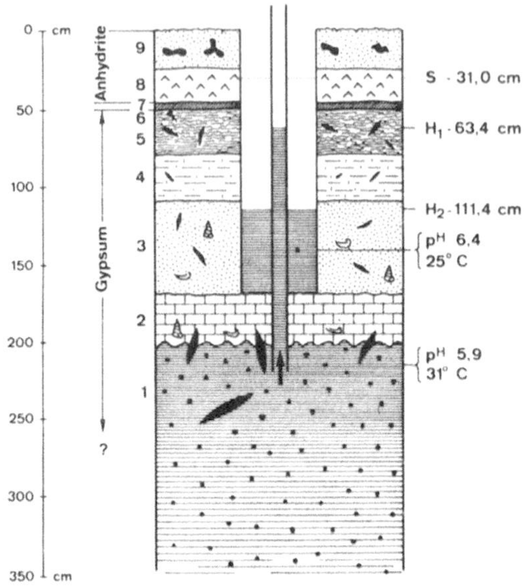

Fig. 4. SEDIMENTARY SEQUENCE AND WATER LEVELS AT LOCALITY 2

9) Brown, coarse, detrital sand with anhydrite nodules	(0- 24 cm)
8) Layered anhydrite	(24- 43 cm)
7) Grey, creamy, carbonate mud	(43- 47 cm
6) Laminated algal mats with small gypsum crystals (3 mm) and some gypsum disks (Ø 4 cm)	(47- 57 cm)
5) Clearly laminated algal mats with some gypsum disks (Ø 8 cm)	(57- 80 cm)
4) Algal laminated, grey carbonate mud	(80-110 cm)
3) Blue-grey, fine, carbonate sand with cerithium shells, mollusk and gypsum disks (Ø 6 cm)	(110-170 cm)
2) Lithified layer, same material as above	(170-200 cm)
1) Blue-grey, coarse sand with large gypsum disks (Ø 20 cm)	(200-350 cm)

S : Height of ancient salt crust

H_1: Manometer height from layer 1

H_2: Water table

In other words, the sediment which maintained such a potential difference should have a permeability of about 500 millidarcies. We are now sending the sample for permeability determination to check this prediction.

Several other holes dug in the sabkhas show similar behavior (Figure 3), although it was not possible to determine the exact differences in manometer height.

The increase in manometer height with depth is exactly that predicted by models 3 and 4. On the inner parts of the sabkha, where the piezometric surface of the deeper horizons is higher than mean sea level, a suply of saline continental groundwater is necessary. A part of the computed model by Freeze and Whitherspoon (1967) for inland basins was adapted to depict sabkha hydrology. This model (Figure 5) shows that groundwater could be transported from a distant source through an extensive permeable substratum, and the water would rise vertically in an area where the ground-surface is a sink. A porous Miocene sandstone formation is present a few meters below the sabkha (Kinsman, 1971), and this formation is probably a recipient of dynamic groundwaters from the distant Oman mountains. The loss of water from evaporation made the sabkha surface a sink. The groundwater moves upward from the Miocene substratum under the vertically oriented hydraulic gradient as the evaporative pumping model in Figure 2d indicated. Williams (1970) applied a similar model of ascending waters to explain the origin of continental evaporites, where inland basins have long been known as sinks for groundwaters. We have now shown that the rising flow-pattern is equally relevant to the genesis of "marine evaporites" under sabkhas.

OUTSTANDING QUESTIONS

Our results so far conform essentially to those predicted by the model of evaporative pumping from a groundwater source (Figure 2d). However, we have not yet determined whether the substratum is higher than the mean sea level in all parts of the Abu Dhabi sabkha. The possibility that sea water also supplies interstitial fluids to sediments on the more coastal parts of the sabkhas must not be excluded. Theoretically, the sea could be the source of water for the permeable substratum in our model if the piezometric surface of that horizon should be depressed below that of mean sea level during periods of extreme drought. In an arid ,oceanic reef-atoll, the sea water must be the source for the permeable substratum could not receive recharge from a distant source. We hope that our studies in the next few years will tell us more about the sources of waters under the sabkha. The relevant question at present is not where the water comes from, but which way it goes.

A second question concerns the material balance. What happens to the salts left behind by the evaporated brines? Halite crust is ubiquitous on the sabkha. However, the amount of halite is far less than it should be after a thousand years of evaporative pumping, unless there have been effective removal mechanisms. We are still studying this question and meanwhile are tempted to quote our friend Kinsman (1969, p. 835):

"Halite forms a crust on much of the sabkha surface but is not an accumulative phase. Some of the halite is blown away by the onshore winds. Perhaps a more important removal mechanism is the flooding of the sabkha by storm water, some of which subsequently flows seaward, removing the dissolved halite in solution."

We would like to emphasize that our work is still in progress. We have been invited by the Editor to submit our preliminary results. However, we still plan to spend two more seasons on the sabkha. What we have found so far has led us to believe that water moves upward under the sabkhas and has permitted us to adopt evaporative-pumping as a working hypothesis. The final word is yet to be written.

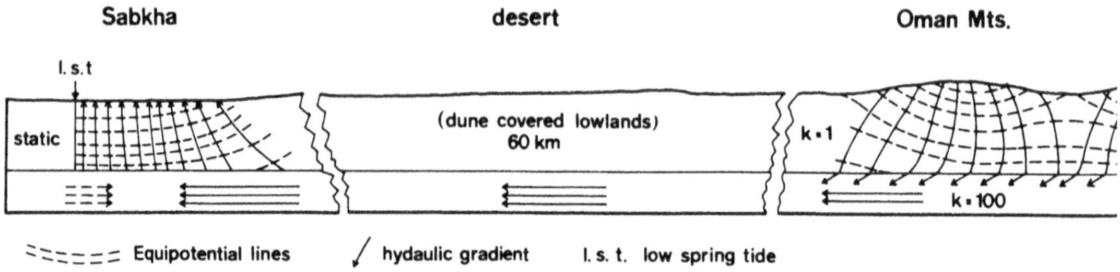

Fig. 5. A model of Sabkha Hydrology; a modified version of the Freeze and Witherspoon (1967) model.

Groundwater from recharge in Oman Mountains is transported through a permeable conduit to the permeable Miocene formation under the sabkha, where the saline groundwater rises to replace evaporative loss. The model requires that the evaporative loss be small in the desert areas between the source and sink; this is possible as the groundwater table must lie rather deep in the desert, where the thick, undersaturated zone serves as a blanket preventing evaporative loss from the permeable conduit. Seawater invasion into the permeable substratum under the more coastal part of the sabkha is probable, especially at times of unusual drought, when the evaporative loss must have been replaced by an inflow from the marine reservoir.

l.s.t. = low spring tide level k = relative permeability

Strontium Geochemistry of Modern and Ancient Calcium Sulphate Minerals

Godfrey P. Butler[1]

ABSTRACT

The study of Sr^{+2} distribution in calcium sulphate rocks has been extensive, but largely descriptive. This paper demonstrates the complexity of Sr^{+2} diagenesis in sedimentary gypsum and anhydrite in several modern evaporite environments. Distribution of Sr^{+2} in these sedimentary sulphates accumulated during the progressive evolution of sea water to a brine is controlled by the solubility products and partition coefficients of aragonite, gypsum, celestite, dolomite, and anhydrite. In arid, supratidal, evaporite environments, this control is modified according to whether the supratidal sediments are dominantly either silici-clastic or carbonate in composition. The absence of dolomitization as in silici-clastic evaporite environments, or presence of dolomitization as in certain carbonate evaporite environments exerts a dominant control on the path of Sr^{+2} diagenesis. In the absence of dolomitization, the mass ratio Sr^{+2}/Ca^{+2} in brines saturated with respect to aragonite, gypsum/anhydrite and celestite rapidly stabilizes at $\sim 2.2 \times 10^{-2}$ and remains at this value into, and probably beyond, halite saturation. Dolomitization in similar brines causes the ratio to fall to a value of less than 2.2×10^{-2}. The magnitude of the departure is dependent on the degree of dolomitization. In carbonate-evaporite sediments which approach 100 % dolomitization, the ratio Sr^{+2}/Ca^{+2} in brines stabilizes between 0.9×10^{-2} and 1.0×10^{-2}. Because the concentration of Sr^{+2} in gypsum/anhydrite is a product of their respective distribution coefficients and the Sr^{+2}/Ca^{+2} ratio in brines, Sr^{+2} in sedimentary sulphates is dependent on the absence or the degree of dolomitization. In other words, Sr^{+2} concentration in sulphates is a function of the nature of their environment of formation.

Mass balance calculations between Sr^{+2} in brines and Sr^{+2} in gypsum/anhydrite from modern evaporite environments permit an evaluation of the average values

1 Esso Production Research Co., P.O.Box 2189, Houston, Texas.

for the partition coefficient of gypsum ($k^{G}_{Sr} \simeq 0.18$) and anhydrite $k^{A}_{Sr} \simeq 0.37$).

Strontium concentrations in anhydrite rocks can be broadly grouped according to the structural type of anhydrite (mosaic vs. laminated). Certain laminated, dolomite free anhydrites have precipitated/equilibrated with brines with estimated Sr^{+2}/Ca^{+2} ratios of $\sim 2.3 \times 10^{-2}$; mosaic anhydrites with dolomite matrix have precipitated/equilibrated with brines with estimated Sr^{+2}/Ca^{+2} ratios of between 0.9×10^{-2} and 1.4×10^{-2}.

Fluid chemistry, and structural type anhydrite-dolomite association in modern evaporite environments suggest 1) that most laminated anhydrites precipitated/equilibrated with brines saturated with respect to gypsum/anhydrite and celestite, and that accumulation probably occurred at a sediment-free brine interface, and 2) that most mosaic anhydrites precipitated/equilibrated with brines saturated with respect to gypsum/anhydrite, celestite and dolomite and that accumulation probably occurred in an arid, supratidal setting.

INTRODUCTION

A considerable volume of data exists in the literature on the distribution of Sr^{+2} in calcium sulphate rocks. A number of authors (Philips, 1947; Müller, 1962, among others) have speculated on the significance of these data, particularly with respect to the origin of anhydrite. These speculations have been made in the absence of data on (1) partition coefficients for Sr^{+2} in gypsum (k^{G}_{Sr}) and anhydrite (k^{A}_{Sr}) and their variation with temperature, and (2) factors which control the ratio Sr^{+2}/Ca^{+2} in brines in evaporite-forming environments. An attempt is made in this paper to provide approximate values for k^{G}_{Sr} and k^{A}_{Sr} from a mass balance of Sr^{+2} in pore brines and Sr^{+2} in gypsum and anhydrite sampled from a variety of modern evaporite environments. These data are used to interpret the distribution of Sr^{+2} in calcium sulphate rocks.

This paper is presented in five parts. The first is a summarized description of the several evaporite environments, and is followed by a discussion of their brine chemistry with particular reference to the controls on the ratio Sr^{+2}/Ca^{+2} in brines. The third part will deal with the experimental techniques for analysis for Sr^{+2} in modern evaporites and presentation of data, which will be followed by the estimation of k^{G}_{Sr} and k^{A}_{Sr} from mass balance calculations. The fifth, and final part, will deal with an interpretation of Sr^{+2} concentrations in calcium sulphate rocks.

BRIEF DESCRIPTION OF EVAPORITE ENVIRONMENTS

Samples of brines and calcium sulphate minerals analyzed for Sr^{+2} were obtained from Abu Dhabi on the Trucial Coast, Baja California, and from Jarvis Island in the south-central Pacific. To avoid repetition, the following is a brief description of these evaporite environments. For detailed descriptions reference should be made to Kinsman, 1966, 1969, Evans et al., 1969, and Butler, 1969a, 1970a, 1971 (Trucial Coast); Kinsman, 1969, Butler, 1970b, and Shearman, 1970 (Baja California); and to Schlanger and Tracey, 1970, for Jarvis Island.

Baja California and Jarvis Island

The Colorado Delta in the N.W. portion of the Gulf of California forms an extensive, clastic, arid, supratidal flat. The suite of evaporite minerals consists of halite, gypsum, anhydrite, and bassinite. Gypsum occurs as 1-2 cm. thick layers within the upper 5-10 cm of coarse, mica-rich arkosic sands which may locally contain an admixture of halite, as small ($\leq$ 2 mm) euhedral and discoidal shaped crystals. Anhydrite is associated invariably with gypsum but its distribution is very patchy. It occurs as small (< 3mm diameter), irregular coalesced nodules, or as isolated blebs of felty crystal aggregates constituting surface, or near-surface layers 1-2 cm thick. Alternatively, it occurs as lath shaped crystals ($\leq 2\mu$ long) disseminated in gypsum layers.

On Jarvis Island gypsum has accumulated a mush up to 0.75 m thick in what is best described as a salt pan environment which is filled by flooding during storms and possibly by seepage through coral rubble rampants around the island. Much of the gypsum has accumulated at a sediment-brine interface. Gypsum crystals are euhedral and range in size from 1 mm to $\leq$ 0.15 mm. Anhydrite is absent.

Abu Dhabi

The island of Abu Dhabi is located in the SE portion of the Trucial Coast. Waters of the Persian Gulf shoal gradually into a coastline of barrier islands and shallow lagoons (see Purser and Evans, in this volume), the latter grading into a low-lying, arid supratidal flat, (or sabkha) whose inner margin is limited by a plateau of Miocene and Pleistocene rocks covered with dune sands (Fig. 1). The Abu Dhabi supratidal flat attains 10 km in width. It is bordered along its seaward margin by living algal mats which are continuous with subsurface Holocene algal stromatolites in the inner parts of the sabkha. The supratidal flat consists essentially of a regressive wedge of Holocene supratidal terrigenous,and intertidal and subtidal marine,algal and carbonate sediments. This sediment wedge, up to 2.7 m thick, is underlaid by uncemented aeolian sands which crop out as a 3 km wide zone along the inner margin of the supratidal flat.

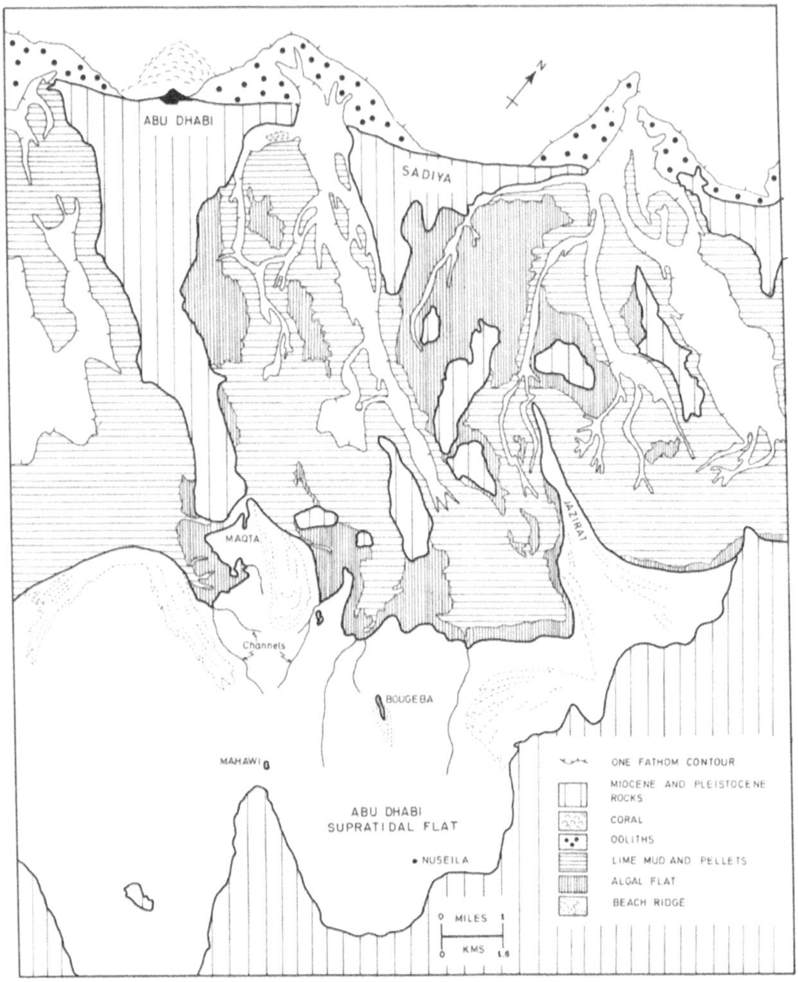

Fig. 1. Location of Abu Dhabi supratidal flat, and interpreted lithofacies distribution in the Abu Dhabi area

A typical stratigraphic sequence through the upper portion of the wedge of marine sediments is as follows: (1) supratidal facies: a terrigenous wedge 0-50 cm thick, (2) upper intertidal facies: commonly an algal facies 15-60 cm thick, and (3) lower intertidal facies: carbonate muds and muddy pellet sands ~ 40 cm thick. These sediments are host to a suite of diagenetic minerals which include gypsum, anhydrite, dolomite, celestite, halite, and bassinite.

Three distinct diagenetic calcium sulphate facies can be observed in the upper portion of the wedge of essentially marine sediments and in the aeolian sand facies when traversing the supratidal flat. The simplified lateral and vertical distribution of these minerals, together with other evaporite mineral species, are summarized in Table 1.

Table 1. Simplified vertical and lateral distribution of evaporites across the Abu Dhabi supratidal flat. The term gypsum describes those crystals of size ≤ 2-3 cm; selenite as crystals of size ≤ 12 cm

Position	Diagenetic Facies	Sediment Facies	Mineral/Structural Description
Low supratidal	Gypsum	Supratidal	Gypsum crystal mush
		Upper intertidal	Tr. gypsum crystals, tr. dolomite
		Low intertidal	Tr. selenite gypsum
Intermediate Supratidal	Anhydrite	Supratidal	Nodular/Mosaic anhydrite
		Upper intertidal	Gypsum + selenite gypsum (≤ 25% dolomite)
		Low intertidal	Selenite gypsum, dolomite (≤ 100% dolomite)
High supratidal	Gypsum	Aeolian	Nodular/Mosaic anhydrite, gypsum

Celestite occurs in trace amounts scattered throughout the sedimentary facies innumerated in Table 1.

PORE FLUID CHEMISTRY

Introduction

The amount of Sr^{+2} entering a mineral such as gypsum precipitated from a homogeneous solution at a given temperature will depend on the value of the partition coefficient defined as

$$k_{Sr}^{Gypsum} = (Sr^{+2}/Ca^{+2})_G \ / \ (Sr^{+2}/Ca^{+2})_L$$

where $(Sr^{+2}/Ca^{+2})_G$ is the ratio of strontium to calcium in gypsum, and $(Sr^{+2}/Ca^{+2})_L$ is the ratio of strontium to calcium in the liquid. Because of the partitioning of Sr^{+2} between solute and precipitated minerals, the ratio Sr^{+2}/Ca^{+2} in pore brines in an evaporite environment plays a dominant role in determining the concentration of Sr^{+2} in precipitated gypsum and anhydrite. The problem is to understand the factors which control the value of Sr^{+2}/Ca^{+2} in evolving brines derived from sea water.

Species of sulphate minerals occurring on Trucial Coast are celestite (C), gypsum (G), and anhydrite (A). A progressively concentrated brine will show values of $(Sr^{+2}/Ca^{+2})_L$ in part dependent on the timing of the precipitation of these minerals related to their respective solubility products (K^C_{SP}, K^G_{SP}, K^A_{SP}) in brines of variable concentration, and on the values of the partition coefficients for gypsum and anhydrite. If dolomitization takes place values of $(Sr^{+2}/Ca^{+2})_L$ will be modified in addition by the solubility product of dolomite (K^D_{SP}) and the partition coefficient of Sr^{+2} for dolomite (k^D_{Sr}).

The following is a discussion of the factors which control the ratio (Sr^{+2}/Ca^{+2}) in brines associated with the supratidal evaporites in the Abu Dhabi supratidal flat with reference, where appropriate, to the evaporite areas of Baja California and Jarvis Island. The chemistry of the supratidal brines is highly complex and,for reasons of clarity and continuity,considerable use will be made of data summaries.

Source of Supratidal Brines

Pore fluids in the Abu Dhabi supratidal flat are supplied from two sources. Lagoon waters are driven across the low supratidal and seaward portion of the intermediate supratidal areas during exceptionally high tides, and during storms (flood recharge). Another source is from mainland groundwaters (continental groundwaters) influxing laterally into the inner margin of the supratidal flat as well as vertically into carbonate sediments in the intermediate supratidal area from the underlying aeolian sands (Butler 1970a). The method of input of marine brines and the effect of continental groundwater input considerably influences both the chemistry of pore fluids and the mineralogy of diagenetic minerals. Influx of dilute continental groundwaters in the inner margin of the supratidal flat, for example, is responsible for the local hydration of anhydrite to gypsum.

Stability of Evaporite Minerals

Approximate brine concentrations and temperatures at which minerals are observed to begin to precipitate from supratidal brines, together with precipitation ranges in terms of brine concentrations, are summarized in Table 2.

Table 2. Approximate observed stability fields of supratidal evaporite minerals

Mineral	Brine Type	Initial Brine Concentration	T°C Range	Precipitation Range	Reference Figure (this paper)
Aragonite	Marine	X 1.2	25-28	X1.2 - X3.4	Fig. 6
Gypsum	Marine	X 3.4	25-35	X3.4 - > X8	Fig. 6
Celestite	Marine	X 3.8	25-28	X3.8 - > X 8	Figs. 4, 5
Anhydrite	Marine	X 7.0	31-35	X7.0 - > X 8	
	Continental	X 7.5	26-30	X7.5 - > X 8	
Dolomite	Marine	X 6	25-28	≥ X 6	Fig. 2
	Marine	X 7	31-35	≥ X 7	

Most of the data presented in Table 2 is self-explanatory from figures presented in subsequent sections, but, because the timing of anhydrite precipitation and dolomitization influences the ratio Sr^{+2}/Ca^{+2} in brines, the stability fields of these two minerals will be described in further detail.

Anhydrite

Sediments in the low supratidal environment are characterized by a mush of displacement gypsum crystals which ranges in thickness from 0-30 cm constituting a zone up to 2.5 km wide immediately inland of the algal flats. Anhydrite is observed firstly within the topmost cm of gypsum mush between 1-200 m inland of the algal mats. It occurs in the form of isolated blebs and small nodules consisting of a felted mass of anhydrite crystals. Five pore brines coexisting with incipient anhydrite range in chlorinity from 130-143‰ with an average chlorinity of 135‰ Cl^-, or a normalized brine concentration of X 7.0 relative to normal seawater. Pore fluid temperature ranges from 31-35°C with an average temperature of ~32°C . (Butler 1969, sample line 4, p. 24; and unpublished data).

Anhydrite in the aeolian sand facies in the high supratidal area is hydrated locally to gypsum by influx of dilute,continental groundwaters. Nineteen pore brines squeezed from coexisting anhydrite and gypsum ranges in chlorinity from 132-153‰ Cl^- with an average chlorinity of 145‰ Cl^- (X7.5). Pore fluid temperatures ranged from 26-30°C, average 28°C (Butler, 1969, sample area 4, p. 87; and unpublished data).

These data for the stability of anhydrite coexisting with gypsum agree fairly closely with the experimental data of Hardie (1967) for gypsum-anhydrite equilibria, as shown below.

This paper	Hardie (1967)
T = 28°C	T = 28°C
Cl^- = 132-153‰	a_{H_2O} = 0.79
Avg. 145‰	Cl^- = 138‰
T = 32°C	T = 32°C
Cl^- = 130-143‰	a_{H_2O} = 0.82
Avg. 135‰	Cl^- = 130‰

Chlorinities were calculated from values of a_{H_2O} using vapor pressure data of Arons and Kientzler (1954).

Dolomite

Lagoon waters have ratio $m_{Mg^{+2}}/m_{Ca^{+2}}$ of 5.3. Across the intertidal and low supratidal areas the ratio in pore brines rises rapidly to values in excess of 35 and then rapidly declines to between 3-4 across the intermediate supratidal area. The variation of the ratio $m_{Mg^{+2}}/m_{Ca^{+2}}$ in brines with increasing brine concentration can be equated to the successive precipitation of aragonite and gypsum and to dolomitization. Mass balance calculations indicate that dolomitization in the supratidal flat proceeds by the following reaction (Butler, 1970a).

$$2CaCO_3 + Mg^{+2} = CaMg\,(CO_3)_2 + Ca^{+2} \quad \text{(Eq. 1)}$$

The Ca^{+2} released by the reaction is precipitated as gypsum or anhydrite. The dolomitization process, most likely, is responsible partly for the nucleation of anhydrite. Initial high Ca^{+2} concentration (Eq. 1) could cause high supersaturation necessary for anhydrite formation (Cruft and Chao, 1970).

The conditions necessary for the initiation of dolomitization are difficult to establish. Figure 2 shows the variation of the ratio $m_{Mg^{+2}}/m_{Ca^{+2}}$ with brine concentration. Values of the ratio between concentration X3.4 - X6.9 occurred in pore brines sampled from the algal flats in the upper intertidal zone. The progressive decrease in some values of $m_{Mg^{+2}}/m_{Ca^{+2}}$ beginning at concentration X6 with increasing brine concentration suggests that in the area where these pore brines were sampled, dolomitization was initiated at concentration of X6 and a $m_{Mg^{+2}}/m_{Ca^{+2}}$ ratio of approximately 18. Temperatures of 25°C - 28°C, and pH values of 6.8 - 7.2 were observed for these particular brines (Butler, 1969; and unpublished data).

The greater portion of pore brines which show reduction in $m_{Mg^{+2}}/m_{Ca^{+2}}$

occur in the low supratidal area (Butler, 1970a, Fig. 10, p. 129). However, sediments in the older, buried, upper intertidal and low intertidal facies locally contain up to 100 ‰ dolomite (Butler, 1971). The variation in the ratio Sr^{+2}/Ca^{+2} in low supratidal brines described in a later section can best be explained by the onset of dolomitization at a brine concentration of $\sim$X7 ($\sim$135‰ Cl^-). Temperatures in these brines ranged from 31-35° (May, 1967) and pH from 7.0-6.6 (Butler, unpublished data). The differential onset of dolomitization in these two sub-environments is probably due to differences in the partial pressure of CO_2 in pore brines.

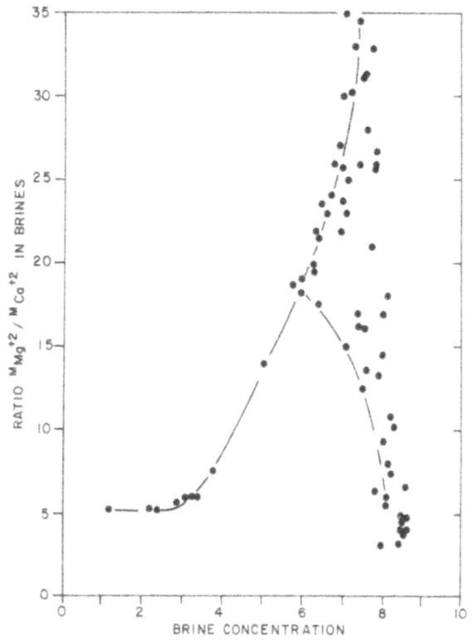

Fig. 2. Variation of magnesium to calcium ratio (mol.) in lagoon-supratidal brines. Note that brine concentration = ratio observed Cl^-‰/19.35 ‰ Cl^-

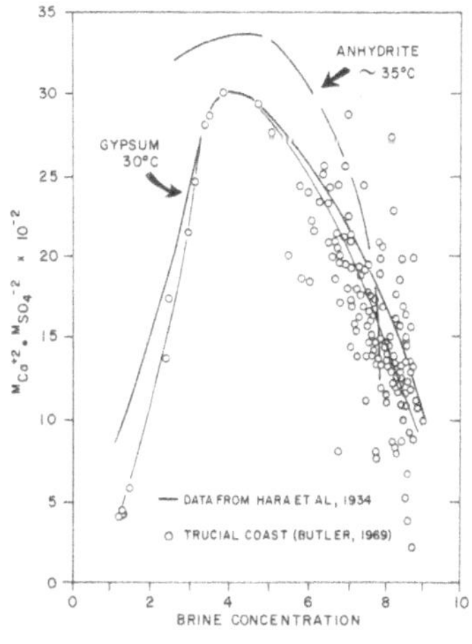

Fig. 3. Ionic product (mol.) in brines saturated with gypsum/anhydrite (experimental) and ionic product in lagoon-supratidal brines (observed)

Calcium Sulphate Ionic Product

The product $^{m}Ca^{+2}$. $^{m}SO_4^{-2}$ in brines closely follow published solubility curves for gypsum and anhydrite (Fig. 3). The product is consequently being buffered by gypsum and anhydrite and the bulk of the brines are in equilibrium with these minerals. Gypsum begins to crystallize from sea water brines at concentration ~X3.4 (Fig. 3). Brines which lie above or below the solubility curve are supersaturated or undersaturated with respect to gypsum or anhydrite. Supersaturated brines are a characteristic of the low supratidal area where increase in brine ionic strength in floodwater is rapid due to the leaching of surface accumulations of halite formed by evaporation of previous marine flooding events (Butler 1970a). Continental brines are, in general, characteristically undersaturated with respect to gypsum and anhydrite.

Strontium Sulphate Ionic Product

The product $^{m}Sr^{+2}$. $^{m}Ca^{+2}$ follows a path similar to that of $^{m}Ca^{+2}$. $^{m}SO_4^{-2}$, indicating that celestite is buffering this product (Fig. 4). Data in Figure 4 suggest that celestite precipitation begins in marine brines at concentrations between X3-X4. Consideration of the ratio Sr^{+2}/Ca^{+2} in brines (below) however, indicates that celestite crystallization commences at ~X3.8. A number of marine flood water-derived brines in the low supratidal area are supersaturated with celestite (Fig. 4). In contrast, continental groundwaters, although generally undersaturated with respect to gypsum/anhydrite, are saturated with respect to celestite.

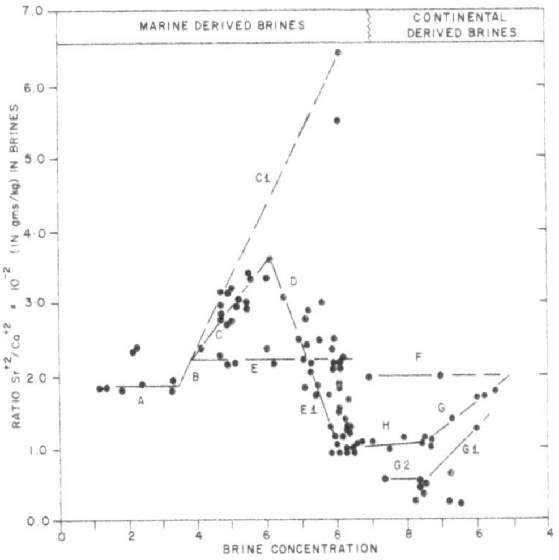

Fig. 4. Strontium X sulphate product (mol.) in lagoon-supratidal brines

Strontium to Calcium Ratio :

Marine Brines

The ratio Sr^{+2}/Ca^{+2} (Sr^{+2} and Ca^{+2} both expressed in gms/kg) varies with brine concentration in a highly complex manner (Fig. 5). The variation is related partly to the type of diagenetic environment from which the brines were sampled. The controls on the distribution of Sr^{+2}/Ca^{+2} in brines are summarized in Table 3. The interpreted controls on the variation of Sr^{+2}/Ca^{+2} with brine concentration with reference to the coded curves in Figure 5 are as follows.

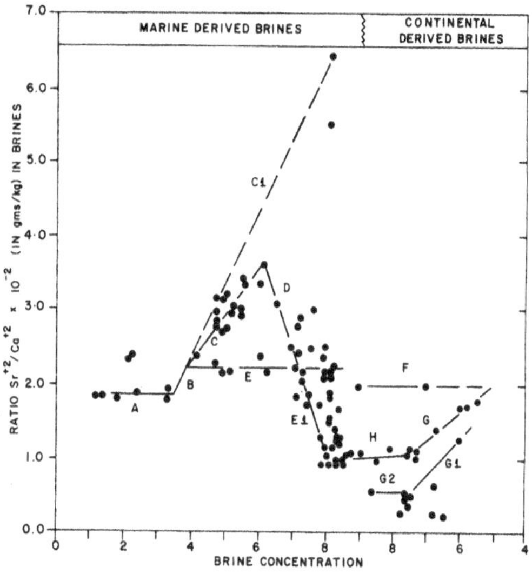

Fig. 5. Strontium to calcium ratio (mass) in lagoon-supratidal brines. Coded curve segments are discussed in text

Curve A - Ratio Sr^{+2}/Ca^{+2} remains constant at $\sim 1.85 \times 10^{-2}$ over the concentration range $\sim$X1.2 to $\sim$X3.4. A plot of normalized Ca^{+2} and SO_4^{-2} concentration versus brine concentration indicates that a trace amount of Ca^{+2} is precipitated as aragonite prior to the onset of gypsum precipitation at $\sim$X3.4 (Fig. 6). For the ratio Sr^{+2}/Ca^{+2} to remain constant during the precipitation of aragonite, every mol. of Ca^{+2} precipitated is accompanied by one mol. of Sr^{+2}. Because these brines remain undersaturated with respect to celestite, these data indicate co-precipitation of Sr^{+2} with aragonite. Consequently, the partition coefficient of Sr^{+2} with respect to aragonite (k_{Sr}^{Ar}) should be $\simeq 1.0$. These samples of lagoon to low upper-intertidal zone brines has a temperature range of 25-28°C (Butler, 1969). A value of $k_{Sr}^{Ar} \simeq 1.0$ at 30°C has been determined experimentally (Kinsman and Holland, 1969). Thus, during the initial concentration of sea water, the ratio Sr^{+2}/Ca^{+2} is controlled by the solubility product of aragonite and $k_{Sr}^{Ar} \simeq 1.0$.

Curve B (Fig. 5) - Ratio Sr^{+2}/Ca^{+2} increases from 1.85×10^{-2} to $\sim 2.2 \times 10^{-2}$ over the brine concentration range X 3.4 (gypsum precipitation) to $\sim$X3.8 (celestite precipitation). For the ratio to increase over this brine concentration range, Sr^{+2}

must be co-precipitated with gypsum, and the value for k^G_{Sr} is < 1. Mass balance calculations between Sr^{+2}/Ca^{+2} in brines and Sr^{+2}/Ca^{+2} in gypsum discussed later in this paper indicate that $k^G_{Sr} \simeq 0.18$.

Curves C and C1 (Fig. 5). Ratio Sr^{+2}/Ca^{+2} increases from $\sim 2.2 \times 10^{-2}$ to 3.6×10^{-2} between concentrations X3.8 - X6.0 (curve C); and rises to 6.4×10^{-2} in flood waters with a range of concentration up to X8 (Curve C1). This continued increase in the value of the ratio Sr^{+2}/Ca^{+2} in continuity with curve B, is attributed to the suppression of celestite precipitation as a result of a high precipitation ratio of gypsum to celestite. A high precipitation ratio of gypsum to celestite is caused by a rapid increase in ionic strength in these algal flat brines by leaching of halite, which is observed to form locally by surface evaporation of tidal waters. Thin (2-10 mm thick) surface deposits of halite are commonly found

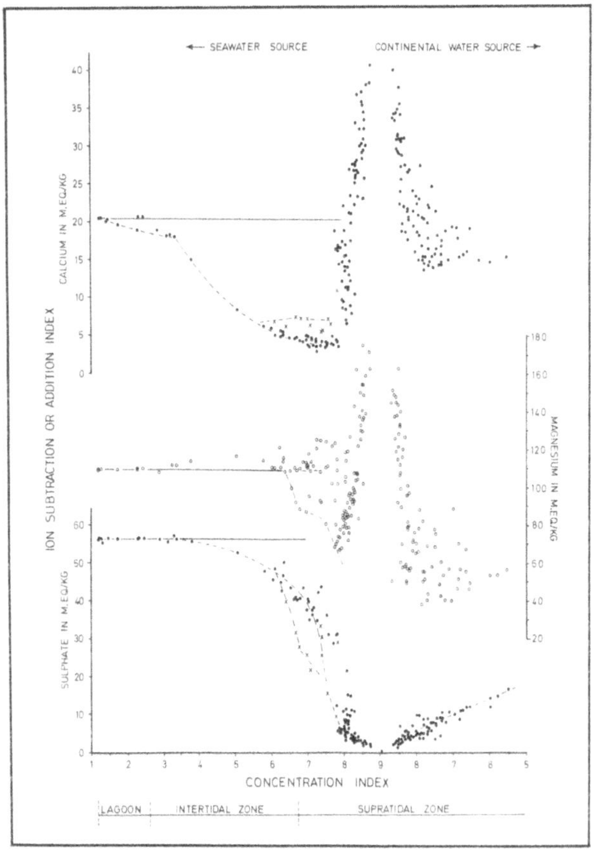

Fig. 6. Concentration of calcium, magnesium, sulphate ions in lagoon-supratidal brines. Ion subtraction/addition index = ratio of ion concentration to brine concentration (index). Calcium and sulphate values (x) indicate selected brines showing subtraction of magnesium

in the inner supratidal areas and seaward portions of the intermediate supratidal areas which are rather frequently flooded by marine waters during storms and exceptionally high tides. Surface crusts of halite are formed by the evaporation of these floodwaters. Subsequent flood waters rapidly become saturated with sodium chloride by the leaching of halite. Brines containing Sr^{+2}/Ca^{+2} ratios of 5.6×10^{-2} at concentrations ~X8 are samples of flood waters taken in the low supratidal area (Fig. 5). Since the solubility of celestite in brines is slightly greater than that of gypsum, rapid precipitation of gypsum from these floodwaters would be expected and the precipitation of celestite suppressed. A sample of flood water which had penetrated into the intermediate supratidal area, however, had a ratio $Sr^{+2}/Ca^{+2} = 2.3 \times 10^{-2}$ at X8.2 and was almost deplete in Ca^{+2}. As would be expected, celestite precipitates from flood water brines when the rate of gypsum precipitation decreases.

High ratios of Sr^{+2}/Ca^{+2} have been reported from other evaporite areas. Brines from Baja California and from solar salt waorks in Newark, California, had ratios ranging from 1.75×10^{-2} to 4.8×10^{-2} with two exceptionally high values of 7×10^{-2} and 12×10^{-2} (Kinsman, 1969, p. 490).

Curve D (Fig. 5). Ratio Sr^{+2}/Ca^{+2} decreases from 3.6×10^{-2} to $\sim 2.2 \times 10^{-2}$ over the brine concentration range X6 to ~X7. This decrease in the ratio in algal flat brines can be attributed either to a high precipitation ratio of celestite to gypsum or to the effect of dolomitization, or to both. The effect of dolomitization is considered more likely for the following reasons:

The data in Figure 3 suggests that dolomitization can begin at brine concentration X6 under a diagenetic environment defined by pH values of between 6.8 and 7.2 and an ${}^{m}Mg^{+2}/{}^{m}Ca^{+2}$ ratio of ~18. In the presence of a brine saturated with respect to gypsum, dolomitization proceeds by

$$2CaCO_3 + Mg^{+2} + {}^{B}Ca^{+2} + {}^{B}SO_4^{-2} + 2H_2O = CaMg(CO_3)_2 + CaSO_4.2H_2O + {}^{B}Ca^{+2} \quad \text{(Eq. 2)}$$

where ${}^{B}Ca^{+2}$ and ${}^{B}SO_4^{-2}$ are calcium and sulphate ions in the brines (Butler, 1970a). Calcium released by dolomitization (see Eq. 1) causes K_{SP}^{G} to be exceeded and equilibrium is maintained by gypsum crystallization. The concentration of calcium in brines (${}^{B}Ca^{+2}$) reacting with aragonite consequently increases with brine concentration, or, in other words, normalized calcium concentration remains constant with brine concentration (compare Ca^{+2} and SO_4^{-2} in brines showing Mg^{+2} reduction, Fig. 6). Consequently, the ratio Sr^{+2}/Ca^{+2} in these brines, initially supersaturated with repsect to celestite, would be expected to decrease sharply at brine concentrations > X6. Dolomitizaiton would be accompanied by crystallization of gypsum and celestite.

The ratio Sr^{+2}/Ca^{+2} in brines from the high intertidal area is controlled essentially by the ratio $K^G_{SP} \cdot K^D_{SP}/K^C_{SP}$ and k^G_{Sr}. Diagenetic dolomite in these sediments contains ~0.066 wt. % Sr^{+2} ($Sr^{+2}/Ca^{+2} \sim 3.0 \times 10^{-3}$) (Butler, 1970a). The ratio Sr^{+2}/Ca^{+2} is thus, in addition, controlled by the value of k^D_{Sr}.

Curve E (Fig. 5). Ratio Sr^{+2}/Ca^{+2} remains essentially constant at ~2,2 $\times 10^{-2}$ between concentrations X3.8 - > X8 in low supratidal (gypsum mush) pore brines, and in chemically-evolved floodwater brines. Along this curve gypsum, celestite, and at concentrations > X7, anhydrite, are being precipitated. For the ratio to remain constant in these brines, Sr^{+2}/Ca^{+2} must be controlled essentially by the ratio K^G_{SP} or K^A_{SP} / K^C_{SP}. Dolomitization has not occurred in these brine sampling localities.

Curve E1 (Fig. 5) - Ratio Sr^{+2}/Ca^{+2} falls from ~2.2 $\times 10^{-2}$ to ~1.0 $\times 10^{-2}$ between concentrations X7 to > X8 in brines in low supratidal and low intermediate supratidal areas. These brines are saturated with respect to celestite and gypsum and anhydrite. The observed decrease in Sr^{+2}/Ca^{+2} can be attributed either to a value of $k^A_{Sr} \simeq >> 1$, or to dolomitization, as in Curve D, or both. Mass balance calculations described in a later section indicate that $k^A_{Sr} \simeq 0.36$. Dolomitization, by equation 2 (above), would thus appear to be the dominant control on the ratio Sr^{+2}/Ca in these brines, in addition to other factors summarized in Table 3.

The bulk of the brines in the intermediate supratidal area contain values of Sr^{+2}/Ca^{+2} of between 0.95×10^{-2} and 1.1×10^{-2} at brine concentrations between ~X8 - ~X8.5, indicating equilibration between brines with carbonate sediment, gypsum, anhydrite, celestite, and dolomite (Fig. 5).

Continental Brines

Continental brines vary widely in concentration. The lowest concentration measured in brines sampled close to the landward margin of the supratidal flat was X5.5. However, brines from an area to the west of the Abu Dhabi supratidal flat in a similar position on the sabkha had concentrations of X9. Concentrations progressively increase seaward from the inner margin of the supratidal flat, indicating progressive evaporation and brine mixing.

Brines derived from continental groundwater influx have a highly variable and complex chemistry indicative of their mixed character. Some brines are initially undersaturated with respect to gypsum and anhydrite whereas others are saturated with respect to these minerals, but most are saturated with respect to celestite. The brines coexist with celestite, gypsum, and anhydrite.

Table 3. Summary of controls on ratio Sr^{+2}/Ca^{+2} in Abu Dhabi supratidal brines

Brine Source	Curve (Fig. 5)	Brine Location	Brine Concn.	Range Sr^{+2}/Ca^{+2} in Brine	Controls on Sr^{+2}/Ca^{+2} in Brine
Marine	A	Lagoon to low upper intertidal	1.2-3.4	1.85×10^{-2}	K_{Sp}^{Ar}, $k_{Sr}^{Ar} \simeq 1.0$
Marine	B	Intermediate upper intertidal (incipient gypsum)	3.4-3.8	1.85×10^{-2} 2.2×10^{-2}	K_{SP}^{G}, $k_{Sr}^{G} \simeq 0.18$
Marine	C	High upper intertidal	3.8-6.0	2.2×10^{-2} 3.6×10^{-2}	K_{SP}^{G}, K_{SP}^{C}, $k_{Sr}^{G} \simeq 0.18$, pptn ratio G/C > > 1
Marine	C1	Low supratidal flood waters (gypsum mush, incipient)	$\leq \sim 8$	6.4×10^{-2}	as C above
Marine	E	Low supratidal (gypsum mush zone)	3.8-> 8	2.2×10^{-2}	K_{SP}^{G}, K_{SP}^{A}, K_{SP}^{C}, k_{Sr}^{G} $\simeq 0.18$, $k_{Sr}^{A} \simeq 0.37$
Marine	E1	Low to intermediate	7.0-8	2.2×10^{-2} $\sim 1.0 \times 10^{-2}$	K_{SP}^{G}, K_{SP}^{A}, K_{SP}^{C}, K_{SP}^{D}, $k_{Sr}^{G} \simeq 0.18$, $k_{Sr}^{A} \simeq 0.37$, k_{Sr}^{D}, ratio $\frac{K_{SP}^{G/A} K_{SP}^{D}}{K_{SP}^{C}}$
Marine	D	High upper intertidal (gypsum incipient)	6.0-7.0	3.6×10^{-2} $\sim 2.2 \times 10^{-2}$	as E1 above
Continental	G,G1	High supratidal	~5.5-7.5	$\sim 2.0 \times 10^{-2}$ 0.5×10^{-2}	K_{SP}^{C}
Continental	H	High supratidal	7.5 -> 8	$\sim 1.1 \times 10^{-2}$	K_{SP}^{C}, K_{SP}^{A}, $k_{Sr}^{AA} \simeq 0.37$
Continental	F	High supratidal	~5.5-9	2.0×10^{-2}	as E above
Continental	G2	High supratidal	7.5-9	$\sim 0.5 \times 10^{-2}$	as E above

The following is a discussion of the ratio Sr^{+2}/Ca^{+2} in these brines with reference to the coded curves in Figure 5. The controls on the ratio are summarized in Table 3.

Curve F - Ratio Sr^{+2}/Ca^{+2} remains constant at $\sim 2.0 \times 10^{-2}$ over the concentration range $\leq$ X5.5 to X9. Curve F parallels curve E in the marine-derived brines, which suggests precipitation of celestite with gypsum, or anhydrite at concentrations > X 7.5. The ratio Sr^{+2}/Ca^{+2} would then be controlled by the ratio $K_{SP}^{G/A}/K_{SP}^{C}$ and k_{Sr}^{G} or k_{Sr}^{A}. Temperatures in these brines ranged from 22°C to 28°C, which are lower than brines from curve E, which ranged from 25°C to 35°C. Brine concentrations at which gypsum and celestite crystallize from continental brines would thus be expected to be slightly lower than in marine brines, explaining the slight difference in the equilibrium value of Sr^{+2}/Ca^{+2} in marine ($\sim 2.2 \times 10^{-2}$) and continental brines ($\sim 2.0 \times 10^{-2}$).

Curves, G, G1, G2 - Ratio of Sr^{+2}/Ca^{+2} decreases from $\sim 2.0 \times 10^{-2}$ to $\lesssim 0.5 \times 10^{-2}$ over the concentration range X5.5 to X9. The product $^{m}Ca^{+2} \cdot {}^{m}SO_4^{-2}$ in these brines generally lies below the solubility curves for gypsum and anhydrite (Fig. 3) and consequently, the brines overall are undersaturated with respect to these minerals. Normalized concentrations of Ca^{+2} are scattered and show no definite trend over the range of brine concentrations considered. Sulphate ions, on the other hand, show an almost linear decrease in concentration (Fig. 6). Traces of celestite are found in these aeolian sands, and the control on the ratio Sr^{+2}/Ca^{+2} is attributed to precipitation of celestite, except in curve G2 where ratio is controlled by K_{SP}^{G} or K_{SP}^{A} / K_{SP}^{C}.

Curve H - Ratio Sr^{+2}/Ca^{+2} remains approximately constant over the brine concentration range X 7.5 to X9.0. Values of $^{m}Ca^{+2} \cdot {}^{m}SO_4^{-2}$ in these brines indicate that they are saturated with respect to gypsum or anhydrite. These brines are similar to those in curves F and E; the ratio in these brines is controlled by the ratio $K_{SP}^{G/A}/K_{SP}^{C}$ and k_{Sr}^{G} or k_{Sr}^{A}.

Two important points emerge from the chemistry of these supratidal brines. Firstly, that in terms of normal, evaporative concentration the ratio Sr^{+2}/Ca^{+2} in marine waters will rise rapidly from 1.85×10^{-2} to $\sim 2.2 \times 10^{-2}$ during the initial stages of evaporative concentration and subsequently remain approximately constant up to, and probably beyond, halite saturation. Secondly, if dolomitization takes place, the ratio Sr^{+2}/Ca^{+2} in marine brines will have values $< 2.2 \times 10^{-2}$.

STRONTIUM IN MODERN CALCIUM SULPHATES

Analytic Technique and Results

Most of the gypsum and anhydrite samples analyzed for Sr^{+2} were associated with carbonate and celestite. Gypsum crystals frequently had carbonate grains enclosed within them and special care had to be taken to exclude these materials before final analysis. Pre-analytical treatment included washing the raw samples with 0.1M HCl , and subjecting the grindings to further acid leaching in a sonic vibrator. Gypsum was dehydrated, ground, and acid leached. Portions of samples were then checked for carbonate and celestite by optical examination using water mounts and by x-ray diffraction. Samples were dissolved in 4.3M HNO_3 and analyzed for Sr^{+2} with a Perkin-Elmer model 290 Atomic Adsorption Spectrophotometer. Analyses are accurate to within ± 0.004 weight percent Sr^{+2} (Appendix). Data for gypsum are reported for $CaSO_4 \cdot 2H_2O$. Wherever possible the original samples were analyzed for crystal size by sieving (gypsum) and optical examination of water mounts (anhydrite) (Appendix). Strontium concentrations in anhydrite and the various species of gypsum are summarized in Table 4.

ESTIMATION OF k_{Sr}^{Gypsum} AND $k_{Sr}^{Anhydrite}$

Published values for k_{Sr}^{G} and k_{Sr}^{A} determined experimentally are sparce and their reliability uncertain. Bonch-Osmolovskaya (1964) determined a values for k_{Sr}^{G} of 0.27 in gypsum precipitated by addition of H_2SO_4 to $CaCl_2$ solution. No temperature data were reported. Purkayastha and Chatterjee (1966) attempted to provide values for k_{Sr}^{G} and k_{Sr}^{A} by precipitating calcium sulphate from H_2SO_4 - $CaCl_2$ solution spiked with Sr^{85}. Average values for $k_{Sr}^{calcium\ sulphate}$ ranged from 0.42 to 0.24 over the temperature range 15°C to 57°C. Unfortunately, the precipitates were not examined, and these authors only assumed that gypsum was precipitated at temperatures < 40°C and anhydrite at temperatures > 40°C. Butler (1970a) assumed as a working hypothesis that $k_{Sr}^{G} \simeq 0.44$ and $k_{Sr}^{A} \simeq 0.50$ but these are no longer valid.

In the following discussion an attempt is made to evaluate average values for k_{Sr}^{G} and k_{Sr}^{A} on the basis of a mass balance between the ratio Sr^{+2}/Ca^{+2} in brines and Sr^{+2} concentration in precipitated gypsum and anhydrite. Note that Sr^{+2}/Ca^{+2} in brines is expressed as a ratio of Sr^{+2} and Ca^{+2} in gms/kg, and that $k_{Sr}^{G/A}$ are related to $(Sr^{+2}/Ca^{+2})_L^{G/A}$ by

$$k_{Sr}^{G} \simeq \frac{Sr^{G}}{0.23(Sr^{+2}/Ca^{+2})_L^{G}}$$

$$k_{Sr}^{A} \simeq \frac{Sr^{A}}{0.294(Sr^{+2}/Ca^{+2})_L^{A}}$$

Table 4. Summary of Sr^{+2} concentration in gypsum and anhydrite

GYPSUM

Location	Type/association	# Samples	Avg. Sr^{+2} wt %	Range Sr^{+2} wt %
Trucial Coast	Discoid (gypsum mush)	14	0.110	0.045-0.282
	Algal	4	0.071	0.360-0.094
	Selenite	6	0.041	0.021-0.057
	Gypsum (aeolian facies)	3	0.041	0.032-0.046
Baja California	Discoid (gypsum mush and layers)	5	0.079	0.071-0.084
Jarvis Island	Discoid (gypsum mush)	6	0.208	0.175-0.264

ANHYDRITE

Location	Type/association	# Samples	Avg. Sr^{+2} wt %	Range Sr^{+2} wt %
Trucial Coast	Nodular/mosaic	36	0.226	0.0880-0.522
Baja California	Nodular/bleb and mosaic	5	0.115	0.099-0.124

where $Sr^{G/A}$ are in weight percent, and the constants 0.23 and 0.294 are the percent formula weights of Ca^{+2} in gypsum and anhydrite necessary to convert $Sr^{G/A}$ into ratio $(Sr/Ca)^{G/A}$.

The ratio Sr^{+2}/Ca^{+2} in brines varies considerably with brine concentration over different portions of the Abu Dhabi supratidal flat. These variations within the stability fields of gypsum and anhydrite, however, tend to be specific to, and characteristic of, the nature of the brine source (marine vs. continental) and of diagenesis within the intertidal and supratidal areas. Equilibrium is maintained between brines and calcium sulphates during crystallization. If there is no re-equilibration between brines and minerals, then concentrations of Sr^{+2} in gypsum or anhydrite precipitated at various brine concentrations in these areas will be proportional to the ratio Sr^{+2}/Ca^{+2} in brines. Actual concentrations of Sr^G or Sr^A will be determined by the product $k_{Sr}^{G/A}(Sr^{+2}/Ca^{+2})_L$. Average values for k_{Sr}^G and k_{Sr}^A can thus be estimated from a mass balance between the average concentration of Sr^G and Sr^A and average values of $(Sr^{+2}/Ca^{+2})_L^G$ and $(Sr^{+2}/Ca^{+2})_L^A$.

Partition Coefficient of Gypsum

The ratio $(Sr^{+2}/Ca^{+2})_L$ remains constant at $\sim 2.2 \times 10^{-2}$ over the concentration range of brines observed in the area of gypsum mush development. Sr^G in 14 discrete size populations of gypsum crystals from this area averaged 0.110 wt. % with a range of 0.045 - 0.282 wt. % Sr^{+2}(Table 4 and Appendix). The average value for Sr^G is weighed by inclusion of sample 50 which contains 0.282 wt. % Sr^{+2}. Omission of this value reduces the average to 0.098 wt. % Sr^{+2} with a range in values for 13 samples of 0.45 - 0.182 wt. % Sr^{+2}. Values of k_{Sr}^G calculated using these two averages and $(Sr^{+2}/Ca^{+2})_L \simeq 2.2 \times 10^{-2}$ are:

Avg. wt. % Sr	k_{Sr}^G
0.098	~0.194
0.110	~0.218

Sample 42 consisted of gypsum crystals which ranged in size from 0.15 -0.5 mm. Rapidly concentrated marine flood waters, or rapidly evaporated brines supersaturated with respect to celestite, would be expected to precipitate a large population of very small gypsum crystals. Such gypsum would be expected to contain a high concentration of Sr^{+2} relative to gypsum precipitated from brines just saturated with respect to gypsum and celestite. Inspection of the surface of silici-clastic sediments in the low to intermediate supratidal area after the retreat of a marine flood event in 1967 showed that the entire surface affected by floodwaters contained discoidal and prismatic gypsum crystals ~0.2 mm in diameter (Butler, 1970a). Six size-discrete samples of crystals from a salt pan environment on Jarvis Island range in size between 0.15-0.5 mm and contain an average Sr^G of 0.208 wt % with a range 0.175-0.264 wt. % (Fig. 7 and Appendix). The ratio Sr^{+2}/Ca^{+2} in one sample of brine was 4.7×10^{-2} at concentration X6.8, indicating supersaturation with respect to celestite and a high precipitation ratio of gypsum/celestite during rapid evaporative concentration. The size and position of samples 42 and 50 (Fig. 7) with respect to the Jarvis Island gypsum samples suggests that these populations of gypsum crystals crystallized from flood waters with a ratio $Sr^{+2}/Ca^{+2} >> 2.2 \times 10^{-2}$. Omission of samples 42 and 50 gives an average Sr^G of 0.90 wt. % with a range for 12 samples of 0.045 - 0.124 wt. %. The calculated value of k_{Sr}^G then becomes ~0.18. Coexisting brine temperatures in the sampling area were 30-34°C with an average temperature of ~33°C. A value of 0.184 for $k_{Sr}^G[({}^{M}Sr^{+2}/{}^{M}Ca^{+2})_G = 0.16 \times 10^{-2}$; $({}^{M}Sr^{+2}/{}^{m}Ca^{+2})_L = 0.86 \times 10^{-2}$ [temperature unspecified] has been found experimentally (?) for gypsum precipitated from seawater brines (Kinsman, 1969; paper presented at SEPM-AAPG meeting, Milwaukee).

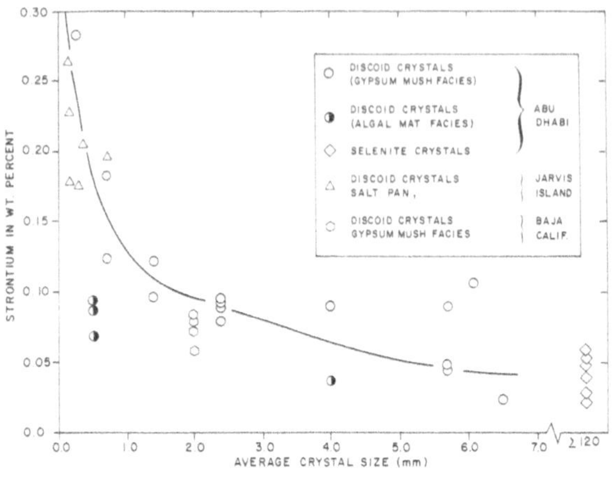

Fig. 7. Strontium in modern gypsum as a function of grain size

A sample of gypsum (gypsum coexisting with anhydrite and formed by dissolution of the anhydrite and reprecipitation of ions as gypsum) in the high supratidal area contained 0.044 wr. % Sr^{+2}, and coexisting squeezed pore fluid had ratio $Sr^{+2}/Ca^{+2} = 1.0 \times 10^{-2}$ at a temperature of 28.0°C. The value for k_{Sr}^{G} calculated in this case is ~0.19.

Supratidal gypsum coexisting with anhydrite from Baja California contains an average Sr^{G} of 0.075 wt. % with a range for five samples of 0.072 - 0.084 wt. % (Table 4 and Appendix). Three samples of brine collected in the area in which these crystals were sampled had ratios Sr^{+2}/Ca^{+2} of 1.82×10^{-2}, 1.94×10^{-2}, and 2.20×10^{-2} at brine concentrations between X7.38 and X7.51. Temperatures of a total of five

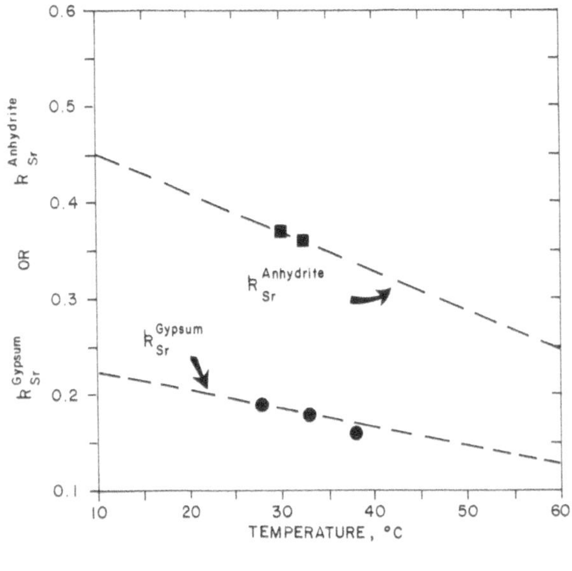

Fig. 8. Estimated variation of k_{Sr}^{G} and k_{Sr}^{A} as a function of temperature

brines in the same area ranged from 31-40°C, with an average of 37°C. k_{Sr}^{G} calculated from an average of these data is ~0.16 at 37°C. These three data points permit an evaluation of the temperature sensitivity of k_{Sr}^{G} (Fig. 8). It is of note that one sample of free brine collected from an old drainage channel immediately seaward of the area contained a ratio Sr^{+2}/Ca^{+2} of 2.5×10^{-2} at concentration X8.

Figure 9 shows the concentration of Sr^{+2} in gypsum which one would expect to find in crystals precipitated from pore brines in various areas of the Abu Dhabi supratidal flat. This figure was constructed using an average value $k_{Sr}^{G} = 0.185$ at 30°C [the average temperature of supratidal sediments (Butler, 1969)] and the average trends in the ratio Sr^{+2}/Ca^{+2} in pore brines with brine concentration from Figure 5. Predicted and observed values of Sr^{G} show close agreement in various types of gypsum precipitated in a number of diagenetic, supratidal sub-environments. Selenite gypsum crystals predominate in intermediate supratidal areas in association with brines of concentrations ≥ ~X8. From Figure 9 (curve H), selenite should contain an average $Sr^{G} \simeq 0.045$ wt. %. Six samples of selenite from the intermediate supratidal area contained average $Sr^{G} = 0.041$ wt. %, range 0.021-0.048 wt. % (Table 4 and Appendix). Similarly, high supratidal gypsum should contain between 0.02 and 0.045 wt. % Sr^{+2} (curves H and G1, Fig. 9). Three samples of these gypsum crystals contained $Sr^{G} = 0.032$, 0.044, and 0.048 wt. % Sr^{+2} (Appendix). Although incipient gypsum crystals develop between algal layers in the upper intertidal zone, they predominate in the older, buried algal layers in the supratidal flat where brine concentrations exceed ~X6 (Butler, 1969, Table 2). Algal gypsum should, then, contain populations of crystals precipitated along curves E, E1, and in part H (Fig. 9). Sr^{G} consequently should range ~0.045 to ~0.095 wt. %. Analysis of four size-discrete samples of algal gypsum gave a range of Sr^{G} of 0.036 to 0.094 wt. %, average 0.071 wt. % (Appendix).

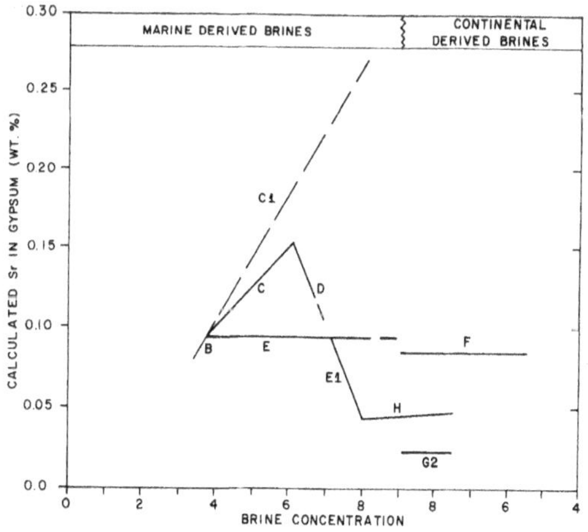

Fig. 9. Calculated Sr^{+2} concentration in gypsum crystallized from supratidal brines. Code and ratio Sr^{+2}/Ca^{+2} in brines as Fig. 5; $k_{Sr}^{G} = 0.18$ (30°C)

Partition Coefficient for Anhydrite

A sample of incipient anhydrite coexisting with gypsum in the zone of gypsum mush development in the Abu Dhabi supratidal flat contained 0.224 wt. % Sr. The associated pore fluid had a ratio of $Sr^{+2}/Ca^{+2} = 2.13 \times 10^{-2}$, concentration X7.2, and temperature of 32.5°C. These data provide a value of $k_{Sr}^{A} \simeq 0.36$ at 32.5°C. Anhydrite begins to precipitate in these low supratidal sediments at a brine concentration of ~X7.0. In brines, the ratio Sr^{+2}/Ca^{+2} decreases from $\sim 2.2 \times 10^{-2}$ to $\sim 1.0 \times 10^{-2}$ over the concentration range X7.0 - ~X8 (curve E1, Fig. 5). The concentration of Sr^{+2} in anhydrites should reflect a continuous series of anhydrite crystals precipitated over this range of Sr^{+2}/Ca^{+2} ratios in the brines. The variation of Sr^{A} with crystal size is interpreted to reflect this sequence of anhydrite precipitation (Fig. 10). For a mass balance calculation it will be assumed that equal amounts of anhydrite are precipitated consecutively for each increment in brine concentration over the range X7 - X8. The average value of Sr^{+2} in anhydrite is 0.226 wt. % with a range for 36 samples of 0.08 - 0.522 wt. %, and the average value in brines for the ratio Sr^{+2}/Ca^{+2} is 1.66×10^{-2}. This average value of Sr^{+2}/Ca^{+2} was determined from Figure 9 using the average of data points picked off at intervals of X0.2 concentration units along curve E1. An average value of $k_{Sr}^{A} \simeq 0.46$ is calculated from these data. As with gypsum, however, there are two distinct populations of anhydrite on the basis of Sr^{+2} concentration. Six samples contain Sr^{+2} ranging between 0.348 - 0.522 wt. % (average 0.45 wt. % Sr^{+2}), 29 samples contain Sr^{+2} ranging between 0.117 - 0.254 wt. % (average 0.180 wt. % Sr^{+2}) and one sample contains 0.08 wt. % Sr^{+2} (Appendix). Using the average Sr^{+2} shown by the bulk of the samples (0.180 wt. %) and ratio $(Sr^{+2}/Ca^{+2})_L = 1.66 \times 10^{-2}$, $k_{Sr}^{A} \simeq 0.37$ which closely agrees with a value for k_{Sr}^{A} of 0.36 determined above. Temperatures in anhydrites throughout the Abu Dhabi supratidal flat range 24-47°C, with an average temperature of ~30°C (February to April, 1964) (Butler, 1969, Table 2). These two data points of $k_{Sr}^{A} \simeq 0.36$ at 32.5°C and $k_{Sr}^{A} \simeq 0.37$ at 30°C allow an evaluation of the temperature variation in the value k_{Sr}^{A} (Fig. 8).

Reasons for anhydrites containing 0.348-0.522 wt. % Sr are difficult to determine. Assuming a value of $k_{Sr}^{A} = 0.37$, these anhydrites formed from brines, supersaturated with celestite, with ratios Sr^{+2}/Ca^{+2} between 3.2×10^{-2} and 4.5×10^{-2}. Information which may have a bearing on the origin of these anhydrites are: (a) these anhydrites tend to be very fine grained (Fig. 10), which would suggest precipitation from rapidly concentrated brines, and (b) these anhydrites occur in the intermediate supratidal flat in areas ancestral to large tidal channels. Flooding in these areas was probably more frequent in the past when the supratidal flat was narrower and because of the decreased tidal and storm fetch damping due to deeper lagoons between the offshore islands. A greater frequency of flooding would increase the supply of brines to the low supratidal area being supersaturated with respect to celestite and having high ratios of Sr^{+2}/Ca^{+2}, as brines between curves C1 and D (Fig. 5). Thus, a distinct possibility arises that anhydrites with Sr^{+2} concentrations between 0.348 - 0.522 wt. % were formed from flood-derived brines supersaturated with respect to celestite.

Figure 11 shows the range in Sr^{A} which one would expect for anhydrite precipitated from pore brines across the supratidal flat. Figure 11 was constructed using an average value of $k_{Sr}^{A} = 0.37$ at 30°C utilising the curves of the ratio Sr^{+2}/Ca^{+2} in brines in Figure 5. Because k_{Sr}^{A} was calculated on the basis of the average Sr^{A}, comparison between observed and calculated Sr^{A} is meaningless. Note, however, the predicted values of Sr^{A} for curves G2 and C1 (Fig. 5). One anhydrite sample from the inner margin of the aeolian sand facies contained 0.08 wt. % Sr, which is close to Sr^{A} predicted from curve G2. Curve G2 would suggest that in some continental-derived brines anhydrites can form with Sr^{A} as low as ~0.06, by long delayed saturation with respect to anhydrite (or gypsum) but prolonged saturation with respect to celestite, which controls the ratio Sr^{+2}/Ca^{+2} in these particular brines. Theoretically, some anhydrites in the low supratidal area,precipitated from flood waters,could contain 0.53 to ~0.69 wt. % Sr (C1, Fig. 11). But since these flood brines tend to stabilize with Sr^{+2}/Ca^{+2} ratios of ~2.2×10^{-2} ($Sr^{A} \simeq 0.19$ wt. % for $k_{Sr}^{A} = 0.37$), anhydrites precipitated from such flood waters could contain between ~0.19 and ~0.69 wt. % Sr.

Surface, or near surface supratidal anhydrite from Baja California contains an average of 0.115 wt. % Sr with a range (for five samples) of 0.099 to 0.124 wt. % Sr. Associated brines had ratios of Sr^{+2}/Ca^{+2} between 1.82×10^{-2} and 2.20×10^{-2} and concentrations between X7.38 and X7.51. Surface temperatures of between 47 and 51°C, (average ~50°C), were recorded in August 1967. At 50°C, k_{Sr}^{A} is estimated to be 0.29 (Fig. 8). Anhydrite precipitated from these brines should have Sr^{+2} concentrations between ~0.16 and ~0.19 wt. % Sr. If, indeed, the anhydrite precipitated from flood waters with ratio Sr^{+2}/Ca^{+2} greater or less than 2.5×10^{-2} (see above section), which seems the more likely case, anhydrite should contain ± 0.21 wt. % Sr. The discrepancy between observed and calculated values of Sr^{A} is considerable. Perhaps the best explanation for this discrepancy is that the anhydrite has developed under conditions which are marginal for its formation and preservation. Brine concentrations are only slightly higher than X 7. Anhydrite crystals are very small (< 2μ) and are intermixed with bassinite while coexisting gypsum crystals show no sign of corrosion. These factors suggest that one is dealing with a disequilibrium mixture of gypsum, anhydrite, and brine, and that the anhydrite is in the process of being dissolved. In the course of dissolution the concentration of Sr^{+2} in the anhydrite would decrease progressively. If this is the case, the concentrations of Sr^{+2} in the anhydrites analyzed reflect progressive stages in the equilibration between anhydrites and pore brines.

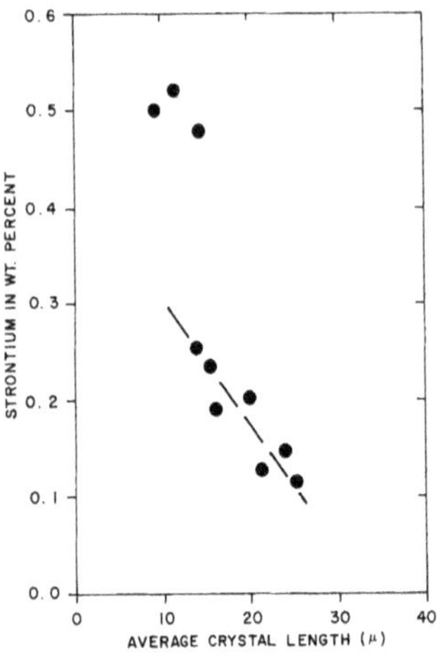

Fig. 10. Strontium in modern anhydrite as a function of crystal length

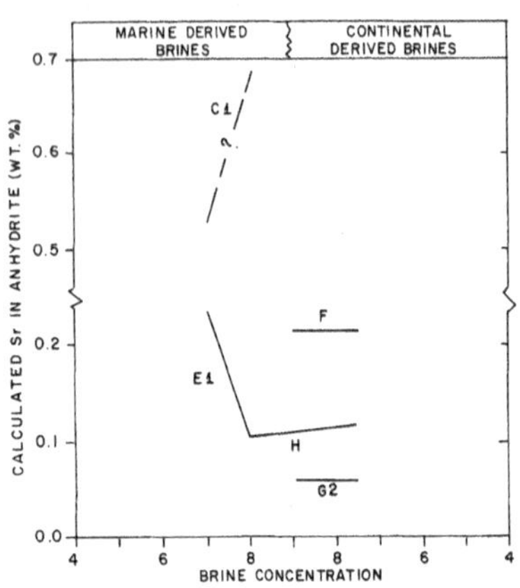

Fig. 11. Calculated Sr^{+2} concentration in anhydrite crystallized from supratidal brines. Code and ratio Sr^{+2}/Ca^{+2} in brines as Fig. 5, k^{A}_{Sr} = 0.37 (30°C)

CALCIUM SULPHATE ROCKS

Two structurally distinct types of anhydrite occur in rocks, mosaic and laminated ("varved"). Examination of modern supratidal flats of the Trucial Coast and Baja California has demonstrated the presence of many of the textures and structures which, in ancient rocks have previously been regarded as the products of primary deposition in relatively deep water,or to late diagenetic changes associated with burial. The occurrence in the Trucial Coast of supratidal nodular and mosaic anhydrites in seams 1-30 cm thick, or as massive beds up to 2.7 m thick in marine and continental supratidal environments (Butler, 1970a) suggests that the mosaic structure may be characteristic of arid supratidal environments. Much of the mosaic anhydrite in the Trucial Coast is closely associated with dolomitized carbonates and with algal stromatolites. Similarly, layered halite rocks presently forming in depressions on supratidal flats in the Gulf of California contain crystals which display a chevron structure essentially similar to that which occurs in many ancient layered salt deposits. Shearman (1970) concludes that the halite layers are not records of annual events but that each one is a product of a long period of reworking of brine pan halite and

Table 5. Summary of Sr^{+2} concentrations in sulphate rocks

Source	Location	Age-Formation	Structural Type	Gypsum/ Anhydrite	# Samples	Avg. Wt. % Sr	Range Wt. % Sr
Dean (1967)	W. Texas Cowden #2 Core	Permian-Castile	Bedded (varved) + calcite	Anhydrite	60	0.241	0.195-0.300
Johnson & Ham (1970) unpub. data	W. Oklahoma	Permian-Cloud Chief	Nodular/Mosaic + matrix dolomite	Anhydrite Gypsum	26 48	0.10 0.086	0.065-0.152 0.047-0.123
		-Blaine	Nodular/Mosaic + matrix dolomite	Anhydrite Gypsum	24 40	0.145 0.100	0.098-0.316 0.045-0.157
	S.W. Oklahoma	Permian-Dog Creek & Blaine	Nodular/Mosaic + matrix dolomite	Anhydrite Gypsum	18 30	0.129 0.038	0.050-0.302 0.001-0.100
Jung & Knitzschke (1960)	Germany	Permian-Zechstein	Bedded (varved) + Dolomite (2-10 %)	Anhydrite	34	0.204	0.13-0.29
Müller (1962)	Germany	Permian-Zechstein					
Table 14, p. 21			Anhydrite with salt	Anhydrite	41	0.246	0.14-0.34
Table 15, p. 22			Anhydrite with salt	Anhydrite	13	0.243	0.18-0.30
in Table 16, p. 22			Bedded (varved) + Dolomite (2-10 %)	Anhydrite	118	0.217	
Holliday (1965)	Spitsbergen	M. Carboniferous Ebbadalen & Nordenskioldbreen	Nodular/Mosaic + matrix dolomite	Anhydrite Gypsum	3 3	0.142 0.068	0.097-0.168 0.068-0.103

groundwater-brine diagenesis. Thus, it is time to take a fresh look at certain ancient evaporite deposits in the light of these observations made in modern arid supratidal flat environments.

A problem remains, however, concerning the origin of laminated anhydrites which may be very thick and extend over wide areas (Anderson and Kirkland, 1966). The essential problem is whether or not laminated anhydrites (or their gypsum precursers) accumulated at a sediment-free brine interface. Because of the uncertainities in the data involved, the following discussion does not attempt to account for the origin of laminated anhydrites. It focuses attention on diagenetic processes, stressing the effects that various reactions can have on fluids from which anhydrite (or its gypsum precurser) may have formed.

Sr^{+2} in Calcium Sulphate Rocks

Table 5 is a summary of Sr^{+2} concentrations in calcium sulphate rocks compiled from literature. The structural classification is after Maiklem et al. (1969). Mosaic, laminated, and anhydrite with halite, contain different concentrations of Sr^{+2}. Employing all the data, mosaic anhydrites contain an average Sr^{+2} concentrate of 0.129 wt. % (range 0.01-0.316 wt. %); laminated 0.217 wt. % (range 0.13-0.30 wt. %); and anhydrite associated with salt, 0.245 wt. % (range 0.14-0.34 wt. %). It should be noted that the total range of Sr^{+2} in modern anhydrites is 0.08-0.522 wt. % with an average of 0.226 wt. %. An interpretation of Sr^{+2} in ancient anhydrites in terms of the original nature of the brine from which they formed is made difficult by a lack of knowledge as to the complexities of their post depositional diagenetic history. Mosaic anhydrites from W and SW Oklahoma, and Spitsbergen are presently either very close to, or on the surface. These anhydrites have a dolomite matrix and are associated with late diagenetic gypsum and celestite. If k^G_{Sr} is taken to be 0.2 at 20°C (Fig. 8) then, using values of Sr^G in Table 5, ratios Sr^{+2}/Ca^{+2} in groundwaters from which gypsum precipitated can be estimated (Table 6).

Pore fluids saturated with respect to celestite, gypsum, and anhydrite should have ratio $Sr^{+2}/Ca^{+2} \simeq 2.2 \times 10^{-2}$ (this paper). Predicted average values of Sr^{+2}/Ca^{+2} in fluids associated with these minerals in rocks shown in Table 6 differ to varying degrees from this value. These data suggest, however, that mosaic anhydrites from W. Oklahoma and from Spitsbergen have recrystallized the least and that the Sr^A average range of 0.100 to 0.145 wt. % Sr^{+2} is probably not much less than that range of Sr^{+2} in the original mosaic anhydrite at deposition. Using $k^A_{Sr} = 0.37$ at 30°C (Fig. 8), the ratio Sr^{+2}/Ca^{+2} in original brines is estimated to have ranged from 0.9×10^{-2} to 1.4×10^{-2}. All the laminated anhydrites and anhydrites with salt listed in Table 6 were sampled from cores/cuttings from depths in excess of 3,000 feet.

Table 6. Sr^{+2} concentration in various gypsum associated with anhydrite, and calculated ratios Sr^{+2}/Ca^{+2} in groundwaters

Location	Avg. Sr^G (wt. %)	Range Sr^G (wt. %)	Avg. $(Sr^{+2}/Ca^{+2})_L$	Range $(Sr^{+2}/Ca^{+2})_L$
W. Oklahoma	0.086	0.047-0.123	1.9×10^{-2}	$1.0\text{-}2.7 \times 10^{-2}$
	0.100	0.045-0.157	2.0×10^{-2}	$0.98\text{-}3.4 \times 10^{-2}$
S.W. Oklahoma	0.038	0.001-0.100	0.8×10^{-2}	$0.2\text{-}2.0 \times 10^{-2}$
Spitsbergen	0.084	0.068-0.103	1.8×10^{-2}	$1.5\text{-}2.2 \times 10^{-2}$

Sr^{+2} concentrations in these anhydrites probably are very close to those in the original anhydrites before burial. Anhydrites associated with dolomite contain lower concentrations of Sr^{+2} than those not associated with dolomite (Table 6). If k_{Sr}^{A} is taken to be 0.37 at 30°C (Fig. 8) then, using values of Sr^A in Table 6, ratios Sr^{+2}/Ca^{+2} in brines from which the anhydrites crystallized can be estimated (Table 7).

Table 7. Sr^{+2} in various laminated anhydrites with or without dolomite association, and calculated Sr^{+2}/Ca^{+2} in brines

Location	Structural Type and Association	Avg. Sr^A (wt. %)	Range Sr^A (wt. %)	Avg. $(Sr^{+2}/Ca^{+2})_L$	Range $(Sr^{+2}/Ca^{+2})_L$
W. Texas	Laminated + calcite	0.241	0.195-0.300	$\sim 2.3 \times 10^{-2}$	$1.8\text{-}2.8 \times 10^{-2}$
Germany	Anhydrite + salt	0.246	0.14-0.34	$\sim 2.3 \times 10^{-2}$	$1.3\text{-}3.2 \times 10^{-2}$
	anhydrite + salt	0.243	0.18-0.30	$\sim 2.3 \times 10^{-2}$	$1.7\text{-}2.8 \times 10^{-2}$
Germany	Laminated + dolomite	0.204	0.130-0.290	1.9×10^{-2}	$1.2\text{-}2.7 \times 10^{-2}$
Germany	Laminated + dolomite	0.217	--	2.0×10^{-2}	--

These data strongly suggest that a relationship exists between the estimated value of the ratio Sr^{+2}/Ca^{+2} in brines from which the anhydrites either formed initially or later equilibrated, and the degree of dolomitization. Sr^{+2} in dolomite-free, laminated anhydrites and anhydrite with salt yields an average estimated Sr^{+2}/Ca^{+2} ratio of $\sim 2.3 \times 10^{-2}$. This ratio is 20% lower (avg. 1.9 - 2.0×10^{-2}) for laminated anhydrites associated with a trace of dolomite (2-10%, Table 5). Mosaic anhydrites with a dolomite matrix yield average ratios of between $\sim 0.9 \times 10^{-2}$ and $\sim 1.4 \times 10^{-2}$.

Marine supratidal brines saturated with respect to celestite, and gypsum/ anhydrite on the Trucial Coast not associated with dolomitization have Sr^{+2}/Ca^{+2} ratios of $\sim 2.2 \times 10^{-2}$. In those associated with dolomitization, however, the ratio ranges from $\sim 2.2 \times 10^{-2}$ to $\sim 1.0 \times 10^{-2}$. Brines with ratio $Sr^{+2}/Ca^{+2} \simeq 1.0 \times 10^{-2}$ appear to be in equilibrium with the assemblage dolomite, celestite, carbonate, and gypsum/anhydrite.

The close agreement between values of Sr^{+2}/Ca^{+2} in Trucial Coast brines and those estimated for certain ancient anhydrite rocks is impressive and surely not coincidental. Dolomitization would be extensive in arid, carbonate, supratidal flats owing to extremes of brine chemistry and temperature. Because of dolomitization, ratio Sr^{+2}/Ca^{+2} in brines would average between $\sim 1.0 \times 10^{-2}$ and $\sim 2.2 \times 10^{-2}$ and Sr^A would average between ~ 0.10 wt. % and approximately 0.24 wt. %. Brines saturated with respect to celestite, gypsum/anhydrite at a sediment-free brine interface would be expected to maintain $Sr^{+2}/Ca^{+2} \simeq 2.2 \times 10^{-2}$. Gypsum precipitated at this interface would contain ~ 0.09 wt. % Sr^{+2} and anhydrite ~ 0.24 wt. % Sr^{+2}. If the idea of actual precipitation of anhydrite at this interface is not relevant, particularly since (a) gypsum rather than anhydrite is the first phase to precipitate from sea water and will continue to do so metastably and (b) anhydrite is difficult to nucleate from brines even though saturated with respect to anhydrite, then on moderate burial gypsum could react with pore brines, dissolve, and the ions reprecipitate as anhydrite. In the absence of dolomitization it is doubtful whether the ratio Sr^{+2}/Ca^{+2} in pore brines would shift appreciably from $\sim 2.2 \times 10^{-2}$ during burial in the presence of gypsum and celestite. If this were the case then anhydrite would still contain ~ 0.24 wt. % Sr^{+2}, but slightly less than 0.24 wt. % if some dolomitization also occurred. It is noteworthy that two samples of brine produced from halite, anhydrite, and potash salts in the Paradox Basin of Utah contained ratios Sr^{+2}/Ca^{+2} of 2.3×10^{-2} and 1.73×10^{-2} (Mayhew and Heylmun, 1966).

The preceding discussion concerns marine derived brines and calcium sulphate minerals formed from marine derived brines. Some brines derived from continental groundwaters in the Abu Dhabi supratidal flat have ratio $Sr^{+2}/Ca^{+2} = 0.55 \times 10^{-2}$ and are saturated with respect to gypsum or anhydrite. These brines are capable of precipitating gypsum or anhydrite with $Sr^G = 0.02$ wt. % and $Sr^A = 0.06$ wt. %. The intriguing possibility arises that a Sr^{+2} concentration of ~ 0.06 wt. % could be characteristic of anhydrites formed in interior playas of desert areas adjacent to coastal, arid, supratidal flats.

ACKNOWLEDGEMENTS

I am indebted to K. S. Johnson of the Oklahoma Geological Survey for the unpublished data on Sr^{+2} concentrations from anhydrites in western and southwestern Oklahoma (Table 3).

APPENDIX

STRONTIUM CONCENTRATIONS IN MODERN GYPSUM AND ANHYDRITE

Values in parenthesis represent size or size range. Values for gypsum and anhydrite are in mm and microns respectively. Samples 38-51 are discoidal gypsum crystals (gypsum mush); 52-55, algal gypsum; 56-61, selenite gypsum; and 62-64, gypsum associated with anhydrite in the aeolian facies.

ABU DHABI

Anhydrite		Gypsum	
Sample #	Wt. % Sr^{+2}	Sample #	Wt. % Sr^{+2}
1	0.192 (38-77)	38	0.048 (≥ 5.7)
2	0.146 (22-25)	39	0.089 (≥ 5.7)
3	0.160	40	0.079 (2.0-2.8)
4	0.166	41	0.123 (1.0-2.0)
5	0.163	42	0.182 (0.5-1.0)
6	0.236	43	0.092 (2.0-2.8)
7	0.117 (22-28)	44	0.091 (2.0-2.8)
8	0.131	45	0.106 (5.7-7.5)
9	0.170	46	0.090 (2.8-5.7)
10	0.080	47	0.093 (2.0-2.8)
11	0.127 (17-30)	48	0.096 (1.0-2.0)
12	0.123	49	0.124 (0.5-1.0)
13	0.194	50	0.282 (0.15-0.5)
14	0.224	51	0.045 (≥ 5.7)
15	0.202 (15-25)		
16	0.235 (13-20)	52	0.068 (≥ 0.5)
17	0.187	53	0.087 (≥ 0.5)
18	0.151	54	0.094 (≥ 0.5)
19	0.195	55	0.036 (2.8-5.7)
20	0.187		
21	0.214	56	0.021
22	0.185	57	0.048
23	0.133	58	0.028
24	0.163	59	0.052
25	0.204	60	0.039
26	0.254 (10-16)	61	0.057
27	0.189		
28	0.221	62	0.032
29	0.192	63	0.044
30	0.191	64	0.048
31	0.157		
32	0.522 (8-23)		
33	0.480 (12-20)		
34	0.453		
35	0.498 (12-20)		
36	0.430		
37	0.348		

BAJA CALIFORNIA

Anhydrite

Sample #	Wt. % Sr^{+2}
65	0.118 ($\leq$ 2)
66	0.122 ($\leq$ 2)
67	0.124 ($\leq$ 2)
68	0.114 ($\leq$ 2)
69	0.099 ($\leq$ 2)

Gypsum

Sample #	Wt. % Sr^{+2}
70	0.084 (~ 2.0)
71	0.073 (~ 2.0)
72	0.068 (~ 2.0)
73	0.079 (~ 2.0)
74	0.071 (~ 2.0)

JARVIS ISLAND

Gypsum

Sample #	Wt. % Sr^{+2}
75	0.228 ($\leq$ 0.15)
76	0.205 (0.15-0.5)
77	0.264 ($\leq$ 0.15)
78	0.197 (0.5-1.0)
79	0.175 (0.15-0.5)
80	0.179 ($\leq$ 0.15)

Sulphur-Isotope Geochemistry of an Arid, Supratidal Evaporite Environment, Trucial Coast

G. P. Butler[1], R. H. Krouse[2], and R. Mitchell[3]

ABSTRACT

In the interpretation of sulphur isotope data from ancient sulphates the effect of sulphur isotope fractionation by sulphate crystallization has been neglected. In the Abu Dhabi supratidal flat, dolomitization and concomitant sulphate precipitation are responsible for depleting > 90 % of the original SO_4^{-2} available in marine-derived pore brines at, or close to, halite saturation. This results in a fractionation of $\Delta^{34}S_c$ of between -3 ‰ and -4 ‰ in precipitated sulphates relative to the source marine waters, which have $\delta S^{34} \simeq +23$ ‰ .

Continental-derived brines in the supratidal flat have SO_4^{-2} with $\delta S^{34} \leq +19$ ‰ . Sulphates crystallized from marine-derived brines are consequently isotopically heavier by up to + 3 ‰ than those crystallized from continental-derived brines.

The effect of bacterial reduction of SO_4^{-2} is recognized in brines and precipitated sulphates, whereby the sulphates have been enriched in S^{34} by as much as 3 ‰.

Fractionation causes values of δS^{34} to increase generally from the base to the top of the anhydrite cycle in the supratidal flat. Sub-cycles of anhydrite caused by periods of erosion within the main anhydrite cycle are identified by a

[1]Esso Production Research Co., P.O. Box 2189, Houston, Texas.

[2]Dept. of Physics, University of Calgary, Calgary, Canada.

[3]Mineralogisk-Geologisk Museum, SARS Gate 1, OSLO 5, Norway.

decrease in δS^{34} values from the base to the top of the sub-cycle. These relationships could allow recognition and segregation of ancient marine anhydrite-dolomite sequences into supratidal cycles.

INTRODUCTION

Sulphur isotope abundance in sulphates have been used to demonstrate that the composition of oceanic sulphate has varied widely through geologic time from $\delta S^{34} = +9$ ‰ in the late Paleozoic to $\simeq +30$ ‰ in the early Paleozoic, with less extreme fluctuations in post Paleozoic time (Holser and Kaplan, 1966). These fluctuations have been attributed, in general, to processes of bacterial reduction of SO_4^{-2} leading to enrichment of S^{34} in SO_4^{-2}, and to massive influx of isotopically light SO_4^{-2} into the oceans from weathered sulphide/sulphate rocks (Thode and Monster, 1965; Holser and Kaplan, 1966). Effects on sulphur isotope distribution in sulphates due to fractionation during sulphate precipitation have been assigned a minor role by these authors.

This paper deals with the sulphur isotope geochemistry of a supratidal environment in the Trucial Coast, the Abu Dhabi supratidal flat, with additional data from Baja, California, and Jarvis Island in the Pacific (Butler, in this volume). As is usually the case when detailed work is done in an area, the validity of data interpretation becomes increasingly less substantiated because of the complex interplay of minor variables. This situation predominates in the interpretation of data from modern evaporite environments. Accordingly, only the highlights of this study will be presented in this paper, which will emphasize the control of isotopic fractionation on δS^{34} distribution in gypsum and anhydrite, and suggest how similar supratidal sulphate cycles may be recognized in ancient evaporite sequences.

SULPHUR ISOTOPE MEASUREMENTS

Sulphate from brines was separated as barium sulphate. These barium sulphates, together with anhydrite and gypsum samples, were reduced to H_2S with a hypophosphorous-hydriodic acid mixture, converted to silver sulphide and then to SO_2 (Thode et al., 1961). The sulphur dioxide was subjected to sulphur isotope ratio analyses by mass spectrometry. $^{34}S/^{32}S$ abundance ratios are expressed as δS^{34} ‰ relative to the Canyon Diablo Meteorite Standard, in the usual manner. For a homogeneous sample, δS^{34} values were reproducible to within ± 0.15 ‰ standard deviation. In some cases, e.g. algal mats, adjacently sampled raw materials displayed fluctuations as high as ± 0.5 ‰ if small sample sizes (< 10 mg sulphur equivalent) were chosen.

BRINES

Marine derived brines are isotopically distinct from continental derived brines in the Abu Dhabi supratidal flat. Marine brines range in δS^{34} from + 23.9 ‰ to $\simeq$ + 19 ‰ ; continental brines from $\simeq \pm$ 19 ‰ to + 17.6 ‰ (Fig. 1). Most inner lagoon brines have $\delta S^{34} \simeq$ + 23 ‰ but three low intertidal pore brines ranged from + 23.4 ‰ to + 23.9 ‰ , possibly indicating slight enrichment in S^{34} by bacterial action (Fig. 1). It is significant that these inner lagoon brines are enriched with S^{34} compared to normal ocean water with $\delta S^{34} \simeq$ + 20 ‰ .

Concentration of SO_4^{-2} in supratidal brines during progressive concentrations is controlled by the solubility products of gypsum, anhydrite and, to a lesser extent, celestite. The overall relationship between SO_4^{-2} concentration and $\delta S^{34}(SO_4^{-2})$ in brines (Fig. 1) strongly suggests that the variation in δS^{34} is due to the preferential fractionation of S^{34} into precipitated sulphate.

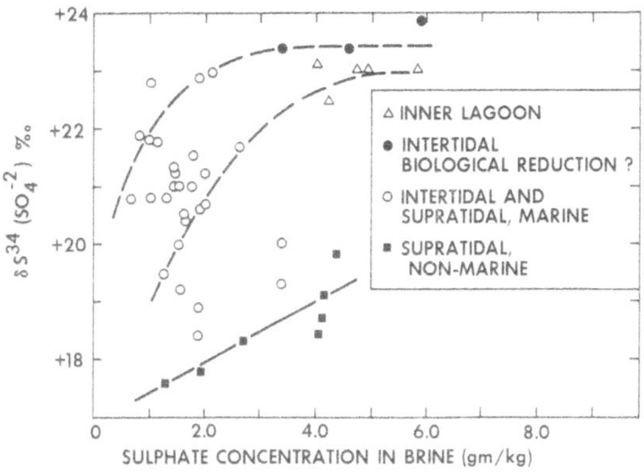

Fig. 1. Sulphur isotope in lagoon-supratidal brines vs. SO_4^{-2} concentration in brines

Holser and Kaplan (1966, Fig. 4, p. 117) suggest that the effect of isotopic fractionation during crystallization becomes significant during the potash phase of precipitation from brines. They assumed that at the beginning of carnallite deposition, 90 % of the sulphate would have been precipitated and, as a result, the fractionation relative to the original sea water would be $\Delta^{34}S_c$ = 2.2 ‰ . Their model is only applicable in the absence of dolomitization. Dolomitization with concomitant calcium sulphate precipitation in the Abu Dhabi supratidal flat caused > 90 % of the SO_4^{-2} to be precipitated in brines at, or near to, halite saturation (Butler, Fig. 6, this volume). Fig. 2 expresses the variation to δS^{34} of SO_4^{-2} in the supratidal flat brines as a function of sulphate precipitation. δS^{34} of SO_4^{-2} was calculated using the relationship (Holser and Kaplan, 1966, p. 116):

$$\ln \frac{r_f}{r_o} = (K - 1) \ln (1 - \frac{m_f}{m_o}) \quad \text{and}$$

$$\delta S^{34} \text{ of } SO_4^{-2} = \frac{r_f - r_o}{r_o} \times 10^3$$

where r_o and r_f are the initial and final S^{34}/S^{32} ratios of SO_4^{-2} in brine; $K^G = 1.100165$ (coefficient of fractionation for gypsum; Thode and Monster, 1965); and values for the ratio m_f/m_o for the weight fraction of sulphate crystallized with increasing brine concentration from the sulphate curve in Butler, Fig. 6 (this volume). δS^{34} for "original" marine brine was taken to be + 23 ‰ (see Fig. 1). Also shown in Fig. 2 is a curve for continental-derived brines assuming the original δS^{34} of SO_4^{-2} to be + 19 ‰ (see Fig. 1). The estimated δS^{34} in precipitated gypsum with respect to brine concentration is given in Fig. 3.

The coefficient of fractionation for anhydrite (K^A) is not known. Anhydrite precipitation in these brines commences at concentrations $\geq$ x 7. If it can be assumed that $K^A = K^G$, then the curves in Fig. 3 can be used to estimate S^{34} in precipitated anhydrite at concentrations $\geq$ x 7.

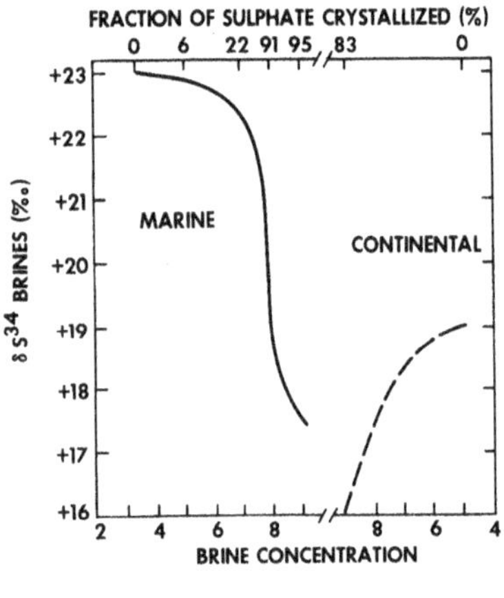

Fig. 2. Estimated δS^{34} of brines as a function of brine concentration

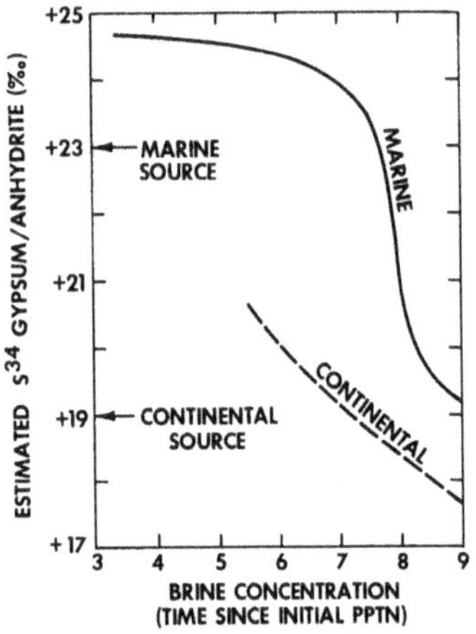

Fig. 3. Estimated sulphur isotope in precipitated sulphates vs. brine concentration (time since initial sulphate precipitation)

ISOTOPIC COMPOSITION OF MODERN GYPSUM AND ANHYDRITE

The estimated ranges of δS^{34} in gypsum and anhydrite agree quite well with the observed ranges in these minerals (Table 1). These data, in general, support the thesis that isotopic fractionation can considerably influence the distribution of δS^{34} in precipitated sulphates when dolomitization occurs.

Because of the basic differences in δS^{34} in marine and continental-derived brines, anhydrites crystallized from the marine brines are isotopically heavier than those from continental brines. The average difference is in the order of + 1.5 ‰ to + 3.0 ‰ (Table 1). The hydration of marine anhydrite to gypsum via a solution phase in a continental brine regime results in the gypsum being isotopically lighter than the "parent" anhydrite (Table 1).

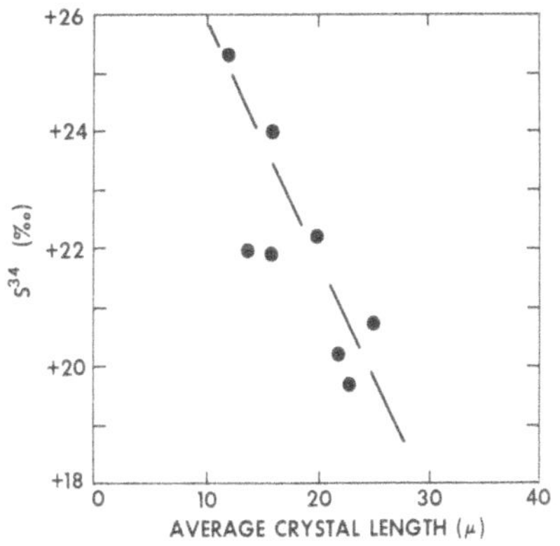

Fig. 4. Sulphur isotope in modern gypsum as a function of crystal grain size

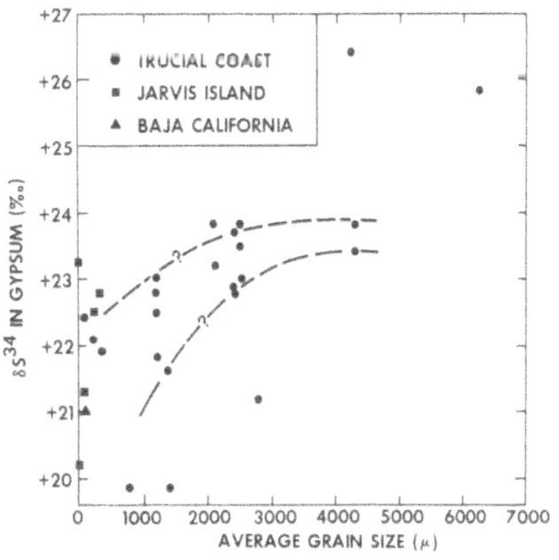

Fig. 5. Sulphur isotope in modern anhydrite as a function of crystal length

As with Sr^{+2} in modern gypsum and anhydrite (Butler, this volume) there is an apparent relationship between crystal size and concentration of δS^{34} in both gypsum and anhydrite (Fig. 4, 5). In fact, the distribution of δS^{34} as a function of gypsum grain size (Fig. 4) is strikingly similar to the distribution shown in Fig. 1.

One possible explanation is that one is seeing the effect of continued crystal precipitation from a brine constantly changing in isotopic composition combined with crystal zonation. Crystal size may simply be a measure of the prevailing isotopic composition of the brines in which the crystals grew. The large ($\simeq$ 4000 μ) gypsum crystals with δS^{34} = +23 to +24 ‰ , for example (Fig. 4), probably have small cores which are isotopically lighter than the surrounding gypsum. Gypsum crystals with $\delta S^{34} \simeq > +24$ ‰ (Fig. 5) probably precipitated from brines which have been enriched in S^{34} by bacterial action. The probability that S^{34} enrichment by bacterial action can be extensive and is reflected by δS^{34} values in anhydrite in particular, will be discussed below.

RECOGNITION OF ANCIENT SUPRATIDAL DOLOMITE-EVAPORITE CYCLES

Diagenetic Model

It is observed in the Abu Dhabi supratidal flat that a) brine concentration generally increased progressively inland across the flat, and that b) anhydrite development generally begins in the surface/near surface sediments in the low supratidal area and occurs progressively at greater depths on passing from the low to the intermediate supratidal area. In general, in the intermediate supratidal area, the youngest anhydrite occurs at depth and the oldest near the surface. To obtain a brine, sea water is evaporated, the degree of brine concentration being proportional to the duration of evaporation. In a simple evaporite system, brine concentration may be considered as a measure of brine age. Although it is realized that the Abu Dhabi supratidal flat is not a passive evaporite system, but rather a dynamic system owing to marine flood recharge (Butler, this volume), brine concentration is probably still somewhat time-dependent because of the high net rate of evaporation (50.4 inches per year; Privett, 1959). This being the case, the abscissa units in Fig. 3 can be considered as time units since initial sulphate precipitation. In the context of observation (b) discussed above, and data in Fig. 3, one would expect anhydrites near the surface of the supratidal flat in the indermediate area to be enriched in S^{34} relative to anhydrite at depth.

Vertical Isotope Trends in Abu Dhabi Supratidal Flat

Figure 6 shows the variation of δS^{34} in anhydrites with depth from the surface in a core taken from the high supratidal area where terrigenous clastic sediments predominate and where pore brines are derived from continental ground waters. Similar depth trends of δS^{34} in anhydrites are shown in Figs. 7, 8. The core sampled for anhydrite in Fig. 7 was taken from the intermediate supratidal area where pore brines are of marine origin. Consequently, the mean δS^{34} value is heavier than in Fig. 6. It should be noted in Fig. 7 that the δS^{34} values of anhydrites denoted by open rectangles have a common carbonate-algal matrix; H_2S gas was detected during coring.

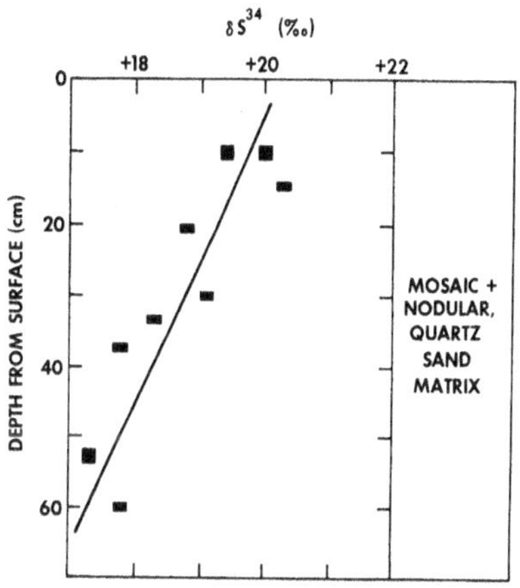

Fig. 6. Depth variation of sulphur isotope in supratidal anhydrites crystallized from continental-derived brines, Note core description

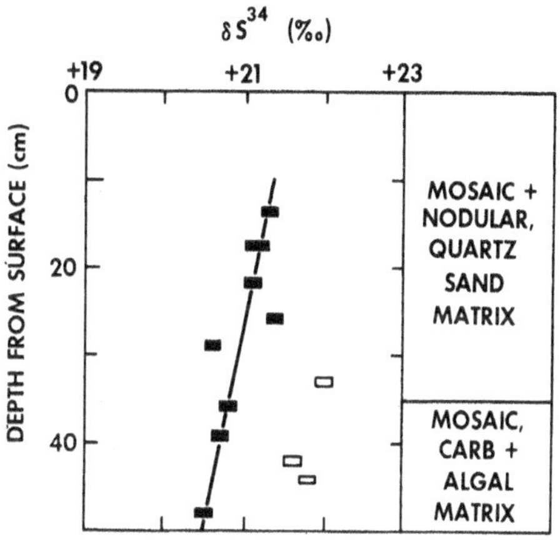

Fig. 7. Depth variation of sulphur isotope in marine supratidal anhydrite cycle. Solid data points = no bacterial action; open = bacterial action. Note core description

These data suggest that these particular anhydrites were precipitated from pore brines which had been enriched in S^{34} by bacterial reduction of SO_4^{-2}.

Similar relationships are illustrated in anhydrites from two cores shown in Fig. 8. Characteristics for these cores are similar to those noted above for the core in Fig. 7. Below 10 cm, values of δS^{34} in anhydrites in both cores can be shown to have a strong inverse relationship with depth consistent with Fig. 7; and that significant departure from this trend occurs in anhydrites sampled from carbonate-algal sediments and the fossil algal mat. The average enrichment in δS^{34} in these samples is $\simeq + 2$ ‰.

The reason for the reversal in δS^{34} values above 10 cm depth in one core (Fig. 8) is uncertain. One possible explanation is that it results from isotopic dilution of pore brines with rain water. An alternative, and preferred explanation, which takes into consideration the marked erosion surface at 10 cm depth (Fig. 8) is that one is dealing with an isotopic system which is unrelated to that below the erosion surface. Flooding by marine water of the supratidal flat during severe storms is a fairly frequent event, the erosion surface being attributed to one of these episodes. Subsequent accumulation of sediment could be by the process of sediment adhesion.

The lower anhydrite ($\delta S^{34} \simeq + 21$ ‰) is visualized as being older than the overlying anhydrite ($\delta S^{34} \simeq + 20$ ‰). This relationship, in the context of progressive isotopic fractionation of the flood-derived pore brines during sediment accumulation and crystallation of anhydrite, could well account for the observed distribution of δS^{34} .

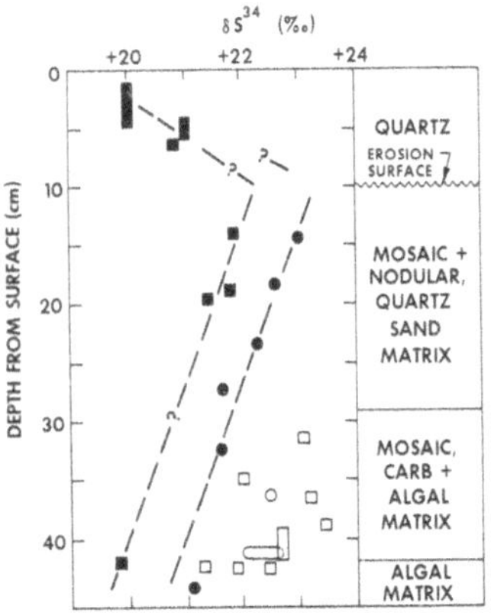

Fig. 8. Depth variations of sulphur isotope in two marine supratidal anhydrite cycles, expressed as squares and circles, respectively. Solid data points = no bacterial action; open = bacterial action

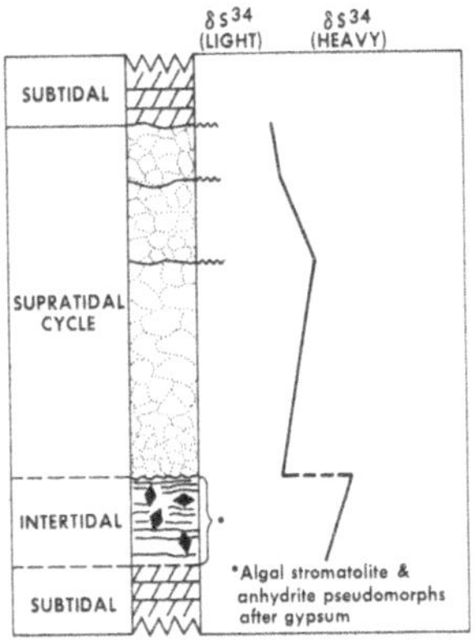

Fig. 9. Note generalized core description. Predicted trend in sulphur isotope in ancient, upper intertidal (algal stromatolite) and supratidal anhydrite-dolomite cycle

A progressive upwards enrichment in S^{34} in anhydrite would appear to define a supratidal cycle in the Abu Dhabi supratidal flat. If our interpretation of the data is valid, sub-cycles caused by erosion within the main cycle would be reflected by progressive enrichment in S^{34} in anhydrite from the top down. In applying these concepts to the segregation of marine, supratidal, evaporite-dolomite cycles in ancient evaporite sequences, the lightest δS^{34} values should be used in attempting to establish trends in order to help eliminate probable effects of bacterial action.

TABLE 1

Summary of Isotope Data, and Comparison between Observed and Estimated δS^{34} in Gypsum and Anhydrite.

Note that all δS^{34} values are positive

Mineral	Type	Fluid Regime	Range $^{o}/oo$	Mean & S.D. $^{o}/oo$	Estimated $^{o}/oo$ (Fig. 3)
GYPSUM					
	Primary	marine	19.9-26.4	22.4 $\pm$ 0.9	19.2-24.7
	via $CaSO_4$	continental	17.9-19.0	18.6 $\pm$ 0.6	17.7-20.6
ANHYDRITE					
	Discrete	marine	19.7-22.6	20.5 $\pm$ 0.8	
	Nodules				19.2-24
	Mosaic	marine	19.2-25.3	22.2 $\pm$ 1.0	
	Nodular	continental	17.3-20.9	19.0 $\pm$ 1.2	17.7-19.1
	Mosaic				

In Figure 9 an attempt in made to characterize trends in δS^{34} which could be expected in an ancient, supratidal, anhydrite-dolomite cycle on the basis of data presented in this paper. A new concept is introduced, however, and it concerns the trend of δS^{34} in sulphates developed in algal stromatolites at the base of the supratidal evaporite cycle.

Gypsum crystals are commonly developed in older, buried algal mats in the inner parts of the Abu Dhabi supratidal flat. As with anhydrite, gypsum crystals have developed initially in the upper algal layers and then occur at progressively deeper horizons within the algal mat inland across the supratidal flat. In terms of the fractionation model outlined in this paper, the top, early formed gypsum crystals should be isotopically heavier than the lower, later formed crystals. Gypsum is the first sulphate mineral to precipitate from supratidal brines, and, thus, these gypsum crystals within the algal mat would be expected to be, on the average, isotopically heavier than the overlying anhydrite. Although overlaid by andhydrite in the intermediate supratidal areas, these gypsum crystals appear stable and in equilibrium with halite saturated pore brines. If they remain stable, on later burial they would be expected to dehydrate to anhydrite with the anhydrite pseudomorphing their observed lenticular to subangular habit. If, under these conditions, the process of gypsum to anhydrite conversion is one of simple dehydration, then the anhydrite should retain the isotopic composition of the parent gypsum producing the trend illustrated in Fig. 9.

Bibliography

A. Bibliography Relating to the Persian Gulf Region

AL-HABEEB, K.H.: Sedimentological investigations of the flood-plain sediments of the middle Euphrates River. Thesis, University of Baghdad (1969).

ARAMCO: Ghawar oil field, Saudi Arabia. Bull. Am. Assoc. Petrol. Geologists 43, 434-454 (1950).

BATE, R.H.: The distribution of Recent Ostracoda in the Abu Dhabi Lagoon, Persian Gulf. Bull. Cent. Rech. Pau-SNPA 5, 239-256 (1971).

BOBEK, H.: Nature and implications of Quaternary climatic changes in Iran. UNESCO proc. XX. Changes in climate, 403-422, Paris (1963).

BOBEK, H.: Zur Kenntnis der Südlichen Lüt. Mitteil. der Österreichischen Geographischen Gesellschaft, Bd. 111, Heft II/III, 155-192 (1969).

BRAMKAMP, R.A., POWERS, R.W.: Two Persian Gulf lagoons. J. Sediment. Petrol. 25, 139-140 (1955).

British Petroleum Co. Ltd.: Geological maps and sections of South-West Persia. Int. Geol. Congr., Proc. 20th Session, Mexico, 1956.

BURINGH, P.: Soil and soil conditions in Iraq. Ministry of Agriculture, 1-322, Baghdad (1960).

BURINGH, P., EDELMAN, C.H.: Some remarks about soils of the alluvial plain of Iraq, south of Baghdad. Neth. J. Agr. Soc. 3, 40-49 (1955).

BUSH, P.R.: Chloride rich brines from sabkha sediments and their possible role in ore formation. Trans. Inst. Miner. Metall. Sect. B. 79, 137-144 (1970).

BUTLER, G.P.: Early diagenesis in the Recent sediments of the Trucial coast of the Persian Gulf. MSc Thesis. Univ. Lond. (1966).

BUTLER, G.P.: Modern evaporite deposition and geochemistry of co-existing brines, the sabkha, Trucial coast, Arabian Gulf. J. Sediment. Petrol. 39, 70-89 (1969).

BUTLER, G.P.: Recent gypsum and anhydrite of the Abu Dhabi sabkha, Trucial Coast: an alternative explanation of origin; in: Third symposium on salt, Vol. 1: Northern Ohio Geol. Soc., p. 120-152 (1970a).

BUTLER, G.P.: Secondary anhydrite from a sabkha, North West Gulf of California, Mexico; in: Third symposium on salt, Vol. 1: Northern Ohio Geol. Soc., p. 120-152 (1970b).

BUTLER, G.P.: Origin and controls on distribution of arid supratidal (sabkha) dolomite, Abu Dhabi, Trucial Coast (abst.) : Bull. Am. Assoc. Petrol. Geologists 55, 332 (1971).

BUTLER, G.P., KENDALL, C.G.St.C., KINSMAN, D.J.J., SHEARMAN, D.J., SKIPWITH, Sir Patrick A.d'E.: Recent anhydrite from the Trucial coast of the Arabian Gulf. Circ. Geol. Soc. Lond. 120, 3 (1965).

CUFF, C.: Lattice disorder in Recent anhydrite and its geological implications. Proc. Geol. Soc. London, No. 1659, 326-330 (1969).

CURTIS, R., EVANS, G., KINSMAN, D.J.J., SHEARMAN, D.J.: Association of dolomite and anhydrite in the Recent sediments of the Persian Gulf. Nature, 197, No. 4868, 679-680 (1963).

DIESTER, L.: Grobfraktionsanalyse von Sedimentkernen aus dem Persischen Golf. Thesis Univ. Kiel (1971).
DIESTER, L.: Zur spätpleistozänen und holozänen Sedimentation im zentralen und östlichen Persischen Golf. "METEOR" Forsch. Ergebnisse, Reihe C, No. 8, 37-83 (1972).
DIESTER, L.: Grobfraktionsanalyse von Sedimentkernen aus dem Persischen Golf. "METEOR" Forsch. Ergebnisse, Reihe C, No. 8, 37-83 (1972).
DIETRICH, G., KRAUSE, G., SEIBOLD, E., VOLLBRECHT, K.: Reisebericht der Indischen Ozean Expedition mit Forschungsschiff "METEOR" 1964-1965. "METEOR" Forsch. Ergebnisse, Reihe A, No. 1, 1-52 (1966).
Dust Storms in Iraq. A review. Geograph. Jour. 43, 121-131.
EINSELE, G., WERNER, F.: Sedimentary processes at the entrance Gulf of Aden / Red Sea. "METEOR" Forsch. Ergebnisse, Reihe C, No. 9 (in press).
ELDER, S.: Umm Shaif oil field. Journ. Inst. Pet. 49, 478, 308-315 (1963).
EMERY, K.O.: Sediments and water of Persian Gulf. Bull. Am. Assoc. Petrol. Geologists 40, 2354-2383 (1956).
ESTEOULE, J., ESTEOULE-CHOUX, J., MELGUEN, M., SEIBOLD, E.: Sur la présence d'attapulgite dans des sédiments récents du Nord-Est du Golfe Persique. C. R. Acad. Sc. Paris 271, 1153-1156.
EVANS, G.: The Recent sedimentary facies of the Persian Gulf region. Phil. Trans. Roy. Soc. London, Ser. A, Vol. 259, No. 1099, 291-298 (1966).
EVANS, G., SHEARMAN, D.J.: Recent celestite from the sediments of the Trucial coast of the Persian Gulf. Nature 202, No. 4930, 385-386 (1964).
EVANS, G., KENDALL, C.G.St.C., SKIPWITH, Sir Patrick A. d'E.: Origin of the coastal flats, the sabkha, of the Trucial coast, Persian Gulf. Nature 202, No. 4934, 579-600 (1964).
EVANS, G., KINSMAN, D.J.J., SHEARMAN, D.J.: A reconnaissance survey of the environment of Recent carbonate sedimentation along the Trucial coast, Persian Gulf. In: Deltaic and shallow marine deposits, ed. by L.M.J.U. van STRAATEN. Elsevier, Amsterdam, p. 129-135 (1964).
EVANS, G., BUSH, P.: Some sedimentological and oceanographic observations on a Persian Gulf lagoon. In: Coastal Lagoons, ed. by A.A. CASTANARES and F.B. PHLEGER. Mem. Simp. Intern. Lagunas Costeras UNAM-UNESCO, Mexico. 155-17o (1969).
EVANS, G., SCHMIDT, V., BUSH, P., NELSON, H.: Stratigraphy and geologic history of the sabkha, Abu Dhabi, Persian Gulf. Sedimentology 12, 145-159 (1969).
EVANS, G.: Coastal and nearshore sedimentation: a comparison of clastic and carbonate deposition. Proc. Geol. Assoc. London 81, 493-508 (1970).
FOX, A.F.: Some problems of petroleum geology in Kuwait. Journ. Inst. Petr. 45, 95-110 (1959).
FUCHS, W., GATTINGER, T.E., HOLZER, H.F.: Explanatory text to the synoptic geologic map of Kuwait; a surface geology of Kuwait and the Neutral Zone. Geological Survey of Austria (1968).
GLENNIE, K.W., EVAMY, B.D.: Dikaka: plants and plant-root structures associated with aeolian sand. Palaeogeog., Palaeoclim., Palaeoecol. 4, 77-87 (1968).
GROOT, de, K.: The chemistry of submarine cement formation at Dohat Hussain in the Persian Gulf. Sedimentology 12, 63-68 (1969).
HAAKE, F.W.: Zur Tiefenverteilung von Miliolinen (Foram.) im Persischen Golf. Pal. Z. 44, 3/4, 196-200 (1970).
HAAS, F.: Shells collected by the Peabody Museum Expedition to the Near East 1950. I. Mollusks from the Persian Gulf. The Nautilus 65, 4, 114-119 (1952).
HARTMANN, M., LANGE, H., SEIBOLD, E., WALGER, E.: Oberflächensedimente im Persischen Golf und Golf von Oman. I. Geologisch-hydrologischer Rahmen und erste sedimentologische Ergebnisse. "METEOR" Forsch. Ergebnisse, Reihe C, No. 4, 1-76 (1971).
HENSON, F.R.S.: Observations on the geology and petroleum occurrences of the Middle East. World Petrol. Congr. Proc. 3rd. The Hague. Section 1, 118-140 (1951).
HOEFS, J., SARNTHEIN, M.: $^{18}O/^{16}O$ Ratios and related temperatures of Recent pteropod shells (Cavolinia longirostris LESUEUR) from the Persian Gulf. Mar. Geol. 10, 11-16 (1971).
HOLM, D.A.: Desert geomorphology in the Arabian Peninsula. Science 132, 1369 (1960).
HOUBOLT, J.J.H.C.: Surface sediments of the Persian Gulf near the Qatar Peninsula. Doctoral thesis, University of Utrecht. Den Haag: Mouton and Co. (1957).

ILLING, L.V., WELLS, A.J., TAYLOR, J.C.M.: Penecontemporary dolomite in the Persian Gulf. In: Dolomitization and Limestone diagenesis, ed. by L.C. PRAY and R.C.MURRAY. S.E.P.M., Spec. Publ. 13, 89-111 (1965).

JESSEN, K., SPARCK, R. (eds.): Danish Scientific Investigations in Iran, parts I-IV. Copenhagen: Munksgaard (1939-1949).

KENDALL, G.G.St.C.: Recent sediments of the Western Khor al Bazam, Abu Dhabi, Trucial Coast. PhD Thesis, University of London (1966).

KENDALL, C.G.St.C., SKIPWITH, Sir Patrick A. d'E.: Recent algal stromatolites of the Khor al Bazam, Southwest Persian Gulf. Abs. Geol. Soc. Am. Spec. Pap. for 1966, 108 (1966).

KENDALL, C.G.St.C., SKIPWITH, Sir Patrick A. d'E.: Recent algal mats of a Persian Gulf Lagoon. J. sediment. Petrol. 38, 1040-1058 (1968).

KENDALL, C.G.St.C., SKIPWITH, Sir Patrick A. d'E.: Holocene shallow-water carbonate and evaporite sediments of Khor al Bazam, Abu Dhabi, Southwest Persian Gulf. Bull. Am. Assoc. Petrol. Geologists 53, 841-869 (1969a).

KENDALL, C.G.St.C., SKIPWITH, Sir Patrick A. d'E.: Geomorphology of a Recent shallow water carbonate province; Khor al Bazam, Trucial Coast, Southwest Persian Gulf. Bull. Geol. Soc. Am. 80, 865-892 (1969b).

KHALAF, M.J.: The climate of Iraq. Bull. Coll. Arts and Sci. 2, 1-51 (1957) Baghdad.

KINSMAN, D.J.J.: Reef coral tolerance of high temperatures and salinities. Nature 202, 1280-1282 (1964).

KINSMAN, D.J.J.: The Recent carbonate sediments near Halat el Bahrani, Trucial coast, Persian Gulf. In: Developments in Sedimentology 1. Deltaic and Shallow Marine Deposits, ed. by L.M.J.U. van STRAATEN, p. 189-192 (1964a).

KINSMAN, D.J.J.: Recent carbonate sedimentation near Abu Dhabi, Trucial Coast, Persian Gulf. PhD Thesis, University of London (1964b).

KINSMAN, D.J.J.: Gypsum and anhydrite of Recent age, Trucial coast, Persian Gulf. Proc. 2nd Salt Symposium, 1. N. Ohio Geol. Soc. Cleveland, Ohio, 302-326 (1966).

KINSMAN, D.J.J.: Huntite from a carbonate - evaporite environment. Am. Miner. 52, 1332-1340 (1967).

KINSMAN, D.J.J.: Modes of formation, sedimentary associations, and diagnostic features of shallow-water and supratidal evaporites. Bull. Am. Assoc. Petrol. Geologists 53, 830-840 (1969).

KINSMAN, D.J.J., PARK, R.K., PATTERSON, R.J.: Sabkhas: Studies in Recent carbonate sedimentation and diagenesis, Persian Gulf. Geol. Soc. Am. Abstracts, 1971 Annual Meetings, 772-774 (1971).

KUKAL, Z., SAADALLAH, A.: Composition and rate of deposition of the Recent dust storm sediments in Iraq. Čas. pro Min. a Geol. 15, 227-234 (1970).

KUKAL, Z., AL-JASSIM, J.: Sedimentology of Pliocene Molasse sediments of the Mesopotamian geosyncline. Sediment. Geol. 5, 57-81 (1971).

LEES, G.M.: The physical geography of South-East Arabia. Geogr. Journ. 121, 441 (1948).

LEES, G.M., FALCON, N.L.: The geographical history of the Mesopotamian plains. Geogr. Journ. 116, 24-39 (1952).

LUTZE, G.F., GRABERT, B., SEIBOLD, E.: Lebendbeobachtungen an Groß-Foraminiferen (Heterostegina) aus dem Persischen Golf. "METEOR" Forsch. Ergebnisse, Reihe C, No. 6, 21-40 (1971).

MARTINI, E.: Nannoplankton und Umlagerungserscheinungen im Persischen Golf und im nördlichen Arabischen Meer. N. Jb. Geol. Pal., Mh. 597-607 (1967).

MELGUEN, M.: Etude de sédiments Pleistocène-Holocène au nordouest du golfe Persique. Analyse de faciès par ordinateur. Thèse, Université de Rennes (1971).

MELVILL, J.C.: The marine Mollusca of the Persian Gulf, Gulf of Oman and North Arabian Sea, as evidenced mainly through collections of Captain F.W. Townsend 1893-1914., etc. Proc. Malacol. Soc., London, 18, 93-117 (1928/1929).

MELVILL, J.C., STANDEN, R.: The Mollusca of the Persian Gulf, Gulf of Oman and Arabian Sea, as evidenced mainly through the collections of Mr. F.W. Townsend, 1893-1900; with description of new species. Part I. Cephalopoda, Gastropoda, Scaphopoda. Proc. Zool. Soc. 327-460 (1901).

MURRAY, J.W.: The Foraminiferida of the Persian Gulf. Part 1. Rosalina adhaerens sp. nov. Ann. Mag. nat. Hist., ser 13, 8, 77-79 (1965a).

MURRAY, J.W.: The Foraminiferida of the Persian Gulf. 2. The Abu Dhabi Region. Palaeogeography, Palaeoclim., Palaeoecol., 1, 307-332 (1965b).

MURRAY, J.W.: The Foraminiferida of the Persian Gulf. 3. The Halat al Bahrani Region. Palaeogeo., Palaeoclim., Palaeoecol. 2, 59-68 (1966a).
MURRAY, J.W.: The Foraminiferida of the Persian Gulf. 4. Khor al Bazam. Palaegeo., Palaeoclim., Palaeoecol. 2, 153-169 (1966b).
MURRAY, J.W.: The Foraminiferida of the Persian Gulf. 5. The shelf of the Trucial Coast. Palaeogeo., Palaeoclim., Palaeoecol. 2, 267-278 (1966c).
MURRAY, J.W.: The Foraminiferida of the Persian Gulf. 6. Living forms in the Abu Dhabi Region. J. Nat. Hist. 4, 55-67 (1970a).
MURRAY, J.W.: The Foraminifera of the hypersaline Abu Dhabi Lagoon, Persian Gulf. Lethaia 3, 51-68 (1970b).
PEERY, K.: Results of the Persian Gulf - Arabian sea oceanographic surveys, 1960-1961. Technical report 176, U.S. Naval Oceanographic Office, Washington, D.C. (1965).
PILKEY, O.H., NOBLE, D.: Carbonate and clay mineralogy of the Persian Gulf. Deep-Sea Research 13, 1-16 (1966).
POEL, G.H. van der: Measurement of some factors concerning the flow of saline water in the sediments on the tidal flats at Qatar, Persian Gulf. Unpubl. Shell Res. (1964).
POWERS, R.W.: Geology of the Arabian Peninsula. Sedimentary geology of Saudi Arabia. Geol. Soc. Am. Prof. Paper 560d (1966).
PRIVETT, D.W.: Monthly charts of evaporation from the Northern Indian Ocean (including the Red Sea and the Persian Gulf). Q.J.R. Met. Soc. 85, 424-478 (1959).
SAMPO, M.: Microfacies and microfossils of the Zagros area, southwestern Iran (from pre-Permian to Miocene). Ed. BRILL, Leiden (1969).
SARNTHEIN, M.: Sedimentologische Merkmale für die Untergrenze der Wellenwirkung im Persischen Golf. Geol. Rundsch. 59/2, 649-666 (1970).
SARNTHEIN, M.: Oberflächensedimente im Persischen Golf und Golf von Oman. II. Quantitative Komponenten Analys der Grobfraktion. "METEOR" Forsch. Ergebnisse, Reihe C, No. 5 (1971).
SARNTHEIN, M.: Recent sedimentation near entrances to adjacent seas: Persian Gulf/Gulf of Oman. Abstracts, VIII. Int. Sed. Congr. (Heidelberg 1971).
SARNTHEIN, M.: Sediments and history of the postglacial transgression in the Persian Gulf and Gulf of Oman. Mar. Geol. 12, 245-266 (1972).
SARNTHEIN, M.: Pteropods and Heteropods in surface sediments of the Persian Gulf. In: The Biology of the Indian Ocean. Symp. Kiel 1971, in press.
SARNTHEIN, M.: Stratigraphic Contamination by Vertical Bioturbation in Holocene Shelf Sediments. 24th Int. Geol. Congr., Proc., Sect. 6, Montreal (1972).
SCHOTT, W., von STACKELBERG, U., ECKHARDT, F.-J., MATTIAT, B., PETER, J., ZOBEL, B.: Geologische Untersuchungen an Sedimenten des indisch-pakistanischen Kontinentalrandes (Arabisches Meer). Geol. Rundsch. 60/1, 264-274 (1970).
SEIBOLD, E.: Nebenmeere im humiden und ariden Klimabereich. Geol. Rundsch. 60/1, 73-105 (1970).
SEIBOLD, E.: Biogenic sedimentation of the Persian Gulf. In: The Biology of the Indian Ocean. Symposium Kiel 1971, in press.
SEIBOLD, E., VOLLBRECHT, K.: Die Bodengestalt des Persischen Golfs. "METEOR" Forsch. Ergebnisse, Reihe C, No. 2, 29-56 (1969).
SEIBOLD, E., ULRICH, J.: Zur Bodengestalt des nordwestlichen Golfs von Oman. "METEOR" Forsch. Ergebnisse, Reihe C, No. 3, 1-14 (1970).
SHEARMAN, D.J.: Recent anhydrite, gypsum, dolomite and halite from the coastal flats of the Arabian shore of the Persian Gulf. Proc. Geol. Soc. Lond. No. 1607, 63 (1963).
SHEARMAN, D.J.: Origin of marine evaporites by diagenesis. Trans. Inst. Miner. Metall., Section B, 75, 208-215 (1966).
SHINN, E.A.: Submarine lithification of Holocene carbonate sediments in the Persian Gulf. Sedimentology 12, 109-144 (1969).
SKIPWITH, P.A. d'E.: Recent carbonate sediments of Eastern Khor al Bazam, Abu Dhabi, Trucial Coast. PhD. Thesis, Imperial College, London (1966).
SUGDEN, W.: Structural analysis and geometrical prediction for change of form with depth of some Arabian plains-type folds. Bull. Am. Assoc. Petrol. Geologists 46, 2213-2228 (1962).
SUGDEN, W.: Some aspects of sedimentation in the Persian Gulf. J. Sediment. Petrol. 33, 355-364 (1963).
SUGDEN, W.: The hydrology of the Persian Gulf and its significance in respect to evaporite deposition. Am. J. Sci. 261, 741-755 (1963).

SUGDEN, W.: Pyrite staining of pellety debris in carbonate sediments from the Middle East and elsewhere. Geol. Mag. 103, 250-256 (1966).
Statistical Abstracts for 1967. Government Press, p. 20, Baghdad.
TAYLOR, J.C.M., ILLING, L.V.: Holocene intertidal calcium carbonate cementation, Qatar, Persian Gulf. Sedimentology 12, 69-107 (1969).
United States Geological Survey (1958-1963), Misc. Geol. Investigations. Maps 1208-209-213-214-215 (Geological maps of the Arabian Peninsula) 1:500.000.
WELLS, A.J.: Recent dolomite in the Persian Gulf. Nature 194, 4825, 274-275 (1962).
ZEIST, W. van, WRIGHT, H.E., Jr.: Preliminary pollen studies at Lake Zeribar, Zagros Mountains, South Western Iran. Science 140, 65-67 (1963).
ZEIST, W. van: Late Quaternary vegetation history of western Iran. Rev. Palaeobot. 2, Palynol. 2, 301-311 (1967).
ZIEGENBEIN, J.: Trübungsmessungen im Persischen Golf und im Golf von Oman. "METEOR" Forsch. Ergebnisse, Reihe A, No. 1, 59-79 (1966).

B. General Bibliography

ADAMS, J.E., FRENZEL, H.N.: Capitan Barrier Reef, Texas and New Mexico. J. Geol. 58, 289-312 (1950).
ADAMS, J.E., RHODES, M.L.: Dolomitisation by seepage refluxion. Bull. Am. Assoc. Petrol. Geologists 44, 1912-1920 (1960).
ALLEN, G., PUJOS-LAMY, A.: Application de l'analyse factorielle (mode Y) à l'étude micropaléontologique d'une carotte marine. C. R. Somm. Soc. Géol. Fr. 7, 257 (1970).
ANDERSON, R.Y., KIRKLAND, D.W.: Intrabasin valve correlation. Bull. Geol. Soc. Am. 77, 241-256 (1966).
ARONS, A.B., KIENTZLER, C.F.: Vapour pressure of sea-salt solutions. Am Geophys. Union Trans. 35, 722-728 (1954).
ARTJUSHKOV, E.V.: The possible genetics and general laws governing the growth of a convection instability in sedimentary rock. Doklady Akad. Nauk SSSR 153, No. 1, 162-165 (1963).
ARTJUSHKOV, E.V.: Basic shapes of convectional structures in sedimentary rock. Doklady Akad. Nauk SSSR 153, No. 2, 412-415 (1963).
BAKER, G., FROSTICK, A.C.: Pisoliths, ooliths, and calcareous growths in limestone caves at Port Campbell, Victoria, Australia. J. Sediment. Petrol. 21, 85-104 (1951).
BARON, G.: Sur la synthèse de la dolomite, application au phénomène de la dolomitisation. Rev. Inst. Franc. Petr. 15, 3-68 (1960).
BATHURST, R.G.C.: Boring algae, micrite envelopes and lithification of molluscan biosparites. Geol. J. 5, 15-32 (1966).
BATHURST, R.G.C.: Oolitic films on low energy carbonate sand grains, Bimini lagoon, Bahamas. Marine Geol. 5, 89-109 (1967).
BATHURST, R.G.C.: Carbonate sediments and their diagenesis. Amsterdam: Elsevier Publishing Co. (1972).
BEALL, A.O.: Textural differentiation within the fine sand grade. J. Geol. 78, 77-93 (1970).
BENZECRI, J.P.: Lecon sur l'analyse statistique des données multidimensionnelle.Public. multigraphiée, Inst. Statistiques, Univ. Paris (1970).
BENZECRI, J.P.: Distance distributionnelle et mêtrique du X^2 en analyse factorielle des correspondances. Public. multigraphiée, Inst. Statistiques, Univ. Paris (1970).
BENZECRI, J.P.: L'analyse des données. Public. multigraphiée, Inst. Statistiques, Univ. Paris (1971).
BLACK, M.: The precipitation of calcium carbonate on the Great Bahama Bank. Geol. Mag. 70, 455-466 (1933).
BODECHTEL, Y., GIERLOFF-EMDEN, H.G.: Weltraumbilder der Erde. S. 176, München: List-Verlag (1969).
BONCH-OSMOLOVSKAYA, K.S.: Dependence of the strontium content in calcium sulphates on the conditions of solid phase formation. Chemical Abstracts 1965, Vol. 63, col. 9667, and Uch. Zap. Nauchn-Issled. Inst. Geol. Arktiki, Regional'n. Geol. 1964 (2), p. 179-187 (1964) (in Russian).
BRIGGS, L.I.: Heavy mineral correlations and provenances. J. Sediment. Petrol. 35, 4, 939-955 (1965).

BUTZER, K.W.: Climatic changes in the arid zones of Africa during early to mid-Holocene times. SAWYER, J.S. (ed.). World climate from 8.000 to 0.B.C., p. 72-83, London (1966).
CORDIER, B.: Analyse factorielle des correspondances. Thèse de 3ème cycle (non publiée), Faculté des Sciences, Univ. Rennes (1965).
CRUFT, E.F., CHAO, P.C.: Nucleation kinetics of the anhydrite-gypsum system. In: Third symposium on salt, Vol. 1; Northern Ohio Geol. Soc. 109-118 (1970).
CULLEN, D.J.: Submarine evidence from New Zealand of a rapid rise in sea level about 11.000 years ago. Paleogeo., Paleoclim., Paleoecol. 3, 289-298 (1967).
CURRAY, J.R.: Late Quaternary sea level. A discussion. Bull. Geol. Soc. Am. 72, 1707-1712 (1961).
DANGEARD, L.: Etude des calacaires par coloration et décalcification. Application à l'étude des calcaires oolitiques. Bull. Soc. Geol. de France, V Série, T VI, 237-245 (1936).
DAVIES, G.R.: Carbonate Bank Sedimentation, Eastern Shark Bay, Western Australia. In: Carbonate Sedimentation and Environments Shark Bay, Western Australia. Am. Assoc. Petrol. Geologists Mem. 13, 85-168 (1970a).
DAVIES, G.R.: Algal-laminated sediments, Gladstone Embayment, Shark Bay, Western Australia. In: Carbonate Sedimentation and Environments Shark Bay, Western Australia. Am. Assoc. Petrol. Geologists, Mem 13, 169-205 (1970b).
DEAN, W.E., Jr.: Petrologic and geochemical variations in the Permian Castile varved anhydrite, Delaware Basin, Texas and New Mexico; Ph.D. Dissertation; Univ. of New Mexico, p. 326 (1967).
DEFFEYES, K.S., LUCIA, F.J., WEYL, D.K.: Dolomitisation of Recent and Plio-Pleistocene sediments by marine evaporite waters on Bonaire, Netherlands Antilles. In: Dolomitization and Limestone Diagenesis. Eds. L.C. PRAY and R.C. MURRAY. S.E.P.M., Spec. Pub. No. 13, 71-88 (1965).
DEN HARTOG, C.: Sea grasses of the World. Verh. K. Ned. Akad. Wet., Afd. Natuurk. Tweede Sectie 59, Pt. 1, 12-38 (1970).
DONAHUE, J.D.: Genesis of oolite and pisolite grains: an energy index. J. Sediment. Petrol. 39, 1399-1411 (1969).
DUNHAM, R.J.: Classification of carbonate rocks according to depositional texture. Am. Assoc. Petrol. Geologists, Mem. 1, 108-120 (1962).
DUNHAM, R.J.: Early vadose silt in Townsend mound (reef), New Mexico and Texas. In: Depositional Environments of carbonate rocks. Ed. G.M. FRIEDMAN. S.E.P.M., Spec. Pub. 14, 139-181 (1969).
DUNHAM, R.J.: Vadose pisolite in the Capitan Reef (Permian), New Mexico and Texas. In: Depositional Environments of carbonate rocks. Ed. G.M. FRIEDMAN. S.E.P.M., Spec. Pub. 14, 182-191 (1969).
EARDLEY, A.J.: Sediments of Great Salt Lake Utah. Bull. Am. Assoc. Petrol. Geologists 22, 1305-1411 (1938).
EMERY, K.O., TRACEY, J.I., LADD, H.S.: Geology of Bikini and nearby atolls. Part I: Geology. Geological Survey of America, Professional paper 260-A (1954).
FABRICIUS, F., KLINGELE, H.: Ultrastrukturen von Ooiden und Oolithen: Zur Genese und Diagenese quartärer Flachwasserkarbonate des Mittelmeeres. Verh. Geol. B.-A. 4, 594-617 (1970).
FAIRBRIDGE, R.W.: The dolomite question. S.E.P.M., Spec. Pub. 5, 125-178 (1957).
FAIRBRIDGE, R.W.: Eustatic changes in sea level. In: Physics and Chemistry of the Earth, Vol. 4. Pergamon Press (1961).
FISHER, R.A., CORBET, A.S., WILLIAMS, C.B.: The relationship between the number of species and the number of individuals in a random sample of an animal population. J. Anim. Ecol. 12, 42-58 (1943).
FREEMAN, T.: Quiet water oolites from Laguna Madre, Texas. J. Sediment. Petrol. 32, 475-483 (1962).
FREEZE, R.A., WITHERSPOON, P.A.: Theoretical analysis of regional ground-water flow. Water Resources Research 3, 623-634 (1967).
FRIEDMAN, G.M.: Occurrence and origin of Quaternary dolomite of Salt Flat, West Texas. J. Sediment. Petrol. 36, 263-267 (1966).
FRIEDMAN, G.M., SANDERS, J.E.: Origin and occurrence of dolostones. In: Carbonate Rocks, Part A. Ed. by G.V. CHILINGAR et al., p. 267-348. New York: Elsevier Publishing Co. (1967).

FRISHMAN, S.A.: Geochemistry of oolites, Baffin Bay, Texas. Thesis, University of Texas (Austin, 1969).
FÜCHTBAUER, H.: Zur Nomenklatur der Sedimentgesteine. Erdöl und Kohle 12, 605-613 (1959).
GLENNIE, K.W.: Desert Sedimentary Environments. Amsterdam: Elsevier Publishing Co. (1970).
GROOT, K., DUYVIS, E.M.: Crystal form of precipitated calcium carbonate as influenced by adsorbed magnesium ions. Nature 212, 183-184 (1966).
HAHNE, C., KIRCHMAYER, M., OTTEMANN, J.: "Höhlenperlen" (cave pearls), besonders aus Bergwerken des Ruhrgebietes. Modellfälle zum Studium diagenetischer Vorgänge an Einzelooiden. N. Jahrb., Geol. Paläontol., Ab. 130, 1-46 (1968).
HARBAUGH, J.W., DEMIRMEN, F.: Application of factor analysis to petrologic variations of Americus Limestone (Lower Permian), Kansas and Oklahoma. The University of Kansas. Special distribution Public. 15 (1964).
HARBAUGH, J.W., MERRIAM, D.F.: Computer applications in stratigraphical analysis. New York (1968).
HARDIE, L.A.: The gypsum-anhydrite equilibrium at one atmosphere pressure. Am. Mineralogist 52, 171-200 (1967).
HARMAN, H.H.: Modern Factor Analysis. Chicago-London (1960).
HARA, R., TANAKA, Y., NAKAMURA, K.: On the calcium sulphate in sea water - 1 Solubilities of dehydrate and anhydrite in sea waters of various concentrations at 0° - 200° C. Tohoku Imperial University Tech. Repts. 11, 87-109 (1934).
HOLLIDAY, D.W.: Secondary gypsum in Middle Carboniferous rocks of Spitsbergen. Geol. Mag. 104, 2, 171-177 (1967).
HOLSER, W.T., KAPLAN, I.R.: Isotope geochemistry of sedimentary sulfates. Chemical Geology 1, 93-135 (1966).
HSÜ, K.J.: Origin of dolomite in sedimentary sequences: a critical analysis. Mineralium Deposita 2, 133-138 (1966).
HSÜ, K.J., SIEGENTHALER, C.: Preliminary experiments on hydrodynamic movement induced by evaporation and their bearing on the dolomite problem. Sedimentology 12, 11-25 (1969).
HUBBERT, M.K.: The theory of groundwater motion. J. Geol. 48, 785-944 (1940).
ILLING, L.V.: Bahaman calcareous sands. Bull. Am. Assoc. Petrol. Geologists 38, 1-95 (1954).
IMBRIE, J., PURDY, E.G.: Classification of modern Bahamian carbonate sediments. HAM, W.E. (ed.). Bull. Am. Assoc. Petrol. Geologists, Mem. 1, 253-272 (1962).
IMBRIE, J.: Factor and vector analysis programs for analyzing geologic data. Technical Report of ONR Task No. 389-135, Evanston, Illinois (1963).
IMBRIE, J., ANDEL, T.H. van: Vector analysis of heavy mineral data. Bull. Geol. Soc. Am. 75, 1131-1156 (1964).
JELGERSMA, S.: Holocene sea level changes in the Netherlands. Dissertation "Ernest van AELST", Maastricht (1961).
JUNG, W., KNITZSCHKE, G.: Kombiniert-feinstratigraphischgeochemische Untersuchungen der Anhydrite des Zechsteins I im SE-Harzvorland. Geologie 9, Nr. 1, 58-72 (1960).
KERR, S.D., THOMPSON, A.: Origin of nodular and bedded anhydrite in Permian shelf sediments, Texas and New Mexico. Bull. Am. Assoc. Petrol. Geologists 42, 1726-1732 (1963).
KELLEY, J.C., WHETTEN, J.T.: Quantitative statistical analyses of Columbia river sediment samples. J. Sediment. Petrol. 39, 3, 1167-1173 (1969).
KINSMAN, D.J.J.: Interpretation of Sr^{2+} concentrations in carbonate minerals and rocks. J. Sediment. Petrol. 39, 486-508 (1969).
KINSMAN, D.J.J., HOLLAND, H.D.: The coprecipitation of cations with $CaCO_3$-IV. The coprecipitation of Sr^{2+} with aragonite between 16° and 96°C. Geochim. Cosmochim. Acta 33, 1-17 (1969).
KLOVAN, J.E.: The use of factor analysis in determining depositional environments from grain-size distributions. J. Sediment. Petrol. 36, 115-125 (1966).
KRUMBEIN, W.C., GRAYBILL, F.A.: An introduction to statistical models in geology. New York-London-Sydney (1965).
KUDRASS, H.R.: Sedimentation am Kontinentalhang vor Portugal und Marokko im Spätpleistozän und Holozän. Unveröffentl. Diss. Univ. Kiel (1972).
LEBART, L., FENELON, J.P.: Statistique et informatique appliquées. Ed. by DUNOD. Paris (1971).

LOGAN, B.W. et al: Classification and environmental significance of algal stromatolites. J. Geol. 72, 68-83 (1964).
LOGAN, B.W. et al: Carbonate sediments and reefs, Yucatan shelf, Mexico. Am. Assoc. Petrol. Geologists, Mem. 11 (1969).
LOREAU, J.-P.: Ultrastructure de la phase carbonatée des oolithes actuelles. C. R. Acad. Sci. 271, 816-819 (1970).
LOREAU, J.-P.: Ultrastructure, nucleation, and crystal growth of natural and experimentally produced aragonite in relation to non-skeletal organic matter. Abstracts, VIII. Int. Sed. Congr. (Heidelberg, 1971).
LOREAU, J.-P.: Explication de la birefringence anormalement faible de l'aragonite des oolithes marines. C. R. Acad. Sci. (in press).
LUCIA, F.J.: Recent sediments and diagenesis of south Bonaire, Netherlands Antilles. J. Sediment. Petrol. 38, 845-858 (1967).
MAIKLEM, W.R., BEBOUT, D.G., GLAISTER, R.P.: Classification of anhydrite - a practical approach. Bull. Canad. Petrol Geology 17, No. 2, 194-233 (1969).
MAYHEW, E.J., HEYLMUM, E.B.: Complex salts and brines of the Paradox Basin. In: Second symposium on salt, Vol. 1.Northern Ohio Geol. Soc., p. 221-235 (1966).
MAXWELL, W.G.J.: Atlas of the Great Barrier Reef. Amsterdam: Elsevier Publishing Co. (1968).
MELGUEN, M.: Exemple de traitement statistique de données sédimentologiques : différenciation de faciès dans un sédiment d'apparence homogène. Bull. de l'Union des Océanographes Français. Paris (1972).
MOORE, R.C. : Meaning of facies. Sedimentary facies in Geologic History. Geol. Soc. Amer. Mem. 39, 1-34 (1949).
MÖRNER, N.A.: Eustatic and climatic changes during the last 15.000 years. Geol. en Mijnbouw 48 (4), 389-399 (1969).
MÜLLER, G.: Zur Geochemie des Strontiums in ozeanen Evaporiten unter besonderer Berücksichtigung der sedimentaren Coelestin Lagerstätte von Hemmelte-West (Süd-Oldenburg) Geologies, Beiheft 35, 1-90 (1962).
MURRAY, R.C.: Origin and diagenesis of gypsum and anhydrite. J. Sediment. Petrol. 34, 512-523 (1964).
MURRAY, R.C.: Hydrology of South Bonaire, N. A. A rock selective dolomitisation model. J. Sediment. Petrol. 39, 1007-1013 (1969).
NESTEROFF, W.D.: De l'origine des oolithes. C. R. Acad. Sci. 242, 1047-1049 (1956).
NEWELL, N.D., RIGBY, J.K.: Geological studies on the Great Bahama Bank. S.E.P.M. Spec. Pub. 5, 15-72 (1957).
NEWELL, N.D. et al: Bahamian oolitic sand. J. Geol. 68, 481-497 (1960).
OSBORNE, R.H.: The American Upper Ordovician standard XI multivariate classification of typical cincinnatian calcarenites. J. Sediment. Petrol. 39, 2, 769-776 (1969).
PAREMANCBLUM, M.: The distribution of heavy minerals and their hydraulic equivalents in sediments of the Mediterranean continental shelf of Israel. J. Sediment. Petrol. 36, 162-174 (1966).
PETERSON, M.N.A., VON DER BORCH, C.C., BIEN, G.S.: Growth of dolomite crystals. Am. J. Sci. 264, 257-272 (1966).
PHILIPS, F.C.: Oceanic salt deposits. Chem. Soc. Quart. Rev. 1, 91-111 (1947).
PHLEGER, F.B.: Sedimentology of Guerrero Negro Lagoon, Baja California, Mexico. In: Submarine Geology and Geophysics. Eds. W.F. WHITTARD and R. BRADSHAW, p. 205-237. London: Butterworths (1965).
PIA. J.: Pflanzen als Gesteinbildner. Berlin: Borntraeger 1926.
PLAS, L. van der, TOBI, A.C.: A chart for judging the reliability of point counting results. Am. J. Sci. 263, 87-90 (1965).
PRICE, W.A.: Environment and formation of the chenier plain. Quaternaria 2, 75-86 (1955).
PURKAYASTHA, B.C., CHATTERJEE, A.: The study of the uptake of strontium tracer by different forms of calcium sulphate. J. Indian Chem. Soc. 43, 687-693 (1966).
PURSER, B.H.: Syn-sedimentary marine lithification of Middle Jurassic limestones in the Paris Basin. Sedimentology 12, 205-230 (1969).
RUSNAK, G.A.: Some observations of Recent oolites. J. Sediment. Petrol. 30, 471-480 (1960).
SCHLANGER, S.O., TRACEY, J.I., Jr.: Dolomitization related to recent emergence of Jarvis Island, Southern Line Islands (Pacific Ocean) (abst.). Geol. Soc. Am., 1970 An. Meetings, 2, n. 7, 676 (1970).

SCHOFIELD, J.C., THOMPSON, H.R.: Post glacial sea levels and isostatic uplift. N.Z.J. Geol. Geophys. 7, 359-370 (1964).
SEIBOLD, E., EXON, N., HARTMANN, M., KÖGLER, F.-C., KRUMM, H., LUTZE, G.F., NEWTON, R.S., WERNER, F.: Marine Geology of Kiel Bay. In: Sedimentology of parts of Central Europe. Guidebook. VIII. Int. Sed. Congr., p. 209-235 (Heidelberg, 1971).
SHEARMAN, D.J.: Recent halite rock, Baja California, Mexico. Institution of Mining and Metallurgy (Newcastle-upon-Tyne). Trans. Sec. B, 79, 155-162 (1970).
SHEARMAN, D.J., SKIPWITH, Sir Patrick A. d'E.: Organic matter in Recent and ancient limestones and its role in their diagenesis. Nature 208, No. 5017, 1310-1311 (1965).
SHEARMAN, D.J., TWYMAN, J., ZAND KARIMI, M.: The genesis and diagenesis of oolites. Proc. Geol. Ass. Lond. 81, 561-575 (1970).
SHEPARD, F.P., MOORE, D.E.: Sedimentary environments differentiated by coarse fraction studies. Bull. Am. Assoc. Petrol. Geologists 38, 1792-1802 (1954).
SHINN, E.A.: Practical significance of birdseye structures in carbonate rocks. J. Sediment. Petrol. 38, 215-223 (1965).
SHINN, E.A.: Burrowing in recent lime sediments of Florida and the Bahamas. J. Paleont. 42, 879-894 (1968).
SHINN, E.A., GINSBURG, R.N., LLOYD, R.M.: Recent supratidal dolomite from Andros Island, Bahamas. In: Dolomitization and Limestone Diagenesis. Eds. L.C. PRAY and R.C. MURRAY. S.E.P.M., Spec. Pub. 13, 112-123 (1965).
SHINN, E.A., LLOYD, R.M., GINSBURG, R.N.: Anatomy of a modern carbonate tidal flat, Andros Island, Bahamas. J. Sediment. Petrol. 39, 1202-1228 (1969).
SITTER, L.U., de: Structural Geology. London: McGraw-Hill (1956).
SMALLEY, I.J., VITA-FINZI, C.: The formation of fine particles in sandy desert and the nature of desert loess. J. Sediment. Petrol. 38, 766-774 (1968).
SORBY, H.C.: The structure and origin of limestones. Proc. Geol. Soc. London 35, 56-95 (1879).
THODE, H.G., MONSTER, J., DUNFORD, H.B.: Sulphur-isotope geochemistry. Geochim. Cosmochim. Acta 25, 159-174 (1961).
THODE, H.G., MONSTER, J.: Sulphur-isotope geochemistry of petroleum, evaporites and ancient seas. Am. Assoc. Petrol. Geol., Mem. 4, 367-377 (1965).
TRICHET, J.: Essai d'explication du dépôt d'aragonite sur des substrats organiques. C. R. Acad. Sci. 265, 1464-1467 (1967).
TRICHET, J.: Recent aragonite deposition on algal substrates during blue-green algae decomposition. Rôle of organic substances. Abstracts, VIII. Int. Sed. Congr. (Heidelberg, 1971).
UEBERLA, K.: Faktorenanalyse. Berlin-Heidelberg-New York: Springer 1968.
USIGLIO, J.: The data of geochemistry. Bull. U.S. Geol. Survey 770 (1924).
VAN STRAATEN, L.M.J.U.: Composition and structure of Recent marine sediments in The Netherlands. Leidse Geol. Meded. XIX (1956).
VAN VEEN, J.: Eb en vloedschaar systemen in de nederlandse getijwateren. Tijdschr. K. Ned. aardrijksk. Genoot. 67, 303-325 (1950).
WEYL, P.K.: The solution behaviour of carbonate minerals in sea water. Univ. Miami. Studies Tropical Oceanog. 5, 178-228 (1967).
WILLIAMS, R.E.: Groundwater flow systems and accumulation of evaporite minerals. Bull. Am. Assoc. Petrol. Geologists 54, 1290-1295 (1969).

International Union
of Geological Sciences,
Series A, Number 3

Ores in Sediments

Edited by G. C. Amstutz and A. J. Bernard

VIII. International Sedimentological Congress
Heidelberg, August 31 – September 3, 1971

Sponsored by the Society of Geology Applied to Mineral Deposits (SGA) and the International Association of Sedimentology

184 figures
VIII, 350 pages. 1973
DM 48,–; US $ 17,80
ISBN 3-540-05712-9
Prices are subject to change without notice

The symposium report "Ores in Sediments" contains some 20 papers on problems of diagenetic history, sedimentary fabrics and paleogeographic extent of metallic and non-metallic deposits in sediments. It is a real milestone in the development of modern theories on mineral deposits in that it maps progress made since the start of the new trend in the early sixties. In addition to a wealth of new observations, almost all contributions propose criteria for the successful exploration of deposits in sediments. The book will be welcome to practically all mining companies and institutions dealing with mineral exploration and research, because 80 % of all metal reserves and most non-metallic ore deposits are stratabound deposits.

■ **Prospectus on request**

Springer-Verlag
Berlin Heidelberg New York
München London Paris Sydney Tokyo Wien

SPRINGER-VERLAG
BERLIN·HEIDELBERG·NEW YORK

A Geochemical and Geophysical Account

Hot Brines and Recent Heavy Metal Deposits in the Red Sea

Edited by **Egon T. Degens** and **David A. Ross.**
Woods Hole Oceanographic Institute,
Woods Hole, Massachusetts.

From the Preface

With 220 figures
(8 in color)
XII, 600 pages. 1969
Atlas format:
8¼ x 11 inches
double spaced
Cloth DM 128,–
US $ 47,40

Prices are subject to change without notice

Cooperative research ventures between oceanographic institutions and nations today frequently start with a series of official meetings, councils, and so forth, followed by several years of research, and finally a group of papers emerging in various technical journals. The study of the Red Sea is an exception to this procedure. It is a good example of the kind of spontaneous cooperation that can occur when individual scientists get excited about a unique problem and work together exchanging samples and data and publishing their final results in a single volume. The problem of the hot holes of the Red Sea required real team-work from scientists of many different disciplines as well as different nationalities.

The papers represent a broad international authorship and cover quite a variety of subject matter. Mathematicians, physicists, chemists, biologists, geologists, oceanographers, lawyers, and economists from numerous countries have participated in this endeavor, and the editors are grateful to them for their enthusiastic support.

■ **Prospectus on request**

GPSR Compliance

The European Union's (EU) General Product Safety Regulation (GPSR) is a set of rules that requires consumer products to be safe and our obligations to ensure this.

If you have any concerns about our products, you can contact us on ProductSafety@springernature.com

In case Publisher is established outside the EU, the EU authorized representative is:

Springer Nature Customer Service Center GmbH
Europaplatz 3
69115 Heidelberg, Germany

Batch number: 10762209

Printed by Printforce, the Netherlands